The

INVISIBLE RAINBOW

The INVISIBLE RAINBOW

A History of
Electricity and Life

Arthur Firstenberg

Chelsea Green Publishing
White River Junction, Vermont
London, UK

Drawings on pages 3 and 159 copyright © 2017 by Monika Steinhoff.
"Two bees" drawing by Ulrich Warnke, used with permission.

Originally published in 2017 by AGB Press, Santa Fe, New Mexico; Sucre, Bolivia.

This paperback edition published by Chelsea Green Publishing, 2020.

Book layout: Jim Bisakowski
Cover design: Ann Lowe

Printed in Canada.
First printing February 2020.
10 9 8 7 6 5 4   22 23 24

Our Commitment to Green Publishing
Chelsea Green sees publishing as a tool for cultural change and ecological steward-
ship. We strive to align our book manufacturing practices with our editorial mission
and to reduce the impact of our business enterprise in the environment. We print
our books and catalogs on chlorine-free recycled paper, using vegetable-based inks
whenever possible. This book may cost slightly more because it was printed on paper
that contains recycled fiber, and we hope you'll agree that it's worth it. *The Invisible
Rainbow* was printed on paper supplied by Marquis that is made of recycled materials
and other controlled sources.

Library of Congress Control Number: 2020930536
ISBN 978-1-64502-009-7 (paperback) | 978-1-64502-010-3 (ebook)

Chelsea Green Publishing
85 North Main Street, Suite 120
White River Junction, VT 05001

Somerset House
London, UK

www.chelseagreen.com

In memory of Pelda Levey—
friend, mentor, and fellow traveler.

Author's Note

FOR EASE OF READING I have kept the endnotes to a minimum. However, all sources referred to in the text can be found in the bibliography at the back of the book, together with other principal works I have consulted. For the convenience of those interested in particular subjects, the literature in the bibliography is organized by chapter, and within some chapters by topic, instead of the usual single alphabetical listing.

A.F.

Contents

Prologue

Once upon a time, the rainbow visible in the sky after a storm represented all the colors there were. Our earth was designed that way. We have a blanket of air above us that absorbs the higher ultraviolets, together with all of the X-rays and gamma rays from space. Most of the longer waves, that we use today for radio communication, were once absent as well. Or rather, they were there in infinitesimal amounts. They came to us from the sun and stars but with energies that were a trillion times weaker than the light that also came from the heavens. So weak were the cosmic radio waves that they would have been invisible, and so life never developed organs that could see them.

The even longer waves, the low-frequency pulsations given off by lightning, are also invisible. When lightning flashes, it momentarily fills the air with them, but they are almost gone in an instant; their echo, reverberating around the world, is roughly ten billion times weaker than the light from the sun. We never evolved organs to see this either.

But our bodies know that those colors are there. The energy of our cells whispering in the radio frequency range is infinitesimal but necessary for life. Every thought, every movement that we make surrounds us with low frequency pulsations, whispers that were first detected in 1875 and are also necessary for life. The electricity that we use today, the substance that we send through wires and broadcast through the air without a thought, was identified around 1700 as a property of life. Only later did scientists learn to extract it and make it move inanimate objects, ignoring—because they could not see—its effects on the living world. It surrounds us today, in all of its colors, at intensities that rival the light from the sun, but we still cannot see it because it was not present at life's birth.

We live today with a number of devastating diseases that do not belong here, whose origin we do not know, whose presence we take for granted and no longer question. What it feels like to be without them is a state of vitality that we have completely forgotten.

"Anxiety disorder," afflicting one-sixth of humanity, did not exist before the 1860s, when telegraph wires first encircled the earth. No hint of it appears in the medical literature before 1866.

Influenza, in its present form, was invented in 1889, along with alternating current. It is with us always, like a familiar guest—so familiar that we have forgotten that it wasn't always so. Many of the doctors who were flooded with the disease in 1889 had never seen a case before.

Prior to the 1860s, diabetes was so rare that few doctors saw more than one or two cases during their lifetime. It, too, has changed its character: diabetics were once skeletally thin. Obese people never developed the disease.

Heart disease at that time was the twenty-fifth most common illness, behind accidental drowning. It was an illness of infants and old people. It was extraordinary for anyone else to have a diseased heart.

Cancer was also exceedingly rare. Even tobacco smoking, in non-electrified times, did not cause lung cancer.

These are the diseases of civilization, that we have also inflicted on our animal and plant neighbors, diseases that we live with because of a refusal to recognize the force that we have harnessed for what it is. The 60-cycle current in our house wiring, the ultrasonic frequencies in our computers, the radio waves in our televisions, the microwaves in our cell phones, these are only distortions of the invisible rainbow that runs through our veins and makes us alive. But we have forgotten.

It is time that we remember.

PART ONE

1. Captured in a Bottle

THE EXPERIMENT OF LEYDEN was a craze that was immense, universal: everywhere you went people would ask you if you had experienced its effects. The year was 1746. The place, any city in England, France, Germany, Holland, Italy. A few years later, America. Like a child prodigy making his debut, electricity had arrived, and the whole Western world turned out to hear his performance.

His midwives—Kleist, Cunaeus, Allamand, and Musschenbroek—warned that they had helped give birth to an *enfant terrible*, whose shocks could take away your breath, boil your blood, paralyze you. The public should have listened, been more cautious. But of course the colorful reports of those scientists only encouraged the crowds.

Pieter van Musschenbroek, professor of physics at the University of Leyden, had been using his usual friction machine. It was a glass globe that he spun rapidly on its axis while he rubbed it with his hands to produce the "electric fluid"—what we know today as static electricity. Hanging from the ceiling by silk cords was an iron gun barrel, almost touching the globe. It was called the "prime conductor," and was normally used to draw sparks of static electricity from the rubbed, rotating glass sphere.

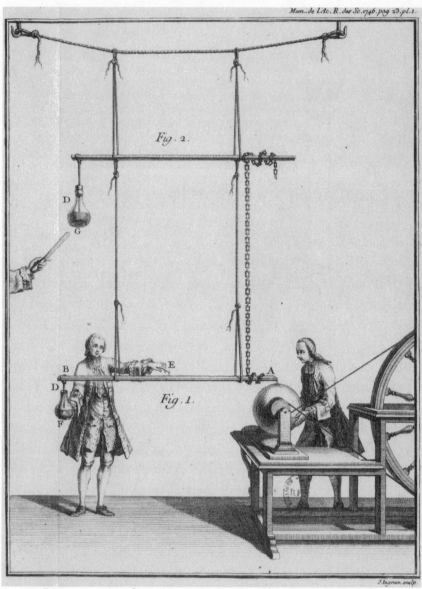

Line engraving from *Mémoires de l'Académie Royale des Sciences*
Plate 1, p. 23, 1746

But electricity, in those early days, was of limited use, because it always had to be produced on the spot and there was no way to store it. So Musschenbroek and his associates designed an ingenious experiment—an experiment that changed the world forever: they attached a wire to the other end of the prime conductor and inserted it in a small glass bottle partly filled with water. They wanted to see if the electric fluid could be stored in a jar. And the attempt succeeded beyond their wildest expectations.

"I am going to tell you about a new but terrible experiment," Musschenbroek wrote to a friend in Paris, "which I advise you never to try yourself, nor would I, who have experienced it and survived by the grace of God, do it again for all the Kingdom of France." He held the bottle in his right hand, and with the other hand he tried to draw sparks from the gun barrel. "Suddenly my right hand was hit with such force, that my whole body shook as though struck by lightning. The glass, although thin, did not break, and my hand was not knocked away, but my arm and whole body were affected more terribly than I can express. In a word, I thought I was done for."[1] His companion in invention, biologist Jean Nicolas Sébastien Allamand, when he tried the experiment, felt a "prodigious blow." "I was so stunned," he said, "that I could not breathe for some moments." The pain along his right arm was so intense that he feared permanent injury.[2]

But only half the message registered with the public. The fact that people could be temporarily or, as we will see, permanently injured or even killed by these experiments became lost in the general excitement that followed. Not only lost, but soon ridiculed, disbelieved, and forgotten. Then as now, it was not socially acceptable to say that electricity was dangerous. Just two decades later, Joseph Priestley, the English scientist who is famous for his discovery of oxygen, wrote his *History and Present State of Electricity*, in which he mocked the "cowardly professor" Musschenbroek, and the "exaggerated accounts" of the first experimenters.[3]

Its inventors were not the only ones who tried to warn the public. Johann Heinrich Winkler, professor of Greek and Latin at Leipzig,

Germany, tried the experiment as soon as he heard about it. "I found great convulsions in my body," he wrote to a friend in London. "It put my blood into great agitation; so that I was afraid of an ardent fever; and was obliged to use refrigerating medicines. I felt a heaviness in my head, as if I had a stone lying upon it. It gave me twice a bleeding at my nose, to which I am not inclined. My wife, who had only received the electrical flash twice, found herself so weak after it, that she could hardly walk. A week after, she received only once the electrical flash; a few minutes after it she bled at the nose."

From their experiences Winkler took away the lesson that electricity was not to be inflicted upon the living. And so he converted his machine into a great beacon of warning. "I read in the newspapers from Berlin," he wrote, "that they had tried these electrical flashes upon a bird, and had made it suffer great pain thereby. I did not repeat this experiment; for I think it wrong to give such pain to living creatures." He therefore wrapped an iron chain around the bottle, leading to a piece of metal underneath the gun barrel. "When then the electrification is made," he continued, "the sparks that fly from the pipe upon the metal are so large and so strong, that they can be seen (even in the day time) and heard at the distance of fifty yards. They represent a beam of lightning, of a clear and compact line of fire; and they give a sound that frightens the people that hear it."

The general public did not react as he planned, however. After reading reports like Musschenbroek's in the proceedings of France's Royal Academy of Sciences, and his own in the *Philosophical Transactions* of the Royal Society of London, eager men and women by the thousands, all over Europe, lined up to give themselves the pleasure of electricity.

Abbé Jean Antoine Nollet, a theologian turned physicist, introduced the magic of the Leyden jar into France. He tried to satisfy the insatiable demands of the public by electrifying tens, hundreds of people at once, having them take each other by the hand so as to form a human chain, arranged in a large circle with the two ends close together. He would place himself at one of the ends, while the person who represented the last link took hold of the bottle. Suddenly the learned abbot, touching with his hand the metal wire inserted in the flask, would complete the circuit and immediately the shock would be felt simultaneously by the whole line. Electricity had become a social affair; the world was possessed, as some observers called it, by "electromania."

The fact that Nollet had electrocuted several fish and a sparrow with the same equipment did not deter the crowds in the least. At Versailles, in the presence of the king, he electrified a company of 240 soldiers of the French Guard holding each other by the hands. He electrified a community of monks at the Carthusian monastery in Paris, stretched out in a circle more than a mile around, each connected to his neighbors by iron wires.

The experience became so popular that the public began to complain of not being able to give themselves the pleasure of an electric shock without having to wait in line or consult a physician. A demand was created for a portable apparatus that everyone could purchase for a reasonable price and enjoy at their leisure. And so the "Ingenhousz bottle" was invented. Enclosed in an elegant-looking case, it was a small Leyden jar joined to a varnished silk ribbon and a rabbit skin with which to rub the varnish and charge the jar.[4]

Electric canes were sold, "priced for all pocketbooks."[5] These were Leyden jars cleverly disguised as walking canes, which you could charge surreptitiously and trick unsuspecting friends and acquaintances into touching.

Then there was the "electric kiss," a form of recreation that even preceded the invention of the Leyden jar but became much more exciting afterwards. Physiologist Albrecht von Haller, at the University of Göttingen, declared incredulously that such parlor games had "taken the place of quadrille." "Could one believe," he wrote, "that a lady's finger, that her whale-bone petticoat, should send forth flashes of true lightning, and that such charming lips could set on fire a house?"

Line engraving c. 1750, reproduced in Jürgen Teichmann,
Vom Bernstein zum Elektron, Deutsches Museum 1982

She was an "angel," wrote German physicist Georg Matthias Bose, with "white-swan neck" and "blood-crowned breasts," who "steals your heart with a single glance" but whom you approach at your peril. He called her "Venus Electrificata" in a poem, published in Latin, French, and German, that became famous throughout Europe:

> If a mortal only touches her hand
> Of such a god-child even only her dress,
> The sparks burn the same, through all of one's limbs,
> As painful as it is, he seeks it again.

Even Benjamin Franklin felt compelled to give instructions: "Let A and B stand on wax; or A on wax and B on the floor; give one of them the electrised phial in hand; let the other take hold of the wire; there will be a small spark; but when their lips approach, they will be struck and shock'd."[6]

Wealthy ladies hosted such entertainment in their homes. They hired instrument makers to craft large, ornate electrical machines that they displayed like pianos. People of more moderate means bought off-the-shelf models that were available in an assortment of sizes, styles, and prices.

Aside from entertainment, electricity, assumed to be related to or identical with the life force, was used primarily for its medical effects. Both electrical machines and Leyden jars found their way into hospitals, and into the offices of doctors wanting to keep up with the times. An even greater number of "electricians" who were not medically trained set up office and began treating patients. One reads of medical electricity being used during the 1740s and 1750s by practitioners in Paris, Montpellier, Geneva, Venice, Turin, Bologna, Leipzig, London, Dorchester, Edinburgh, Shrewsbury, Worcester, Newcastle-Upon-Tyne, Uppsala, Stockholm, Riga, Vienna, Bohemia, and The Hague.

The famous French revolutionary and doctor Jean-Paul Marat, also a practitioner of electricity, wrote a book about it titled *Mémoire sur l'électricité médicale* ("Memoir on Medical Electricity").

Franklin treated patients with electricity in Philadelphia—so many of them that static electric treatments later became known, in the nineteenth century, as "franklinization."

John Wesley, the founder of the Methodist Church, published a 72-page tract in 1759 titled *Desideratum; or, Electricity Made Plain and Useful*. He called electricity "the noblest Medicine yet known in the World," to be used in diseases of the nervous system, skin, blood, respiratory system, and kidneys. "A person standing on the ground," he felt obliged to add, "cannot easily kiss an electrified person standing on the rosin."[7] Wesley himself electrified thousands of people at the headquarters of the Methodist movement and at other locations around London.

And it wasn't just prominent individuals who were setting up shop. So many non-medical people were buying and renting machines for medical use that London physician James Graham wrote, in 1779: "I tremble with apprehension for my fellow creatures, when I see in almost every street in this great metropolis a barber – a surgeon – a tooth-drawer – an apothecary, or a common mechanic turned electrical operator."[8]

Since electricity could initiate contractions of the uterus, it became a tacitly understood method of obtaining abortions. Francis Lowndes, for example, was a London electrician with an extensive practice who advertised that he treated poor women gratis "for amenorrhea."[9]

Even farmers began testing electricity on their crops and proposing it as a means of improving agricultural production, as we will see in chapter 6.

The use of electricity on living beings in the eighteenth century was so widespread in Europe and America that a wealth of valuable knowledge was collected about its effects on people, plants, and animals, knowledge that has been entirely forgotten, that is far more extensive and detailed than what today's doctors are aware of, who see daily, but without recognition, its effects on their patients, and who do not even know such knowledge ever existed. This information is both formal and informal—letters from individuals describing their experiences;

accounts written up in newspapers and magazines; medical books and treatises; papers read at meetings of scientific societies; and articles published in newly founded scientific journals.

As early as the 1740s, ten percent of all articles published in the *Philosophical Transactions* were related to electricity. And during the last decade of that century, fully seventy percent of all articles on electricity in the prestigious Latin journal, *Commentarii de rebus in scientis naturali et medicina gestis*, had to do with its medical uses and its effects on animals and people.[10]

But the floodgates were wide open, and the torrent of enthusiasm about electricity rushed on unhindered, and would continue to do so during the coming centuries, sweeping caution against the rocks, crushing hints of danger like so many bits of driftwood, obliterating whole tracts of knowledge and reducing them to mere footnotes in the history of invention.

2. The Deaf to Hear, and the Lame to Walk

A BURMESE ELEPHANT has the same set of genes whether it toils in a logging camp or runs free in the forest. But its DNA will not tell you the details of its life. In the same way, electrons cannot tell us what is most interesting about electricity. Like elephants, electricity has been forced to bear our burdens and move great loads, and we have worked out more or less precisely its behavior while in captivity. But we must not be fooled into believing we know everything important about the lives of its wild cousins.

What is the source of thunder and lightning, that causes clouds to become electrified and discharge their fury upon the earth? Science still does not know. Why does the earth have a magnetic field? What makes combed hair frizzy, nylon cling, and party balloons stick to walls? This most common of all electrical phenomena is still not well understood. How does our brain work, our nerves function, our cells communicate? How is our body's growth choreographed? We are still fundamentally ignorant. And the question raised in this book—"What is the effect of electricity on life?"—is one that modern science doesn't even ask. Science's only concern today is to keep human exposure below a level that will cook your cells. The effect of nonlethal electricity is something mainstream science no longer wants to know. But in the eighteenth century, scientists not only asked the question, but began to supply answers.

Early friction machines were capable of being charged to about ten thousand volts—enough to deliver a stinging shock, but not enough, then or now, to be thought dangerous. By way of comparison, a person can accumulate thirty thousand volts on their body in walking across a synthetic carpet. Discharging it stings, but won't kill you.

A one-pint Leyden jar could deliver a more powerful shock, containing about 0.1 joules of energy, but still about a hundred times less than what is thought to be hazardous, and thousands of times less than shocks that are routinely delivered by defibrillators to revive people who are in cardiac arrest. According to mainstream science today, the sparks, shocks, and tiny currents used in the eighteenth century should have had no effects on health. But they did.

Imagine you were a patient in 1750 suffering from arthritis. Your electrician would seat you in a chair that had glass legs so that it was well insulated from the ground. This was done so that when you were connected to the friction machine, you would accumulate the "electric fluid" in your body instead of draining it into the earth. Depending on the philosophy of your electrician, the severity of your disease, and your own tolerance for electricity, there were a number of ways to "electrize" you. In the "electric bath," which was the most gentle, you would simply hold in your hand a rod connected to the prime conductor, and the machine would be cranked continuously for minutes or hours, communicating its charge throughout your body and creating an electrical "aura" around you. If this was done gently enough, you would feel nothing—just as a person who shuffles their feet on a carpet can accumulate a charge on their body without being aware of it.

After you were thus "bathed," the machine would be stopped and you might be treated with the "electric wind." Electricity discharges most easily from pointed conductors. Therefore a grounded, pointed metal or wooden wand would be brought toward your painful knee and you would again feel very little—perhaps the sensation of a small breeze as the charge that had built up in your body slowly dissipated through your knee into the grounded wand.

For a stronger effect, your electrician might use a wand with a rounded end, and instead of a continuous current draw actual sparks from your ailing knee. And if your condition were severe—say your leg was paralyzed—he could charge up a small Leyden jar and give your leg a series of strong shocks.

Electricity was available in two flavors: positive, or "vitreous" electricity, obtained by rubbing glass, and negative, or "resinous" electricity, originally obtained by rubbing sulfur or various resins. Your electrician would most likely treat you with positive electricity, as it was the variety normally found on the surface of the body in a state of health.

The goal of electrotherapy was to stimulate health by restoring the electrical equilibrium of the body where it was out of balance. The idea was certainly not new. In another part of the world, the use of natural electricity had been developed to a fine art over thousands of years. Acupuncture needles, as we will see in chapter 9, conduct atmospheric electricity into the body, where it travels along precisely mapped pathways, returning to the atmosphere through other needles that complete the circuit. By comparison electrotherapy in Europe and America, although similar in concept, was an infant science, using instruments that were like sledgehammers.

European medicine in the eighteenth century was full of sledgehammers. If you went to a conventional doctor for your rheumatism, you might expect to be bled, purged, vomited, blistered, and even dosed with mercury. It's easy to understand that going to an electrician instead might seem a very attractive alternative. And it remained attractive until the beginning of the twentieth century.

After more than half a century of unceasing popularity, electrotherapy fell temporarily out of favor during the early 1800s in reaction to certain cults, one of which had grown up in Europe around Anton Mesmer and his so-called "magnetic" healing, and another in America around Elisha Perkins and his "electric" tractors—three-inch-long metallic pencils with which one made passes over a diseased part of the body. Neither man used actual magnets or electricity at all, but they gave both those methods, for a while, a bad name. By mid-century electricity was again mainstream, and in the 1880s ten thousand American physicians were administering it to their patients.

Electrotherapy finally fell permanently out of favor in the early twentieth century, perhaps, one suspects, because it was incompatible with what was then going on in the world. Electricity was no longer a subtle force that had anything to do with living things. It was a dynamo, capable of propelling locomotives and executing prisoners, not curing patients. But sparks delivered by a friction machine, a century and a half before the world was wired, carried quite different associations.

There is no doubt that electricity sometimes cured diseases, both major and minor. The reports of success, over almost two centuries, were sometimes exaggerated, but they are too numerous and often too detailed and well-attested to dismiss them all. Even in the early 1800s, when electricity was not in good repute, reports continued to emerge that cannot be ignored. For example, the London Electrical Dispensary, between September 29, 1793, and June 4, 1819, admitted 8,686 patients for electrical treatment. Of these, 3,962 were listed as "cured," and another 3,308 as "relieved" when they were discharged— an 84 percent success rate.[1]

Although the main focus of this chapter will be on effects that are not necessarily beneficial, it is important to remember why eighteenth century society was enthralled with electricity, just as we are today. For almost three hundred years the tendency has been to chase its benefits and dismiss its harms. But in the 1700s and 1800s, the daily use of electricity in medicine was a constant reminder, at least, that electricity was intimately connected with biology. Here in the West, electricity as a biological science remains in its infancy today, and even its cures have been long forgotten. I will recall just one of them.

Making the Deaf Hear

In 1851, the great neurologist Guillaume Benjamin Duchenne de Boulogne achieved renown for something for which he is least remembered today. A well-known figure in the history of medicine, he was certainly no quack. He introduced modern methods of physical examination that are still in use. He was the first physician ever to take a biopsy from a living person for the purpose of diagnosis. He published the first accurate clinical description of polio. A number of diseases that he identified are named for him, most notably Duchenne muscular dystrophy. He is remembered for all those things. But in his own time he was the somewhat unwilling center of attention for his work with the deaf.

Duchenne knew the anatomy of the ear in great detail, in fact it was for the purpose of elucidating the function of the nerve called the chorda tympani, which passes through the middle ear, that he asked a few deaf people to volunteer to be the subjects of electrical experiments. The incidental and unexpected improvement in their hearing caused Duchenne to be inundated with requests from within the deaf community to come to Paris for treatments. And so he began to minister to large numbers of people with nerve deafness, using the same apparatus that he had designed for his research, which fit snugly into the ear canal and contained a stimulating electrode.

His procedure, to a modern reader, might seem unlikely to have had any effect at all: he exposed his patients to pulses of the feeblest possible current, spaced half a second apart, for five seconds at a time. Then he gradually increased the current strength, but never to a painful level, and never for more than five seconds at a time. And yet by this means he restored good hearing, in a matter of days or weeks, to a 26-year-old man who had been deaf since age ten, a 21-year-old man who had been deaf since he had measles at age nine, a young woman recently made deaf by an overdose of quinine, given for malaria, and numerous others with partial or complete hearing loss.[2]

Fifty years earlier, in Jever, Germany, an apothecary named Johann Sprenger became famous throughout Europe for a similar reason. Though he was denounced by the director of the Institute for the Deaf and Dumb in Berlin, he was besieged by the deaf themselves with requests for treatment. His results were attested in court documents, and his methods were adopted by contemporary physicians. He himself was reported to have fully or partially restored hearing to no less than forty deaf and hard of hearing individuals, including some deaf from birth. His methods, like Duchenne's, were disarmingly simple and gentle. He made the current weaker or stronger according to the sensitivity of his patient, and each treatment consisted of brief pulses of electricity spaced one second apart for a total of four minutes per ear. The electrode was placed on the tragus (the flap of cartilage in front of the ear) for one minute, inside the ear canal for two minutes, and on the mastoid process behind the ear for one minute.

And fifty years before Sprenger, Swedish physician Johann Lindhult, writing from Stockholm, reported the full or partial restoration of hearing, during a two-month period, to a 57-year-old man who had been deaf for thirty-two years; a youth of twenty-two, whose hearing loss was recent; a seven-year-old girl, born deaf; a youth of twenty-nine, hard of hearing since age eleven; and a man with hearing loss and tinnitus of the left ear. "All patients," wrote Lindhult, "were treated with gentle electricity, either the simple current or the electric wind."

Lindhult, in 1752, was using a friction machine. Half a century later, Sprenger used galvanic currents from an electric pile, forerunner of today's batteries. Half a century after that, Duchenne used alternating current from an induction coil. British surgeon Michael La Beaume, similarly successful, used a friction machine in the 1810s and galvanic currents later on. What they all had in common was their insistence on keeping their treatments brief, simple, and painless.

Seeing and Tasting Electricity

Aside from attempting to cure deafness, blindness, and other diseases, early electricians were intensely interested in whether electricity could be directly perceived by the five senses—another question about which modern engineers have no interest, and modern doctors have no knowledge, but whose answer is relevant to every modern person who suffers from electrical sensitivity.

When he was still in his early twenties, the future explorer Alexander von Humboldt lent his own body to the elucidation of this mystery. It would be several years before he left Europe on the long voyage that was to propel him far up the Orinoco River and to the top of Mount Chimborazo, collecting plants as he went, making systematic observations of the stars and the earth and the cultures of Amazonian peoples. Half a century would pass before he would begin work on his five-volume *Kosmos*, an attempt to unify all existing scientific knowledge. But as a young man supervising mining operations in the Bayreuth district of Bavaria, the central question of his day occupied his spare time.

Is electricity really the life force, people were asking? This question, gnawing gently at the soul of Europe since the days of Isaac Newton, had suddenly become insistent, forcing itself out of the lofty realms of philosophy and into dinnertime discussions around the tables of ordinary people whose children would have to live with the chosen answer. The electric battery, which produced a current from the contact of dissimilar metals, had just been invented in Italy. Its implications were huge: friction machines—bulky, expensive, unreliable, subject to atmospheric conditions—might no longer be necessary. Telegraph systems, already designed by a few visionaries, might now be practical. And questions about the nature of the electric fluid might come closer to being answered.

In the early 1790s, Humboldt threw himself into this research with enthusiasm. He wished, among other things, to determine whether he could perceive this new form of electricity with his own eyes, ears, nose, and taste buds. Others were doing similar experiments—Alessandro Volta in Italy, George Hunter and Richard Fowler in England, Christoph Pfaff in Germany, Peter Abilgaard in Denmark—but none more thoroughly or diligently than Humboldt.

Consider that today we are accustomed to handling nine-volt batteries with our hands without a thought. Consider that millions of us are walking around with silver and zinc, as well as gold, copper, and other metals in the fillings in our mouths. Then consider the following experiment of Humboldt's, using a single piece of zinc, and one of silver, that produced an electric tension of about a volt:

"A large hunting dog, naturally lazy, very patiently let a piece of zinc be applied against his palate, and remained perfectly tranquil while another piece of zinc was placed in contact with the first piece and with his tongue. But scarcely one touched his tongue with the silver, than he showed his aversion in a humorous manner: he contracted his upper lip convulsively, and licked himself for a very long time; it sufficed afterwards to show him the piece of zinc to remind him of the impression he had experienced and to make him angry."

The ease with which electricity can be perceived, and the variety of the sensations, would be a revelation to most doctors today. When Humboldt touched the top of his own tongue with the piece of zinc, and its point with the piece of silver, the taste was strong and bitter. When he moved the piece of silver underneath, his tongue burned. Moving the zinc further back and the silver forward made his tongue feel cold. And when the zinc was moved even further back he became nauseated and sometimes vomited—which never happened if the two metals were the same. The sensations always occurred as soon as the zinc and silver pieces were placed in metallic contact with each other.[3]

A sensation of sight was just as easily elicited, by four different methods, using the same one-volt battery: by applying the silver "armature" on one moistened eyelid and the zinc on the other; or one in a nostril and the other on an eye; or one on the tongue and one on an eye; or even one on the tongue and one against the upper gums. In each case, at the moment the two metals touched each other, Humboldt saw a flash of light. If he repeated the experiment too many times, his eyes became inflamed.

In Italy, Volta, the inventor of the electric battery, succeeded in eliciting a sensation of sound, not with one pair of metals, but with thirty, attached to electrodes in each ear. With the metals he originally used in his "pile," using water as an electrolyte, this may have been about a twenty-volt battery. Volta heard only a crackling sound which could have been a mechanical effect on the bones of his middle ears, and he did not repeat the experiment, fearing that the shock to his brain might be dangerous.[4] It remained for German physician Rudolf Brenner, seventy years later, using more refined equipment and smaller currents, to demonstrate actual effects on the auditory nerve, as we will see in chapter 15.

Speeding up the Heart and Slowing it Down

Back in Germany, Humboldt, armed with the same single pieces of zinc and silver, turned his attention next to the heart. Together with his older brother Wilhelm, and supervised by well-known physiologists, Humboldt removed the heart of a fox and prepared one of its nerve fibers so that the armatures could be applied to it without touching the heart itself. "At each contact with the metals the pulsations of the heart were clearly changed; their speed, but especially their force and their elevation were augmented," he recorded.

The brothers next experimented on frogs, lizards, and toads. If the dissected heart beat 21 times in a minute, after being galvanized it beat 38 to 42 times in a minute. If the heart had stopped beating for five minutes, it restarted immediately upon contact with the two metals.

Together with a friend in Leipzig, Humboldt stimulated the heart of a carp that had almost stopped beating, pulsing only once every four minutes. After massaging the heart proved to have no effect, galvanization restored the rate to 35 beats per minute. The two friends kept the heart beating for almost a quarter of an hour by repeated stimulation with a single pair of dissimilar metals.

On another occasion, Humboldt even managed to revive a dying linnet that was lying feet up, eyes closed on its back, unresponsive to the prick of a pin. "I hastened to place a small plate of zinc in its beak and a small piece of silver in its rectum," he wrote, "and I immediately established a communication between the two metals with an iron rod. What was my astonishment, when at the moment of contact the bird opened its eyes, raised itself on its feet and beat its wings. It breathed again for six or eight minutes and then calmly died."[5]

Nobody proved that a one-volt battery could restart a human heart, but scores of observers before Humboldt had reported that electricity increased the human pulse rate—knowledge that is not possessed by doctors today. German physicians Christian Gottlieb Kratzenstein[6] and Carl Abraham Gerhard,[7] German physicist Celestin Steiglehner,[8] Swiss physicist Jean Jallabert,[9] French physicians François Boissier de Sauvages de la Croix,[10] Pierre Mauduyt de la Varenne,[11] and Jean-Baptiste Bonnefoy,[12] French physicist Joseph Sigaud de la Fond,[13] and Italian physicians Eusebio Sguario[14] and Giovan Giuseppi Veratti[15] were just a few of the observers who reported that the electric bath increased the pulse rate by anywhere from five to thirty beats per minute, when positive electricity was used. Negative electricity had the opposite effect. In 1785, Dutch pharmacist Willem van Barneveld conducted 169 trials on 43 of his patients—men, women, and children aged nine to sixty—finding an average five percent increase in the pulse rate when the person was bathed with positive electricity, and a three percent *decrease* in the pulse rate when the person was bathed with negative electricity.[16] When positive sparks were drawn the pulse increased by twenty percent.

But these were only averages: no two individuals reacted the same to electricity. One person's pulse always increased from sixty to ninety beats per minute; another's always doubled; another's pulse became much slower; another reacted not at all. Some of van Barneveld's subjects reacted in a manner opposite to the majority: a negative charge always accelerated their pulse, while a positive charge slowed it down.

"Istupidimento"

Observations of these kinds came quickly and abundantly, so that by the end of the eighteenth century a basic body of knowledge had been built up about the effects of the electric fluid—usually the positive variety—on the human body. It increased both the pulse rate, as we have seen, and the strength of the pulse. It augmented all of the secretions of the body. Electricity caused salivation, and made tears to flow, and sweat to run. It caused the secretion of ear wax, and nasal mucus. It made gastric juice flow, stimulating the appetite. It made milk to be let down, and menstrual blood to issue. It made people urinate copiously and move their bowels.

Most of these actions were useful in electrotherapy, and would continue to be so until the early twentieth century. Other effects were purely unwanted. Electrification almost always caused dizziness, and sometimes a sort of mental confusion, or "istupidimento," as the Italians called it.[17] It commonly produced headaches, nausea, weakness, fatigue, and heart palpitations. Sometimes it caused shortness of breath, coughing, or asthma-like wheezing. It often caused muscle and joint pains, and sometimes mental depression. Although electricity usually caused the bowels to move, often with diarrhea, repeated electrification could result in constipation.

Electricity caused both drowsiness and insomnia.

Humboldt, in experiments on himself, found that electricity increased blood flow from wounds, and caused serum to flow copiously out of blisters.[18] Gerhard divided one pound of freshly drawn blood into two equal parts, placed them next to each other, and electrified one of them. The electrified blood took longer to clot.[19] Antoine Thillaye-Platel, pharmacist at the Hôtel-Dieu, the famous hospital in Paris, agreeing, said that electricity is contraindicated in cases of hemorrhage.[20] Consistent with this are numerous reports of nosebleeds from electrification. Winkler and his wife, as already mentioned, got nosebleeds from the shock of a Leyden jar. In the 1790s, Scottish physician and anatomist Alexander Monro, who is remembered for discovering the function of the lymphatic system, got nosebleeds from just a one-volt battery, whenever he tried to elicit the sensation of light in his eyes. "Dr. Monro was so excitable by galvanism that he bled from the nose when, having the zinc very gently inserted in his nasal fossae, he put it in contact with an armature applied to his tongue. The hemorrhage always took place at the moment when the lights appeared." This was reported by Humboldt.[21] In the early 1800s, Conrad Quensel, in Stockholm, reported that galvanism "frequently" caused nosebleeds.[22]

Line engraving from Abbé Nollet, *Recherches sur les Causes Particulières des Phénomènes Électriques*, Paris: Frères Guérin, 1753

Abbé Nollet proved that at least one of these effects—perspiration—occurred merely from being in an electric field. Actual contact with the friction machine wasn't even necessary. He had electrified cats, pigeons, several kinds of songbirds, and finally human beings. In carefully controlled repeatable experiments, accompanied by modern-looking data tables, he had demonstrated measurable weight loss in all of his electrified subjects, due to an increase in evaporation from their skin. He had even electrified five hundred houseflies in a gauze-covered jar for four hours and found that they too had lost extra weight—4 grains more than their non-electrified counterparts in the same amount of time.

Then Nollet had the idea to place his subjects on the floor underneath the electrified metal cage instead of in it, and they still lost as much, and even a bit more weight than when they were electrified themselves. Nollet had also observed an acceleration in the growth of seedlings sprouted in electrified pots; this too occurred when the pots were only placed on the floor beneath. "Finally," wrote Nollet, "I made a person sit for five hours on a table near the electrified metal cage." The young woman lost 4½ drams more weight than when she had actually been electrified herself.[23]

Nollet was thus the first person, back in 1753, to report significant biological effects from exposure to a DC electric field—the kind of field that according to mainstream science today has no effect whatsoever. His experiment was later replicated, using a bird, by Steiglehner, professor of physics at the University of Ingolstadt, Bavaria, with similar results.[24]

Table 1 lists the effects on humans, reported by most early electricians, of an electric charge or small currents of DC electricity. Electrically sensitive people today will recognize most if not all of them.

Table 1 - Effects of Electricity as Reported in the Eighteenth Century

Therapeutic and neutral effects

Change in pulse rate
Sensations of taste, light,
    and sound
Increase of body temperature
Pain relief
Restoration of muscle tone
Stimulation of appetite
Mental exhilaration
Sedation
Perspiration
Salivation
Secretion of ear wax
Secretion of mucus
Menstruation, uterine
    contraction
Lactation
Lacrimation
Urination
Defecation

Non-therapeutic effects

Dizziness
Nausea
Headaches
Nervousness
Irritability
Mental confusion
Depression
Insomnia
Drowsiness
Fatigue
Weakness
Numbness and tingling
Muscle and joint pains
Muscle spasms and cramps
Backache
Heart palpitations
Chest pain
Colic
Diarrhea
Constipation
Nosebleeds, hemorrhage
Itching
Tremors
Seizures
Paralysis
Fever
Respiratory infections
Shortness of breath
Coughing
Wheezing and asthma attacks
Eye pain, weakness, and fatigue
Ringing in the ears
Metallic taste

3. Electrical Sensitivity

"I HAVE ALMOST ENTIRELY given up the electrical experiments." The author of these words, in referring to his own inability to tolerate electricity, wrote them not in the modern era of alternating currents and radio waves, but in the mid-eighteenth century when all there was was static electricity. French botanist Thomas-François Dalibard confided his reasons to Benjamin Franklin in a letter dated February 1762. "First, the different electrical shocks have so strongly attacked my nervous system that I am left with a convulsive tremor in my arm so that I can scarcely bring a glass to my mouth; and if I now were to touch one electrical spark I would be unable to sign my name for 24 hours. Another thing that I notice is that it is almost impossible for me to seal a letter because the electricity of the Spanish wax, communicating itself to my arm, increases my tremor."

Dalibard was not the only one. Benjamin Wilson's 1752 book, *A Treatise on Electricity*, helped promote the popularity of electricity in England, but he himself did not fare so well by it. "Upon repeating those shocks often for several weeks together," he wrote, "I at last was weakened so much that a very small quantity of electric matter in the vial would shock me to a great degree, and cause an uncommon pain. So that I was obliged to desist from trying any more." Even rubbing a glass globe with his hand—the basic electrical machine of his day—gave him "a very violent headache."[1]

The man who authored the first book in German devoted solely to electricity, *Neu-Entdeckte Phænomena von Bewunderns-würdigen Würckungen der Natur* ("Newly Discovered Phenomena of the Wonderful Workings of Nature," 1744), became gradually paralyzed on one side of his body. Called the first electrical martyr, Johann Doppelmayer, professor of mathematics at Nuremberg, stubbornly persisted in his researches and died of a stroke in 1750 after one of his electrical experiments.[2]

These were just three of the earliest casualties—three scientists who helped birth an electrical revolution in which they themselves could not participate.

Even Franklin developed a chronic neurological illness that began during the period of his electrical researches and that recurred periodically for the rest of his life. Although he also suffered from gout, this other problem worried him more. Writing on March 15, 1753, of a pain in his head, he said, "I wish it were in my foot, I think I could bear it better." One recurrence lasted for the better part of five months while he was in London in 1757. He wrote to his doctor about "a giddiness and a swimming in my head," "a humming noise," and "little faint twinkling lights" that disturbed his vision. The phrase "violent cold," appearing often in his correspondence, was usually accompanied by mention of that same pain, dizziness, and problems with his eyesight.[3] Franklin, unlike his friend Dalibard, never recognized a connection to electricity.

Jean Morin, professor of physics at the Collège Royale de Chartres, and author, in 1748, of *Nouvelle Dissertation sur l'Électricité* ("New Dissertation on Electricity"), thought that it was never healthy to expose oneself to electricity in any form, and to illustrate his point he described an experiment conducted not with a friction machine but with his pet cat. "I stretched out a large cat on the cover of my bed," he recounted. "I rubbed it, and in the darkness I saw sparks fly." He continued this for more than half an hour. "A thousand tiny fires flew here and there, and continuing the friction, the sparks grew until they seemed like spheres or balls of fire the size of a hazelnut... I brought my eyes near one ball, and I immediately felt a lively and painful stinging in my eyes; there was no shock in the rest of my body; but the pain was followed by a faintness that made me fall to the side, my strength failed me, and I battled, so to speak, against passing out, I fought against my own weakness from which I did not recover for several minutes."[4]

Such reactions were by no means confined to scientists. What is known to few doctors today was known universally to all eighteenth-century electricians, and to the nineteenth-century electrotherapists who followed them: electricity had side effects and some individuals were enormously and unaccountably more sensitive to it than others. "There are persons," wrote Pierre Bertholon, a physicist from Languedoc, in 1780, "on whom artificial electricity made the greatest impression; a small shock, a simple spark, even the electric bath, feeble as it is, produced profound and lasting effects. I found others in whom strong electrical operations seemed not to cause any sensation at all... Between these two extremes are many nuances that correspond to the diverse individuals of the human species."[5]

Sigaud de la Fond's numerous experiments with the human chain never produced the same results twice. "There are people for whom electricity can be unfortunate and very harmful," he declared. "This impression being relative to the disposition of the organs of those who experience it and of the sensitivity or irritability of their nerves, there are probably not two persons in a chain composed of many, who experience strictly the same degree of shock."[6]

Mauduyt, a physician, proposed in 1776 that "the face of the constitution depends in great part on the communication between the brain, the spinal cord and the different parts by means of the nerves. Those in whom this communication is less free, or who experience the nervous illness, are then more affected than others."[7]

Few other scientists made any attempt to explain the differences. They simply reported them as fact—a fact as ordinary as that some people are fat and some thin, some tall and some short—but a fact that one had to take into account if one were going to offer electricity as a treatment, or otherwise expose people to it.

Even Abbé Nollet, popularizer of the human chain and electricity's leading missionary, reported this variability in the human condition from the very beginning of his campaign. "Pregnant women especially, and delicate persons," he wrote in 1746, "should not be exposed to it." And later: "Not all persons are equally appropriate to the experiments of electricity, be it for exciting that virtue, be it for receiving it, be it finally for feeling its effects."[8]

British physician William Stukeley, in 1749, was already so familiar with the side effects of electricity that he observed, after an earthquake at London on March 8 of that year, that some felt "pains in their joints, rheumatism, sickness, headach, pain in their back, hysteric and nervous disorders... *exactly as upon electrification*; and to some it has proved fatal."[9] He concluded that electrical phenomena must play an important role in earthquakes.

And Humboldt was so amazed by the extraordinary human variability that he wrote, in 1797: "It is observed that susceptibility to electrical irritation, and electrical conductivity, differ as much from one individual to another, as the phenomena of living matter differ from those of dead material."[10]

The term "electrical sensitivity," in use again today, reveals a truth but conceals a reality. The truth is that not everyone feels or conducts electricity to the same degree. In fact if most people were aware of how vast the spectrum of sensitivity really is, they would have reason to be as astonished as Humboldt was, and as I still am. But the hidden reality is that however great the apparent differences between us, electricity is still part and parcel of our selves, as necessary to life as air and water. It is as absurd to imagine that electricity doesn't affect someone because he or she is not aware of it, as to pretend that blood doesn't circulate in our veins when we are not thirsty.

Today, people who are electrically sensitive complain about power lines, computers, and cell phones. The amount of electrical energy being deposited into our bodies incidentally from all this technology is far greater than the amount that was deposited deliberately by the machines available to electricians during the eighteenth and early nineteenth centuries. The average cell phone, for example, deposits about 0.1 joule of energy into your brain every second. For a one-hour phone call, that's 360 joules. Compare that to a maximum of only 0.1 joule from the complete discharge of a one-pint Leyden jar. Even the 30-element electric pile which Volta attached to his ear canals could not have delivered more than 150 joules in an hour, even if all the energy were absorbed by his body.

Consider also that a static charge of thousands of volts accumulates on the surface of computers screens—both old desktop computers and new wireless laptops—whenever they are in use, and that part of this charge is deposited on the surface of your body when you sit in front of one. This is probably less charge than was provided by the electric bath, but no one was subjected to the electric bath for forty hours a week.

Electrotherapy is indeed an anachronism. In the twenty-first century we are all engaged in it whether we like it or not. Even if occasional use was once beneficial to some, perpetual bombardment is not likely to be so. And modern researchers trying to determine the biological effects of electricity are a bit like fish trying to determine the impact of water. Their eighteenth century predecessors, before the world was inundated, were in a much better position to record its effects.

The second phenomenon pointed out by Humboldt has equally profound implications for both modern technology and modern medicine: not only were some people more sensitive to its effects than others, but individuals differed extremely in their ability to *conduct* electricity and in their tendency to accumulate a charge on the surface of their body. Some people could not help gathering a charge wherever they went, simply by moving and breathing. They were walking spark generators, like the Swiss woman whom the Scottish writer Patrick Brydone heard about in his travels. Her sparks and shocks, he wrote, were "strongest in a clear day, or during the passage of thunder-clouds, when the air is known to be replete with that fluid."[11] Something was physiologically different about such individuals.

And, conversely, human non-conductors were found, people who conducted electricity so poorly, even when their hands were well moistened, that their presence in a human chain interrupted the flow of current. Humboldt performed many experiments of this type with so-called "prepared frogs." When the person at one end of a chain of eight people grasped a wire connected to the sciatic nerve of a frog while the person at the other end grasped a wire connected to its thigh muscle, the completing of the circuit made the muscle convulse. But not if there was a human non-conductor anywhere in the chain. Humboldt himself interrupted the chain one day when he was running a fever and was temporarily a non-conductor. He also could not elicit the flash of light in his eyes with the current on that day.[12]

In the *Transactions of the American Philosophical Society* for 1786 is a report along the same lines by Henry Flagg about experiments that took place at Rio Essequibo (now Guyana), in which a many-personed chain grasped the two ends of an electric eel. "If someone was present who constitutionally was not apt to receive the impression of the electric fluid," wrote Flagg, "that person did not receive the shock at the moment of contact with the fish." Flagg mentioned one such woman who, like Humboldt, had a mild fever at the time of the experiment.

This led some eighteenth century scientists to postulate that both electrical sensitivity and electrical conductivity were indicators of one's overall state of health. Bertholon observed that a Leyden bottle drew feebler sparks more slowly from a patient who was running a fever than an identical bottle did from a healthy person. During episodes of the chills, the opposite was true: the patient then seemed to be a super-conductor and the sparks drawn from him or her were stronger than normal.

According to Benjamin Martin, "a person who has the small-pox cannot be electrified by any means whatever."[13]

But despite the above observations, neither electrical sensitivity nor electrical conductivity were reliable indicators of either good health or bad. Most often they seemed to be random attributes. Musschenbroek, for example, in his *Cours de Physique*, mentioned three individuals whom he was never, at any time, able to electrify at all. One was a vigorous, healthy 50-year-old man; the second a healthy, pretty 40-year-old mother of two; and the third a 23-year-old paralyzed man.[14]

Age and sex seemed to be factors. Bertholon thought that electricity had a greater effect on mature young men than on infants or the elderly.[15] French surgeon Antoine Louis agreed. "A man of twenty-five years," he wrote, "is electrified more easily than a child or an old person."[16] According to Sguario, "women generally are electrized more easily, and in a better manner, than men, but in one or the other sex a fiery and sulfurous temperament better than others, and youths better than old people."[17] According to Morin, "adults and persons with a more robust temperament, more hot-blooded, more fiery, are also more susceptible to the movement of this substance."[18] These early observations that vigorous young adults are in some way more susceptible to electricity than others may seem surprising. But we will see later the importance of this observation to the public health problems of the modern era, including especially the problem of influenza.

To illustrate in some detail the typical reactions of electrically sensitive people, I have chosen Benjamin Wilson's report on the experiences of his servant, who volunteered to be electrified in 1748 when he was twenty-five years of age. Wilson, being electrically sensitive himself, was naturally more attentive to these effects than some of his colleagues. Present-day electrically sensitive people will recognize most of the effects, including the after-effects that lasted for days.

"After the first and second experiments," wrote Wilson, "he complained of his spirits being depressed, and of being a little sick. Upon making the fourth experiment, he became very warm, and the veins of his hands and face swelled to a great degree. The pulse beat more than ordinary quick, and he complained of a violent oppression at his heart (as he called it) which continued along with the other symptoms near four hours. Upon uncovering his breast, it appeared to be much inflamed. He said that his head ached violently, and that he felt a pricking pain in his eyes and at his heart; and a pain in all his joints. When the veins began to swell, he complained of a sensation which he compared to that arising from strangling, or a stock tying too tight about the neck. Six hours after the making of the experiments most of these complaints left him. The pain in his joints continued till the next day, at which time he complained of weakness, and was very apprehensive of catching cold. On the third day he was quite recovered.

"The shocks he received were trifling," Wilson added, "compared with those which are commonly received by most persons when they join hands to compleat the circuit for amusement."[19]

Morin, who stopped subjecting himself to electricity before 1748, also highlighted its ill effects in some detail. "Persons who are electrified on resin cakes, or on a wool cushion, often become like asthmatics," he observed. He reported the case of a young man of thirty who, after being electrified, suffered from a fever for thirty-six hours and a headache for eight days. He denounced medical electricity, concluding from his own experiments on people with rheumatism and gout that "all left suffering much more than before." "Electricity brings with it symptoms to which it is not prudent to expose oneself," he said, "because it is not always easy to repair the damage." He especially disapproved of the medical use of the Leyden jar, telling the story of a man with eczema on his hand who, receiving a shock from a small jar containing only two ounces of water, was rewarded with a pain in his hand that endured more than a month. "He was not so eager after that," said Morin, "to be the whipping boy for the electrical phenomena."[20]

Whether electricity did more good than harm was not a trivial issue for people who lived at that time.

Morin, who was electrically sensitive, and Nollet, who was not, came to loggerheads over the future of our world, there at the dawn of the electrical era. Their debate played out very publicly in the books and magazines of their time. Electricity was, first and foremost, known to be a property of living things and to be necessary for life. Morin thought of electricity as a kind of atmosphere, an exhalation that surrounded material bodies, including living bodies, and communicated itself to others by proximity. He was frightened by Nollet's notion that electricity might instead be a substance that flowed in a direction from one place to another, that could not flow out unless more of it flowed in from somewhere else, a substance that humanity had now captured and could send anywhere in the world at will. The debate began in 1748, just two years after the invention of the Leyden jar.

"It would be easy," prophesied Nollet with amazing accuracy, "to make a great number of bodies feel the effects of electricity at the same time, without moving them, without inconveniencing them, even if they are at very considerable distances; because we know that this virtue is transmitted with enormous ease to a distance by chains or by other contiguous bodies; some metal pipes, some iron wires stretched far away... a thousand other means even easier, that ordinary industry could invent, would not fail to put these effects within reach of the whole world, and to extend the use of it as far as one would wish."[21]

Morin was shocked. What would become of the bystanders, he immediately thought? "The living bodies, the spectators, would quickly lose that spirit of life, that principle of light and of fire that animates them... To put the whole universe, or at least a sphere of immense size in play, in action, in movement for a simple crackling of a little electrical spark, or for the formation of a luminous halo five to six inches long at the end of an iron bar, that would be truly to create a great commotion for no good reason. To make the electrical material penetrate in the interior of the densest metals, and then to make it radiate out with no obvious cause; that is perhaps to speak of good things; but the whole world will not agree."[22]

Nollet responded with sarcasm: "Truly, I don't know if the whole universe must feel thus the experiments that I make in a small corner of the world; how will this flowing material that I cause to come toward my globe from nearby, how will its flow be felt in China, for example? But that would be of great consequence! Hey! What would become, as Mr. Morin remarks so well, of the living bodies, of the spectators!"[23]

Like other prophets who have shouted warnings instead of praise for new technologies, Morin was not the most popular scientist of his time. I have even seen him condemned by one modern historian as a "pompous critic," a "gladiator" who "rose against" the electrical visionary Nollet.[24] But the differences between the two men were in their theories and conclusions, not their facts. The side effects of electricity were known to everyone, and continued to be so until the dawn of the twentieth century.

George Beard and Alphonso Rockwell's authoritative 1881 textbook on *Medical and Surgical Electricity* devoted ten pages to these phenomena. The terms they used were "electro-susceptibility," referring to those who were easily injured by electricity, and "electro-sensibility," referring to those who sensed electricity to an extraordinary degree. One hundred and thirty years after Morin's first warnings, these physicians said: "There are individuals whom electricity always injures, the only difference in the effect on them between a mild and a severe application being, that the former injures less than the latter. There are patients upon whom all electrotherapeutic skill and experience are wasted; their temperaments are not *en rapport* with electricity. It matters not what may be the special disease or symptoms of disease from which they suffer—paralysis, or neuralgia, or neurasthenia, or hysteria, or affections of special organs—the immediate and the permanent effects of galvanization or faradization, general or localized, are evil and only evil." The symptoms to watch out for were the same as in the previous century: headache and backache; irritability and insomnia; general malaise; excitation or increase of pain; over-excitation of the pulse; chills, as though the patient were catching a cold; soreness, stiffness, and dull aching; profuse perspiration; numbness; muscle spasms; light or sound sensitivity; metallic taste; and ringing in the ears.

Electro-susceptibility runs in families, said Beard and Rockwell, and they made the same observations about gender and age that early electricians had made: women, on average, were a little more susceptible to electricity than men, and active adults between twenty and fifty bore electricity more poorly than at other ages.

Like Humboldt, they were also astonished by the people who were *insensitive* to the electrical energy. "It should be added," they said, "that some persons are *indifferent* to electricity—they can bear almost any strength of either current very frequently and for long applications, without experiencing any effect either good or evil. Electricity may be poured over them in limitless measures; they may be saturated with it, and they may come out from the applications not a whit better or worse." They were frustrated that there was no way to predict whether a person was *en rapport* with electricity or not. "Some women," they observed, "even those who are exquisitely delicate, can bear enormous doses of electricity, while some men who are very hardy can bear none at all."[25]

Obviously electricity is not, as so many modern doctors would have it—those who recognize that it affects our health at all—an ordinary kind of stressor, and it is a mistake to assume that one's vulnerability to it is an indicator of one's state of health.

Beard and Rockwell did not give any estimates of the numbers of people not *en rapport* with electricity, but in 1892, otologist Auguste Morel reported that twelve percent of healthy subjects had a low threshold for at least the auditory effects of electricity. In other words, twelve percent of the population was, and presumably still is, able in some way to hear unusually low levels of electric current.

Weather Sensitivity
Unlike electrical sensitivity per se, the study of human sensitivity to the weather has a venerable history going back five thousand years in Mesopotamia, and possibly as long in China and Egypt. In his treatise on *Airs, Waters and Places*, written about 400 B.C., Hippocrates said that the human condition is largely determined by the climate of the place where one lives, and its variations. This is a discipline that, however much ignored and underfunded, is mainstream. And yet the name of this science, "biometeorology," hides an open secret: some thirty percent of any population, no matter their ethnic origin, are weather sensitive and therefore, according to some textbooks in that field, electrically sensitive.[26]

The International Society for Biometeorology was founded in 1956 by Dutch geophysicist Solco Tromp with headquarters in, appropriately, Leyden, the city that launched the electrical age over two centuries before. And for the next forty years—until cell phone companies began to put pressure on researchers to repudiate an entire, long-established scientific discipline[27]—bioelectricity and biomagnetism were the subjects of intensive research and were the focus of one of the Society's ten permanent Study Groups. In 1972, an International Symposium was held in the Netherlands on the "Biological Effects of Natural Electric, Magnetic and Electromagnetic Fields." In 1985, the Fall issue of the *International Journal of Biometeorology* was devoted entirely to papers on the effects of air ions and atmospheric electricity.

"We do great injustice to the electrosensitive patients," wrote Felix Gad Sulman, "when we treat them as psychiatric patients." Sulman was a medical doctor at Hadassah University Medical Center in Jerusalem, and chair of the Medical School's Bioclimatology Unit. In 1980, he published a 400-page monograph titled *The Effects of Air Ionization, Electric Fields, Atmospherics and Other Electric Phenomena on Man and Animal*. Sulman, together with fifteen colleagues in other medical and technical fields, had studied 935 weather-sensitive patients over a period of fifteen years. One of their most riveting findings was that eighty percent of these patients could predict weather changes twelve to forty-eight hours before they happened. "The 'prophetic' patients were all sensitive to the electrical changes preceding the arrival of a weather change," Sulman wrote. "They reacted by serotonin release to ions and atmospherics which naturally arrive with the speed of electricity—before the slow pace of the weather winds."[28]

Weather sensitivity had emerged from within the walls of centuries of imprecise medical hearsay and was being exposed to the light of rigorous laboratory analysis. But this put the field of biometeorology on a collision course with an emerging technological dynamo. For if a third of the earth's population are that sensitive to the gentle flow of ions and the subtle electromagnetic whims of the atmosphere, what must the incessant rivers of ions from our computer screens, and the turbulent storms of emissions from our cell phones, radio towers, and power lines be doing to us all? Our society is refusing to make the connection. In fact, at the 19th International Congress of Biometeorology held September 2008 in Tokyo, Hans Richner, professor of physics at the Swiss Federal Institute of Technology, stood up and actually told his colleagues that because cell phones are not dangerous, and their electromagnetic fields are so much stronger than those from the atmosphere, therefore decades of research were wrong and biometeorologists should not study human interactions with electric fields any more.[29] In other words, since we are all using cell phones, therefore we have to presume that they're safe, and so all the effects on people, plants and animals from mere atmospheric fields that have been reported in hundreds of laboratories could not have happened! It is no wonder that long-time biometeorological researcher Michael Persinger, professor at Laurentian University in Ontario, says that the scientific method has been abandoned.[30]

But in the eighteenth century, electricians did make the connection. The reactions of their patients to the friction machine shed new light on an ancient mystery. The problem was framed by Mauduyt. "Men and animals," he explained, "experience a sort of weakness and languor on stormy days. This depression reaches its highest degree at the moment preceding the storm, it diminishes shortly after the storm has burst, and especially when a certain quantity of rain has fallen; it dissipates and terminates with it. This fact is well known, important, and has occupied physicians for a long time without their being able to find a sufficient explanation."[31]

The answer, said Bertholon, was now at hand: "Atmospheric electricity and artificial electricity depend on one and the same fluid that produces various effects relative to the animal economy. A person who is insulated and electrized by the bath represents one who stands on the earth when it is electrified to excess; both are filled to overabundance with the electric fluid. It is accumulated around them in the same fashion."[32] The electric circuit created by a machine was a microcosm of the grand circuit created by the heavens and the earth.

Italian physicist Giambatista Beccaria described the global electrical circuit in surprisingly modern terms (see chapter 9). "Previous to rain," he wrote, "a quantity of electrical matter escapes out of the earth, in some place where there was a redundancy of it; and ascends to the higher regions of the air... The clouds that bring rain diffuse themselves from over those parts of the earth which abound with the electric fire, to those parts which are exhausted of it; and, by letting fall their rain, restore the equilibrium between them."[33]

Eighteenth century scientists were not the first to discover this. The Chinese model, formulated in the *Yellow Emperor's Classic of Internal Medicine*, written in the fourth century B.C., is similar. In fact, if one understands that "Qi" is electricity, and that "Yin" and "Yang" are negative and positive, the language is almost identical: "The pure Yang forms the heaven, and the turbid Yin forms the earth. The Qi of the earth ascends and turns into clouds, while the Qi of the heaven descends and turns into rain."[34]

Famous weather sensitive—and therefore electrically sensitive—individuals have included Lord Byron, Christopher Columbus, Dante, Charles Darwin, Benjamin Franklin, Goethe, Victor Hugo, Leonardo da Vinci, Martin Luther, Michelangelo, Mozart, Napoleon, Rousseau, and Voltaire.[35]

4. The Road Not Taken

DURING THE 1790s, European science faced an identity crisis. For centuries, philosophers had been speculating about the nature of four mysterious substances that animated the world. They were light, electricity, magnetism, and caloric (heat). Most thought the four fluids were somehow related to one another, but it was electricity that was most obviously connected with life. Electricity alone breathed motion into nerves and muscles, and pulsations into the heart. Electricity boomed from the heavens, stirred winds, tossed clouds, pelted the earth with rain. Life was movement, and electricity made things move.

Electricity was "an electric and elastic spirit" by which "all sensation is excited, and the members of animal bodies move at the command of the will, namely, by the vibrations of this spirit, mutually propagated along the solid filaments of the nerves, from the outward organs of sense to the brain, and from the brain into the muscles."[1] So spoke Isaac Newton in 1713, and for the next century few disagreed.

Electricity was:

> "an element that is to us more intimate than the very
> air that we breathe."
>
> *Abbé Nollet, 1746*[2]

"the principle of animal functions, the instrument of will and the vehicle of sensations."

French physicist Marcelin Ducarla-Bonifas, 1779[3]

"that fire necessary to all bodies and which gives them life… that is both attached to known matter and yet apart from it."

Voltaire, 1772[4]

"one of the principles of vegetation; it's what fertilizes our fields, our vines, our orchards, and what brings fecundity to the depths of the waters."

Jean-Paul Marat, M.D., 1782[5]

"the Soul of the Universe" that "produces and sustains Life thro-out all Nature, as well in Animals as in Vegetables"

John Wesley, founder of the Methodist Church, 1760.[6]

Then came Luigi Galvani's stunning announcement that simply touching a brass hook to an iron wire would cause a frog's leg to contract. A modest professor of obstetrics at the Institute of Sciences of Bologna, Galvani thought this proved something about physiology: each muscle fiber must be something like an organic Leyden jar. The metallic circuit, he reasoned, released the "animal electricity" that was manufactured by the brain and stored in the muscles. The function of the nerves was to discharge that stored electricity, and the dissimilar metals, in direct contact with the muscle, somehow mimicked the natural function of the animal's own nerves.

But Galvani's countryman, Alessandro Volta, held an opposing, and at that time heretical opinion. The electric current, he claimed, came not from the animal, but from the dissimilar metals themselves. The convulsions, according to Volta, were due entirely to the external stimulus. Furthermore, he proclaimed, "animal electricity" did not even exist, and to try to prove it he made his momentous demonstration that the electric current could be produced by the contact of different metals alone, without the intervention of the animal.

The combatants represented two different ways of looking at the world. Galvani, trained as a physician, sought his explanations in biology; the metals, to him, were an adjunct to a living organism. Volta, the self-taught physicist, saw precisely the opposite: the frog was only an extension of the non-living metallic circuit. For Volta, the contact of one conductor with another was a sufficient cause, even for the electricity within the animal: muscles and nerves were nothing more than moist conductors, just another kind of an electric battery.

Their dispute was a clash not just between scientists, not just between theories, but between centuries, between mechanism and spirit, an existential struggle that was ripping the fabric of western civilization in the late 1790s. Hand weavers were shortly to rise in revolt against mechanical looms, and they were destined to lose. The material, in science as in life, was displacing and obscuring the vital.

Volta, of course, won the day. His invention of the electric battery gave an enormous boost to the industrial revolution, and his insistence that electricity had nothing to do with life also helped steer its direction. This mistake made it possible for society to harness electricity on an industrial scale—to wire the world, even as Nollet had envisioned—without worrying about the effects such an enterprise might have on biology. It permitted people to begin to disregard the accumulated knowledge gained by eighteenth century electricians.

Eventually, one learns if one reads the textbooks, Italian physicists Leopoldo Nobili and Carlo Matteucci, and then a German physiologist named Emil du Bois-Reymond, came along and proved that electricity did after all have something to do with life, and that nerves and muscles were not just moist conductors. But the mechanistic dogma was already entrenched, resisting all attempts to properly restore the marriage between life and electricity. Vitalism was permanently relegated to religion, to the realm of the insubstantial, divorced forever from the domain of serious investigative science. The life force, if it existed, could not be subjected to experiment, and it certainly could not be the same stuff that turned electric motors, lit light bulbs, and traveled thousands of miles on copper wires. Yes, electricity had finally been discovered in nerves and muscles, but its action was only a by-product of the journeys of sodium and potassium ions across membranes and the flight of neurotransmitters across synapses. Chemistry, that was the thing, the fertile, seemingly endless scientific soil that nurtured all biology, all physiology. Long-range forces were banished from life.

The other, even more significant change that occurred after 1800 is that gradually people even forgot to wonder what the nature of electricity was. They began to build a permanent electrical edifice, whose tentacles snaked everywhere, without noticing, or thinking about, its consequences. Or, rather, they recorded its consequences in minute detail without ever making the connection to what they were building.

5. Chronic Electrical Illness

In 1859, THE CITY of London underwent an astonishing metamorphosis. A tangle of electric wires, suddenly and inescapably, was brought to the streets, shops, and residential rooftops of its two and a half million inhabitants. I will let one of the most famous English novelists, who was an eyewitness, begin the story.

"About twelve years ago," wrote Charles Dickens, "when the tavern fashion of supplying beer and sandwiches at a fixed price became very general, the proprietor of a small suburban pothouse reduced the system to an absurdity by announcing that he sold a glass of ale and an electric shock for fourpence. That he really traded in this combination of science and drink is more than doubtful, and his chief object must have been to procure an increase of business by an unusual display of shopkeeping wit. Whatever motive he had to stimulate his humor, the fact should certainly be put upon record that he was a man considerably in advance of his age. He was probably not aware that his philosophy in sport would be made a science in earnest in the space of a few years, any more than many other bold humorists who have been amusing on what they know nothing about. The period has not yet arrived when the readers of Bishop Wilkin's famous discourse upon aerial navigation will be able to fly to the moon, but the hour is almost at hand when the fanciful announcement of the beer-shop keeper will represent an every-day familiar fact. A glass of ale and an electric shock will shortly be sold for fourpence, and the scientific part of the bargain will be something more useful than a mere fillip to the human nerves. It will be an electric shock that sends a message across the house-tops through the web of wires to any one of a hundred and twenty district telegraph stations, that are to be scattered amongst the shopkeepers all over the town.

"The industrious spiders have long since formed themselves into a commercial company, called the London District Telegraph Company (limited), and they have silently, but effectively, spun their trading web. One hundred and sixty miles of wire are now fixed along parapets, through trees, over garrets, round chimney-pots, and across roads on the southern side of the river, and the other one hundred and twenty required miles will soon be fixed in the same manner on the northern side. The difficulty decreases as the work goes on, and the sturdiest Englishman is ready to give up the roof of his castle in the interests of science and the public good, when he finds that many hundreds of his neighbors have already led the way."

English citizens did not necessarily welcome the prospect of electric wires being attached to their homes. "The British householder has never seen a voltaic battery kill a cow," wrote Dickens, "but he has heard that it is quite capable of such a feat. The telegraph is worked, in most cases, by a powerful voltaic battery, and therefore the British householder, having a general dread of lightning, logically keeps clear of all such machines." Nonetheless, Dickens tells us, the agents of the London District Telegraph Company persuaded nearly three thousand five hundred property owners to lend their rooftops as resting places for the two hundred and eighty miles of wires that were crisscrossing all of London, and that were shortly to drop into the shops of grocers, chemists, and tavern-keepers all over the city.[1]

A year later, the electrical web above London homes became even more densely woven when the Universal Private Telegraph Company opened its doors. In contrast to the first company, whose stations accepted only public business, Universal rented telegraph facilities to individuals and businesses for private use. Cables containing up to a hundred wires each formed the backbone of the system, each wire departing from its companions at the nearest approach to its destination. By 1869, this second company had strung more than two thousand five hundred miles of cable, and many times as much wire over the heads and under the feet of Londoners, to serve about fifteen hundred subscribers scattered throughout the city.

A similar transformation was occurring more or less everywhere in the world. The rapidity and intensity with which this happened is not appreciated today.

The systematic electrification of Europe had begun in 1839 with the opening of the magnetic telegraph on the Great Western Railway between West Drayton and London. The electrification of America began a few years later, when Samuel Morse's first telegraph line marched from Baltimore to Washingon in 1844 along the Baltimore and Ohio Railroad. Even earlier, electric doorbells and annunciators began decorating homes, offices, and hotels, the first complete system having been installed in 1829 in Boston's Tremont House, where all hundred and seventy guest rooms were connected by electric wires to a system of bells in the main office.

Electric burglar alarms were available in England by 1847, and soon afterwards in the United States.

By 1850, telegraph lines were under construction on every continent except Antarctica. Twenty-two thousand miles of wire had been energized in the United States; four thousand miles were advancing through India, where "monkeys and swarms of large birds" were alighting on them"[2]; one thousand miles of wire were spreading in three directions from Mexico City. By 1860, Australia, Java, Singapore, and India were being joined undersea. By 1875, thirty thousand miles of submarine cable had demolished oceanic barriers to communication, and the tireless weavers had electrified seven hundred thousand miles of copper web over the surface of the earth—enough wire to encircle the globe almost thirty times.

And the traffic of electricity accelerated even more than the number of wires, as first duplexing, then quadriplexing, then automatic keying meant that current flowed at all times—not just when messages were being sent—and that multiple messages could be sent over the same wire at the same time, at a faster and faster rate.

Almost from the beginning, electricity became a presence in the average urban dweller's life. The telegraph was never just an adjunct to railroads and newspapers. In the days before telephones, telegraph machines were installed first in fire and police stations, then in stock exchanges, then in the offices of messenger services, and soon in hotels, private businesses, and homes. The first municipal telegraph system in New York City was built by Henry Bentley in 1855, connecting fifteen offices in Manhattan and Brooklyn. The Gold and Stock Telegraph Company, incorporated in 1867, supplied instantaneous price quotations from the Stock, Gold, and other Exchanges telegraphically to hundreds of subscribers. In 1869, the American Printing Telegraph Company was created to provide private telegraph lines to businesses and individuals. The Manhattan Telegraph Company was organized in competition two years later. By 1877, the Gold and Stock Telegraph Company had acquired both those companies and was operating 1,200 miles of wire. By 1885, the industrious spiders linking almost thirty thousand homes and businesses had to spin webs over New York even more intricate than the ones over Dickens' London.

In the midst of this transformation, a slender, slightly deaf clergyman's son wrote the first clinical histories of a previously unknown disease that he was observing in his neurology practice in New York City. Dr. George Miller Beard was only three years out of medical school. Yet his paper was accepted and published, in 1869, in the prestigious *Boston Medical and Surgical Journal*, later renamed the *New England Journal of Medicine*.

A self-assured young man, possessed of a serenity and hidden sense of humor that attracted people to him, Beard was a sharp observer who, even so early in his career, was not afraid to break new medical ground. Although he was sometimes ridiculed by his elders for his novel ideas, one of his colleagues was to say many years after his death that Beard "never said an unkind word against anyone."[3] Besides this new disease, he also specialized in electrotherapy and hypnotherapy, both of which he was instrumental in restoring to good repute, half a century after the death of Mesmer. In addition, Beard contributed to the knowledge of the causes and treatment of hay fever and seasickness. And in 1875 he collaborated with Thomas Edison in investigating an "etheric force" that Edison had discovered, which was able to travel through the air, causing sparks in nearby objects without a wired circuit. Beard correctly surmised, a decade before Hertz and two decades before Marconi, that this was high frequency electricity, and that it might one day revolutionize telegraphy.[4]

George Miller Beard, M.D.
(1839-1883)

As far as the new disease that he described in 1869, Beard did not guess its cause. He simply thought it was a disease of modern civilization, caused by stress, that was previously uncommon. The name he gave it, "neurasthenia," just means "weak nerves." Although some of its symptoms resembled other diseases, neurasthenia seemed to attack at random and for no reason and no one was expected to die from it. Beard certainly didn't connect the disease with electricity, which was actually his preferred treatment for neurasthenia—when the patient could tolerate it. When he died in 1883, the cause of neurasthenia, to everyone's frustration, had still not been identified. But in a large portion of the world where the term "neurasthenia" is still in everyday use among doctors—and the term is used in most of the world outside of the United States—electricity is recognized today as one of its causes. And the electrification of the world was undoubtedly responsible for its appearance out of nowhere during the 1860s, to become a pandemic during the following decades.

Today, when million-volt power lines course through the countryside, twelve-thousand-volt lines divide every neighborhood, and sets of thirty-ampere circuit breakers watch over every home, we tend to forget what the natural situation really is. None of us can begin to imagine what it would feel like to live on an unwired earth. Not since the presidency of James Polk have our cells, like puppets on invisible strings, been given a second's rest from the electric vibrations. The gradual increase in voltage during the past century and a half has been only a matter of degree. But the sudden overwhelming of the earth's own nurturing fields, during the first few decades of technological free-for-all, had a drastic impact on the very character of life.

In the earliest days telegraph companies, in countryside and in cities, built their lines with only one wire, the earth itself completing the electric circuit. None of the return current flowed along a wire, as it does in electrical systems today; all of it traveled through the ground along unpredictable paths.

Twenty-five-foot-high wooden poles supported the wires on their journeys between towns. In cities, where multiple telegraph companies competed for customers and space was at a premium, forests of overhead wires tangled their way between housetops, church steeples, and chimneys, to which they attached themselves like vines. And from those vines hung electric fields that blanketed the streets and byways and the spaces within the homes to which they clung.

The historical numbers provide a clue to what happened. According to George Prescott's 1860 book on the *Electric Telegraph*, a typical battery used for a 100-mile length of wire in the United States was "fifty cups of Grove," or fifty pairs of zinc and platinum plates, which provided an electric potential of about 80 volts.[5] In the earliest systems, the current only flowed when the telegraph operator pressed the sending key. There were five letters per word and, in the Morse alphabet, an average of three dots or dashes per letter. Therefore, if the operator was proficient and averaged thirty words per minute, she pressed the key at a rhythm of 7.5 strokes per second. This is the very near the fundamental resonant frequency (7.8 Hz) of the biosphere, to which all living things, as we will see in chapter 9, are tuned, and whose average strength—about a third of a millivolt per meter—is given in textbooks. It is easy to calculate, using these simple assumptions, that the electric fields beneath the earliest telegraph wires were up to 30,000 times stronger than the natural electric field of the earth at that frequency. In reality the rapid interruptions in telegraph keying also produced a wide range of radio frequency harmonics, which also traveled along the wires and radiated through the air.

The magnetic fields can also be estimated. Based on the values for electrical resistance for wires and insulators as given by Samuel Morse himself,[6] the amount of current on a typical long-distance wire varied from about 0.015 ampere to 0.1 ampere, depending on the length of the line and the weather. Since the insulation was imperfect, some current escaped down each telegraph pole into the earth, a flow which increased when it rained. Then, using the published value of 10^{-8} gauss for the magnetic field of the earth at 8 Hz, one may calculate that the magnetic field from a single early telegraph wire would have exceeded the earth's natural magnetic field at that frequency for a distance of two to twelve miles on either side of the line. And since the earth is not uniform, but contains underground streams, iron deposits, and other conductive paths over which the return current would travel, exposure of the population to these new fields varied widely.

In cities, each wire carried about 0.02 ampere and exposure was universal. The London District Telegraph Company, for example, commonly had ten wires together, and the Universal Private Telegraph Company had up to one hundred wires together, strung above the streets and rooftops over a large part of town. Although the apparatus and alphabet of London District differed from those used in America, the current through its wires fluctuated at a similar rate—about 7.2 vibrations per second if the operator transmitted 30 words per minute.[7] And Universal's dial telegraph was a hand-cranked magneto-electric machine that actually sent alternating current through the wires.

One enterprising scientist, professor of physics John Trowbridge at Harvard University, decided to put to the test his own conviction that signals riding on telegraph wires that were grounded at both ends were escaping from their appointed paths and could easily be detected at remote locations. His test signal was the clock at the Harvard Observatory, which transmitted time signals four miles by wire from Cambridge to Boston. His receiver was a newly-invented device—a telephone—connected to a length of wire five hundred feet long and grounded to the earth at both ends. Trowbridge found that by tapping the earth in this way he could clearly hear the ticking of the observatory clock up to a mile from the observatory at various points not in the direction of Boston. The earth was being massively polluted with stray electricity, Trowbridge concluded. Electricity originating in the telegraph systems of North America should even be detectable on the other side of the Atlantic Ocean, he said after doing some calculations. If a powerful enough Morse signal, he wrote, were sent from Nova Scotia to Florida over a wire that was grounded at both ends, someone on the coast of France should be able to hear the signal by tapping the earth using his method.

A number of historians of medicine who have not dug very deep have asserted that neurasthenia was not a new disease, that nothing had changed, and that late nineteenth and early twentieth century high society was really suffering from some sort of mass hysteria.[8]

A list of famous American neurasthenes reads like a Who's Who of literature, the arts, and politics of that era. They included Frank Lloyd Wright, William, Alice and Henry James, Charlotte Perkins Gilman, Henry Brooks Adams, Kate Chopin, Frank Norris, Edith Wharton, Jack London, Theodore Dreiser, Emma Goldman, George Santayana, Samuel Clemens, Theodore Roosevelt, Woodrow Wilson, and a host of other well-known figures.

Historians who think they have found neurasthenia in older text-books have been confused by changes in medical terminology, changes that have prevented an understanding of what happened to our world a hundred and fifty years ago. For example, the term "nervous" was used for centuries without the connotations given to it by Freud. It simply meant, in today's language, "neurological." George Cheyne, in his 1733 book, *The English Malady*, applied the term "nervous disorder" to epilepsy, paralysis, tremors, cramps, contractions, loss of sensation, weakened intellect, complications of malaria, and alcoholism. Robert Whytt's 1764 treatise on "nervous disorders" is a classic work on neurology. It can be confusing to see gout, tetanus, hydrophobia, and forms of blindness and deafness called "nervous disorders" until one realizes that the term "neurological" did not replace "nervous" in clinical medicine until the latter half of the nineteenth century. "Neurology," at that time, meant what "neuroanatomy" means today.

Another source of confusion for a modern reader is the old use of the terms "hysterical" and "hypochondriac" to describe neurological conditions of the body, not the mind. The "hypochondria" were the abdominal regions and "hystera," in Greek, was the uterus; as Whytt explained in his treatise, hysterical and hypochondriac disorders were those neurological diseases that were believed to have their origins in the internal organs, "hysterical" traditionally being applied to women's diseases and "hypochondriac" to men's. When the stomach, bowels and digestion were involved, the illness was called hypochondriac or hysterical depending on the patient's sex. When the patient had seizures, blackouts, tremors, or palpitations, but the internal organs were not affected, the illness was called simply "nervous."

Confounding this confusion still further were the Draconian treatments that were standard medical practice until well into the nineteenth century, which themselves often caused serious neurological problems. These were based on the humoral theory of medicine as set forth by Hippocrates in the fifth century B.C. For thousands of years all sickness was believed to be caused by an imbalance of "humors"—the four humors being phlegm, yellow bile, black bile, and blood—so that the goal of medical treatment was to strengthen the deficient humors and drain off those that were in excess. Therefore all medical complaints, major and minor, were subject to treatment by some combination of purging, vomiting, sweating, bleeding, medicines, and dietary prescriptions. And the drugs were liable to be neurotoxic, preparations containing heavy metals such as antimony, lead, and mercury being frequently prescribed.

By the early nineteenth century, some doctors had begun to question the humoral theory of disease, but the term "neurology" had not yet acquired its modern meaning. During this time the realization that many illnesses were still being called "hysterical" and "hypochondriac" when there was nothing wrong with the uterus or internal organs led a number of physicians to try out new names for diseases of the nervous system. In the eighteenth century Pierre Pomme's "vaporous conditions" included cramps, convulsions, vomiting, and vertigo. Some of these patients had total suppression of urine, spitting of blood, fevers, smallpox, strokes, and other illnesses that sometimes took their lives. When the disease didn't kill them the frequent bleedings often did. Thomas Trotter's book, *A View of the Nervous Temperament*, written in 1807, included cases of worms, chorea, tremors, gout, anemia, menstrual disorders, heavy metal poisonings, fevers, and convulsions leading to death. A series of later French doctors tried out names like "proteiform neuropathy," "nervous hyperexcitability," and "the nervous state." Claude Sandras' 1851 *Traité Pratique des Maladies Nerveuses* ("Practical Treatise on Nervous Diseases") is a conventional textbook on neurology. Eugène Bouchut's 1860 book on "l'état nerveux" ("the nervous state") contained many case histories of patients suffering from the effects of blood-letting, tertiary syphilis, typhoid fever, miscarriage, anemia, paraplegia, and other acute and chronic illnesses of known causes, some lethal. Beard's neurasthenia is not to be found.

In fact, the first description anywhere of the disease to which Beard called the world's attention is in Austin Flint's textbook of medicine published in New York in 1866. A professor at the Bellevue Hospital Medical College, Flint devoted two brief pages to it and gave it almost the same name Beard was to popularize three years later. Patients with "nervous asthenia," as he called it, "complain of languor, lassitude, want of buoyancy, aching of the limbs, and mental depression. They are wakeful during the night, and enter upon their daily pursuits with a sense of fatigue."[9] These patients did not have anemia or any other evidence of organic disease. They also did not die of their disease; on the contrary, as Beard and others were later to also observe, they seemed to be protected from ordinary acute illnesses and lived, on average, longer than others.

These first publications were the beginning of an avalanche. "More has been written about neurasthenia in the course of the last decade," wrote Georges Gilles de la Tourette in 1889, "than on epilepsy or hysteria, for example, during the last century."[10]

The best way to familiarize the reader with both the disease and its cause is to introduce another prominent New York City physician who herself suffered from it—though by the time she told her story the American medical profession had been trying to find the cause of neurasthenia for nearly half a century and, not finding one, had concluded that the illness was psychosomatic.

Dr. Margaret Abigail Cleaves, born in the territory of Wisconsin, had graduated from medical school in 1879. She had first worked at the State Hospital for the Insane at Mt. Pleasant, Iowa, and from 1880 to 1883 had served as chief physician to the female patients of the

Margaret Abigail Cleaves, M.D.
(1848-1917)

Pennsylvania State Lunatic Hospital. In 1890 she had moved to the big city, where she had opened a private practice in gynecology and psychiatry. It was not until 1894, at the age of 46, that she was diagnosed with neurasthenia. What was new was her heavy exposure to electricity: she had begun to specialize in electrotherapy. Then, in 1895, she opened the New York Electro-Therapeutic Clinic, Laboratory, and Dispensary, and within a matter of months experienced what she termed her "complete break."

The details, written down over time in her *Autobiography of a Neurasthene*, describe the classic syndrome presented nearly half a century earlier by Beard. "I knew neither peace nor comfort night nor day," she wrote. "There remained all the usual pain of nerve trunks or peripheral nerve endings, the exquisite sensitiveness of the body, the inability to bear a touch heavier than the brush of a butterfly's wing, the insomnia, lack of strength, the recurrence of depression of spirits, the inability to use my brain at my study and writing as I wished."

"It was with the greatest difficulty," she wrote on another occasion, "to even use knife and fork at the table, while the routine carving was an impossibility."

Cleaves had chronic fatigue, poor digestion, headaches, heart palpitations and tinnitus. She found the sounds of the city unbearable. She smelled and tasted "phosphorus." She became so sensitive to the sun that she lived in darkened rooms, able to go outdoors only at night. She gradually lost her hearing in one ear. She became so affected by atmospheric electricity that, by her sciatica, her facial pain, her intense restlessness, her feeling of dread, and her sensation "of a crushing weight bowing me to the earth," she could predict with certainty 24 to 72 hours in advance that the weather was going to change. "Under the influence of oncoming electrical storms," she wrote, "my brain does not function."[11]

And yet through it all, suffering until the end of her life, she was dedicated to her profession, exposing herself day in and day out to electricity and radiation in their various forms. She was a founding and very active officer of the American Electro-Therapeutic Association. Her textbook on *Light Energy* taught the therapeutic uses of sunlight, arc light, incandescent light, fluorescent light, X-rays, and radioactive elements. She was the first physician to use radium to treat cancer.

How could she not have known? And yet it was easy. In her day as in ours, electricity did not cause disease, and neurasthenia—it had finally been decided—resided in the mind and emotions.

Other related illnesses were described in the late nineteenth and early twentieth centuries, occupational diseases suffered by those who worked in proximity to electricity. "Telegrapher's cramp," for example, called by the French, more accurately, "mal télégraphique" ("telegraphic sickness") because its effects were not confined to the muscles of the operator's hand. Ernest Onimus described the affliction in Paris in the 1870s. These patients suffered from heart palpitations, dizziness, insomnia, weakened eyesight, and a feeling "as though a vice were gripping the back of their head." They suffered from exhaustion, depression, and memory loss, and after some years of work a few descended into insanity. In 1903, Dr. E. Cronbach in Berlin gave case histories for seventeen of his telegraphist patients. Six had either excessive perspiration or extreme dryness of hands, feet, or body. Five had insomnia. Five had deteriorating eyesight. Five had tremors of the tongue. Four had lost a degree of their hearing. Three had irregular heartbeats. Ten were nervous and irritable both at work and at home. "Our nerves are shattered," wrote an anonymous telegraph worker in 1905, "and the feeling of vigorous health has given way to a morbid weakness, a mental depression, a leaden exhaustion... Hanging always between sickness and health, we are no longer whole, but only half men; as youths we are already worn out old men, for whom life has become a burden... our strength prematurely sapped, our senses, our memory dulled, our impressionability curtailed." These people knew the cause of their illness. "Has the release of electrical power from its slumber," asked the anonymous worker, "created a danger for the health of the human race?"[12] In 1882, Edmund Robinson encountered similar awareness among his telegraphist patients from the General Post Office at Leeds. For when he suggested treating them with electricity, they "declined trying anything of the kind."

Long before that, an anecdote from Dickens could have served as a warning. He had toured St. Luke's Hospital for Lunatics. "We passed a deaf and dumb man," he wrote, "now afflicted with incurable madness." Dickens asked what employment the man had been in. "'Aye,' says Dr. Sutherland, 'that is the most remarkable thing of all, Mr. Dickens. He was employed in the transmission of electric-telegraph messages.'" The date was January 15, 1858.[13]

Telephone operators, too, often suffered permanent injury to their health. Ernst Beyer wrote that out of 35 telephone operators that he had treated during a five-year period, not a single one had been able to return to work. Hermann Engel had 119 such patients. P. Bernhardt had over 200. German physicians routinely attributed this illness to electricity. And after reviewing dozens of such publications, Karl Schilling, in 1915, published a clinical description of the diagnosis, prognosis, and treatment of illness caused by chronic exposure to electricity. These patients typically had headaches and dizziness, tinnitus and floaters in the eyes, racing pulse, pains in the region of the heart, and palpitations. They felt weak and exhausted and were unable to concentrate. They could not sleep. They were depressed and had anxiety attacks. They had tremors. Their reflexes were elevated, and their senses were hyper-acute. Sometimes their thyroid was hyperactive. Occasionally, after long illness, their heart was enlarged. Similar descriptions would come throughout the twentieth century from doctors in the Netherlands, Belgium, Denmark, Austria, Italy, Switzerland, the United States, and Canada.[14] In 1956, Louis Le Guillant and his colleagues reported that in Paris "there is not a single telephone operator who doesn't experience this nervous fatigue to one degree or another." They described patients with holes in their memory, who couldn't carry on a conversation or read a book, who fought with their husbands for no reason and screamed at their children, who had abdominal pains, headaches, vertigo, pressure in their chest, ringing in their ears, visual disturbances, and weight loss. A third of their patients were depressed or suicidal, almost all had anxiety attacks, and over half had disturbed sleep.

As late as 1989, Annalee Yassi reported widespread "psychogenic illness" among telephone operators in Winnipeg, Manitoba and St. Catharines, Ontario, and in Montreal, Bell Canada reported that 47 percent of its operators complained of headaches, fatigue, and muscular aches related to their work.

Then there was "railway spine," a misnamed illness that was investigated as early as 1862 by a commission appointed by the British medical journal *Lancet*. The commissioners blamed it on vibrations, noise, speed of travel, bad air, and sheer anxiety. All those factors were present, and no doubt contributed their share. But there was also one more that they did not consider. Because by 1862, every rail line was sandwiched between one or more telegraph wires running overhead and the return currents from those lines coursing beneath, a portion of which flowed along the metal rails themselves, upon which the passenger cars rode. Passengers and train personnel commonly suffered from the same complaints later reported by telegraph and telephone operators: fatigue, irritability, headaches, chronic dizziness and nausea, insomnia, tinnitus, weakness, and numbness. They had rapid heart beat, bounding pulse, facial flushing, chest pains, depression, and sexual dysfunction. Some became grossly overweight. Some bled from the nose, or spat blood. Their eyes hurt, with a "dragging" sensation, as if they were being pulled into their sockets. Their vision and their hearing deteriorated, and a few became gradually paralyzed. A decade later they would have been diagnosed with neurasthenia—as many railroad employees later were.

The most salient observations made by Beard and the late nineteenth century medical community about neurasthenia are these:

It spread along the routes of the railroads and telegraph lines.

It affected both men and women, rich and poor, intellectuals and farmers.

Its sufferers were often weather sensitive.

It sometimes resembled the common cold or influenza.

It ran in families.

It seized most commonly people in the prime of their life, ages 15 to 45 according to Beard, 15 to 50 according to Cleaves, 20 to 40 according to H. E. Desrosiers,[15] 20 to 50 according to Charles Dana.

It lowered one's tolerance for alcohol and drugs.

It made people more prone to allergies and diabetes.

Neurasthenes tended to live longer than average.

And sometimes—a sign whose significance will be discussed in chapter 10—neurasthenes passed reddish or dark brown urine.

It was the German physician Rudolf Arndt who finally made the connection between neurasthenia and electricity. His patients who could not tolerate electricity intrigued him. "Even the weakest galvanic current," he wrote, "so weak that it scarcely deflected the needle of a galvanometer, and was not perceived in the slightest by other people, bothered them in the extreme." He proposed in 1885 that "electrosensitivity is characteristic of high-grade neurasthenia." And he prophesied that electrosensitivity "may contribute not insubstantially to the elucidation of phenomena that now seem puzzling and inexplicable."

He wrote this in the middle of an intense, unrelenting haste to wire the whole world, driven by an unquestioning embrace of electricity, even an adoration, and he wrote it as though he knew he was risking his reputation. A large obstacle to the proper study of neurasthenia, he suggested, was that people who were less sensitive to electricity did not take its effects at all seriously: instead, they placed them in the realm of superstition, "lumped together with clairvoyance, mind-reading and mediumship."[16]

That obstacle to progress confronts us still today.

The Renaming

In December 1894, an up-and-coming Viennese psychiatrist wrote a paper whose influence was enormous and whose consequences for those who came after have been profound and unfortunate. Because of him, neurasthenia, which is still the most common illness of our day, is accepted as a normal element of the human condition, for which no external cause need be sought. Because of him, environmental illness, that is, illness caused by a toxic environment, is widely thought not to exist, its symptoms automatically blamed on disordered thoughts and out-of-control emotions. Because of him, we are today putting millions of people on Xanax, Prozac, and Zoloft instead of cleaning up their environment. For over a century ago, at the dawn of an era that blessed the use of electricity full throttle not just for communication but for light, power, and traction, Sigmund Freud renamed neurasthenia "anxiety neurosis" and its crises "anxiety attacks." Today we call them also "panic attacks."

The symptoms listed by Freud, in addition to anxiety, will be familiar to every doctor, every "anxiety" patient, and every person with electrical sensitivity:

Irritability
Heart palpitations, arrhythmias, and chest pain
Shortness of breath and asthma attacks
Perspiration
Tremor and shivering
Ravenous hunger
Diarrhea
Vertigo
Vasomotor disturbances (flushing, cold extremities, etc.)
Numbness and tingling
Insomnia
Nausea and vomiting
Frequent urination
Rheumatic pains
Weakness
Exhaustion

Freud ended the search for a physical cause of neurasthenia by reclassifying it as a mental disease. And then, by designating almost all cases of it as "anxiety neurosis," he signed its death warrant. Although he pretended to leave neurasthenia as a separate neurosis, he didn't leave it many symptoms, and in Western countries it has been all but forgotten. In some circles it persists as "chronic fatigue syndrome," a disease without a cause that many doctors believe is also psychological and that most don't take seriously. Neurasthenia survives in the United States only in the common expression, "nervous breakdown," whose origin few people remember.

In the International Classification of Diseases (ICD-10), there is a unique code for neurasthenia, F48.0, but in the version used in the United States (ICD-10-CM), F48.0 has been removed. In the American version, neurasthenia is only one among a list of "other non-psychotic mental disorders" and is almost never diagnosed. Even in the Diagnostic and Statistical Manual (DSM-V), the official system for assigning codes to mental diseases in American hospitals, there is no code for neurasthenia.

It was a death warrant only in North America and Western Europe, however. Half the world still uses neurasthenia as a diagnosis in the sense intended by Beard. In all of Asia, Eastern Europe, Russia and the former Soviet Republics, neurasthenia is today the most common of all psychiatric diagnoses as well as one of the most frequently diagnosed diseases in general medical practice.[17] It is often considered a sign of chronic toxicity.[18]

In the 1920s, just as the term was being abandoned in the West, it was first coming into use in China.[19] The reason: China was just beginning to industrialize. The epidemic that had begun in Europe and America in the late nineteenth century had not yet reached China at that time.

In Russia, which began to industrialize along with the rest of Europe, neurasthenia became epidemic in the 1880s.[20] But nineteenth century Russian medicine and psychology were heavily influenced by neurophysiologist Ivan Sechenov, who emphasized external stimuli and environmental factors in the workings of the mind and body. Because of Sechenov's influence, and that of his pupil Ivan Pavlov after him, the Russians rejected Freud's redefinition of neurasthenia as anxiety neurosis, and in the twentieth century Russian doctors found a number of environmental causes for neurasthenia, prominent among which are electricity and electromagnetic radiation in their various forms. And as early as the 1930s, because they were looking for it and we weren't, a new clinical entity was discovered in Russia called "radio wave sickness," which is included today, in updated terms, in medical textbooks throughout the former Soviet Union and ignored to this day in Western countries, and to which I will return in later chapters.

In its early stages the symptoms of radio wave sickness are those of neurasthenia.

As living beings, not only do we possess a mind and a body, but we also have nerves that join the two. Our nerves are not just conduits for the ebb and flow of electric fluid from the universe, as was once believed, nor are they just an elaborate messenger service to deliver chemicals to muscles, as is currently thought. Rather, as we will see, they are both. As a messenger service, the nervous system can be poisoned by toxic chemicals. As a network of fine transmission wires, it can easily be damaged or unbalanced by a great or unfamiliar electric load. This has effects on both mind and body that we know today as anxiety disorder.

6. The Behavior of Plants

WHEN I FIRST ENCOUNTERED the works of Sir Jagadis Chunder Bose, I was stunned. The son of a public official in East Bengal, Bose was educated in Cambridge, where he received a degree in natural science that he took back to his home country. A genius in both physics and botany, he had an extraordinary eye for detail as well as a unique talent for designing precision measuring equipment. With an intuition that all living things share the same fundamentals, this man built elegant machinery that could magnify the movements of ordinary plants one hundred million times, while recording such movements automatically, and he proceeded in this

Sir Jagadis Chunder Bose
(1858-1937)

way to study the *behavior* of plants in the same manner that zoologists study the behavior of animals. In consequence, he was able to locate the *nerves* of plants—not just unusually active plants like Mimosa and Venus fly trap, but "normal" plants—and he actually dissected them out and proved that they generate action potentials like any animal's nerves. He performed conduction experiments on the nerves of ferns in the same way physiologists do with the sciatic nerves of frogs.

Bose also located pulsating cells in a plant's stem which he showed are responsible for pumping the sap, which have special electrical properties, and he built what he called a magnetic sphygmograph that magnified the pulsations ten million times and measured changes in sap pressure.

I was astonished, because you can search botanical textbooks today without finding so much as a hint that plants have anything like a heart and a nervous system. Bose's books, including *Plant Response* (1902), *The Nervous Mechanism of Plants* (1926), *Physiology of the Ascent of Sap* (1923), and *Plant Autographs and Their Revelations* (1927), languish in the archives of research libraries.

But Bose did more than just find the nerves of plants. He demonstrated the effects of electricity and radio waves on them, and he obtained similar results with sciatic nerves of frogs, proving the exquisite sensitivity of all living things to electromagnetic stimuli. His expertise in these areas was beyond question. He had been appointed Officiating Professor of Physics at the Presidency College in Calcutta in 1885. He made contributions in the field of solid-state physics, and is credited with the invention of the device—called a coherer—that was used to decode the first wireless message sent across the Atlantic Ocean by Marconi. In fact, Bose had given a public demonstration of wireless transmission in a lecture hall in Calcutta in 1895, more than a year before Marconi's first demonstration on Salisbury Plain in England. But Bose took out no patents, and sought no publicity for his invention of the radio. Instead he gave up those technical pursuits to devote the rest of his life to the more humble study of plant behavior.

In applying electricity to plants, Bose built on a tradition that was already a century and a half old.

The first to electrify a plant with a friction machine was a Dr. Mainbray of Edinburgh, who connected two myrtle trees to a machine throughout October 1746; the two trees sent out new branches and buds that autumn as though it were springtime. The following October,

Abbé Nollet, having received this news, conducted the first of a series of more rigorous experiments in Paris. In addition to Carthusian monks and soldiers of the French guard, Nollet was electrifying mustard seeds as they sprouted in tin bowls back in his laboratory. The electrified sprouts grew four times as tall as normal, but with stems that were weaker and more slender.[1]

That December, around Christmas time, Jean Jallabert electrified jonquil, hyacinth, and narcissus bulbs in carafes of water.[2] The following year Georg Bose electrified plants at Wittenberg,[3] and Abbé Menon at Angers,[4] and for the rest of the eighteenth century plant growth demonstrations were *de rigeur* among scientists studying frictional electricity. The energized plants sprouted earlier, grew faster and longer, opened their flowers sooner, sent out more leaves, and were generally—but not always—sturdier.

Jean-Paul Marat even watched electrified lettuce seeds germinate in the month of December when the ambient temperature was two degrees above freezing.[5]

Giambattista Beccaria in Turin was the first, in 1775, to suggest the use of these effects for the benefit of agriculture. Soon afterwards Francesco Gardini, also in Turin, stumbled upon the opposite effect: plants deprived of the natural atmospheric field did not grow as well. A network of iron wires had been stretched over the ground for the purpose of detecting atmospheric electricity. But the wires happened to run above part of a monastery's garden, shielding it from the atmospheric fields that the wires were measuring. For the three years that the wire net had been in place, the gardeners tending that section had complained that their harvests of fruits and seeds were fifty to seventy percent less than in the rest of their gardens. So the wires were removed, and production returned to normal. Gardini drew a remarkable inference. "Tall plants," he said, "have a harmful influence on the development of plants that grow at their base, not only by depriving them of light and heat, but also because they absorb atmospheric electricity at their expense."[6]

In 1844, W. Ross was the first of many to apply electricity to a field of crops, using a one-volt battery much like the one from which Humboldt had so successfully elicited sensations of light and taste, only larger. He buried a copper plate five feet by fourteen inches at one end of a row of potatoes, a zinc plate two hundred feet away at the other end, and connected the two plates with a wire. And in July he harvested potatoes averaging two and a half inches in diameter from the electrified row, versus only one-half inch from the untreated row.[7]

In the 1880s, Professor Selim Lemström of the University of Helsingfors in Finland conducted large-scale experiments on crops with a friction machine, suspending over his crops a network of pointed wires connected to the positive pole of the machine. Over a period of years he found that electricity stimulated the growth of some crops—wheat, rye, barley, oats, beets, parsnips, potatoes, celeriac, beans, leeks, raspberries, and strawberries—while it stunted the growth of peas, carrots, kohlrabi, rutabagas, turnips, cabbages, and tobacco.

And in 1890, Brother Paulin, Director of the Agricultural Institute at Beauvais, France, invented what he called a "géomagnétifère" to draw down atmospheric electricity like Benjamin Franklin had once done with his kite. Perched atop a tall pole 40 to 65 feet high was an iron collecting rod, terminating in five pointed branches. Four such poles were planted on every hectare of land, and the electricity collected by them was carried to the soil and distributed to the crops by means of underground wires.

According to contemporary newspaper accounts the effect was visually startling. Like supercrops, all of the potato plants within a sharply delineated ring were greener, taller, and "twice as vigorous" as the surrounding plants. The yield of potatoes within the electrified areas was fifty to seventy percent greater than outside them. Repeated in a vineyard, the experiment produced grape juice with seventeen percent more sugar, and wine with an exceptional alcohol content. Further trials in fields of spinach, celery, radishes, and turnips were just as impressive. Other farmers, using similar apparatus, improved their yields of wheat, rye, barley, oats, and straw.[8]

All these experiments with frictional electricity, feeble electric batteries, and atmospheric fields might make one suspect that it doesn't take very much current to affect a plant. But until the end of the nineteenth century the experiments lacked precision, and accurate measurements were not available.

Which brings me back to Jagadis Chunder Bose.

In 1859, Eduard Pflüger had formulated a simple model of how electric currents affect animal nerves. If two electrodes are attached to a nerve and the current is suddenly turned on, the negative electrode, or cathode, momentarily stimulates the section of nerve near it, while the positive electrode, or anode, has a deadening effect. The reverse occurs at the moment the current is broken. The cathode, said Pflüger, increases excitability at "make," and decreases excitability at "break," while the anode does just the opposite. While the current is flowing and not changing, supposedly nervous activity is not affected whatsoever by the current. Pflüger's Law, formulated a century and a half ago, is widely believed until the present day, and is the basis for modern electrical safety codes that are designed to prevent shocks at "make" or "break" of circuits but that do not prevent low-level continuous currents from being induced in the body because they are presumed to be of no consequence.

Unfortunately Pflüger's Law is not true and Bose was the first to prove it. One problem with Pflüger's Law is that it was based on experiments using relatively strong electric currents, on the order of one milliampere (a thousandth of an ampere). But, as Bose showed, it is not even correct at those levels.[9] Experimenting on himself in much the same way Humboldt had done a century before, Bose applied an electromotive force of 2 volts to a skin wound, and to his surprise the cathode, both at make, *and as long as the current flowed*, made the wound much more painful. The anode, both at make and while the current flowed, soothed the wound. But exactly the opposite occurred when he applied a much lower voltage. At a third of a volt, the cathode soothed and anode irritated.

After experimenting on his own body, Bose, being a botanist, tried a similar experiment on a plant. He took a twenty-centimer length of the nerve of a fern, and applied an electromotive force of only a tenth of a volt across the ends. This sent a current of about three ten-millionths of an ampere through the nerve, or about one thousand times less than the range of currents most modern physiologists and makers of safety regulations are used to thinking about. Again, at this low level of current, Bose found precisely the reverse of Pflüger's Law: the anode stimulated the nerve and the cathode made it less responsive. Evidently, in plants as well as in animals, electricity could have exactly opposite effects depending on the strength of the current.

Still Bose was not satisfied, because under certain circumstances the effects did not consistently follow either pattern. Maybe, suspected Bose, Pflüger's model was not only wrong but simplistic. He speculated that the applied currents were actually altering the conductivity of the nerves and not just the threshold of their response. Bose questioned the received wisdom that nervous functioning was a neat all-or-nothing response based only on chemicals in a watery solution.

His ensuing experiments confirmed his suspicions spectacularly. Contrary to existing theories—existing still today in the twenty-first century—of how nerves function, a constantly applied electric current, even though tiny, profoundly altered the conductivity of the animal and plant nerves Bose tested. If the applied current was in the same direction as nervous impulses, the speed of the impulses became slower and, in the animal, the muscular response to stimulation became weaker. If the applied current was in the opposite direction, nervous impulses traveled faster and muscles responded more vigorously. By manipulating the magnitude and direction of the applied current, Bose found that he could control nerve conduction at will, in animals and in plants, making nerves more or less sensitive to stimulation, or even blocking conduction altogether. And after the current was turned off, a rebound effect was observed. If a given amount of current depressed conduction, the nerve became hypersensitive after it was turned off, and remained so for a period of time. In one experiment a brief current of 3 microamperes—3 millionths of an ampere—produced nervous hypersensitivity for 40 seconds.

An incredibly tiny current was all that was needed: in plants, one microampere, and in animals a third of a microampere, was enough to slow or speed up nerve impulses by about twenty percent.[10] This is about the amount of current that would flow through your hand if you touched both ends of a one-volt battery, or that would flow through your body if you slept under an electric blanket. It is much less than the currents that are induced in your head when you talk on a cell phone. And, as we will see, it requires even less current to affect growth than to affect nerve activity.

In 1923, Vernon Blackman, an agricultural researcher at Imperial College in England, found in field experiments that electric currents averaging less than one milliampere (one thousandth of an ampere) per *acre* increased the yields of several types of crops by twenty percent. The current passing through each plant, he calculated, was only about 100 picoamperes—that's 100 trillionths of an ampere, about a thousand times less than the currents Bose had found were necessary to stimulate or deaden nerves.

But the field results were inconsistent. So Blackman took his experiments into the laboratory where both exposure and growth conditions could be precisely controlled. Barley seeds were sprouted in glass tubes, and at varying heights above each plant was a metal point charged to about 10,000 volts by a DC power supply. The current flowing through each plant was measured precisely with a galvanometer, and Blackman found that a maximal increase in growth was obtained with a current of only 50 picoamperes, applied for just one hour per day. Increasing the time of application diminished the effect. Increasing the current to a tenth of a microampere was always harmful.

In 1966, Lawrence Murr and colleagues at Pennsylvania State University, experimenting on sweet corn and bush beans, verified Blackman's finding that currents around one microampere inhibited growth and damaged leaves. They then took these experiments one step farther: they undertook to discover the *smallest* current that would affect growth. And they found that any current greater than *one qua-drillionth of an ampere* would stimulate plant growth.

In his radio experiments, Bose used a device he called a magnetic crescograph, which recorded the growth rate of plants, magnified ten million times.[11] Remember that Bose was also an expert in wireless technology. When he set up a radio transmitter at one end of his property, and a plant attached to a receiving aerial at the other end, two hundred meters away, he found that even a brief radio transmission changed a plant's growth rate within a few seconds. The broadcast frequency, implied from his description, was about 30 MHz. We are not told what the power was. However, Bose recorded that a "feeble stimulus" produced an immediate acceleration of growth, and that "moderate" radio energy retarded growth. In other experiments he proved that exposure to radio waves slowed the ascent of sap.[12]

Bose's conclusions, drawn in 1927, were striking and prophetic. "The perceptive range of the plant," he wrote, "is inconceivably greater than ours; it not only perceives, but also responds to the different rays of the vast aetherial spectrum. Perhaps it is as well that our senses are limited in their range. For life would otherwise be intolerable under the constant irritation of these ceaseless waves of space-signalling to which brick walls are quite transparent. Hermetically-sealed metal chambers would then have afforded us the only protection."[13]

7. Acute Electrical Illness

ON MARCH 10, 1876, seven famous words sent an even greater avalanche of wires cascading down over an already tangled world: "Mr. Watson, come here, I want you."

As though living in a desert that was waiting to be planted and watered, millions of people heard and heeded the call. For although in 1879 only 250 people owned telephones in all of New York City, just ten years later, from that same soil, fertilized by an idea, dense forests of telephone poles were sprouting eighty and ninety feet tall, bearing up to thirty cross-branches each. Each tree in these electric groves supported up to three hundred wires, obscuring the sun and darkening the avenues below.

The Blizzard of 1888, New York City
Courtesy of the Museum of the City of New York

Calvert and German Streets, Baltimore, Maryland, circa 1889.
From E. B. Meyer, Underground Transmission and Distribution,
McGraw-Hill, N.Y., 1916

The electric light industry was conceived at roughly the same time.
One hundred and twenty-six years after a few Dutch pioneers taught
their eager pupils how to store a small quantity of electric fluid in a glass
jar, the Belgian Zénobe Gramme gave to the descendants of those
pioneers the knowledge, so to speak, of how to remove that jar's lid.
His invention of the modern dynamo made possible the generation of
virtually unlimited quantities of electricity. By 1875, dazzling carbon
arc lamps were lighting outdoor public spaces in Paris and Berlin. By
1883, wires carrying two thousand volts were trailing across residential
rooftops in the West End of London. Meanwhile, Thomas Edison had
invented a smaller and gentler lamp, the modern incandescent, that was
more suitable for bedrooms and kitchens, and in 1881 on Pearl Street
in New York City he built the first of hundreds of central stations supplying direct current (DC) electric power to outlying customers. Thick

wires from these stations soon joined their thinner comrades, strung between high branches of the spreading electric groves shading streets in cities across America.

And then another species of invention was planted alongside: alternating current (AC). Although many, including Edison, wanted to eradicate the invader, to pull it out by the roots as being too dangerous, their warnings were to no avail. By 1885, the Hungarian trio of Károly Zipernowsky, Otis Bláthy, and Max Déri had designed a complete AC generation and distribution system and began installing these in Europe.

In the United States, George Westinghouse adopted the AC system in the spring of 1887 and the "battle of the currents" escalated, Westinghouse vying with Edison for the future of our world. In one of the last salvos of that brief war, on page 16 of its January 12, 1889 issue, *Scientific American* published the following challenge:

> The direct and alternating current advocates are engaged in active attack upon each other on the basis of the relative harmfulness of the two systems. One engineer has suggested a species of electric duel to settle the matter. He proposes that he shall receive the direct current while his opponent shall receive the alternating current. Both are to receive it at the same voltage, and it is to be gradually increased until one succumbs, and voluntarily relinquishes the contest.

The State of New York settled the matter by adopting the electric chair as its new means of executing murderers. Yet, although alternating current was the more dangerous, it won the duel which was even then playing out not between individual combatants, but between commercial interests. Long-distance suppliers of electricity had to find economical ways to deliver ten thousand times more power through the average wire than had previously been necessary. Using the technology available at that time, direct current systems could not compete.

From these beginnings electrical technology, having been carefully sowed, fertilized, watered, and nurtured, shot skyward and outward toward and beyond every horizon. It was Nikola Tesla's invention of the polyphase AC motor, patented in 1888, enabling industries to use alternating current not just for lighting but for power, that provided the last necessary ingredient. In 1889, quite suddenly, the world was being electrified on a scale that could scarcely have been conceived when Dr. George Beard first described a disease called neurasthenia. The telegraph had "annihilated space and time," many had said at the time. But twenty years later the electric motor made the telegraph look like a child's toy, and the electric locomotive was poised to explode onto the countryside.

In early 1888, just thirteen electric railways had operated in the United States on a total of forty-eight miles of track, and a similar number in all of Europe. So spectacular was the growth of this industry that by the end of 1889, roughly a thousand miles of track had been electrified in the United States alone. In another year that number again tripled.

Eighteen eighty-nine is the year manmade electrical disturbances of the earth's atmosphere took on a global, rather than local, character. In that year the Edison General Electric Company was incorporated, and the Westinghouse Electric Company was reorganized as the Westinghouse Electric and Manufacturing Company. In that year Westinghouse acquired Tesla's alternating current patents and put them to use in its generating stations, which grew to 150 in number in 1889, and to 301 in 1890. In the United Kingdom, amendment of the Electric Lighting Act in 1888 eased regulations on the electric power industry and made central power station development commercially feasible for the first time. And in 1889, the Society of Telegraph Engineers and Electricians changed its name to the now more appropriate Institution of Electrical Engineers. In 1889, sixty-one producers in ten countries were manufacturing incandescent lamps, and American and European companies were installing plants in Central and South America. In that year *Scientific American* reported that "so far as we know, every city in the United States is provided with arc and incandescent illumination, and the introduction of electric lighting is rapidly extending to the smaller towns."[1] Also in that year, Charles Dana, writing in the *Medical Record*, reported on a new class of injuries, previously produced only by lightning. They were due, he said, to "the extraordinary increase now going on in the practical application of electricity, nearly $100,000,000 being already invested in lights and power alone." In 1889, most historians agree, the modern electrical era opened.

And in 1889, as if the heavens had suddenly opened as well, doctors in the Americas, Europe, Asia, Africa, and Australia were overwhelmed by a flood of critically ill patients suffering from a strange disease that seemed to have come like a thunderbolt from nowhere, a disease that many of these doctors had never seen before. That disease was influenza, and that pandemic lasted four continuous years and killed at least one million people.

Influenza Is an Electrical Disease

Suddenly and inexplicably, influenza, whose descriptions had remained consistent for thousands of years, changed its character in 1889. Flu had last seized most of England in November 1847, over half a century earlier. The last flu epidemic in the United States had raged in the winter of 1874–1875. Since ancient times, influenza had been known as a capricious, unpredictable disease, a wild animal that came from nowhere, terrorized whole populations at once without warning and without a schedule, and disappeared as suddenly and mysteriously as it had arrived, not to be seen again for years or decades. It behaved unlike any other illness, was thought not to be contagious, and received its name because its comings and goings were said to be governed by the "influence" of the stars.

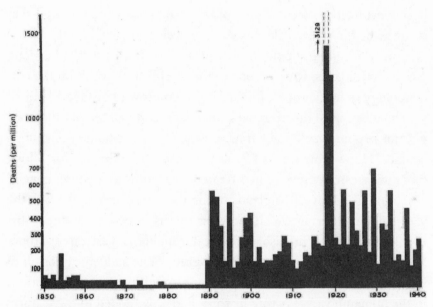

Influenzal deaths per million in England and Wales, 1850-1940[2]

But in 1889 influenza was tamed. From that year forward it would be present always, in every part of the world. It would vanish mysteriously as before, but it could be counted on to return, at more or less the same time, the following year. And it has never been absent since.

Like "anxiety disorder," influenza is so common and so seemingly familiar that a thorough review of its history is necessary to unmask this stranger and convey the enormity of the public health disaster that occurred one hundred and thirty years ago. It's not that we don't know enough about the influenza virus. We know more than enough. The microscopic virus associated with this disease has been so exhaustively studied that scientists know more about its tiny life cycle than about any other single microorganism. But this has been a reason to ignore many unusual facts about this disease, including the fact that it is not contagious.

In 2001, Canadian astronomer Ken Tapping, together with two British Columbia physicians, were the latest scientists to confirm, yet again, that for at least the last three centuries influenza pandemics have been most likely to occur during peaks of solar magnetic activity—that is, at the height of each eleven-year sun cycle.

Such a trend is not the only aspect of this disease that has long puzzled virologists. In 1992, one of the world's authorities on the epidemiology of influenza, R. Edgar Hope-Simpson, published a book in which he reviewed the essential known facts and pointed out that they did not support a mode of transmission by direct human-to-human contact. Hope-Simpson had been perplexed by influenza for a long time, in fact ever since he had treated its victims as a young general practitioner in Dorset, England, during the 1932–1933 epidemic—the very epidemic during which the virus that is associated with the disease in humans was first isolated. But during his 71-year career Hope-Simpson's questions were never answered. "The sudden explosion of information about the nature of the virus and its antigenic reactions in the human host," he wrote in 1992, had only "added to the features calling for explanation."[3]

Why is influenza seasonal? he still wondered. Why is influenza almost completely absent except during the few weeks or months of an epidemic? Why do flu epidemics end? Why don't out-of-season epidemics spread? How do epidemics explode over whole countries at once, and disappear just as miraculously, as if suddenly prohibited? He could not figure out how a virus could possibly behave like this. Why does flu so often target young adults and spare infants and the elderly? How is it possible that flu epidemics traveled at the same blinding speed in past centuries as they do today? How does the virus accomplish its so-called "vanishing trick"? This refers to the fact that when a new strain of the virus appears, the old strain, between one season and the next, has vanished completely, all over the world at once. Hope-Simpson listed twenty-one separate facts about influenza that puzzled him and that seemed to defy explanation if one assumed that it was spread by direct contact.

He finally revived a theory that was first put forward by Richard Shope, the researcher who isolated the first flu virus in pigs in 1931, and who also did not believe that the explosive nature of many outbreaks could be explained by direct contagion. Shope, and later Hope-Simpson, proposed that the flu is not in fact spread from person to person, or pig to pig, in the normal way, but that it instead remains latent in human or swine carriers, who are scattered in large numbers throughout their communities until the virus is reactivated by an environmental trigger of some sort. Hope-Simpson further proposed that the trigger is connected to seasonal variations in solar radiation, and that it may be electromagnetic in nature, as a good many of his predecessors during the previous two centuries had suggested.

When Hope-Simpson was young and beginning his practice in Dorset, a Danish physician named Johannes Mygge, at the end of a long and distinguished career, had just published a monograph in which he too showed that influenza pandemics tended to occur during years of maximum solar activity, and further that the yearly number of cases of flu in Denmark rose and fell with the number of sunspots. In an era in which epidemiology was becoming nothing more than a search for microbes, Mygge admitted, and knew already from hard experience, that "he who dances out of line risks having his feet stomped on."[4] But he was certain that influenza had something to do with electricity, and he had come to this conviction in the same way I did: from personal experience.

In 1904 and 1905, Mygge had kept a careful diary of his health for nine months, and he later compared it to records of the electrical potential of the atmosphere, which he had recorded three times a day for ten years as part of another project. It turned out that his incapacitating migraine-like headaches, which he had always known were connected to changes in the weather, almost always fell on the day of, or one day before, a sudden severe rise or drop in the value of the atmospheric voltage.

But headaches were not the only effects. On the days of such electrical turmoil, almost without exception, his sleep was broken and unrestful and he was bothered with dizziness, irritable mood, a feeling of confusion, buzzing sensations in his head, pressure in his chest, and an irregular heartbeat, and sometimes, he wrote, "my condition had the character of a threatening influenza attack, which in every case was not essentially different from the onset of an actual attack of that illness."[5]

Others who have connected influenza with sunspots or atmospheric electricity include John Yeung (2006), Fred Hoyle (1990), J. H. Douglas Webster (1940), Aleksandr Chizhevskiy (1936), C. Conyers Morrell (1936), W. M. Hewetson (1936), Sir William Hamer (1936), Gunnar Edström (1935), Clifford Gill (1928), C. M. Richter (1921), Willy Hellpach (1911), Weir Mitchell (1893), Charles Dana (1890), Louise Fiske Bryson (1890), Ludwig Buzorini (1841), Johann Schönlein (1841), and Noah Webster (1799). In 1836, Heinrich Schweich observed that all physiological processes produce electricity, and proposed that an electrical disturbance of the atmosphere may prevent the body from discharging it. He repeated the then-common belief that the accumulation of electricity within the body causes the symptoms of influenza. No one has yet disproven this.

It is of interest that between 1645 and 1715, a period astronomers call the Maunder Minimum, when the sun was so quiet that virtually no sunspots were to be seen and no auroras graced polar nights—during which, according to native Canadian tradition, "the people were deserted by the lights from the sky,"[6] —there were also no worldwide pandemics of flu. In 1715, sunspots reappeared suddenly after a lifetime's absence. In 1716, the famous English astronomer Sir Edmund Halley, at sixty years of age, published a dramatic description of the northern lights. It was the first time he had ever seen them. But the sun was still not fully active. As though it had woken up after a long sleep, it stretched its legs, yawned, and lay down again after displaying only half the number of sunspots that it shows us today at the peak of each eleven-year solar cycle. It wasn't until 1727 that the sunspot number surpassed 100 for the first time in over a century. And in 1728 influenza arrived in waves over the surface of the earth, the first flu pandemic in almost a hundred and fifty years. More universal and enduring than any in previously recorded history, that epidemic appeared on every continent, became more violent in 1732, and by some reports lasted until 1738, the peak of the next solar cycle.[7] John Huxham, who practiced medicine in Plymouth, England, wrote in 1733 that "scarce any one had escaped it." He added that there was "a madness among dogs; the horses were seized with the catarrh before mankind; and a gentleman averred to me, that some birds, particularly the sparrows, left the place where he was during the sickness."[8] An observer in Edinburgh reported that some people had fevers for sixty continuous days, and that others, not sick, "died suddenly."[9] By one estimate, some two million people worldwide perished in that pandemic.[10]

If influenza is primarily an electrical disease, a response to an electrical disturbance of the atmosphere, then it is not contagious in the ordinary sense. The patterns of its epidemics should prove this, and they do. For example, the deadly 1889 pandemic began in a number of widely scattered parts of the world. Severe outbreaks were reported in May of that year simultaneously in Bukhara, Uzbekistan; Greenland; and northern Alberta.[11] Flu was reported in July in Philadelphia[12] and in Hillston, a remote town in Australia,[13] and in August in the Balkans.[14] This pattern being at odds with prevailing theories, many historians have pretended that the 1889 pandemic didn't "really" start until it had seized the western steppes of Siberia at the end of September and that it then spread in an orderly fashion from there outward throughout the rest of the world, person to person by contagion. But the trouble is that the disease still would have had to travel faster than the trains and ships of the time. It reached Moscow and St. Petersburg during the third or fourth week of October, but by then, influenza had already been reported in Durban, South Africa[15] and Edinburgh, Scotland.[16] New Brunswick, Canada,[17] Cairo,[18] Paris,[19] Berlin,[20] and Jamaica[21] were reporting epidemics in November; London, Ontario on December 4;[22] Stockholm on December 9;[23] New York on December 11;[24] Rome on December 12;[25] Madrid on December 13;[26] and Belgrade on December 15.[27] Influenza struck explosively and unpredictably, over and over in waves until early 1894. It was as if something fundamental had changed in the atmosphere, as if brush fires were being ignited by some unknown vandal randomly, everywhere in the world.

One observer in East Central Africa, which was struck in September 1890, asserted that influenza had never before appeared in that part of Africa at all, not within the memory of the oldest living inhabitants.[28]

"Influenza," said Dr. Benjamin Lee of the Pennsylvania State Board of Health, "spreads like a flood, inundating whole sections in an hour… It is scarcely conceivable that a disease which spreads with such astonishing rapidity, goes through the process of re-development in each person infected, and is only communicated from person to person or by infected articles."[29]

Influenza works its caprice not only on land, but at sea. With today's speed of travel this is no longer obvious, but in previous centuries, when sailors were attacked with influenza weeks, or even months, out of their last port of call, it was something to remember. In 1894, Charles Creighton described fifteen separate historical instances where entire ships or even many ships in a naval fleet were seized by the illness far from landfall, as if they had sailed into an influenzal fog, only to discover, in some cases, upon arriving at their next port, that influenza had broken out on land at the same time. Creighton added one report from the contemporary pandemic: the merchantship "Wellington" had sailed with its small crew from London on December 19, 1891, bound for Lyttelton, New Zealand. On the 26th of March, after over three months at sea, the captain was suddenly shaken by intense febrile illness. Upon arriving at Lyttelton on April 2, "the pilot, coming on board found the captain ill in his berth, and on being told the symptoms at once said, 'It is the influenza: I have just had it myself.'"[30]

An 1857 report was so compelling that William Beveridge included it in his 1975 textbook on influenza: "The English warship Arachne was cruising off the coast of Cuba 'without any contact with land.' No less than 114 men out of a crew of 149 fell ill with influenza and only later was it learnt that there had been outbreaks in Cuba at the same time."[31]

The speed at which influenza travels, and its random and simultaneous pattern of spread, has perplexed scientists for centuries, and has been the most compelling reason for some to continue to suspect atmospheric electricity as the cause, despite the known presence of an extensively studied virus. Here is a sampling of opinion, old and modern:

> Perhaps no disease has ever been observed to affect so many people in so short a time, as the Influenza, almost a whole city, town, or neighborhood becoming affected in a few days, indeed much sooner than could be supposed to spread from contagion.
>
> Mercatus relates, that when it prevailed in Spain, in 1557, the greatest part of the people were seized in one day.
>
> Dr. Glass says, when it was rife in Exeter, in 1729, two thousand were attacked in one night.
>
> Shadrach Ricketson, M.D. (1808),
A Brief History of the Influenza[32]

> The simple fact is to be recollected that this epidemic affects a whole region in the space of a week; nay, a whole continent as large as North America, together with all the West Indies, in the course of a few weeks, where the inhabitants over such vast extent of country, could not, within so short a lapse of a time, have had the least communication or intercourse whatever. This fact alone is sufficient to put all idea of its being propagated by contagion from one individual to another out of the question.
>
> Alexander Jones, M.D. (1827),
Philadelphia Journal of the
Medical and Physical Sciences[33]

> Unlike cholera, it outstrips in its course the speed of human intercourse.
>
> Theophilus Thompson, M.D. (1852),
Annals of Influenza or Epidemic

*Catarrhal Fever in Great Britain from
1510 to 1837*[34]

Contagion alone is inadequate to explain the sudden outbreak of the disease in widely distant countries at the same
time, and the curious way in which it has been known to
attack the crews of ships at sea, where communication with
infected places or persons was out of the question.

Sir Morell Mackenzie, M.D. (1893),
Fortnightly Review[35]

Usually influenza travels at the same speed as man but at
times it apparently breaks out simultaneously in widely separated parts of the globe.

Jorgen Birkeland (1949),
Microbiology and Man[36]

[Before 1918] there are records of two other major epidemics of influenza in North America during the past two
centuries. The first of these occurred in 1789, the year in
which George Washington was inaugurated President. The
first steamboat did not cross the Atlantic until 1819, and the
first steam train did not run until 1830. Thus, this outbreak
occurred when man's fastest conveyance was the galloping
horse. Despite this fact, the influenza outbreak of 1789 spread
with great rapidity; many times faster and many times farther
than a horse could gallop.

James Bordley III, M.D. and
A. McGehee Harvey, M.D. (1976),
*Two Centuries of American Medicine,
1776–1976*[37]

Flu virus may be communicated from person to person in droplets of moisture from the respiratory tract. However, direct communication cannot account for simultaneous outbreaks of influenza in widely separated places.

> Roderick E. McGrew (1985),
> *Encyclopedia of Medical History*[38]

Why have epidemic patterns in Great Britain not altered in four centuries, centuries that have seen great increases in the speed of human transport?

> John J. Cannell, M.D. (2008),
> "On the Epidemiology of Influenza,"
> in *Virology Journal*

The role of the virus, which infects only the respiratory tract, has baffled some virologists because influenza is not only, or even mainly, a respiratory disease. Why the headache, the eye pain, the muscle soreness, the prostration, the occasional visual impairment, the reports of encephalitis, myocarditis, and pericarditis? Why the abortions, stillbirths, and birth defects?[39]

In the first wave of the pandemic of 1889 in England, neurological symptoms were most often prominent and respiratory symptoms absent.[40] Most of Medical Officer Röhring's 239 flu patients at Erlangen, Bavaria, had neurological and cardiovascular symptoms and no respiratory disease. Nearly one-quarter of the 41,500 cases of flu reported in Pennsylvania as of May 1, 1890 were classified as primarily neurological and not respiratory.[41] Few of David Brakenridge's patients in Edinburgh, or Julius Althaus' patients in London, had respiratory symptoms. Instead they had dizziness, insomnia, indigestion, constipation, vomiting, diarrhea, "utter prostration of mental and bodily strength," neuralgia, delirium, coma, and convulsions. Upon recovery many were left with neurasthenia, or even paralysis or epilepsy. Anton Schmitz published an article titled "Insanity After Influenza" and concluded that influenza was primarily an epidemic nervous disease. C. H. Hughes called influenza a "toxic neurosis." Morell Mackenzie agreed:

> In my opinion the answer to the riddle of influenza is poisoned nerves... In some cases it seizes on that part of (the nervous system) which governs the machinery of respiration, in others on that which presides over the digestive functions; in others again it seems, as it were, to run up and down the nervous keyboard, jarring the delicate mechanism and stirring up disorder and pain in different parts of the body with what almost seems malicious caprice... As the nourishment of every tissue and organ in the body is under the direct control of the nervous system, it follows that anything which affects the latter has a prejudicial effect on the former; hence it is not surprising that influenza in many cases leaves its mark in damaged structure. Not only the lungs, but the kidneys, the heart, and other internal organs and the nervous matter itself may suffer in this way.[42]

Insane asylums filled up with patients who had had influenza, people suffering variously from profound depression, mania, paranoia, or hallucinations. "The number of admissions reached unprecedented proportions," reported Albert Leledy at the Beauregard Lunatic Asylum, at Bourges, in 1891. "Admissions for the year exceed those of any previous year," reported Thomas Clouston, superintending physician of the Royal Edinburgh Asylum for the Insane, in 1892. "No epidemic of any disease on record has had such mental effects," he wrote. In 1893, Althaus reviewed scores of articles about psychoses after influenza, and the histories of hundreds of his own and others' patients who had gone insane after the flu during the previous three years. He was perplexed by the fact that the majority of psychoses after influenza were developing in men and women in the prime of their life, between the ages of 21 and 50, that they were most likely to occur after only mild or slight cases of the disease, and that more than one-third of these people had not yet regained their sanity.

The frequent lack of respiratory illness was also noted in the even deadlier 1918 pandemic. In his 1978 textbook Beveridge, who had lived through it, wrote that half of all influenza patients in that pandemic did not have initial symptoms of nasal discharge, sneezing, or sore throat.[43]

The age distribution is also wrong for contagion. In other kinds of infectious diseases, like measles and mumps, the more aggressive a strain of virus is and the faster it spreads, the more rapidly adults build up immunity and the younger the population that gets it every year. According to Hope-Simpson, this means that between pandemics influenza should be attacking mainly very young children. But influenza keeps on stubbornly targeting adults; the average age is almost always between twenty and forty, whether during a pandemic or not. The year 1889 was no exception: influenza felled preferentially vigorous young adults in the prime of their life, as if it were maliciously choosing the strongest instead of the weakest of our species.

Then there is the confusion about animal infections, which are so much in the news year after year, scaring us all about catching influenza from swine or birds. But the inconvenient fact is that throughout history, for thousands of years, all sorts of animals have caught the flu at the same time as humans. When the army of King Karlmann of Bavaria was seized by influenza in 876 A.D., the same disease also decimated the dogs and the birds.[44] In later epidemics, up to and including the twentieth century, illness was commonly reported to break out among dogs, cats, horses, mules, sheep, cows, birds, deer, rabbits, and even fish at the same time as humans.[45] Beveridge listed twelve epidemics during the eighteenth and nineteenth centuries in which horses caught the flu, usually one or two months before the humans. In fact, this association was considered so reliable that in early December 1889, Symes Thompson, observing flu-like illness in British horses, wrote to the *British Medical Journal* predicting an imminent outbreak in humans, a forecast which shortly proved true.[46] During the 1918–1919 pandemic, monkeys and baboons perished in great numbers in South Africa and Madagascar, sheep in northwest England, horses in France, moose in northern Canada, and buffalo in Yellowstone.[47] There is no mystery here. We are not catching the flu from animals, nor they from us. If influenza is caused by abnormal electromagnetic conditions in the atmosphere, then it affects all living things at the same time, including living things that don't share the same viruses or live closely with one another.

The obstacle to unmasking the stranger that is influenza is the fact that it is two different things. Influenza is a virus and it is also a clinical illness. The confusion comes about because since 1933, human influenza has been defined by the organism that was discovered in that year, and not by clinical symptoms. If an epidemic strikes, and you come down with the same disease as everyone else, but an influenza virus can't be isolated from your throat and you don't develop antibodies to one, then you are said not to have influenza. But the fact is that although influenza viruses are associated in some way with disease epidemics, they have never been shown to cause them.

Seventeen years of surveillance by Hope-Simpson in and around the community of Cirencester, England, revealed that despite popular belief, influenza is not readily communicated from one person to another within a household. Seventy percent of the time, even during the "Hong Kong flu" pandemic of 1968, only one person in a household would get the flu. If a second person had the flu, both often caught it on the same day, which meant that they did not catch it from each other. Sometimes different minor variants of the virus were circulating in the same village, even in the same household, and on one occasion two young brothers who shared a bed had different variants of the virus, proving that they could not have caught it from each other, or even from the same third person.[48] William S. Jordan, in 1958, and P. G. Mann, in 1981, came to similar conclusions about the lack of spread within families.

Another indication that something is wrong with prevailing theories is the failure of vaccination programs. Although vaccines have been proven to confer some immunity to particular strains of flu virus, several prominent virologists have admitted over the years that vaccination has done nothing to stop epidemics and that the disease still behaves just as it did a thousand years ago.[49] In fact, after reviewing 259 vaccination studies from the *British Medical Journal* spanning 45 years, Tom Jefferson recently concluded that influenza vaccines have had essentially no impact on any real outcomes, such as school absences, working days lost, and flu-related illnesses and deaths.[50]

The embarrassing secret among virologists is that from 1933 until the present day, there have been *no* experimental studies proving that influenza—either the virus or the disease—is ever transmitted from person to person by normal contact. As we will see in the next chapter, all efforts to experimentally transmit it from person to person, even in the middle of the most deadly disease epidemic the world has ever known, have failed.

8. Mystery on the Isle of Wight

IN 1904 THE BEES began to die.

From this quiet island, 23 miles long and 13 miles wide, lying off England's southern coast, one looks across the English Channel toward the distant shores of France. In the preceding decade two men, one on each side of the Channel, one a physician and physicist, the other an inventor and entrepreneur, had occupied their minds with a newly discovered form of electricity. The work of each man had very different implications for the future of our world.

At the westernmost end of the Isle of Wight, near offshore chalk formations called The Needles, in 1897, a handsome young man named Giuglielmo Marconi erected his own "needle," a tower as tall as a twelve-story building. It supported the antenna for what became the world's first permanent radio station. Marconi was liberating electricity, vibrating at close to a million cycles per second, from its confining wires, and was broadcasting it freely through the air itself. He did not stop to ask if this was safe.

A few years earlier, in 1890, a well-known physician, director of the Laboratory of Biological Physics at the Collège de France in Paris, had already begun investigations bearing on the important question Marconi was not asking: how does electricity of high frequencies affect living organisms? A distinguished presence in physics as well as medicine, Jacques-Arsène d'Arsonval is remembered today for his many contributions in both fields. He devised ultra-sensitive meters to measure magnetic fields, and equipment to measure heat production and respiration in animals; made improvements to the microphone and the telephone; and created a new medical specialty called darsonvalization, which is still practiced today in the nations of the former Soviet Bloc. In the West it has evolved into diathermy, which is the therapeutic use of radio waves to produce heat within the body. But darsonvalization is the use of radio waves medicinally at low power, without generating heat, to produce the kinds of effects d'Arsonval discovered in the early 1890s.

Jacques-Arsène d'Arsonval
(1851-1940)

He had first observed that electrotherapy, as then practiced, was not producing uniform results, and he wondered if this was because of lack of precision in the form of the electricity being applied. He therefore designed an induction machine capable of putting out perfectly smooth sine waves, "without jerks or teeth,"[1] that would not be injurious to the patient. When he tested this current on human subjects he found, as he had predicted, that at therapeutic doses it caused no pain, yet had potent physiological effects.

"We have seen that with very steady sine waves, nerve and muscle are not stimulated," he wrote. "The passage of the current nevertheless is responsible for profound modification of metabolism as shown by the consumption of a greater amount of oxygen and the production of considerably more carbon dioxide. If the shape of the wave is changed, each electrical wave will produce a muscular contraction."[2] D'Arsonval had already discovered the reason, 125 years ago, why today's digital technologies, whose waves have nothing but "jerks and teeth," are causing so much illness.

D'Arsonval next experimented with alternating currents of high frequency. Using a modification of the wireless apparatus devised a few years earlier by Heinrich Hertz, he exposed humans and animals to currents of 500,000 to 1,000,000 cycles per second, applied either by direct contact or indirectly by induction from a distance. They were close to the frequencies Marconi was soon going to broadcast from the Isle of Wight. In no case did the subject's body temperature increase. But in every case his subject's blood pressure fell significantly, without—in the case of human subjects at least—any conscious sensation. D'Arsonval measured the same changes in oxygen consumption and carbon dioxide production as with low frequency currents. These facts proved, he wrote, "that the currents of high frequency penetrate deeply into the organism."[3]

These early results should have made anyone experimenting with radio waves think twice before exposing the whole world to them indiscriminately—should have at least made them cautious. Marconi, however, was unfamiliar with d'Arsonval's work. Largely self-educated, the inventor had no inkling of radio's potential dangers and no fear of it. Therefore when he powered up his new transmitter on the island he had no suspicion that he might be doing himself or anyone else any harm.

If radio waves are dangerous, Marconi, of all people in the world, should have suffered from them. Let us see if he did.

As early as 1896, after a year and a half of experimenting with radio equipment in his father's attic, the previously healthy 22-year-old youth began running high temperatures which he attributed to stress. These fevers were to recur for the rest of his life. By 1900 his doctors were speculating that perhaps he had unknowingly had rheumatic fever as a child. By 1904 his bouts of chills and fevers had become so severe that it was thought they were recurrences of malaria. At that time he was occupied with building a permanent super-high-power radio link across the Atlantic Ocean between Cornwall, England and Cape Breton Island, Nova Scotia. Because he thought that longer distances required longer waves, he suspended tremendous wire net aerials, occupying acres of land, from multiple towers hundreds of feet tall on both sides of the ocean.

On March 16, 1905, Marconi married Beatrice O'Brien. In May, after their honeymoon, he took her to live in the station house at Port Morien on Cape Breton, surrounded by twenty-eight huge radio towers in three concentric circles. Looming over the house, two hundred antenna wires stretched out from a center pole like the spokes of a great umbrella more than one mile in circumference. As soon as Beatrice settled in, her ears began to ring.

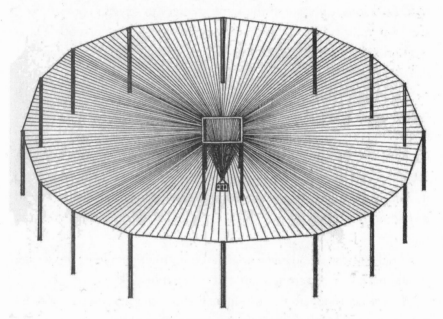

From: W. J. Baker, *A History of the Marconi Company*,
St. Martin's Press, N.Y., 1971

After three months there she was ill with severe jaundice. When Marconi took her back to England it was to live underneath the other monster aerial, at Poldhu Bay in Cornwall. She was pregnant all this time, and although she moved to London before giving birth, her child had spent most of its nine months of fetal life bombarded with powerful radio waves and lived only a few weeks, dying of "unknown causes."

At about the same time Marconi himself collapsed completely, spending much of February through May of 1906 feverish and delirious.

Between 1918 and 1921, while engaged in designing short wave equipment, Marconi suffered from bouts of suicidal depression.

In 1927, during the honeymoon he took with his second wife Maria Cristina, he collapsed with chest pains and was diagnosed with a severe heart condition. Between 1934 and 1937, while helping to develop microwave technology, he suffered as many as nine heart attacks, the final one fatal at age 63.

Bystanders sometimes tried to warn him. Even at his first public demonstration on Salisbury Plain in 1896, there were spectators who later sent him letters describing various nerve sensations they had experienced. His daughter Degna, reading them much later while doing research for the biography of her father, was particular taken by one letter, from a woman "who wrote that his waves made her feet tickle." Degna wrote that her father received letters of this sort frequently. When, in 1899, he built the first French station in the coastal town of Wimereux, one man who lived close by "burst in with a revolver," claiming that the waves were causing him sharp internal pains. Marconi dismissed all such reports as fantasy.

In what may have been an even more ominous warning, Queen Victoria of England, in residence at Osborne House, her estate at the north end of the Isle of Wight, suffered a cerebral hemorrhage and died on the evening of January 22, 1901, just as Marconi was firing up a new, more powerful transmitter twelve miles away. He was hoping to communicate with Poldhu the next day, 300 kilometers distant, twice as far as any previously recorded radio broadcast, and he did. On January 23 he sent a telegram to his cousin Henry Jameson Davis, saying "Completely successful. Keep information private. Signed William."

And then there were the bees.

In 1901, there were already two Marconi stations on the Isle of Wight—Marconi's original station, which had been moved to Niton at the south end of the island next to St. Catherine's Lighthouse, and the Culver Signal Station run by the Coast Guard at the east end on

Culver Down. By 1904, two more had been added. According to an article published in that year by Eugene P. Lyle in *World's Work* magazine, four Marconi stations were now operating on the small island, communicating with a steadily growing number of naval and commercial ships of many nations, steaming through the Channel, that were equipped with similar apparatus. It was the greatest concentration of radio signals in the world at that time.

In 1906, the Lloyd's Signal Station, half a mile east of St. Catherine's Lighthouse, also acquired wireless equipment. At this point the bee situation became so severe that the Board of Agriculture and Fisheries called in biologist Augustus Imms of Christ's College, Cambridge, to investigate. Ninety percent of the honey bees had disappeared from the entire island for no apparent reason. The hives all had plenty of honey. But the bees could not even fly. "They are often to be seen crawling up grass stems, or up the supports of the hive, where they remain until they fall back to the earth from sheer weakness, and soon afterwards die," he wrote. Swarms of healthy bees were imported from the mainland, but it was of no use: within a week the fresh bees were dying off by the thousands.

In coming years "Isle of Wight disease" spread like a plague throughout Great Britain and into the rest of the world, severe losses of bees being reported in parts of Australia, Canada, the United States, and South Africa.[4] The disease was also reported in Italy, Brazil, France, Switzerland, and Germany. Although for years one or another parasitic mite was blamed, British bee pathologist Leslie Bailey disproved those theories in the 1950s and came to regard the disease itself as a sort of myth. Obviously bees had died, he said, but not from anything contagious.

Over time, Isle of Wight disease took fewer and fewer bee lives as the insects seemed to adapt to whatever had changed in their environment. Places that had been attacked first recovered first.

Then, in 1917, just as the bees on the Isle of Wight itself appeared to be regaining their former vitality, an event occurred that changed the electrical environment of the rest of the world. Millions of dollars of United States government money were suddenly mobilized in a crash program to equip the Army, Navy, and Air Force with the most modern communication capability possible. The entry of the United States into the Great War on April 6, 1917, stimulated an expansion of radio broadcasting that was as sudden and rapid as the 1889 expansion of electricity.

Again it was the bees that gave the first warning.

"Mr. Charles Schilke of Morganville, Monmouth County, a bee-keeper with considerable experience operating about 300 colonies reported a great loss of bees from the hives in one of his yards located near Bradevelt," read one report, published in August 1918.[5] "Thousands of dead were lying and thousands of dying bees were crawling about in the vicinity of the hive, collecting in groups on bits of wood, on stones and in depressions in the earth. The affected bees appeared to be practically all young adult workers about the age when they would normally do the first field work, but all ages of older bees were found. No abnormal condition within the hive was noticed at this time."

This outbreak was confined to Morganville, Freehold, Milhurst, and nearby areas of New Jersey, just a few miles seaward from one of the most powerful radio stations on the planet, the one in New Brunswick that had just been taken over by the government for service in the war. A 50,000-watt Alexanderson alternator had been installed in February of that year to supplement a less efficient 350,000-watt spark apparatus. Both provided power to a mile-long aerial consisting of 32 parallel wires supported by 12 steel towers 400 feet tall, broadcasting military communications across the ocean to the command in Europe.

Radio came of age during the First World War. For long distance communications there were no satellites, and no shortwave equipment. Vacuum tubes had not yet been perfected. Transistors were decades into the future. It was the era of immense radio waves, inefficient aerials the size of small mountains, and spark gap transmitters that scattered radiation like buckshot all over the radio spectrum to interfere with everyone else's signals. Oceans were crossed by brute force, three

hundred thousand watts of electricity being supplied to those mountains to achieve a radiated power of perhaps thirty thousand. The rest
was wasted as heat. Morse code could be sent but not voice. Reception
was sporadic, unreliable.

Few of the great powers had had a chance to establish overseas
communication with their colonies before war intervened in 1914.
The United Kingdom had two ultra-powerful stations at home, but no
radio links with a colony. The first such link was still under construction near Cairo. France had one powerful station at the Eiffel Tower,
and another at Lyon, but no links with any of its overseas colonies.
Belgium had a powerful station in the Congo State, but blew up its
home station at Brussels after war broke out. Italy had one powerful station in Eritrea, and Portugal had one in Mozambique and one
in Angola. Norway had one ultrapotent transmitter, Japan one, and
Russia one. Only Germany had made much progress in building an
Imperial Chain, but within months after the declaration of war, all of
its overseas stations—at Togo, Dar-es-Salaam, Yap, Samoa, Nauru,
New Pomerania, Cameroon, Kiautschou, and German East Africa—
were destroyed.[6]

Radio, in short, was in its faltering infancy, still crawling, its attempts
to walk hindered by the onset of the European War. During 1915 and
1916, the United Kingdom made progress in installing thirteen long-
range stations in various parts of the world in order to keep in contact
with its navy.

When the United States entered the war in 1917, it changed the
terrain in a hurry. The United States Navy already had one giant
transmitter at Arlington, Virginia and a second at Darien, in the
Canal Zone. A third, in San Diego, began broadcasting in May 1917,
a fourth, at Pearl Harbor, on October 1 of that year, and a fifth, at
Cavite, the Philippines, on December 19. The Navy also took over and
upgraded private and foreign-owned stations at Lents, Oregon; South
San Francisco, California; Bolinas, California; Kahuku, Hawaii; Heeia
Point, Hawaii; Sayville, Long Island; Tuckerton, New Jersey; and New

Brunswick, New Jersey. By late 1917, thirteen American stations were sending messages across two oceans.

Fifty more medium and high powered radio stations ringed the United States and its possessions for communication with ships. To equip its ships the Navy manufactured and deployed over ten thousand low, medium, and high powered transmitters. By early 1918, the Navy was graduating over four hundred students per week from its radio operating courses. In the short course of a year, between April 6, 1917 and early 1918, the Navy built and was operating the world's largest radio network.

America's transmitters were far more efficient than most of those built previously. When a 30-kilowatt Poulson arc was installed at Arlington in 1913, it was found to be so much superior to the 100-kilowatt spark apparatus there that the Navy adopted the arc as its preferred equipment and ordered sets with higher and higher ratings. A 100-kilowatt arc was installed at Darien, a 200-kilowatt arc in San Diego, 350-kilowatt arcs at Pearl Harbor and Cavite. In 1917, 30-kilowatt arcs were being installed on Navy ships, outclassing the transmitters on most ships of other nations.

Still, the arc was basically only a spark gap with electricity flowing across it continuously instead of in bursts. It still sprayed the airway with unwanted harmonics, transmitted voices poorly, and was not reliable enough for continuous day and night communication. So the Navy tried out its first high-speed alternator, the one it inherited at New Brunswick. Alternators did not have spark gaps at all. Like fine musical instruments, they produced pure continuous waves that could be sharply tuned, and modulated for crystal clear voice or telegraphic communication. Ernst Alexanderson, who designed them, also designed an antenna to go with them that increased radiation efficiency sevenfold. When tested against the 350-kilowatt timed spark at the same station, the 50-kilowatt alternator proved to have a bigger range.[7] So in February 1918, the Navy began to rely on the alternator to handle continuous communications with Italy and France.

In July 1918, another 200-kilowatt arc was added to the system the Navy had taken over at Sayville. In September 1918, a 500-kilowatt arc went on the air at a new naval station at Annapolis, Maryland. Meanwhile the Navy had ordered a second, more powerful alternator for New Brunswick, of 200-kilowatt capacity. Installed in June, it too went on the air full time in September. New Brunswick immediately became the most powerful station in the world, outclassing Germany's flagship station at Nauen, and was the first that transmitted both voice and telegraphic messages across the Atlantic Ocean clearly, continuously, and reliably. Its signal was heard over a large part of the earth.

The disease that was called Spanish influenza was born during these months. It did not originate in Spain. It did, however, kill tens of millions all over the world, and it became suddenly more fatal in September of 1918. By some estimates the pandemic struck more than half a billion people, or a third of the world's population. Even the Black Death of the fourteenth century did not kill so many in so short a period of time. No wonder everyone is terrified of its return.

A few years ago researchers dug up four bodies in Alaska that had lain frozen in the permafrost since 1918 and were able to identify RNA from an influenza virus in the lung tissue of one of them. This was the monster germ that was supposed to have felled so many in the prime of their lives, the microbe that so resembles a virus of pigs, against whose return we are to exercise eternal vigilance, lest it decimate the world again.

But there is no evidence that the disease of 1918 was contagious.

The Spanish influenza apparently originated in the United States in early 1918, seemed to spread around the world on Navy ships, and first appeared on board those ships and in seaports and Naval stations. The largest early outbreak, laying low about 400 people, occurred in February in the Naval Radio School at Cambridge, Massachusetts.[8] In March, influenza spread to Army camps where the Signal Corps was being trained in the use of the wireless: 1,127 men contracted influenza in Camp Funston, in Kansas, and 2,900 men in the Oglethorpe camps in Georgia. In late March and April, the disease spread to the civilian population, and around the world.

Mild at first, the epidemic exploded with death in September, everywhere in the world at once. Waves of mortality traveled with astonishing speed over the global ocean of humanity, again and again until their force was finally spent three years later.

Its victims were often sick repeatedly for months at a time. One of the things that puzzled doctors the most was all of the bleeding. Ten to fifteen percent of flu patients seen in private practice,[9] and up to forty percent of flu patients in the Navy[10] suffered from nosebleeds, doctors sometimes describing the blood as "gushing" from the nostrils.[11] Others bled from their gums, ears, skin, stomach, intestines, uterus, or kidneys, the most common and rapid route to death being hemorrhage in the lungs: flu victims drowned in their own blood. Autopsies revealed that as many as one-third of fatal cases had also hemorrhaged into their brain,[12] and occasionally a patient appeared to be recovering from respiratory symptoms only to die of a brain hemorrhage.

"The regularity with which these various hemorrhages appeared suggested the possibility of there being a change in the blood itself," wrote Drs. Arthur Erskine and B. L. Knight of Cedar Rapids, Iowa in late 1918. So they tested the blood from a large number of patients with influenza and pneumonia. "In every case tested without a single exception," they wrote, "the coagulability of the blood was lessened, the increase in time required for coagulation varying from two and one-half to eight minutes more than normal. Blood was tested as early as the second day of infection, and as late as the twentieth day of convalescence from pneumonia, with the same results… Several local physicians also tested blood from their patients, and, while our records are at this time necessarily incomplete, we have yet to receive a report of a case in which the time of coagulation was not prolonged."

This is consistent not with any respiratory virus, but with what has been known about electricity ever since Gerhard did the first experiment on human blood in 1779. It is consistent with what is known about the effects of radio waves on blood coagulation.[13] Erskine and Knight saved their patients not by fighting infection, but by giving them large doses of calcium lactate to facilitate blood clotting.

Another astonishing fact that makes no sense if this pandemic was infectious, but that makes good sense if it was caused by radio waves, is that instead of striking down the old and the infirm like most diseases, this one killed mostly healthy, vigorous young people between the ages of eighteen and forty—just as the previous pandemic had done, with a little less vehemence, in 1889. This, as we saw in chapter 5, is the same as the predominant age range for neurasthenia, the chronic form of electrical illness. Two-thirds of all influenza deaths were in this age range.[14] Elderly patients were rare.[15] One doctor in Switzerland wrote that he "knew of no case in an infant and no severe case in persons over 50," but that "one robust person showed the first symptoms at 4 p.m. and died before 10 the next morning."[16] A reporter in Paris went so far as to say that "*only* persons between 15 and 40 years of age are affected."[17]

The prognosis was better if you were in poor physical condition. If you were undernourished, physically handicapped, anemic, or tuberculous, you were much less likely to get the flu and much less likely to die from it if you did.[18] This was such a common observation that Dr. D. B. Armstrong wrote a provocative article, published in the *Boston Medical and Surgical Journal*, titled "Influenza: Is it a Hazard to Be Healthy?" Doctors were seriously discussing whether they were actually giving their patients a death sentence by advising them to keep fit!

The flu was reported to be even more fatal for pregnant women.

A further peculiarity that had doctors scratching their heads was that in most cases, after the patients' temperature had returned to normal, their pulse rate fell below 60 and remained there for a number of days. In more serious cases the pulse rate fell to between 36 and 48, an indication of heart block.[19] This too is puzzling for a respiratory virus, but will make sense when we learn about radio wave sickness.

Patients also regularly lost some of their hair two to three months after recovering from the flu. According to Samuel Ayres, a dermatologist at Massachusetts General Hospital in Boston, this was an almost daily occurrence, most of these patients being young women. This is not an expected after-effect of respiratory viruses either, but hair loss has been widely reported from exposure to radio waves.[20]

Yet another puzzling observation was that so few patients in 1918 had sore throats, runny noses, or other initial respiratory symptoms.[21] But neurological symptoms, just as in the pandemic of 1889, were rampant, even in mild cases. They ranged from insomnia, stupor, dulled perceptions, unusually heightened perceptions, tingling, itching, and impairment of hearing to weakness or partial paralysis of the palate, eyelids, eyes, and various other muscles.[22] The famous Karl Menninger reported on 100 cases of psychosis triggered by influenza, including 35 of schizophrenia, that he saw during a three-month period.[23]

Although the infectious nature of this illness was widely assumed, masks, quarantines, and isolation were all without effect.[24] Even in an isolated country like Iceland the flu spread universally, in spite of the quarantining of its victims.[25]

The disease seemed to spread impossibly fast. "There is no reason to suppose that it traveled more rapidly than persons could travel [but] it has appeared to do so," wrote Dr. George A. Soper, Major in the United States Army.[26]

But most revealing of all were the various heroic attempts to prove the infectious nature of this disease, using volunteers. All these attempts, made in November and December 1918 and in February and March 1919, failed. One medical team in Boston, working for the United States Public Health Service, tried to infect one hundred healthy volunteers between the ages of eighteen and twenty-five. Their efforts were impressive and make entertaining reading:

"We collected the material and mucous secretions of the mouth and nose and throat and bronchi from cases of the disease and transferred this to our volunteers. We always obtained this material in the same way. The patient with fever, in bed, had a large, shallow, traylike arrangement before him or her, and we washed out one nostril with some sterile salt solutions, using perhaps 5 c.c., which is allowed to run into the tray; and that nostril is blown vigorously into the tray. This is repeated with the other nostril. The patient then gargles with some of the solution. Next we obtain some bronchial mucus through cough-ing, and then we swab the mucous surface of each nares and also the mucous surface of the throat... Each one of the volunteers... received 6 c.c. of the mixed stuff that I have described. They received it into each nostril; received it in the throat, and on the eye; and when you think that 6 c.c. in all was used, you will understand that some of it was swallowed. None of them took sick."

In a further experiment with new volunteers and donors, the salt solution was eliminated, and with cotton swabs, the material was trans-ferred directly from nose to nose and from throat to throat, using donors in the first, second, or third day of the disease. "None of these volunteers who received the material thus directly transferred from cases took sick in any way... All of the volunteers received at least two, and some of them three 'shots' as they expressed it."

In a further experiment 20 c.c. of blood from each of five sick donors were mixed and injected into each volunteer. "None of them took sick in any way."

"Then we collected a lot of mucous material from the upper respi-ratory tract, and filtered it through Mandler filters. This filtrate was injected into ten volunteers, each one receiving 3.5 c.c. subcutaneously, and none of these took sick in any way."

Then a further attempt was made to transfer the disease "in the natural way," using fresh volunteers and donors: "The volunteer was led up to the bedside of the patient; he was introduced. He sat down alongside the bed of the patients. They shook hands, and by instruc-tions, he got as close as he conveniently could, and they talked for five minutes. At the end of the five minutes, the patient breathed out as hard as he could, while the volunteer, muzzle to muzzle (in accordance with his instructions, about 2 inches between the two), received this expired breath, and at the same time was breathing in as the patient breathed out... After they had done this for five times, the patient coughed directly into the face of the volunteer, face to face, five dif-ferent times... [Then] he moved to the next patient whom we had selected, and repeated this, and so on, until this volunteer had had that sort of contact with ten different cases of influenza, in different stages of the disease, mostly fresh cases, none of them more than three days old... None of them took sick in any way."

"We entered the outbreak with a notion that we knew the cause of the disease, and were quite sure we knew how it was transmitted from person to person. Perhaps," concluded Dr. Milton Rosenau, "if we have learned anything, it is that we are not quite sure what we know about the disease."[27]

Earlier attempts to demonstrate contagion in horses had met with the same resounding failure. Healthy horses were kept in close contact with sick ones during all stages of the disease. Nose bags were kept on horses that had nasal discharges and high temperatures. Those nose bags were used to contain food for other horses which, however, stubbornly remained healthy. As a result of these and other attempts, Lieutenant Colonel Herbert Watkins-Pitchford of the British Army Veterinary Corps wrote in July 1917 that he could find no evidence that influenza was ever spread directly from one horse to another.

The other two influenza pandemics of the twentieth century, in 1957 and 1968, were also associated with milestones of electrical technology, pioneered once again by the United States.

Radar, first used extensively during World War II, was deployed on a spectacular scale by the United States during the mid-1950s, as it sought to surround itself with a triple layer of protection that would detect any nuclear attack. The first and smallest barrier was the 39 stations of the Pinetree Line, which kept vigil from coast to coast across southern Canada and from Nova Scotia northward to Baffin Island. This line, completed in 1954, was the roots, as it were, for a huge tree of surveillance that grew between 1956 and 1958, whose branches spread across mid- and high-latitude Canada, sent shoots into Alaska, and drooped down over the Atlantic and Pacific Oceans to guard the United States on east, west, and north. When it was complete hundreds of radar domes, resembling golf balls the size of buildings, littered the Canadian landscape from ocean to ocean, and from the American border to the Arctic.

The Mid-Canada Line, extending 2,700 miles from Hopedale, Labrador to Dawson Creek, British Columbia, consisted of 98 powerful Doppler radars 30 miles apart and roughly 300 miles north of the Pinetree Line. Construction of the first station began on October 1, 1956, and the completed system was dedicated on January 1, 1958.

The 58 stations of the Distant Early Warning or DEW Line kept their frozen watch roughly along the 69th parallel, 200 miles north of the Arctic Circle, in a chain extending from Baffin Island to the Northwest Territories and across Alaska. Each main site, of which there were 33, had two pulsed transmitters, one controlling a pencil beam for long-range precision tracking, the other a wider beam for general surveillance. Each beam had a peak power of 500 kilowatts, so that each site had a maximum peak capacity of one million watts. The frequency was between 1220 and 1350 MHz. The other twenty-five "gap-filler" stations had continuous wave Dopplers rated at 1 kilowatt and operated at 500 MHz. Construction began in 1955 and the completed system was dedicated on July 31, 1957.

The DEW Line extended down into the Atlantic and Pacific Oceans in lines of Navy ships—four in the Atlantic and five in the Pacific—supplemented by fleets of Lockheed aircraft that cruised in twelve- to fourteen-hour shifts at 3,000 to 6,000 feet in altitude. The radar-bearing ships and planes of the Atlantic Barrier were based in Maryland and Newfoundland and patrolled the waters out to the Azores. Atlantic operations began testing on July 1, 1956, and were fully deployed one year later. The Pacific Barrier, based in Hawaii and Midway, scanned the ocean off western North America and patrolled roughly from Midway to Kodiak Island. Its first two ships were assigned to Pearl Harbor in 1956, and the Barrier became fully operational on July 1, 1958.

In addition, three "Texas Towers," equipped with long-range radars, were placed about 100 miles off the Atlantic coast and affixed to the ocean floor. The first, 110 miles east of Cape Cod, began operation in December 1955, while the third, 84 miles southeast of New York Harbor, was activated in early summer 1957.

Finally, every one of the 195 initial radar sites blanketing Canadian skies had to be able to send surveillance data from mostly very remote locations, and so high power radio transmitters were added to each site, typically operating in the microwave spectrum between 600 and 1000 MHz, with broadcast powers of up to 40 kilowatts. These used a technology called "tropospheric scatter." Huge antennas the shape of curved billboards aimed their signals above the distant horizon so as to bounce them off particles in the lower atmosphere six miles above the earth, and thereby reach a receiver hundreds of miles away.

Another complete network of such antennas, called the White Alice Communications System, was installed throughout Alaska at the same time. The first ones were put into service on November 12, 1956, and the complete system was dedicated on March 26, 1958.

The "Asian" influenza pandemic began about the end of February 1957 and lasted for more than a year. The bulk of the mortality occurred in the fall and winter of 1957-1958.

A decade later the United States launched the world's first constellation of military satellites into orbit at an altitude of about 18,000 nautical miles, right in the heart of the outer Van Allen radiation belt. Called the Initial Defense Communication Satellite Program (IDCSP), its 28 satellites became operational after the last eight were launched on June 13, 1968. The "Hong Kong" flu pandemic began in July 1968 and lasted until March 1970.

Although there had already been a few satellites in space, they had all been launched one at a time during the 1960s, and at the beginning of 1968 there had been a total of only 13 operating satellites orbiting above the earth. In one fell swoop the IDCSP not only more than tripled the number, but placed them in the middle of the most vulnerable layer of the earth's magnetosphere.

In each case—in 1889, 1918, 1957, and 1968—the electrical envelope of the earth, which will be described in the next chapter, and to which we are all attached by invisible strings, was suddenly and profoundly disturbed. Those for whom this attachment was strongest, whose roots were most vital, whose life's rhythms were tuned most closely to the accustomed pulsations of our planet—in other words, vigorous, healthy young adults, and pregnant women—those were the individuals who most suffered and died. Like an orchestra whose conductor has suddenly gone mad, their organs, their living instruments, no longer knew how to play.

9. Earth's Electric Envelope

A

All things by immortal power,
Near or far,
Hiddenly
To each other linked are,
That thou canst not stir a flower
Without troubling of a star.

Francis Thompson, in
The Mistress of Vision

When i look at a flower, what I see is not the same as what a honey bee sees, who comes to drink its nectar. She sees beautiful patterns of ultraviolet that are invisible to me, and she is blind to the color red. A red poppy is ultraviolet to her. A cinquefoil flower, which looks pure yellow to me, is to her purple, with a yellow center luring her to its nectar. Most white flowers are blue-green to her eye.

When I look upon the night sky, the stars appear as points of color twinkling through earth's atmosphere. Everywhere else, except for the moon and a few planets, is blackness. But it is the blackness of illusion.

If you could see all the colors in the world, including the ultra-violets that honeybees can see, the infrareds that snakes can see, the low electric frequencies that catfish and salamanders can see, the radio waves, the X-rays, the gamma rays, the slow galactic pulsations, if you could see everything that is really there in its myriad shapes and hues, in all of its blinding glory, instead of blackness you'd see form and motion everywhere, day and night.

Almost all of the matter in the universe is electrically charged, an endless sea of ionized particles called plasma, named after the contents of living cells because of the unpredictable, life-like behavior of electrified matter. The stars we see are made of electrons, protons, bare atomic nuclei, and other charged particles in constant motion. The space between the stars and galaxies, far from being empty, teems with electrically charged subatomic particles, swimming in vast swirling electromagnetic fields, accelerated by those fields to near-light speeds. Plasma is such a good conductor of electricity, far better than any metals, that filaments of plasma—invisible wires billions of light-years long—transport electromagnetic energy in gigantic circuits from one part of the universe to another, shaping the heavens. Under the influence of electromagnetic forces, over billions of years, cosmic whirlpools of matter collect along these filaments, like beads on a string, evolving into the galaxies that decorate our night sky. In addition, thin sheaths of electric current called double layers, like the membranes of biological cells, divide intergalactic space into immense compartments, each of which can have different physical, chemical, electrical, and magnetic properties. There may even, some speculate, be matter on one side of a double layer and antimatter on the other. Enormous electric fields prevent the different regions of space from mixing, just as the integrity of our own cells is preserved by the electric fields of the membranes surrounding them.

Our own Milky Way, in which we live, a medium-sized spiral galaxy one hundred thousand light-years across, rotates around its center once every two hundred and fifty million earth years, generating around itself a galactic-size magnetic field. Filaments of plasma five hundred light-years long, generating additional magnetic fields, have been photographed looping out of our galactic center.

Our sun, also made of plasma, sends out an ocean of electrons, protons, and helium ions in a steady current called the solar wind. Blowing at three hundred miles per second, it bathes the earth and all of the planets before diffusing out into the plasma between the stars.

The earth, with its core of iron, rotates on its axis in the electric fields of the solar system and the galaxy, and as it rotates it generates its own magnetic field that traps and deflects the charged particles of the solar wind. They wrap the earth in an envelope of plasma called the magnetosphere, which stretches out on the night side of the planet into a comet-like tail hundreds of millions of miles long. Some of the particles from the solar wind collect in layers we call the Van Allen belts, where they circulate six hundred to thirty-five thousand miles above our heads. Driven along magnetic lines of force toward the poles, the electrons collide with oxygen and nitrogen atoms in the upper atmosphere. These fluoresce to produce the northern and southern lights, the aurorae borealis and australis, that dance in the long winter nights of the high latitudes.

The sun also bombards our planet with ultraviolet light and X-rays. These strike the air fifty to two hundred and fifty miles above us, ionizing it, freeing the electrons that carry electric currents in the upper atmosphere. This, the earth's own layer of plasma, is called the ionosphere.

The earth is also showered with charged particles from all directions called cosmic rays. These are atomic nuclei and subatomic particles that travel at velocities approaching the speed of light. From within the earth comes radiation emitted by uranium and other radioactive elements. Cosmic rays from space and radiation from the rocks and soil provide the small ions that carry the electric currents that surround us in the lower atmosphere.

In this electromagnetic environment we evolved.

We all live in a fairly constant vertical electric field averaging 130 volts per meter. In fair weather, the ground beneath us has a negative charge, the ionosphere above us has a positive charge, and the potential difference between ground and sky is about 300,000 volts. The most spectacular reminder that electricity is always playing around and through us, bringing messages from the sun and stars, is, of course, lightning. Electricity courses through the sky far above us, explodes downward in thunderstorms, rushes through the ground beneath us, and flows gently back up through the air in fair weather, carried by small ions. All of this happens continuously, as electricity animates the entire earth; about one hundred bolts of lightning, each delivering a trillion watts of energy, strike the earth every second. During thunderstorms the electric tension in the air around us can reach 4,000 volts per meter and more.

When I first learned about the global electrical circuit, twenty-five years ago, I drew the following sketch to help me think about it.

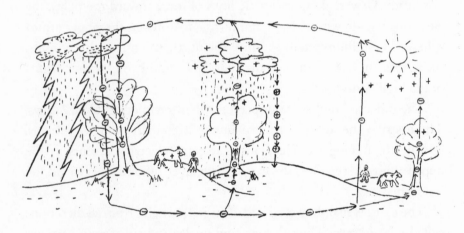

Living organisms, as the drawing indicates, are part of the global circuit. Each of us generates our own electric fields, which keep us vertically polarized like the atmosphere, with our feet and hands negative with respect to our spine and head. Our negative feet walk on the negative ground, as our positive heads point to the positive sky. The complex electric circuits that course gently through our bodies are completed by ground and sky, and in this very real way the earth and sun, the Great Yin and the Great Yang of the *Yellow Emperor's Classic*, are energy sources for life.

It is not widely appreciated that the reverse is also true: not only does life need the earth, but the earth needs life. The atmosphere, for example, exists only because green things have been growing for billions of years. Plants created the oxygen, all of it, and very likely the nitrogen too. Yet we fail to treat our fragile cushion of air as the irreplaceable treasure that it is, more precious than the rarest diamond. Because for every atom of coal or oil that we burn, for every molecule of carbon dioxide that we produce from them, we destroy forever one molecule of oxygen. The burning of fossil fuels, of ancient plants that once breathed life into the future, is really the undoing of creation.

Electrically, too, life is essential. Living trees rise hundreds of feet into the air from the negatively charged ground. And because most raindrops, except in thunderstorms, carry positive charge down to earth, trees attract rain out of the clouds, and the felling of trees contributes electrically towards a loss of rainfall where forests used to stand.

"As for men," said Loren Eiseley, "those myriad little detached ponds with their own swarming corpuscular life, what were they but a way that water has of going about beyond the reach of rivers?"[1] Not only we, but especially trees, are the earth's way of watering the desert. Trees increase evaporation and lower temperatures, and the currents of life speeding through their sap are continuous with the sky and the rain.

We are all part of a living earth, as the earth is a member of a living solar system and a living universe. The play of electricity across the galaxy, the magnetic rhythms of the planets, the eleven-year cycle of sunspots, the fluctuations in the solar wind, thunder and lightning upon this earth, biological currents within our bodies—the one depends upon all the others. We are like tiny cells in the body of the universe. Events on the other side of the galaxy affect all life here on earth. And it is perhaps not too far-fetched to say that any dramatic change in life on earth will have a small but noticeable effect on the sun and stars.

B

When the City and South London Electric Railway began operating in 1890, it interfered with delicate instruments at the Royal Observatory at Greenwich four and a half miles away.[2] Little did the physicists there know that electromagnetic waves from that and every other electric railway were also radiating into space and altering the earth's magnetosphere, a fact that would not be discovered until decades later. To understand its significance for life, let us return first to the story of lightning.

The house we live in, which is the biosphere, the roughly 55-mile-high space filled with air that wraps around the earth, is a resonant cavity that rings like a gong every time a lightning bolt strikes. In addition to maintaining the static electric field of 130 volts per meter in which we all stand and walk, and in which birds fly, lightning sets the biosphere ringing at particular low frequency tones—8 beats per second (or Hz), 14, 20, 26, 32, and so forth. These tones are named for Winfried Schumann, the German physicist who predicted their existence, and who, with his student Herbert König, proved their constant presence in the atmosphere in 1953.

It so happens that in a state of awake relaxation, our brains tune in to these precise frequencies. The dominant pattern of a human electroencephalogram, from before birth through adulthood—the well-known alpha rhythm, ranging from 8 to 13 Hz, or 7 to 13 Hz in a newborn—is bounded by the first two Schumann resonances. An old part of the brain called the limbic system, which is involved in emotions, and in long-term memory, produces theta waves, of 4 to 7 Hz, which are bounded above by the first Schumann resonance. The theta rhythm is more prominent in young children, and in adults in meditation. These same frequencies, alpha and theta, with surprisingly little variation, pulsate, so far as is known, in all animals. In a state of relaxation, dogs show an alpha rhythm, identical to ours, of 8 to 12 Hz. In cats the range is slightly wider, from 8 to 15 Hz. Rabbits, guinea pigs, goats and cows, frogs, birds, and reptiles all show nearly the same frequencies.[3]

Schumann's student König was so impressed by the resemblances these atmospheric waves bear to the electrical oscillations of the brain that he conducted a series of experiments with far-reaching implications. The first Schumann resonance, he wrote, is so completely identical to the alpha rhythm that even an expert is hard pressed to tell the difference between the tracings from the brain and the atmosphere. König did not think this was a coincidence. The first Schumann resonance appears during fair weather, he noted, in calm, balanced conditions, just as the alpha rhythm appears in the brain in a calm, relaxed state. The delta rhythm, on the other hand, which consists of irregular, higher amplitude waves around 3 Hz, appears in the atmosphere under disturbed, unbalanced weather conditions, and in the brain in disturbed or disease states—headaches, spastic conditions, tumors, and so forth.

In an experiment involving nearly fifty thousand people attending a Traffic Exhibition in Munich in 1953, König was able to prove that these latter types of disturbed waves, when present in the atmosphere, significantly slow human reactions times, while the 8 Hz Schumann waves do just the opposite. The larger the Schumann signal in the atmosphere, the quicker people's reactions were on that day. König then duplicated these effects in the laboratory: an artificial field of 3 Hz (delta range) slowed human reactions, while an artificial field of 10 Hz (alpha range) accelerated them. König also noted that during the 3 Hz exposure some of his subjects complained of headaches, fatigue, tightness in their chest, or sweating from their palms.[4]

In 1965, James R. Hamer published the results of experiments along these same lines that he had conducted for Northrop Space Laboratories, in an article which he titled "Biological Entrainment of the Human Brain by Low Frequency Radiation." Like König, he showed that frequencies above 8 Hz quickened reaction times, while lower frequencies had the opposite effect. But he went further. He proved that the human brain could distinguish between frequencies that differed only slightly from each other—but only if the signal was weak enough. When he reduced the signal strength to 0.0038 volts per meter, which is close to the value of the earth's own fields, 7½ Hz had a significantly different effect than 8½ Hz, and 9½ Hz than 10½ Hz.

Lightning is not yet done with its repertoire. In addition to the static field that we walk in and the low frequencies that speak to our brains, lightning also provides us with a steady symphony of higher frequencies called atmospherics, or just "sferics," which reach thousands of cycles per second. They sound like twigs snapping if you listen to them on a very low frequency (VLF) radio, and usually originate in thunderstorms that may, however, be thousands of miles away. Other sounds, called whistlers, resembling the descending tones of a slide whistle, often originate in thunderstorms on the opposite end of the earth. Their falling tones are produced during the long journey these waves have taken as they are guided along magnetic field lines into outer space and back to earth in the opposite hemisphere. These waves may even bounce back and forth many times from one end of the earth to another, resulting in trains of whistles that seemed so unworldly when they were first discovered in the 1920s that they generated newspaper articles with not-so-inappropriate titles like "Voices From Outer Space."[5]

Among the other sounds one may hear, especially at higher latitudes, originating somewhere in the electrical environment of our planet, are a steady hiss, and a "dawn chorus," so named because of its resemblance to chirping birds. Both of these sounds rise and fall gently every 10 seconds or so with the slow pulsations of the earth's magnetic field.

This VLF symphony bathes our nervous system. Its frequencies, ranging roughly from 200 to 30,000 Hz, span the range of our auditory system and also, as König observed, include the frequencies of the impulses that our brains send to our muscles. The effect our VLF environment has on our well-being was resoundingly demonstrated by Reinhold Reiter in 1954 when he tabulated the results of a number of population studies that he and his colleagues had conducted in

Germany, involving about one million people. Births, deaths, suicides, rapes, work injuries, traffic accidents, human reaction times, amputees' pains, and complaints of people with brain injuries all rose significantly on days with strong VLF sferics.[6]

Our VLF environment regulates biological rhythms in both humans and animals. Golden hamsters, which have been popular pets since the 1930s, live in the wild near Aleppo, Syria where, every winter for about three months, they go in and out of hibernation. But scientists who have tried to use hamsters as a subject for hibernation studies in the laboratory have been puzzled by their inability to trigger hibernation in these animals by exposing them to prolonged cold, reducing hours of daylight, or controlling any other known environmental factor.[7]

In the mid-1960s, climatologists Wolfgang Ludwig and Reinhard Mecke took a different approach. They kept a hamster during the winter in a Faraday cage, shielded from all natural electromagnetic waves, and without any alteration of temperature or hours of daylight. At the beginning of the fourth week they introduced the natural out-door atmospheric frequencies by means of an antenna, whereupon the hamster promptly fell asleep. During the following two months, the researchers were able to put the animal into and out of hibernation by introducing, or removing, either the natural outdoor frequencies, or artificial VLF fields that imitated the natural winter pattern. Then, at the beginning of the thirteenth week of the experiment, the frequencies in the enclosure were changed so as to imitate the natural summer pattern, and within half an hour, as if panicked by the sudden change in season, the animal woke up and began a "movement storm," running day and night for an entire week until the experiment was terminated. In repetitions of this experiment on other hamsters, the researchers found that this high level of activity could not be induced unless the state of hibernation had been triggered first. The artificial fields they used were extremely weak—as small as 10 millivolts per meter for the electric field and 26.5 microamperes per meter for the magnetic field.

One way to find out if the earth's natural fields are as important to people as to hamsters would be to place human subjects in a completely shielded room for a few weeks and see what happens. Which is exactly what behavioral physiologist Rütger Wever did at the Max Planck Institute in Germany. In 1967 he had an underground building constructed containing two isolation chambers. Both were carefully shielded against outside light and sound, and one was shielded also against electromagnetic fields. During the next two decades hundreds of people had their sleep cycles, body temperature, and other internal rhythms monitored while they lived in one or the other of these rooms, usually for a month at a time. Wever found that even without any variation in light and darkness, and without any clocks or time cues, the body's sleep cycle and internal rhythms remained close to 24 hours, so long as the earth's natural electromagnetic fields were present. However, when those fields were excluded, the body's rhythms usually became longer, erratic, and desynchronized with each other. The average "free-running" sleep cycle was 25 hours, but in individual cases was as short as 12 hours and as long as 65 hours. Variations in body temperature, potassium excretion, speed of mental processes, and other rhythms drifted at their own separate rates, completely different from one another, and no longer coinciding with the sleep-wake cycle at all. But as soon as an artificial 10 Hz signal—close to the first Schumann resonance—was introduced into the shielded room, the body's rhythms all immediately resynchronized to a 24-hour period.

C

Life, residing between heaven and earth, partakes of both polarities. As we will see in the next chapter, the distribution of electric charge in living beings has been measured and mapped externally. In plants this was done by professor of anatomy Harold Saxton Burr, at Yale University, and in animals by orthopedic surgeon Robert O. Becker, at the State University of New York, Upstate Medical Center, Syracuse. The areas of greatest positive voltage in animals are the center of the head, the heart, and the lower abdomen, and in trees the crown.

The places of greatest negative voltage, in trees, are the roots, and in animals, the four feet and the end of the tail. These are the places where the global electrical circuit enters and leaves the body on its way between heaven and earth. And the channels through which the electricity travels *inside* living beings, distributing the electricity of heaven and earth to every organ, were precisely mapped several thousand years ago, and are part of a body of knowledge that we know today as Chinese acupuncture. It was written down in the *Huangdi Neijing*, the *Yellow Emperor's Classic of Internal Medicine*, between 500 and 300 B.C. The very names of key acupuncture points reveal an understanding that the circuitry of the body is continuous with that of earth and sky. Kidney 1, for example, the point underneath the foot, in the center of the sole, is known in Chinese as yong quan, meaning "bubbling spring," because earth energy bubbles up into the feet through these points and climbs up the legs into the rest of the body toward the heavens. Governing Vessel 20, the point on top of the head, in the center, is called bai hui, the "hundred convergences." This is also the "thousand petal lotus" of Indian traditions, the place where the energy of heaven descends into our body toward the earth, and the flows of our body converge and reach toward the sky.

But not until the 1950s did scientists, beginning with Yoshio Nakatani in Japan and Reinhold Voll in Germany, begin to actually measure the electrical conductivity of acupuncture points and meridians, and to finally translate the word "qi" (formerly spelled "chi") into modern language: it means "electricity."

Hsiao-Tsung Lin is a professor of chemical and material science at National Central University in Taiwan. The qi that flows through our meridians, he tells us, is an electrical current that brings both power and information to our cells, current whose source is both internal and external. Every acupuncture point has a double function: as an amplifier for the internal electrical signals, boosting their strength as they travel along the meridians; and as an antenna that receives electromagnetic signals from the environment. The dantians, or energy centers of Chinese medicine, located in the head, heart, and abdomen—equivalent to the chakras of Indian tradition—are electromagnetic oscillators that resonate at particular frequencies, and that communicate with the meridians and regulate their flow. They have capacitance and inductance like oscillators in any electronic circuitry. The body, says Lin, is a super-complex electromagnetic oscillation network, enormously intricate and delicate.

In 1975, Becker and his colleagues at Upstate Medical Center found that, in general, acupuncture points are not only places of low resistance, but of high potential, averaging five millivolts higher than the surrounding skin. They also found that the path of a meridian, at least on the surface of the body, has significantly greater conductivity and lower electrical resistance than nearby skin.

As a result of the work of Nakatani, Voll, Becker, and others, electroacupuncture, using microampere currents, has taken its place alongside traditional acupuncture, and commercial point locators, which find acupuncture points by measuring the electrical conductivity of the skin, have come into use among nontraditional practitioners here in the West.[8] In China, electroacupuncture devices have been in use since 1934. They are a tacit acknowledgement that the body is an electrical instrument, and that its health or sickness depends on the proper distribution and balance of the electrical energies that constantly flow around and through us. But ironically they also prevent that scientific knowledge from becoming true knowledge, for to substitute artificial electricity for atmospheric electricity in replenishing the body is to forget that the electricity of the air is there, nourishing us and giving us life.

At the Shanghai University of Traditional Chinese Medicine, the Fujian Institute of Traditional Chinese Medicine, and elsewhere in China, scientists continue to confirm that the substance that flows in our meridians is electricity, and that electricity is not only a force that moves locomotives, but is the incredibly complex and delicate stuff of life. Typically, the electrical resistance of an acupuncture point is two to six times lower than the resistance of the surrounding skin, and its capacitance—its ability to store electrical energy—is five times as great.[9] Commercial point locators do not always work, because some-times—depending on the internal state of the individual—an acupunc-ture point can have a higher resistance than its surroundings. But the meridians always respond in an active and nonlinear way to electrical stimulation, and they react, say modern researchers, exactly like an electrical circuit.[10]

The physical structures of the conductive points and meridians have been tentatively identified. In the 1960s, a North Korean physi-cian, Bong Han Kim, published detailed photographs of an entire net-work of tiny corpuscles, and threadlike structures that connect them, that exist throughout the body in our skin, in our internal organs and nervous system, and in and around our blood vessels. These ducts, he found, were electrically conductive and the fluid within them, surpris-ingly, contained large amounts of DNA. Their electrical pulsations were considerably slower than the heartbeat: in the skin of a rabbit, the pulsation rate was between 10 and 20 per minute. The pathways of the superficial ducts in the skin matched the classical pathways of the acupuncture meridians. The reason Kim succeeded in identifying this system is that he worked only on living animals, because the ducts and corpuscles, almost transparent to begin with, disappear shortly after death. He stained the living tissue with an unspecified blue dye that was absorbed only by this network of ducts and corpuscles. Kim's book, *On the Kyungrak System*, was published in Pyongyang in 1963. The reason his work has been so completely ignored has partly to do with his relations with the North Korean government—Kim was expunged from official records in 1966, and rumor has it that he committed sui-cide—and partly with the fact that the outside world does not *want* to find physical proof of our electrical nature. But in the mid-1980s, Jean-Claude Darras, a French physician working in the nuclear medi-cine department at Necker Hospital in Paris, replicated some of Kim's experiments. He injected a radioactive dye containing technetium-99 into various acupuncture points on the feet of volunteers, and found that the dye migrated precisely along the meridian pathways of classi-cal acupuncture, just as Kim had found.[11]

In 2002, Kwang-Sup Soh, who had already been investigating the electromagnetic properties of acupuncture meridians, headed up a team at Seoul National University in South Korea, which looked for and found most of the threadlike duct system described by Kim. A breakthrough came in November 2008 with the discovery that trypan blue, a dye that was previously known to stain only dead cells, if injected into living tissue, will stain *only* the nearly invisible threads and corpuscles they had painstakingly begun to identify. The "primo vascular system," as it was now called, suddenly became a subject of research in other centers in South and North Korea, as well as in China, Europe, Japan, and the United States. The ducts and corpuscles of this system were found, just as Kim had described, resting on the surface of and penetrating inside the internal organs, floating inside the large blood and lymphatic vessels, winding along the outside of major blood vessels and nerves, traveling inside the brain and spinal cord, and following the paths of the known meridians within the deep layers of the skin.[12] When the surface of the skin was stained with the dye, only points along the meridians absorbed it.[13] In September 2010, at the First International Symposium of Primo Vascular System, held in Jecheon, Korea, Satoru Fujiwara, retired professor of anatomy at Osaka City University, Japan, reported tentative success at surgically identifying a superficial primo node—an acupuncture point—in the skin of a rabbit's abdomen.[14] And in 2015, researchers at Seoul National University used a commercially available staining kit to reveal a threadlike vessel running just beneath the abdominal skin of anesthetized living rats.[15] The vessel, colored dark blue from the stain, followed the pathway of the acupuncture meridian called the conception vessel, and connected discrete corpuscles corresponding in location to the known acupuncture points on that meridian. The fine structure of this system of nodes and ducts was revealed by electron microscopy. The staining process, they noted, takes less than ten minutes.

D

In the early 1970s, atmospheric physicists finally woke up to the fact that the earth's magnetic field was highly disturbed. Not all of those whistlers, hiss, chorus, lion roars, and other colorful sounds they had been listening to for half a century were caused by nature! This discovery came about as a result of efforts to *deliberately* alter the earth's electromagnetic environment—efforts that have culminated, today, in the operation of Project HAARP, located in Gakona, Alaska (see chapter 16).

Under contract with the Office of Naval Research, scientists at Stanford University's Radioscience Laboratory had built a 100-kilowatt VLF transmitter at Siple Station, Antarctica, broadcasting in the 1.5 to 16 kHz range. The purposes of the 13-mile-long antenna that stretched over the frozen ice, according to Robert Helliwell, one of the members of the Stanford team, included "control of the ionosphere, control of the radiation belts and new methods of v.l.f. and u.l.f. communication."[16] It had been discovered accidentally in 1958 that VLF transmissions originating on the earth interact with particles in the magnetosphere, stimulating them to emit new VLF waves, which can then be received at the opposite end of the earth. The purpose of the Stanford project was to do this deliberately—to inject sufficient quantities of very low frequency energy into the magnetosphere so that it would not only trigger new waves, but that these triggered waves might in turn cause electrons to rain out of the earth's radiation belts into the atmosphere, altering the properties of the ionosphere for military purposes. A primary goal of the Department of Defense was to devise a method of stimulating the ionosphere to emit VLF (very low frequency), ELF (extra low frequency), or even ULF (ultra low frequency) waves in order to communicate with submarines submerged beneath the oceans.[17] The VLF transmitter at Siple, and a VLF receiver in northern Quebec, at Roberval, were part of this early research.

The data they collected were surprising. First, the signal received in Quebec, immediately after transmission from Antarctica, was larger than expected. The waves broadcast from Antarctica were not only triggering new emissions from particles in the magnetosphere, but were being amplified more than a thousandfold in the magnetosphere before returning to earth and being received in Quebec. Only half a watt of broadcast power was required in order to be detected near the opposite pole of the earth after being relayed from the magnetosphere.[18] The second surprise was that Roberval was receiving frequencies that were unrelated to the frequencies that originated at Siple, but that were instead multiples of 60 Hz. The Siple signal had been altered, on its journey through outer space, to bear the imprint of the electric power grid.

Since those first discoveries, scientists have learned a great deal about this form of pollution, now known as "power line harmonic radiation." It appears that harmonics from all of the world's power grids leak continuously into the magnetosphere, where they are greatly amplified as they bounce back and forth between the northern and southern hemisphere, generating their own rising and falling whistlers just like radiation from lightning.

But there is a fundamental difference. Before 1889, whistlers and other lightning-triggered sounds played continuously over the entire range of the terrestrial instrument. Today the music is stilted, dulled, often confined to multiples of 50 or 60 Hz. Every component of the natural symphony has been radically altered. The "dawn chorus" is quieter on Sundays than on other days of the week, and the starting frequencies of most chorus emissions are power line harmonics.[19] "It seems likely that the entire hiss band is caused by power line radiation," wrote Helliwell in 1975. And the natural, slow pulsations of the earth's magnetic field, below 1 Hz, which are also important to all life, are strongest on weekends, evidently because they are being suppressed by radiation from the power grid, and this radiation is stronger on weekdays.[20] Antony Fraser-Smith, also at Stanford, by analyzing geomagnetic activity data collected since 1868, showed that this is not a new phenomenon but has been happening since the first use of alternating current, and has been increasing over time.[21] Data collected between 1958 and 1992 showed that Pc 1 activity, representing geomagnetic pulsations between 0.2 and 5 Hz, has been fifteen to twenty percent greater on weekends than in the middle of the week.[22]

The structure of the Van Allen radiation belts seems also to have been altered. What the Department of Defense had wanted to do intentionally was apparently already being done massively by the world's electric power grids. Why, physicists had long wondered, are there two electron-filled radiation belts around the earth, an inner and an outer, separated by a layer that is virtually empty of electrons? This "electron slot," some think, is continually drained of its electrons by their interaction with radiation from power lines.[23] These electrons, in turn, rain down over the earth, modifying the electrical properties of the atmosphere.[24] Not only may this increase the frequency of thunderstorms,[25] but it may shift the values of the Schumann resonances to which all living things are attuned.[26]

In short, the electromagnetic environment of the entire earth is radically different today from what it was before 1889. Satellite observations show that radiation originating from power lines often overwhelms natural radiation from lightning.[27] Power line radiation is so intense that atmospheric scientists lament their inability to do fundamental research: there is almost nowhere left on earth, or even in space, where a VLF receiver can be used to study natural phenomena.[28]

Under natural conditions, as they existed before 1889, intense VLF activity, leading to electron rain and the shifting of the Schumann resonances, occurred only during geomagnetic storms. Today, the magnetic storm never ends.

E

Influenza

If the atmosphere is, at times, electrified
beyond the degree which is usual, and necessary to preserve the body in a due state of
excitement, the nerves must be too highly
excited, and under a continued operation of
undue stimulus, become extremely irritable,
and subject to debility.

NOAH WEBSTER, *A Brief History of
Epidemic and Pestilential Diseases,*
1799, p. 38

A large, rapid, qualitative change in the earth's electromagnetic environment has occurred six times in history.

In 1889, power line harmonic radiation began. From that year forward the earth's magnetic field bore the imprint of power line frequencies and their harmonics. In that year, exactly, the natural magnetic
activity of the earth began to be suppressed. This has affected all life
on earth. The power line age was ushered in by the 1889 pandemic of
influenza.

In 1918, the radio era began. It began with the building of hundreds
of powerful radio stations at LF and VLF frequencies, the frequencies
guaranteed to most alter the magnetosphere. The radio era was ushered in by the Spanish influenza pandemic of 1918.

In 1957, the radar era began. It began with the building of hundreds
of powerful early warning radar stations that littered the high latitudes
of the northern hemisphere, hurling millions of watts of microwave
energy skyward. Low-frequency components of these waves rode on

magnetic field lines to the southern hemisphere, polluting it as well. The radar era was ushered in by the Asian flu pandemic of 1957.

In 1968, the satellite era began. It began with the launch of dozens of satellites whose broadcast power was relatively weak. But since they were already in the magnetosphere, they had as big an effect on it as the small amount of radiation that managed to enter it from sources on the ground. The satellite era was ushered in by the Hong Kong flu pandemic of 1968.

The other two mileposts of technology—the beginning of the wireless era and the activation of the High Frequency Active Auroral Research Program (HAARP)—belong to very recent times and will be discussed later in this book.

10. Porphyrins and the Basis of Life

> I see little hope to be able to explain the
> subtle difference between a normal and a
> sick cell as long as we do not understand the
> basic difference between a cat and a stone.
>
> ALBERT SZENT-GYÖRGYI

STRANGELY ENOUGH, "porphyrin" is not a household word. It is not a sugar, fat, or protein, nor is it a vitamin, mineral, or hormone. But it is more basic to life than any other of life's components, because without it we would not be able to breathe. Plants could not grow. There would not be any oxygen in the atmosphere. Wherever energy is transformed, wherever electrons flow, there look for porphyrins. When electricity alters nerve conduction, or interferes with the metabolism of our cells, porphyrins are centrally involved.

As I write this chapter, a dear friend has just died. For the last seven years she had had to live without electricity, hardly ever seeing the sun. She seldom ventured out in the daytime; when she did, she covered herself from head to foot in thick leather clothing, a broad-brimmed leather hat hiding her face, and glasses bearing two layers of dark lenses concealing her eyes. A former dancer who loved music, nature, and the outdoors, Bethany was virtually abandoned by a world in which she no longer belonged.

Her condition, probably caused by her years of work for a computer company, was a classic example of an illness that has been known to medicine only since 1891, its emergence at that time being one of the side effects of the sudden worldwide expansion of electrical technology. Its connection with electricity was discovered a century later. Although it is now considered an extremely rare genetic disease, affecting as few as one person in fifty thousand, porphyria was originally thought to affect as many as ten percent of the population. Its supposed rarity is due in large part to the ostrich-like behavior of the medical profession after World War II.

In the late 1940s, medical practitioners were staring at an impossible contradiction. Most synthetic chemicals were known poisons. But one of the legacies of the war was the ability to manufacture products from petroleum, easily and cheaply, to substitute for almost every consumer product imaginable. Now, thanks to the fledgling petrochemical industry, bringing us "Better Living Through Chemistry," synthetic chemicals were going to be literally everywhere. We were going to be wearing them, sleeping on them, washing our clothes, our hair, our dishes, and our homes with them, bathing in them, insulating our houses with them, carpeting our floors with them, spraying our crops, our lawns, and our pets with them, preserving our food with them, coating our cookware with them, packing our groceries in them, moisturizing our skin with them, and perfuming our bodies with them.

The medical profession had two choices. It could have attempted to study the health effects, singly and in combination, of the hundreds of thousands of new chemicals that were kaleidoscoping over our world, a virtually impossible task. The attempt itself would have put the profession on a collision course with the mushrooming petrochemical industry, threatening the banning of most new chemicals and the strangling of the economic boom of the next two decades.

The other alternative was for the profession to bury its collective head in the sand and pretend that the world's population was not actually going to become poisoned.

Environmental medicine was born as a medical specialty in 1951, founded by Dr. Theron Randolph.[1] It had to be created: the scale of the poisoning was too great to go completely ignored. The sheer numbers of sickened patients, abandoned by mainstream medicine, produced an urgent need for practitioners trained to recognize at least some of the effects of the new chemicals and to treat the resulting diseases. But the specialty was ignored by the mainstream as though it didn't exist, its practitioners ostracized by the American Medical Association. When I attended medical school from 1978 to 1982, environmental medicine wasn't even on the curriculum. Chemical sensitivity, the unfortunate name that has been given to the millions of poisoned patients, was never mentioned in school. Neither was porphyria, arguably a more appropriate name. It still isn't mentioned, not in any medical school in the United States.

Heightened sensitivity to chemicals, we recall, was first described by New York physician George Miller Beard, who considered it a symptom of a new disease. The initial electrification of society through telegraph wires brought with it the constellation of health complaints known as neurasthenia, two of which were a tendency to develop allergies and a drastically reduced tolerance for alcohol and drugs.

By the late 1880s, insomnia, another prominent symptom of neurasthenia, had become so rampant in western civilization that the sale of sleeping pills and potions became big business, with new formulations coming on the market almost every year. Bromides, paraldehyde, chloral, amyl hydrate, urethane, hypnol, somnal, cannabinon, and other hypnotics flew off pharmacists' shelves to satisfy the frustrated urge to sleep—and the addiction that so often followed the long term use of these drugs.

In 1888, one more drug was added to the list. Sulfonal was a sleeping medication that had a reputation for its prompt effect, its non-addictive nature, and its relative lack of side effects. There was just one problem, which only became widely known after three years of its popularity: it killed people.

But its effects were quirky, unexpected. Nine people could take sulfonal, even in large doses and for a long time, with no untoward effects, but the tenth person, sometimes after only a few or even one small dose, would become critically ill. He or she would typically be confused, so weak as to be unable to walk, constipated, with pain in the abdomen, sometimes with a skin rash, and reddish urine often described as the color of port wine. The reactions were idiosyncratic, liable to affect almost any organ, and the patients were apt to die of heart failure without warning. Between four and twenty percent of the general population were reported to be subject to such side effects from taking sulfonal.[2]

During the ensuing decades the chemistry of this surprising disease was worked out.

Porphyrins are light-sensitive pigments that play pivotal roles in the economy of both plants and animals, and in the ecology of planet Earth. In plants a porphyrin bound to magnesium is the pigment called chlorophyll, that makes plants green and is responsible for photosynthesis. In animals an almost identical molecule bound to iron is the pigment called heme, the essential part of hemoglobin that makes blood red and enables it to carry oxygen. It is also the essential part of myoglobin, the protein that makes muscles red and delivers oxygen from our blood to our muscle cells. Heme is also the central component of cytochrome c and cytochrome oxidase, enzymes that are contained in every cell of every plant, animal and bacterium, that transport electrons from nutrients to oxygen so that our cells can extract energy. And heme is the main component of the cytochrome P-450 enzymes in our liver that detoxify environmental chemicals for us by oxidizing them.

In other words, porphyrins are the very special molecules that interface between oxygen and life. They are responsible for the creation, maintenance, and recycling of all of the oxygen in our atmosphere: they make possible the release of oxygen from carbon dioxide by plants, the extraction of oxygen back out of the air by both plants and animals, and the use of that oxygen by living things to burn carbohydrates, fats, and proteins for energy. The high reactivity of these molecules, which makes them transformers of energy, and their affinity for heavy metals, also makes them toxic when they accumulate in excess in the body, as happens in the disease called porphyria—a disease that is not really a disease at all, but a genetic trait, an inborn sensitivity to environmental pollution.

Our cells manufacture heme from a series of other porphyrins and porphyrin precursors in a series of eight steps, catalyzed by eight different enzymes. Like workers on an assembly line, each enzyme has to work at the same rate as all the others in order to keep up with the demand for the final product, heme. A slowdown by any one enzyme creates a bottleneck, and the porphyrins and precursors that accumulate behind the bottleneck get deposited all over the body, causing disease. Or if the first enzyme is working harder than the rest, it produces precursors faster than the enzymes down the line can handle, with the same result. Their accumulation in the skin can cause mild to disfiguring skin lesions, and mild to severe light sensitivity. Their accumulation in the nervous system causes neurological illness, and their accumulation in other organs causes corresponding illness. And when excess porphyrins spill into the urine, it takes on the color of port wine.

Because porphyria is assumed to be so rare, it is almost always misdiagnosed as some other disease. It is fairly called "the little imitator" because it can affect so many organs and mimic so many other conditions. Since patients usually feel so much sicker than they look, they are sometimes wrongly thought to have psychiatric disorders and too often wind up on mental wards. And since most people don't carefully examine their own urine, they usually fail to notice its reddish hue, particularly since the color may be evident only during severe disabling attacks.

The enzymes of the heme pathway are among the most sensitive elements of the body to environmental toxins. Porphyria, therefore, is a response to environmental pollution and was indeed extremely rare in an unpolluted world. Except for one severe, disfiguring congenital form, of which only a few hundred cases are known in the world, porphyrin enzyme deficiencies do not normally cause disease at all. Human beings are genetically diverse, and in times past most people with relatively lower levels of one or more porphyrin enzymes were simply more sensitive to their environment. In an unpolluted world this was a survival advantage, allowing the possessors of this trait to easily avoid places and things that might do them harm. But in a world in which toxic chemicals are inescapable, the porphyrin pathway is to some degree always stressed, and only those with high enough enzyme levels tolerate the pollution well. Sensitivity has become a curse.

Because of the way it was discovered, and the lack of synthetic chemicals in the environment at that time, porphyria became known as a rare disease that was triggered in genetically susceptible people by certain drugs, such as sulfonal and barbiturates, which these patients had to avoid. It was not until another century had passed, in the early 1990s, that Dr. William E. Morton, professor of occupational and environmental medicine at Oregon Health Sciences University, realized that because ordinary synthetic chemicals were far more widespread in the modern environment than pharmaceuticals, they had to be the most common triggers of porphyric attacks. Morton proposed that the controversial disease called multiple chemical sensitivity (MCS) was in most cases identical with one or more forms of porphyria. And when he began testing his MCS patients he found that, indeed, 90 percent of them were deficient in one or more porphyrin enzymes. He then investigated a number of their family trees, looking for the same trait, and succeeded in demonstrating a genetic basis for MCS—something no one had attempted before because MCS had never before been connected to a testable biological marker.[3] Morton also found that most people with electrical sensitivity had porphyrin enzyme deficiencies, and that electrical and chemical sensitivities appeared to be manifestations of the same disease. Porphyria, Morton showed, is not the extremely rare illness it is currently thought to be, but has to affect at least five to ten percent of the world's population.[4]

Morton was courageous, because the rare-disease world of porphyria had come to be dominated by a handful of clinicians who controlled virtually all research and scholarship in their small, inbred field. They tended to diagnose porphyria only during acute attacks with severe neurological symptoms and to exclude cases of milder, smoldering illness. They generally would not make the diagnosis unless porphyrin excretion in urine or stool was at least five to ten times normal. "This makes no sense," wrote Morton in 1995, "and would be analogous to restricting the diagnosis of diabetes mellitus to those who have ketoacidosis or restricting the diagnosis of coronary artery disease to those who have myocardial infarction."[5]

The higher numbers reported by Morton agree with the numbers reported over a century ago—the proportion of the population that became ill when they took the sleeping medication sulfonal. They are consistent with the finding, in the 1960s, of "mauve factor," a lavender-staining chemical, not only in the urine of patients diagnosed with porphyria, but in the urine of five to ten percent of the general population.[6] Mauve factor was eventually identified as a breakdown product of porphobilinogen, one of the porphyrin precursors.[7] Morton also found, in agreement with recent reports from England, the Netherlands, Germany, and Russia, that persistent neurological problems occur during the chronic, smoldering phase of every type of porphyria—even those types which were previously supposed to cause only skin lesions.[8]

Hans Günther, the German doctor who, in 1911, gave porphyria its name, stated that "such individuals are neuropathic and suffer from insomnia and nervous irritability."[9] Morton has brought us back to the original view of porphyria: it is not only a fairly common disease but exists most often in a chronic form with comparatively mild symptoms. And its principal cause is the synthetic chemicals and electromagnetic fields that pollute our modern environment.

Porphyrins are central to our story not only because of a disease named porphyria, which affects a few percent of the population, but because of the part porphyrins play in the modern epidemics of heart disease, cancer, and diabetes, which affect half the world, and because their very existence is a reminder of the role of electricity in life itself, a role which a few courageous scientists have slowly elucidated.

As a child, Albert Szent-Györgyi (pronounced approximately like "Saint Georgie") hated books and needed a tutor's help to pass his exams. But later, having graduated from Budapest Medical School in 1917, he went on to become one of the world's greatest geniuses in the field of biochemistry. In 1929 he discovered Vitamin C, and during the next few years he worked out most of the steps in cellular respiration, a system now known as the Krebs cycle. For these two discoveries he was awarded the Nobel Prize in Physiology or Medicine in 1937. He then spent the next two decades figuring out how muscles function. After emigrating to the United States and settling at Woods Hole, Massachusetts, he received the Albert Lasker Award of the American Heart Association in 1954 for his work on muscles.

Albert Szent-Györgyi, M.D., Ph.D. (1893-1986)

But perhaps his greatest insight is one for which he is least known, although he devoted almost half his life to the subject. For on March 12, 1941, in a lecture delivered in Budapest, he boldly stood up before his peers and suggested to them that the discipline of biochemistry was obsolete and should be brought into the twentieth century. Living organisms, he told them, were not simply bags of water in which molecules floated like tiny billiard balls, forming chemical bonds with other billiard balls with which they happened to collide. Quantum theory, he said, had made such old ideas invalid; biologists needed to study solid state physics.

In his own specialty, although he had worked out the structures of the molecules involved in muscular contraction, he could not begin to fathom why they had those particular structures, nor how the molecules communicated with one another to coordinate their activities.

He saw such unsolved problems everywhere he looked in biology. "One of my difficulties within protein chemistry," he bluntly told his colleagues, "was that I could not imagine how such a protein molecule can 'live.' Even the most involved protein structural formula looks 'stupid,' if I may say so."

The phenomena that had forced Szent-Györgyi to face these questions were the porphyrin-based systems of life. He pointed out that in plants, 2,500 chlorophyll molecules form a single functional unit, and that in dim light at least 1,000 chlorophyll molecules have to cooperate simultaneously in order to split one molecule of carbon dioxide and create one molecule of oxygen.

He spoke about the "enzymes of oxidation"—the cytochromes in our cells—and wondered, again, how the prevailing model could be correct. How could a whole series of large protein molecules be arranged geometrically so that electrons could wander directly from one to the other in a precise sequence? "Even if we could devise such an arrangement," he said, "it would still be incomprehensible how the energy liberated by the passing of an electron from one substance to the other, *viz.*, from one iron atom to the other, could do anything useful."

Szent-Györgyi proposed that organisms are alive because thousands of molecules form single systems with shared energy levels, such as physicists were describing in crystals. Electrons don't have to pass directly from one molecule to another, he said; instead of being attached to only one or two atoms, electrons are mobile, belong to the whole system, and transmit energy and information over large distances. In other words, the stuff of life is not billiard balls but liquid crystals and semiconductors.

Szent-Györgyi's sin was not that he was incorrect. He wasn't. It was his failure to respect the old animosity. Electricity and life were long divorced; the industrial revolution had been running full bore for a century and a half. Millions of miles of electric wires clothed the earth, exhaling electric fields that permeated all living things. Thousands of radio stations blanketed the very air with electromagnetic oscillations that one could not avoid. Skin and bones, nerves and muscles were not allowed to be influenced by them. Proteins were not permitted to be semiconductors. The threat to industry, economics, and modern culture would be too great.

So biochemists continued to think of proteins, lipids, and DNA as though they were little marbles drifting in a watery solution and colliding with one another at random. They even thought of the nervous system this way. When forced to, they admitted parts of quantum theory, but only on a limited basis. Biological molecules were still only permitted to interact with their immediate neighbors, not to act at a distance. It was okay to acknowledge modern physics only that much, like opening a small hole in a dam for knowledge to leak through one drop at a time, while the main structure is reinforced lest a flood demolish it.

Old knowledge about chemical bonds and enzymes in a water solution must now coexist with new models of electron transport chains. It was necessary to invent these to explain phenomena that were most central to life: photosynthesis and respiration. Large porphyrin-containing protein molecules no longer had to move and physically interact with one another in order for anything useful to happen. These molecules could stay put and electrons could shuttle between them instead. Biochemistry was becoming that much more alive. But it still had a long way to go. For even in the new models, electrons were constrained to move only, like little messenger boys, between one protein molecule and its immediate neighbor. They could cross the street, so to speak, but they couldn't travel down a highway to a distant town. Organisms were still pictured essentially as bags of water containing very complex solutions of chemicals.

The laws of chemistry had explained a lot about metabolic processes, and electron transport now explained even more, but there was not yet an organizing principle. Elephants grow from tiny embryos, which grow from single brainless cells. Salamanders regenerate perfect limbs. When we are cut, or break a bone, cells and organs throughout our body mobilize and coordinate their activities to repair the damage.

How does the information travel? How, borrowing Szent-Györgyi's words, do protein molecules "live"?

Despite Szent-Györgyi's sin, his predictions have proven correct. Molecules in cells do not drift at random to collide with one another. Most are firmly anchored to membranes. The water inside cells is highly structured and does not resemble the free-flowing liquid that sloshes around in a glass before you drink it. Piezoelectricity, a property of crystals that makes them useful in electronic products, that transforms mechanical stress into electrical voltages and vice versa, has been found in cellulose, collagen, horn, bone, wool, wood, tendon, blood vessel walls, muscle, nerve, fibrin, DNA, and every type of protein examined.[10] In other words—something most biologists have been denying for two centuries—electricity is essential to biology.

Szent-Györgyi was not the first to challenge conventional thinking. It was Otto Lehmann, already in 1908, who, noticing the close resemblance between the shapes of known liquid crystals and many biological structures, proposed that the very basis of life was the liquid crystalline state. Liquid crystals, like organisms, had the ability to grow from seeds; to heal wounds; to consume other substances, or other crystals; to be poisoned; to form membranes, spheres, rods, filaments and helical structures; to divide; to "mate" with other forms, resulting in offspring that had characteristics of both parents; to transform chemical energy into mechanical motion.

After Szent-Györgyi's daring Budapest lecture, others pursued his ideas. In 1949, Dutch researcher E. Katz explained how electrons could move through a semiconducting chlorophyll crystal during photosynthesis. In 1955, James Bassham and Melvin Calvin, working for the U.S. Atomic Energy Commission, elaborated on this theory. In 1956, William Arnold, at Oak Ridge National Laboratory, confirmed experimentally that dried chloroplasts—the particles in green plants that contain chlorophyll—have many of the properties of semiconductors. In 1959, Daniel Eley, at Nottingham University, proved that dried proteins, amino acids, and porphyrins are indeed semiconductors. In 1962, Roderick Clayton, also at Oak Ridge, found that photosynthetic tissues in living plants behave like semiconductors. In 1970, Alan Adler, at the New England Institute, showed that thin films of porphyrins do also. In the 1970s, biochemist Freeman Cope, at the United States Naval Air Development Center in Warminster, Pennsylvania, emphasized the importance of solid state physics for a true understanding of biology, as did biologist Allan Frey, the most active American researcher into the effects of microwave radiation on the nervous system at that time. Ling Wei, professor of electrical engineering at the University of Waterloo in Ontario, stated baldly that a nerve axon is an electrical transmission line and that its membrane is an ionic transistor. He said that the equivalent circuitry "can be found in any electronics book today," and that "one can easily derive the nerve behavior from semiconductor physics." When he did so, his equations predicted some of the properties of nerves that were, and still are, puzzling to physiologists.

In 1979, a young professor of bioelectronics at the University of Edinburgh published a book titled *Dielectric and Electronic Properties of Biological Materials*. The earlier work of Eley and Arnold had been criticized because the activation energies they had measured—the amount of energy necessary to make proteins conduct electricity—seemed to be too large. Supposedly there was not enough energy available in living organisms to lift electrons into the conduction band. Proteins might be made to conduct electricity in the laboratory, said the critics, but this could not happen in the real world. Eley and Arnold, however, had done all their work on dried proteins, not living ones. The young professor, Ronald Pethig, pointed out the obvious: water is essential to life, and proteins become more conductive if you added water to them. In fact, studies had shown that adding only 7.5 percent water increased the conductivity of many proteins ten thousandfold or more! Water, he proposed, is an electron donor that "dopes" proteins and turns them into good semiconductors.

The electronic role of living water had already been noted by others. Physiologist Gilbert Ling, realizing that cell water is a gel and not a liquid, developed his theory of the electronic nature of cells in 1962. More recently, Gerald Pollack, professor of bioengineering at the University of Washington, has taken up this line of investigation. He was inspired by Ling when they met at a conference in the mid-1980s. Pollack's most recent book, *The Fourth Phase of Water: Beyond Solid, Liquid, and Vapor*, was published in 2011.

The late geneticist Mae-Wan Ho, in London, has clothed Szent-Györgyi's ideas in garments that all can see. She developed a technique using a polarizing microscope that displayed, in vivid color, the interference patterns generated by the liquid crystalline structures that make up living creatures. The first animal she put under her microscope was a tiny worm—a fruit fly larva. "As it crawls along, it weaves its head from side to side flashing jaw muscles in blue and orange stripes on a magenta background," she wrote in 1993 in her book, *The Rainbow and the Worm: The Physics of Organisms*. She and many others have urged that the liquid crystalline properties of our cells and tissues not only teach us about our chemistry, but have something special to tell us about life itself.

Włodzimierz Sedlak, pursuing Szent-Györgyi's ideas in Poland, developed the discipline of bioelectronics within the Catholic University of Lublin during the 1960s. Life, he said, is not only a collection of organic compounds undergoing chemical reactions, but those chemical reactions are coordinated with electronic processes that take place in an environment of protein semiconductors. Other scientists working at the same university are continuing to develop this discipline theoretically and experimentally today. Marian Wnuk has focused on porphyrins as key to the evolution of life. He states that the principal function of porphyrin systems is an electronic one. Józef Zon, head of the Department of Theoretical Biology at the University, has focused on the electronic properties of biological membranes.

Oddly enough, the use of porphyrins in electronic products instructs us about biology. Adding thin films of porphyrins to commercially available photovoltaic cells increases the voltage, current, and total power output.[11] Prototype solar cells based on porphyrins have been produced,[12] as have organic transistors based on porphyrins.[13]

The properties that make porphyrins suitable in electronics are the same properties that make us alive. As everyone knows, playing with fire is dangerous; oxidation releases tremendous energy quickly and violently. How, then, do living organisms make use of oxygen? How do we manage to breathe and metabolize our food without being destroyed in a conflagration? The secret lies in the highly pigmented, fluorescent molecule called porphyrin. Strong pigments are always efficient energy absorbers, and if they are also fluorescent, they are also good energy transmitters. As Szent-Györgyi taught us in his 1957 book, *Bioenergetics*, "fluorescence thus tells us that the molecule is capable of accepting energy and does not dissipate it. These are two qualities any molecule must have to be able to act as an energy transmitter."[14]

Porphyrins are more efficient energy transmitters than any other of life's components. In technical terms, their ionization potential is low, and their electron affinity high. They are therefore capable of transmitting large amounts of energy rapidly in small steps, one low-energy electron at a time. They can even transmit energy electronically from oxygen to other molecules, instead of dissipating that energy as heat and burning up. That's why breathing is possible. On the other side of the great cycle of life, porphyrins in plants absorb the energy of sunlight and transport electrons that change carbon dioxide and water into carbohydrates and oxygen.

Porphyrins, the Nervous System, and the Environment

There is one more place these surprising molecules are found: in the nervous system, the organ where electrons flow. In fact, in mammals, the central nervous system is the *only* organ that shines with the red fluorescent glow of porphyrins when examined under ultraviolet light. These porphyrins, too, perform a function that is basic to life. They occur, however, in a location where one might least expect to find them—not in the neurons themselves, the cells that carry messages from our five senses to our brain, but in the myelin sheaths that envelop them—the sheaths whose role has been almost totally neglected by researchers and whose breakdown causes one of the most common and least understood neurological diseases of our time: multiple sclerosis. It was orthopedic surgeon Robert O. Becker who, in the 1970s, discovered that myelin sheaths are really electrical transmission lines.

In a state of health the myelin sheaths contain primarily two types of porphyrins—coproporphyrin III and protoporphyrin—in a ratio of two to one, complexed with zinc. The exact composition is crucial. When environmental chemicals poison the porphyrin pathway, excess porphyrins, bound to heavy metals, build up in the nervous system as in the rest of the body. This disrupts the myelin sheaths and changes their conductivity which, in turn, alters the excitability of the nerves they surround. The entire nervous system becomes hyperreactive to stimuli of all kinds, including electromagnetic fields.

The cells surrounding our nerves were hardly even studied until recently. In the nineteenth century, anatomists, finding no apparent function for them, supposed that they must have only a "nutritive" and "supportive" role, protecting the "real" nerves that they surrounded. They named them glial cells after the Greek word for "glue." The discovery of the action potential, which transmits signals along each neuron, and of neurotransmitters, the chemicals that carry signals from one neuron to the next, had ended the discussion. From then on, glial cells were thought to be little more than packing material. Most biologists ignored the fact, discovered by German physician Rudolf Virchow in 1854, that myelin is a liquid crystal. They did not think it was relevant.

However, working from the 1960s to the early 1980s and author, in 1985, of *The Body Electric*, Becker found quite another function for the myelin-containing cells and took another step toward restoring electricity to its proper role in the functioning of living things.

When he began his research in 1958, Becker was simply looking for a solution to orthopedists' greatest unsolved problem: nonunion of fractures. Occasionally, despite the best medical care, a bone would refuse to heal. Surgeons, believing that only chemical processes were at work, simply scraped the fracture surfaces, devised complicated plates and screws to hold the bone ends rigidly together, and hoped for the best. Where this did not work, limbs had to be amputated. "These approaches seemed superficial to me," Becker recalled. "I doubted that we would ever understand the failure to heal unless we truly understood healing itself."[15]

Becker began to pursue the ideas of Albert Szent-Györgyi, thinking that if proteins were semiconductors, maybe bones were too, and maybe electron flow was the secret to the healing of fractures. Ultimately he proved that this was correct. Bones were not just made of collagen and appatite, as he was taught in medical school; they were also doped with tiny amounts of copper, much as silicon wafers in computers are doped with tiny amounts of boron or aluminum. The presence of greater or lesser amounts of metal atoms regulates the electrical conductivity of the circuitry—in bones as in computers. With this understanding, Becker designed machines that delivered miniscule electric currents—as small as 100 trillionths of an ampere—to fractured bones to stimulate the healing process, with great success: his devices were the forerunners of machines that are used today by orthopedic surgeons in hospitals throughout the world.

Becker's work on the nervous system is less well known. As already mentioned, the functioning of neurons had been worked out, up to a point, in the nineteenth century. They transmit enormous amounts of information to and from the brain at high speed, including data about one's environment, and instructions to one's muscles. They do this via the familiar action potential and neurotransmitters. And since the action potential is an all-or-nothing event, neuron signaling is an on-off digital system like today's computers. But Becker thought that this could not explain the most important properties of life; there had to be a slower, more primitive, and more sensitive analog system that regulates growth and healing, that we inherited from lower forms of life—a system that might be related to the acupuncture meridians of Chinese medicine, which western medicine also made no attempt to understand.

A number of researchers before Becker, among them Harold Saxton Burr at Yale, Lester Barth at Columbia, Elmer Lund at the University of Texas, Ralph Gerard and Benjamin Libet at the University of Chicago, Theodore Bullock at U.C.L.A., and William Burge at the University of Illinois, had measured DC voltages on the surfaces of living organisms, both plants and animals, and embryos. Most biologists paid no attention. After all, certain DC currents, called "currents of injury," were well known, and were thought to be well understood. They had been discovered by Carlo Matteucci as long ago as the 1830s. Biologists had assumed, for a century, that these currents were meaningless artifacts, caused simply by ions leaking out of wounds. But when, in the 1930s and 1940s, a growing number of scientists, using better techniques, began to find DC voltages on all surfaces of all living things, and not just on the surfaces of wounds, a few began to wonder whether those "currents of injury" just might be a bit more important than they had learned in school.

The accumulated work of these scientists showed that trees,[16] and probably all plants, are polarized electrically, positive to negative, from leaves to roots, and that animals are similarly polarized from head to feet. In humans potential differences of up to 150 millivolts or more could sometimes be measured between one part of the body and another.[17]

Becker was the first to map the charge distribution in an animal in some detail, accomplishing this with salamanders in 1960. The places of greatest positive voltage, he found, as measured from the back of the animal, were the center of the head, the upper spine over the heart, and the lumbosacral plexus at the lower end of the spine, while the places of greatest negative voltage were the four feet and the end of the tail. In addition, the head of an alert animal was polarized from back to front, as though an electric current were always flowing in one direction through the middle of its brain. However, when an animal was anesthetized the voltage diminished as the anesthetic took effect, and then the head reversed polarity when the animal lost consciousness. This suggested to him a novel method of inducing anesthesia, and when Becker tried it, it worked like a charm. In the salamander, at least, passing an electric current of only 30 millionths of an ampere from front to back through the center of its head caused the animal to become immediately unconscious and unresponsive to pain. When the current was turned off, the animal promptly woke up. He observed the same back-to-front polarity in alert humans, and the same reversal during sleep and anesthesia.[18]

While Becker did not try it himself, even tinier electric currents have been used in psychiatry to put humans to sleep since about 1950 in Russia, Eastern Europe, and Asian countries that were once part of the Soviet Union. In these treatments, current is sent from front to back through the midline of the head, reversing the normal polarity of the brain, just as Becker did with his salamanders. The first publications describing this procedure specified short pulses of 10 to 15 microamperes each, 5 to 25 times per second, which gave an average current of only about 30 billionths of an ampere. Although larger currents will cause immediate unconsciousness in a human, just like in a salamander, those tiny currents are all that is necessary to put a person to sleep. This technique, called "electrosleep," has been used for over half a century to treat mental disorders, including manic-depressive illness and schizophrenia, in that part of the world.[19]

The normal electrical potentials of the body are also necessary for the perception of pain. The abolition of pain in a person's arm, for example, whether caused by a chemical anesthetic, hypnosis, or acupuncture, is accompanied by a reversal of electrical polarity in that arm.[20]

By the 1970s it had become clear to the researchers who were looking into such things that the DC potentials they were measuring played a key role in organizing living structures. They were necessary for growth and development.[21] They were also needed for regeneration and healing.

Tweedy John Todd demonstrated as long ago as 1823 that a salamander cannot regenerate a severed leg if you destroy that leg's nerve supply. So for a century and a half, scientists searched for the chemical signal that must be transmitted by nerves to trigger growth. No one ever found one. Finally, embryologist Sylvan Meryl Rose, in the mid-1970s at Tulane University, proposed that maybe there was no such chemical, and that the long-sought signal was purely electrical. Could the currents of injury, he asked, that had previously been considered mere artifacts, themselves play a central role in healing?

Rose found that they did. He recorded the patterns of the currents in the wound stumps of salamanders as they regenerated their severed limbs. The end of the stump, he found, was always strongly positive during the first few days after injury, then reversed polarity to become strongly negative for the next couple of weeks, finally reestablishing the weakly negative voltage found on all healthy salamander legs. Rose then found that salamanders would regenerate their legs normally, even without a nerve supply, provided he carefully duplicated, with an artificial source of current, the electrical patterns of healing that he had observed. Regeneration would not take place if the polarity, magnitude, or sequence of currents were not correct.

Once having established that the signals that trigger regeneration are electrical and not chemical in nature, these scientists were in for yet another surprise. For the DC potentials of the body that, as we have seen, are necessary not just for regeneration but for growth, healing, pain perception, and even consciousness, seemed to be generated not in the "real" nerves but in the myelin-containing cells that surround them—the cells that also contain porphyrins. Proof came by accident while Becker was again working on the problem of why some bone fractures fail to mend. Since he had already learned that nerves were essential to healing, he tried, in the early 1970s, to create an animal model for fractures that do not heal by severing the nerve supply to a series of rats' legs before breaking them.

To his surprise, the leg bones still healed normally—with a six-day delay. Yet six days was not nearly enough time for a rat to regenerate a severed nerve. Could bones be an exception, he wondered, to the rule that nerves are needed for healing? "Then we took a more detailed look at the specimens," wrote Becker. "We found that the Schwann cell sheaths were growing across the gap during the six-day delay. As soon as the perineural sleeve was mended, the bones began to heal normally, indicating that at least the healing, or output, signal, was being carried by the sheath rather than the nerve itself. The cells that biologists had considered merely insulation turned out to be the real wires."[22] It was the Schwann cells, Becker concluded—the myelin-containing glial cells—and not the neurons they surrounded, that carried the currents that determined growth and healing. And in a much earlier study Becker had already shown that the DC currents that flow along salamander legs, and presumably along the limbs and bodies of all higher animals, are of semiconducting type.[23]

Which brings us full circle. The myelin sheaths—the liquid crystalline sleeves surrounding our nerves—contain semiconducting porphyrins,[24] doped with heavy metal atoms, probably zinc.[25] It was Harvey Solomon and Frank Figge who, in 1958, first proposed that these porphyrins must play an important role in nerve conduction. The implications of this are especially important for people with chemical and electromagnetic sensitivities. Those of us who, genetically, have relatively less of one or more porphyrin enzymes, may have a "nervous temperament" because our myelin is doped with slightly more zinc than our neighbors' and is more easily disturbed by the electromagnetic fields (EMFs) around us. Toxic chemicals and EMFs are therefore synergistic: exposure to toxins further disrupts the porphyrin pathway, causing the accumulation of more porphyrins and their precursors, rendering the myelin and the nerves they surround still more sensitive to EMFs. According to more recent research, a large excess of porphyrin precursors can prevent the synthesis of myelin and break apart the myelin sheaths, leaving the neurons they surround naked and exposed.[26]

The true situation is undoubtedly more complex than this, but to put all the pieces correctly together will require researchers who are willing to step outside our cultural blinders and acknowledge the existence of electrical transmission lines in the nervous systems of animals. Already, mainstream science has taken the first step by finally acknowledging that glial cells are much more than packing material.[27] In fact, a discovery by a team of researchers at the University of Genoa is currently revolutionizing neurology. Their discovery is related to breathing.[28]

Everyone knows that the brain consumes more oxygen than any other organ, and that if a person stops breathing, the brain is the first organ to die. What the Italian team confirmed in 2009 is that as much as ninety percent of that oxygen is consumed not by the brain's nerve cells, but by the myelin sheaths that surround them. Traditional wisdom has it that the consumption of oxygen for energy takes place only in tiny bodies inside cells called mitochondria. That wisdom has now been turned on its head. In the nervous system, at least, most of the oxygen appears to be consumed in the multiple layers of fatty substance called myelin, which contain no mitochondria at all, but which forty-year-old research showed contains non-heme porphyrins and is semiconducting. Some scientists are even beginning to say that the myelin sheath is, in effect, itself a giant mitochondrion, without which the huge oxygen needs of our brain and nervous system could never be met. But to truly make sense of this collection of facts will also require the recognition that both the neurons, as Ling Wei proposed, and the myelin sheaths that envelop them, as Robert Becker proposed, work together to form a complex and elegant electrical transmission line system, subject to electrical interference just like transmission lines built by human engineers.

The exquisite sensitivity of even the normal nervous system to electromagnetic fields was proven in 1956 by zoologists Carlo Terzuolo and Theodore Bullock—and then ignored by everyone since. In fact, even Terzuolo and Bullock were astonished by the results. Experimenting on crayfish, they found that although a substantial amount of electric current was needed to cause a previously silent nerve to fire, incredibly tiny currents could cause an already firing nerve to alter its firing rate tremendously. A current of only 36 billionths of an ampere was enough to increase or decrease a nerve's rate of firing by five to ten percent. And a current of 150 billionths of an ampere—thousands of times less than is widely assumed, still today, by developers of modern safety codes, to have any biological effect whatever—would actually double the rate of firing, or silence the nerve altogether. Whether it increased or decreased the activity of the nerve depended only on the direction in which the current was applied to the nerve.

The Zinc Connection

The role of zinc was discovered in the 1950s by Henry Peters, a porphyrinologist at the University of Wisconsin Medical School. Like Morton after him, Peters was impressed by the number of people who seemed to have mild or latent porphyria, and thought the trait was far more prevalent that was commonly believed.[29]

Peters discovered that his porphyria patients who had neurological symptoms were excreting very large amounts of zinc in their urine—up to 36 times normal. In fact, their symptoms correlated better with the levels of zinc in their urine than with the levels of porphyrins they were excreting. With this information, Peters did the most logical thing: in scores of patients, he tried chelation to reduce the body's load of zinc, and it worked! In patient after patient, when courses of treatment with BAL or EDTA had reduced the level of zinc in their urine to normal, their illness resolved, and the patient remained symptom-free for up to several years.[30] Contrary to conventional wisdom, which assumes that zinc deficiency is common and should be supplemented, Peters' patients, because of their genetics and their polluted environment, were actually zinc-poisoned—as at least five to ten percent of the population, with hidden porphyria, may also be.

For the next forty years Peters found tremendous resistance to his idea that zinc toxicity was at all common, but growing evidence is now accumulating that this is so. Large amounts of zinc are in fact entering our environment, our homes, and our bodies from industrial processes, galvanized metals, and even the fillings in our teeth. Zinc is in denture cream and in motor oil. There is so much zinc in automobile tires that their constant erosion makes zinc one of the main components of road dust—which washes into our streams, rivers, and reservoirs, eventually getting into our drinking water.[31] Wondering whether this was perhaps poisoning us all, a group of scientists from Brookhaven

National Laboratory, the United States Geological Survey, and several universities raised rats on water supplemented with a low level of zinc. By three months of age, the rats already had memory deficits. By nine months of age, they had elevated levels of zinc in their brains.[32] In a human experiment, pregnant women in a slum area of Bangladesh were given 30 milligrams of zinc daily, in the expectation that this would benefit the mental development and motor skills of their babies. The researchers found just the opposite.[33] In a companion experiment, a group of Bangladeshi infants were given 5 milligrams of zinc daily for five months, with the same surprising result: the supplemented infants scored more poorly on standard tests of mental development.[34] And a growing body of literature shows that zinc supplements worsen Alzheimer's disease,[35] and that chelation therapy to reduce zinc improves cognitive functioning in Alzheimer's patients.[36] An Australian team who examined autopsy specimens found that Alzheimer's patients had double the amount of zinc in their brains as people without Alzheimer's, and that the more severe the dementia, the higher the zinc levels.[37]

Nutritionists have long been misled by using blood tests to judge the body's stores of zinc; scientists are finding out that blood levels are not reliable, and that unless you are severely malnourished there is no relation between the amount of zinc in your diet and the level of zinc in your blood.[38] In some neurological diseases, including Alzheimer's disease, it is common to have high levels of zinc in the brain while having normal or low levels of zinc in the blood.[39] In a number of diseases including diabetes and cancer, urinary zinc is high while blood zinc is low.[40] It appears that the kidneys respond to the body's total load of zinc, and not to the levels in the blood, so that blood levels can become low, not because of a zinc deficiency but because the body is overloaded with zinc and the kidneys are removing it from the blood as fast as they can. It also appears to be much more difficult than we used to think for people to become deficient by eating a zinc-poor diet; the body is amazingly capable of compensating for even extremely low levels of dietary zinc by increasing intestinal absorption and decreasing excretion through urine, stool, and skin.[41] While the recommended dietary allowance for adult males is 11 milligrams per day, a man can take in as little as 1.4 milligrams of zinc a day and still maintain homeostasis and normal levels of zinc in the blood and tissues.[42] But a person who increases his or her daily intake beyond 20 milligrams may risk toxic effects in the long term.

Canaries in the Mine

In our cells, the manufacture of heme from porphyrins can be inhibited by a large variety of toxic chemicals, and not—so far as we know—by electricity. But we will see in the coming chapters that electromagnetic fields interfere with the most important job that this heme is supposed to do for us: enabling the combustion of our food by oxygen so that we can live and breathe. Like rain on a campfire, electromagnetic fields douse the flames of metabolism. They reduce the activity of the cytochromes, and there is evidence that they do so in the simplest of all possible ways: by exerting a force that alters the speed of the electrons being transported along the chain of cytochromes to oxygen.

Every person on the planet is affected by this invisible rain that penetrates into the fabric of our cells. Everyone has a slower metabolism, is less alive, than if those fields were not there. We will see how this slow asphyxiation causes the major diseases of civilization: cancer, diabetes, and heart disease. There is no escape. Regardless of diet, exercise, lifestyle, and genetics, the risk of developing these diseases is greater for every human being and every animal than it was a century and a half ago. People with a genetic predisposition simply have a greater risk than everyone else, because they have a bit less heme in their mitochondria to start with.

In France, liver cancer was found to be 36 times as frequent in people carrying a gene for porphyria as in the general population.[43] In Sweden and Denmark the rate was 39 times as high, and the lung cancer rate triple the general rate.[44] Chest pain, heart failure, high blood pressure, and EKGs suggestive of oxygen starvation are well known in porphyria.[45] Porphyria patients with normal coronary arteries often die of heart arrhythmias[46] or heart attacks.[47] Glucose tolerance tests and insulin levels are usually abnormal.[48] In one study, 15 of 36 porphyria patients had diabetes.[49] The protean manifestations of this disease, capable of affecting almost any organ, are widely blamed on impaired cellular respiration due to a deficiency of heme.[50] Indeed, no porphyrin expert has offered a better explanation.

The five to ten percent of the population who have lower porphyrin enzyme levels are the so-called canaries in the coal mine, whose songs of warning, however, have been tragically ignored. They are the people who came down with neurasthenia in the last half of the nineteenth century when telegraph wires swept the world; the victims of sleeping pills in the late 1880s, of barbiturates in the 1920s, and of sulfa drugs in the 1930s; the men, women, and children with multiple chemical sensitivity, poisoned by the soup of chemicals that have rained on us since World War II; the abandoned souls with electrical sensitivity left behind by the computer age, forced into lonely exile by the inescapable radiation of the wireless revolution.

In Part Two of this book we will see just how extensively the general population of the world has been affected as a result of the failure to heed their warnings.

PART TWO

PART TWO

11. Irritable Heart

ON THE FIRST DAY OF AUTUMN, 1998, Florence Griffith Joyner, former Olympic track gold medalist, died in her sleep at the age of thirty-eight when her heart stopped beating. That same fall, Canadian ice hockey player Stéphane Morin, age twenty-nine, died of sudden heart failure during a hockey game in Germany, leaving behind a wife and newborn son. Chad Silver, who had played on the Swiss national ice hockey team, also age twenty-nine, died of a heart attack. Former Tampa Bay Buccaneers nose tackle Dave Logan collapsed and died from the same cause. He was forty-two. None of these athletes had any history of heart disease.

A decade later, responding to mounting alarm among the sports community, the Minneapolis Heart Institute Foundation created a National Registry of Sudden Deaths in Athletes. After combing through public records, news reports, hospital archives, and autopsy records, the Foundation identified 1,049 American athletes in thirty-eight competitive sports who had suffered sudden cardiac arrest between 1980 and 2006. The data confirmed what the sports community already knew. In 1980, heart attacks in young athletes were rare: only nine cases occurred in the United States. The number rose gradually but steadily, increasing about ten percent per year, until 1996, when the number of cases of fatal cardiac arrest among athletes suddenly doubled. There were 64 that year, and 66 the following year. In the last year of the study, 76 competitive athletes died when their hearts gave out, most of them under eighteen years of age.[1]

The American medical community was at a loss to explain it. But in Europe, some physicians thought they knew the answer, not only to the question of why so many young athletes' hearts could no longer stand the strain of exertion, but to the more general question of why so many young people were succumbing to diseases from which only old people used to die. On October 9, 2002, an association of German doctors specializing in environmental medicine began circulating a document calling for a moratorium on antennas and towers used for mobile phone communications. Electromagnetic radiation, they said, was causing a drastic rise in both acute and chronic diseases, prominent among which were "extreme fluctuations in blood pressure," "heart rhythm disorders," and "heart attacks and strokes among an increasingly younger population."

Three thousand physicians signed this document, named the Freiburger Appeal after the German city in which it was drafted. Their analysis, if correct, could explain the sudden doubling of heart attacks among American athletes in 1996: that was the year digital cell phones first went on sale in the United States, and the year cell phone companies began building tens of thousands of cell towers to make them work.

Although I knew about the Freiburger Appeal and the profound effects electricity could have on the heart, when I first conceived this book I did not intend to include a chapter on heart disease, for I was still in denial despite the abundant evidence.

We recall from chapter 8 that Marconi, the father of radio, had ten heart attacks after he began his world-changing work, including the one that killed him at the young age of 63.

"Anxiety disorder," which is rampant today, is most often diagnosed from its cardiac symptoms. Many suffering from an acute "anxiety attack" have heart palpitations, shortness of breath, and pain or pressure in the chest, which so often resemble an actual heart attack that hospital emergency rooms are visited by more patients who turn out to have nothing more than "anxiety" than by patients who prove to have something wrong with their hearts. And yet we recall from chapter 6 that "anxiety neurosis" was an invention of Sigmund Freud, a renaming of a disease formerly called neurasthenia, that became prevalent only in the late nineteenth century following the building of the first electrical communication systems.

Radio wave sickness, described by Russian doctors in the 1950s, includes cardiac disturbances as a prominent feature.

Not only did I know all this, but I myself have suffered for thirty-five years from palpitations, abnormal heart rhythm, shortness of breath, and chest pain, related to exposure to electricity.

Yet when my friend and colleague Jolie Andritzakis suggested to me that heart disease itself had appeared in the medical literature for the first time at the beginning of the twentieth century and that I should write a chapter about it, I was taken by surprise. In medical school I had had it so thoroughly drilled into me that cholesterol is the main cause of heart disease that I had never before questioned the wisdom that bad diet and lack of exercise are the most important factors contributing to the modern epidemic. I had no doubt that electromagnetic radiation could cause heart *attacks*. But I did not yet suspect that it was responsible for heart *disease*.

Then another colleague, Dr. Samuel Milham, muddied the waters some more. Milham is an M.D. and an epidemiologist, retired from the Washington State Department of Health. He wrote an article in 2010, followed by a short book, suggesting that the modern epidemics of heart disease, diabetes, and cancer are largely if not entirely caused by electricity. He included solid statistics to back up these assertions.

I decided to dive in.

I first became aware of Milham's work in 1996, when I was asked to help with a national lawsuit against the Federal Communications Commission. I was still living in Brooklyn, and knew only that the telecommunications industry was promising a "wireless revolution." The industry wanted to place a cell phone in the hands of every American, and in order to make those devices work in the urban canyons of my home town they were applying for permission to erect thousands of microwave antennas close to street level throughout New York. Advertisements for the newfangled phones were beginning to appear on radio and television, telling the public why they needed such things and that they would make ideal Christmas gifts. I did not have any idea how radically the world was about to change.

Then came a phone call from David Fichtenberg, a statistician in Washington State, who told me the FCC had just released human exposure guidelines for microwave radiation, and asked if I wanted to join a nationwide legal challenge against them. The new guidelines, I came to find out, had been written by the cell phone industry itself and did not protect people from any of the effects of microwave radiation except one: being cooked like a roast in a microwave oven. None of the known effects of such radiation, apart from heat—effects on the heart, nervous system, thyroid gland, and other organs—were taken into consideration.

Worse, Congress had passed a law that January that actually made it illegal for cities and states to regulate this new technology on the basis of health. President Clinton had signed it on February 8. The industry, the FCC, Congress, and the President were conspiring to tell us that we should all feel comfortable holding devices that emit microwave radiation directly against our brains, and that we should all get used to living in close quarters with microwave towers, because they were coming to a street near you whether you liked it or not. A giant biological experiment had been launched, and we were all going to be unwitting guinea pigs.

Except that the outcome was already known. The research had been done, and the scientists who had done it were trying to tell us what the new technology was going to do to the brains of cell phone users, and to the hearts and nervous systems of people living in the vicinity of cell towers—which one day soon was going to be everybody.

Samuel Milham, Jr. was one of those researchers. He had not done any of the clinical or experimental research on individual humans or animals; such work had been done by others in previous decades.

Samuel Milham, M.D., M.P.H

Milham is an epidemiologist, a scientist who proves that the results obtained by others in the laboratory actually happen to masses of people living in the real world. In his early studies he had shown that electricians, power line workers, telephone linesmen, aluminum workers, radio and TV repairmen, welders, and amateur radio operators—those whose work exposed them to electricity or electromagnetic radiation—died far more often than the general public from leukemia, lymphoma, and brain tumors. He knew that the new FCC standards were inadequate, and he made himself available as a consultant to those who were challenging them in court.

In recent years, Milham turned his skills to the examination of vital statistics from the 1930s and 1940s, when the Roosevelt administration made it a national priority to electrify every farm and rural community in America. What Milham discovered surprised even him. Not only cancer, he found, but also diabetes and heart disease seemed to be directly related to residential electrification. Rural communities that had no electricity had little heart disease—until electric service began. In fact, in 1940, country folk in electrified regions of the country were suddenly dying of heart disease four to five times as frequently as those who still lived out of electricity's reach. "It seems unbelievable that mortality differences of this magnitude could go unexplained for over 70 years after they were first reported," wrote Milham.[2] He speculated that early in the twentieth century nobody was looking for answers.

But when I began reading the early literature I found that *everyone* was looking for answers. Paul Dudley White, for example, a well-known cardiologist associated with Harvard Medical School, puzzled over the problem in 1938. In the second edition of his textbook, *Heart Disease*, he wrote in amazement that Austin Flint, a prominent physician practicing internal medicine in New York City during the last half of the nineteenth century, had not encountered a single case of angina pectoris (chest pain due to heart disease) for one period of five years. White was provoked by the tripling of heart disease rates in his home state of Massachusetts since he had begun practicing in 1911. "As a cause of death," he wrote, "heart disease has assumed greater and greater proportions in this part of the world until now it leads all other causes, having far outstripped tuberculosis, pneumonia, and malignant disease." In 1970, at the end of his career, White was still unable to say why this was so. All he could do was wonder at the fact that coronary heart disease—disease due to clogged coronary arteries, which is the most common type of heart disease today—had once been so rare that he had seen almost no cases in his first few years of practice. "Of the first 100 papers I published," he wrote, "only two, at the end of the 100, were concerned with coronary heart disease."[3]

Heart disease had not, however, sprung full-blown from nothing at the turn of the twentieth century. It had been relatively uncommon but not unheard of. The vital statistics of the United States show that rates of heart disease had begun to rise long before White graduated from medical school. The modern epidemic actually began, quite suddenly, in the 1870s, at the same time as the first great proliferation of telegraph wires. But that is to jump ahead of myself. For the evidence that heart disease is caused primarily by electricity is even more extensive than Milham suspected, and the mechanism by which electricity damages the heart is known.

To begin with, we need not rely only on historical data for evidence supporting Milham's proposal, for electrification is still going on in a few parts of the world.

From 1984 to 1987, scientists at the Sitaram Bhartia Institute of Science and Research decided to compare rates of coronary heart disease in Delhi, India, which were disturbingly high, with rates in rural areas of Gurgaon district in Haryana state 50 to 70 kilometers away. Twenty-seven thousand people were interviewed, and as expected, the researchers found more heart disease in the city than in the country. But they were surprised by the fact that virtually all of the supposed risk factors were actually greater in the rural districts.

City dwellers smoked much less. They consumed fewer calories, less cholesterol, and much less saturated fat than their rural counterparts. Yet they had five times as much heart disease. "It is clear from the present study," wrote the researchers, "that the prevalence of coronary heart disease and its urban-rural differences are not related to any particular risk factor, and it is therefore necessary to look for other factors beyond the conventional explanations."[4] The most obvious factor that these researchers did not look at was electricity. For in the mid-1980s the Gurgaon district had not yet been electrified.[5]

In order to make sense of these kinds of data it is necessary to review what is known—and what is still not known—about heart disease, electricity, and the relationship between the two.

My Hungarian grandmother, who was the main cook in my family while I was growing up, had arteriosclerosis (hardening of the arteries). She fed us the same meals she cooked for herself and, at the advice of her doctor, they were low in fat. She happened to be a marvelous cook, so after I left home I continued eating in a similar style because I was hooked on the taste. For the past thirty-eight years I have also been a vegetarian. I feel healthiest eating this way, and I believe that it is good for my heart.

However, soon after I began to do research for this chapter, a friend gave me a book to read titled *The Cholesterol Myths*. It was published in 2000 by Danish physician Uffe Ravnskov, a specialist in internal medicine and kidney disease and a retired family practice doctor living in Lund, Sweden. I resisted reading it, because Ravnskov is not unbiased: he thinks vegetarians are pleasure-avoiding stoics who heroically deny themselves the taste of proper food in the mistaken belief that this will make them live longer.

Ignoring his prejudices, I eventually read Ravnskov's book and found it well-researched and thoroughly referenced. It demolishes the idea that people are having more heart attacks today because they are stuffing themselves with more animal fat than their ancestors did. On its surface, his thesis is contrary to what I was taught as well as to my own experience. So I obtained copies of many of the studies he quoted, and read them over and over until they finally made sense in light of what I knew about electricity. The most important thing to keep in mind is that the early studies did not have the same outcome as research being done today, and that there is a reason for this difference. Even recent studies from different parts of the world do not always agree with each other, for the same reason.

Ravnskov, however, has become something of an icon among portions of the alternative health community, including many environmental physicians who are now prescribing *high*-fat diets—emphasizing animal fats—to their severely ill patients. They are misreading the medical literature. The studies that Ravnskov relied on show unequivocally that some factor other than diet is responsible for the modern scourge of heart disease, but they also show that cutting down on dietary fat in today's world helps to prevent the damage caused by that other factor. Virtually every large study done since the 1950s in the industrialized world—agreeing with what I was taught in medical school—has shown a direct correlation between cholesterol and heart disease.[6] And every study comparing vegetarians to meat eaters has found that vegetarians today have both lower cholesterol levels and a reduced risk of dying from a heart attack.[7]

Ravnskov speculated that this is because people who eat no meat are also more health-conscious in other ways. But the same results have been found in people who are vegetarians only for religious reasons. Seventh Day Adventists all abstain from tobacco and alcohol, but only about half abstain from meat. A number of large long-term studies have shown that Adventists who are also vegetarians are two to three times less likely to die from heart disease.[8]

Perplexingly, the very early studies—those done in the first half of the twentieth century—did not give these kinds of results and did not show that cholesterol was related to heart disease. To most researchers, this has been an insoluble paradox, contradicting present ideas about diet, and has been a reason for the mainstream medical community to dismiss the early research.

For example, people with the genetic trait called familial hyper-cholesterolemia have extremely high levels of cholesterol in their blood—so high that they sometimes have fatty growths on their joints and are prone to gout-like attacks in toes, ankles, and knees caused by cholesterol crystals. In today's world these people are prone to dying young of coronary heart disease. However, this was not always so. Researchers at Leiden University in the Netherlands traced the ancestors of three present-day individuals with this disorder until they found a pair of common ancestors who lived in the late eighteenth century. Then, by tracing all descendants of this pair and screening all living descendants for the defective gene, they were able to identify 412 individuals who either had definitely carried the gene and passed it on, or who were siblings who had a fifty percent chance of carrying it. They found, to their amazement, that before the 1860s people with this trait had a fifty percent *lower* mortality rate than the general population. In other words, cholesterol seemed to have had protective value and people with very high cholesterol levels lived longer than average. Their mortality rate, however, rose steadily during the late nineteenth century until it equaled the rate of the general population in about 1915. The mortality of this subgroup continued rising during the twentieth century, reaching double the average during the 1950s and then leveling off somewhat.[9] One can speculate, based on this study, that before the 1860s cholesterol did not cause coronary heart disease, and there is other evidence that this is so.

In 1965, Leon Michaels, working at the University of Manitoba, decided to see what historical documents revealed about fat consumption in previous centuries when coronary heart disease was extremely rare. What he found also contradicted current wisdom and convinced him that there must be something wrong with the cholesterol theory. One author in 1696 had calculated that the wealthier half of the English population, or about 2.7 million people, ate an amount of flesh yearly averaging 147.5 pounds per person—more than the national average for meat consumption in England in 1962. Nor did the consumption of animal fats decline at any time before the twentieth century. Another calculation made in 1901 had shown that the servant-keeping class of England consumed, on average, a much larger amount of fat in 1900 than they did in 1950. Michaels did not think that lack of exercise could explain the modern epidemic of heart disease either, because it was among the idle upper classes, who had never engaged in manual labor, and who were eating much *less* fat than they used to, that heart disease had increased the most.

Then there was the incisive work of Jeremiah Morris, Professor of Social Medicine at the University of London, who observed that in the first half of the twentieth century, coronary heart disease had increased while coronary atheroma—cholesterol plaques in the coronary arteries—had actually *decreased*. Morris examined the autopsy records at London Hospital from the years 1908 through 1949. In 1908, 30.4 percent of all autopsies in men aged thirty to seventy showed advanced atheroma; in 1949, only 16 percent. In women the rate had fallen from 25.9 percent to 7.5 percent. In other words, cholesterol plaques in coronary arteries were far less common than before, but they were contributing to more disease, more angina, and more heart attacks. By 1961, when Morris presented a paper about the subject at Yale University Medical School, studies conducted in Framingham, Massachusetts[10] and Albany, New York[11] had established a connection between cholesterol and heart disease. Morris was sure that some other, unknown environmental factor was also important. "It is tolerably certain," he told his audience, "that more than fats in the diet affect blood lipid levels, more than blood lipid levels are involved in atheroma formation, and more than atheroma is needed for ischemic heart disease."

That factor, as we will see, is electricity. Electromagnetic fields have become so intense in our environment that we are unable to metabolize fats the way our ancestors could.

Whatever environmental factor was affecting human beings in America during the 1930s and 1940s was also affecting all the animals in the Philadelphia Zoo.

The Laboratory of Comparative Pathology was a unique facility founded at the zoo in 1901. And from 1916 to 1964, laboratory director Herbert Fox and his successor, Herbert L. Ratcliffe, kept complete records of autopsies performed on over thirteen thousand animals that had died in the zoo.

During this period, arteriosclerosis increased an astonishing ten- to twenty-fold among all species of mammals and birds. In 1923, Fox had written that such lesions were "exceedingly rare," occurring in less than two percent of animals as a minor and incidental finding at autopsy.[12] The incidence rose rapidly during the 1930s, and by the 1950s arteriosclerosis was not only occurring in young animals, but was often the cause of their death rather than just a finding on autopsy. By 1964, the disease occurred in one-quarter of all mammals and thirty-five percent of all birds.

Coronary heart disease appeared even more suddenly. In fact, before 1945 the disease did not exist in the zoo.[13] And the first heart attacks ever recorded in zoo animals occurred ten years later, in 1955. Arteriosclerosis had been occurring with some regularity since the 1930s in the aorta and other arteries, but not in the coronary arteries of the heart. But sclerosis of the coronary arteries now increased so rapidly among both mammals and birds that by 1963, over 90 percent of all mammals and 72 percent of all birds that died in the zoo had coronary disease, while 24 percent of the mammals and 10 percent of the birds had had heart attacks. And a majority of the heart attacks were occurring in young animals in the first half of their expected life spans. Arteriosclerosis and heart disease were now occurring in 45 families of mammals and 65 families of birds residing in the zoo—in deer and in antelope; in prairie dogs and squirrels; in lions, and tigers, and bears; and in geese, storks, and eagles.

Diet had nothing to do with these changes. The increase in arteriosclerosis had begun well before 1935, the year that more nutritious diets were introduced throughout the zoo. And coronary disease did not make its appearance until ten years later, yet the animals' diets were the same at all times between 1935 and 1964. The population density, for mammals at least, remained about the same during all fifty years, as did the amount of exercise they got. Ratcliffe tried to find the answer in social pressures brought about by breeding programs that were begun in 1940. He thought that psychological stresses must be affecting the animals' hearts. But he could not explain why, more than two decades later, coronary disease and heart attacks were continuing to increase, spectacularly, throughout the zoo, and among all species, whether or not they were being bred. Nor could he explain why sclerosis of artieries outside the heart had increased during the 1930s, nor why, thousands of miles away, researchers were finding arteriosclerosis in 22 percent of the animals in the London Zoo in 1960,[14] and a similar number in the Zoo of Antwerp, Belgium in 1962.[15]

The element that increased most spectacularly in the environment during the 1950s when coronary disease was exploding among humans and animals was radio frequency (RF) radiation. Before World War II, radio waves had been widely used for only two purposes: radio communication, and diathermy, which is their therapeutic use in medicine to heat parts of the body.

Suddenly the demand for RF generating equipment was unquenchable. While the use of the telegraph in the Civil War had stimulated its commercial development, and the use of radio in World War I had done the same for that technology, the use of radar in World War II spawned scores of new industries. RF oscillators were being mass produced for the first time, and hundreds of thousands of people were being exposed to radio waves on the job—radio waves that were now used not only in radar, but in navigation; radio and television broadcasting; radio astronomy; heating, sealing and welding in dozens of industries; and "radar ranges" for the home. Not only industrial workers, but the entire population, were being exposed to unprecedented levels of RF radiation.

For reasons having more to do with politics than science, history took opposite tracks on opposite sides of the world. In Western Bloc countries, science went deeper into denial. It had buried its head, ostrich-like, in the year 1800, as we saw in chapter 4, and now simply piled on more sand. When radar technicians complained of headaches, fatigue, chest discomfort, and eye pain, and even sterility and hair loss, they were sent for a quick medical exam and some blood work. When nothing dramatic turned up, they were ordered back to work.[16] The attitude of Charles I. Barron, medical director of the California division of Lockheed Aircraft Corporation, was typical. Reports of illness from microwave radiation "had all too often found their way into lay publications and newspapers," he said in 1955. He was addressing representatives of the medical profession, the armed forces, various academic institutions, and the airline industry at a meeting in Washington, DC. "Unfortunately," he added, "the publication of this information within the past several years coincided with the development of our most powerful airborne radar transmitters, and considerable apprehension and misunderstanding has arisen among engineering and radar test personnel." He told his audience that he had examined hundreds of Lockheed employees and found no difference between the health of those exposed to radar and those not exposed. However, his study, which was subsequently published in the *Journal of Aviation Medicine*, was tainted by the same see-no-evil attitude. His "unexposed" control population were actually Lockheed workers who were exposed to radar intensities of less than 3.9 milliwatts per square centimeter—a level that is almost four times the legal limit for exposure of the general public in the United States today. Twenty-eight percent of these "unexposed" employees suffered from neurological or cardiovascular disorders, or from jaundice, migraines, bleeding, anemia, or arthritis. And when Barron took repeated blood samples from his "exposed" population—those who were exposed to *more* than 3.9 milliwatts per square centimeter—the majority had a significant drop in their red cell count over time, and a significant increase in their white cell count. Barron dismissed these findings as "laboratory errors."[17]

The Eastern Bloc experience was different. Workers' complaints were considered important. Clinics dedicated entirely to the diagnosis and treatment of workers exposed to microwave radiation were established in Moscow, Leningrad, Kiev, Warsaw, Prague, and other cities. On average, about fifteen percent of workers in these industries became sick enough to seek medical treatment, and two percent became permanently disabled.[18]

The Soviets and their allies recognized that the symptoms caused by microwave radiation were the same as those first described in 1869 by American physician George Beard. Therefore, using Beard's terminology, they called the symptoms "neurasthenia," while the disease that caused them was named "microwave sickness" or "radio wave sickness."

Intensive research began at the Institute of Labor Hygiene and Occupational Diseases in Moscow in 1953. By the 1970s, the fruits of such investigations had produced thousands of publications.[19] Medical textbooks on radio wave sickness were written, and the subject entered the curriculum of Russian and Eastern European medical schools. Today, Russian textbooks describe effects on the heart, nervous system, thyroid, adrenals, and other organs.[20] Symptoms of radio wave exposure include headache, fatigue, weakness, dizziness, nausea, sleep disturbances, irritability, memory loss, emotional instability, depression, anxiety, sexual dysfunction, impaired appetite, abdominal pain, and digestive disturbances. Patients have visible tremors, cold hands and feet, flushed face, hyperactive reflexes, abundant perspiration, and brittle fingernails. Blood tests reveal disturbed carbohydrate metabolism and elevated triglycerides and cholesterol.

Cardiac symptoms are prominent. They include heart palpitations, heaviness and stabbing pains in the chest, and shortness of breath after exertion. The blood pressure and pulse rate become unstable. Acute exposure usually causes rapid heartbeat and high blood pressure, while chronic exposure causes the opposite: low blood pressure and a heart-beat that can be as slow as 35 to 40 beats per minute. The first heart sound is dulled, the heart is enlarged on the left side, and a murmur is heard over the apex of the heart, often accompanied by premature beats and an irregular rhythm. The electrocardiogram may reveal a blockage of electrical conduction within the heart, and a condition known as left axis deviation. Signs of oxygen deprivation to the heart muscle—a flattened or inverted T wave, and depression of the ST interval—are extremely frequent. Congestive heart failure is sometimes the ulti-mate outcome. In one medical textbook published in 1971, the author, Nikolay Tyagin, stated that in his experience only about fifteen percent of workers exposed to radio waves had normal EKGs.[21]

Although this knowledge has been completely ignored by the American Medical Association and is not taught in any American med-ical school, it has not gone unnoticed by some American researchers.

Trained as a biologist, Allan H. Frey became interested in micro-wave research in 1960 by following his curiosity. Employed at the General Electric Company's Advanced Electronics Center at Cornell University, he was already exploring how electrostatic fields affect an animal's nervous system, and he was experimenting with the biological effects of air ions. Late that year, while attending a conference, he met a technician from GE's radar test facility at Syracuse, who told Frey that he could hear radar. "He was rather surprised," Frey later recalled, "when I asked if he would take me to a site and let me hear the radar. It seemed that I was the first person he had told about hearing radars who did not dismiss his statement out of hand."[22] The man took Frey to his work site near the radar dome at Syracuse. "And when I walked around there and climbed up to stand at the edge of the pulsating beam, I could hear it, too," Frey remembers. "I could hear the radar going *zip-zip-zip*."[23]

This chance meeting determined the future course of Frey's career. He left his job at General Electric and began doing full-time research into the biological effects of microwave radiation. In 1961, he published his first paper on "microwave hearing," a phenomenon that is now fully recognized although still not fully explained. He spent the next two decades experimenting on animals to determine the effects of microwaves on behavior, and to clarify their effects on the auditory system, the eyes, the brain, the nervous system, and the heart. He discovered the blood-brain barrier effect, an alarming damage to the protective shield that keeps bacteria, viruses, and toxic chemicals out of the brain—damage that occurs at levels of radiation that are much lower than what is emitted by cell phones today. He proved that nerves, when firing, emit pulses of radiation themselves, in the infrared spectrum. All of Frey's pioneering work was funded by the Office of Naval Research and the United States Army.

When scientists in the Soviet Union began reporting that they could modify the rhythm of the heart at will with microwave radiation, Frey took a special interest. N. A. Levitina, in Moscow, had found that she could either speed up an animal's heart rate or slow it down, depending on which part of the animal's body she irradiated. Irradiating the back of an animal's head quickened its heart rate, while irradiating the back of its body, or its stomach, slowed it down.[24]

Frey, in his laboratory in Pennsylvania, decided to take this research one step farther. Based on the Russian results and his knowledge of physiology he predicted that if he used brief pulses of microwave radiation, synchronized with the heartbeat and timed to coincide precisely with the beginning of each beat, he would cause the heart to speed up, and might disrupt its rhythm.

It worked like magic. He first tried the experiment on the isolated hearts of 22 different frogs. The heart rate increased every time. In half the hearts, arrhythmias occurred, and in some of the experiments the heart stopped. The pulse of radiation was most damaging when it occurred exactly one-fifth of a second after the beginning of each beat. The average power density was only six-tenths of a microwatt per square centimeter—roughly ten thousand times weaker than the radiation that a person's heart would absorb today if he or she kept a cell phone in a shirt pocket while making a call.

Frey conducted the experiments with isolated hearts in 1967. Two years later, he tried the same thing on 24 live frogs, with similar though less dramatic results. No arrhythmias or cardiac arrests occurred, but when the pulses of radiation coincided with the beginning of each beat, the heart speeded up significantly.[25]

The effects Frey demonstrated occur because the heart is an electrical organ and microwave pulses interfere with the heart's pacemaker. But in addition to these direct effects, there is a more basic problem: microwave radiation, and electricity in general, starves the heart of oxygen because of effects at the cellular level. These cellular effects were discovered, oddly enough, by a team that included Paul Dudley White. In the 1940s and 1950s, while the Soviets were beginning to describe how radio waves cause neurasthenia in workers, the United States military was investigating the same disease in military recruits.

The job that was assigned to Dr. Mandel Cohen and his associates in 1941 was to determine why so many soldiers fighting in the Second World War were reporting sick because of heart symptoms. Although their research spawned a number of shorter articles in medical journals, the main body of their work was a 150-page report that has been long forgotten. It was written for the Committee of Medical Research of the Office of Scientific Research and Development—the office that was created by President Roosevelt to coordinate scientific and medical research related to the war effort. The only copy I located in the United States was on a single deteriorating roll of microfilm buried in the Pennsylvania storage facility of the National Library of Medicine.[26]

Unlike their predecessors since the time of Sigmund Freud, this medical team not only took these anxiety-like complaints seriously, but looked for and found physical abnormalities in the majority of these patients. They preferred to call the illness "neurocirculatory asthenia," rather than "neurasthenia," "irritable heart," "effort syndrome," or "anxiety neurosis," as it had variously been known since the 1860s. But the symptoms confronting them were the same as those first described by George Miller Beard in 1869 (see chapter 5). Although the focus of this team was the heart, the 144 soldiers enrolled in their study also had respiratory, neurological, muscular, and digestive symptoms. Their average patient, in addition to having heart palpitations, chest pains, and shortness of breath, was nervous, irritable, shaky, weak, depressed, and exhausted. He could not concentrate, was losing weight, and was troubled by insomnia. He complained of headaches, dizziness, and nausea, and sometimes suffered from diarrhea or vomiting. Yet standard laboratory tests—blood work, urinalysis, X-rays, electrocardiogram, and electroencephalogram—were usually "within normal limits."

Cohen, who directed the research, brought to it an open mind. Raised in Alabama and educated at Yale, he was then a young professor at Harvard Medical School who was already challenging delivered wisdom and lighting one of the earliest sparks of what would eventually be a revolution in psychiatry. For he had the courage to call Freudian psychoanalysis a cult back in the 1940s when its practitioners were asserting control in every academic institution, capturing the imagination of Hollywood, and touching every aspect of American culture.[27]

Mandel Ettelson Cohen
(1907-2000)

Paul White, one of the two chief investigators—the other was neurologist Stanley Cobb—was already familiar with neurocirculatory asthenia from his civilian cardiology practice, and thought, contrary to Freud, that it was a genuine physical disease. Under the leadership of these three individuals, the team confirmed that this was indeed the case. Using the techniques that were available in the 1940s, they accomplished what no one in the nineteenth century, when the epidemic began, had been able to do: they demonstrated conclusively that neurasthenia had a physical and not a psychological cause. And they gave the medical community a list of objective signs by which the illness could be diagnosed.

Most patients had a rapid resting heart rate (over 90 beats per minute) and a rapid respiratory rate (over 20 breaths per minute), as well as a tremor of the fingers and hyperactive knee and ankle reflexes. Most had cold hands, and half the patients had a visibly flushed face and neck.

It has long been known that people with disorders of circulation have abnormal capillaries that can be most easily seen in the nail fold—the fold of skin at the base of the fingernails. White's team routinely found such abnormal capillaries in their patients with neurocirculatory asthenia.

They found that these patients were hypersensitive to heat, pain and, significantly, to electricity—they reflexively pulled their hands away from electric shocks of much lower intensity than did normal healthy individuals.

When asked to run on an inclined treadmill for three minutes, the majority of these patients could not do it. On average, they lasted only a minute and a half. Their heart rate after such exercise was excessively fast, their oxygen consumption during the exercise was abnormally low and, most significantly, their ventilatory efficiency was abnormally low. This means that they used less oxygen, and exhaled less carbon dioxide, than a normal person even when they breathed the same amount of air. To compensate, they breathed more air more rapidly than a healthy person and were still not able to continue running because their bodies were still not using enough oxygen.

A fifteen-minute walk on the same treadmill gave similar results. All subjects were able to complete this easier task. However, on average, the patients with neurocirculatory asthenia breathed fifteen percent more air per minute than healthy volunteers in order to consume the same amount of oxygen. And although, by breathing faster, the patients with neurocirculatory asthenia managed to consume the same amount of oxygen as the healthy volunteers, they had twice as much lactic acid in their blood, indicating that their cells were not using that oxygen efficiently.

Compared to healthy individuals, people with this disorder were able to extract less oxygen from the same amount of air, and their cells were able to extract less energy from the same amount of oxygen. The researchers concluded that these patients suffered from a defect of aerobic metabolism. In other words, something was wrong with their mitochondria—the powerhouses of their cells. The patients correctly complained that they could not get enough air. This was starving all of their organs of oxygen and causing both their heart symptoms and their other disabling complaints. Patients with neurocirculatory asthenia were consequently unable to hold their breath for anything like a normal period of time, even when breathing oxygen.[28]

During the five years of Cohen's team's study, several types of treatment were attempted with different groups of patients: oral testosterone; massive doses of vitamin B complex; thiamine; cytochrome *c*; psychotherapy; and a course of physical training under a professional trainer. None of these programs produced any improvement in symptoms or endurance.

"We conclude," wrote the team in June 1946, "that neurocirculatory asthenia is a condition that actually exists and has not been invented by patients or medical observers. It is not malingering or simply a mechanism aroused during war time for purposes of evading military service. The disorder is quite common both as a civilian and as a service problem."[29] They objected to Freud's term "anxiety neurosis" because anxiety was obviously a result, and not a cause, of the profound physical effects of not being able to get enough air.

In fact, these researchers virtually disproved the theory that the disease was caused by "stress" or "anxiety." It was not caused by hyperventilation.[30] Their patients did not have elevated levels of stress hormones—17-ketosteroids—in their urine. A twenty-year follow-up study of civilians with neurocirculatory asthenia revealed that these people typically did not develop any of the diseases that are supposed to be caused by anxiety, such as high blood pressure, peptic ulcer, asthma, or ulcerative colitis.[31] However, they did have abnormal electrocardiograms that indicated that the heart muscle was being starved of oxygen, and that were sometimes indistinguishable from the EKGs of people who had actual coronary artery disease or actual structural damage to the heart.[32]

The connection to electricity was provided by the Soviets. Soviet researchers, during the 1950s, 1960s, and 1970s, described physical signs and symptoms and EKG changes, caused by radio waves, that were identical to those that White and others had first reported in the 1930s and 1940s. The EKG changes indicated both conduction blocks and oxygen deprivation to the heart.[33] The Soviet scientists—in agreement with Cohen and White's team—concluded that these patients were suffering from a defect of aerobic metabolism. Something was wrong with the mitochondria in their cells. And they discovered what that defect was. Scientists that included Yury Dumanskiy, Mikhail Shandala, and Lyudmila Tomashevskaya, working in Kiev, and F. A. Kolodub, N. P. Zalyubovskaya and R. I. Kiselev, working in Kharkov, proved that the activity of the electron transport chain—the mitochondrial enzymes that extract energy from our food—is diminished not only in animals that are exposed to radio waves,[34] but in animals exposed to magnetic fields from ordinary electric power lines.[35]

The first war in which the electric telegraph was widely used—the American Civil War—was also the first in which "irritable heart" was a prominent disease. A young physician named Jacob M. Da Costa, visiting physician at a military hospital in Philadelphia, described the typical patient.

"A man who had been for some months or longer in active service," he wrote, "would be seized with diarrhoea, annoying, yet not severe enough to keep him out of the field; or, attacked with diarrhoea or fever, he rejoined, after a short stay in hospital, his command, and again underwent the exertions of a soldier's life. He soon noticed that he could not bear them as formerly; he got out of breath, could not keep up with his comrades, was annoyed with dizziness and palpitation, and with pain in his chest; his accoutrements oppressed him, and all this though he appeared well and healthy. Seeking advice from the surgeon of the regiment, it was decided that he was unfit for duty, and he was sent to a hospital, where his persistently quick acting heart confirmed his story, though he looked like a man in sound condition."[36]

Exposure to electricity in this war was universal. When the Civil War broke out in 1861, the east and west coasts had not yet been linked, and most of the country west of the Mississippi was not yet served by any telegraph lines. But in this war, every soldier, at least on the Union side, marched and camped near such lines. From the attack on Fort Sumter on April 12, 1861, until General Lee's surrender at Appomattux, the United States Military Telegraph Corps rolled out 15,389 miles of telegraph lines on the heels of the marching troops, so that military commanders in Washington could communicate instantly with all of the troops at their encampments. After the war all of these temporary lines were dismantled and disposed of.[37]

"Hardly a day intervened when General Grant did not know the exact state of facts with me, more than 1,500 miles off as the wires ran," wrote General Sherman in 1864. "On the field a thin insulated wire may be run on improvised stakes, or from tree to tree, for six or more miles in a couple of hours, and I have seen operators so skillful that by cutting the wire they would receive a message from a distant station with their tongues."[38]

Because the distinctive symptoms of irritable heart were encountered in every army of the United States, and attracted the attention of so many of its medical officers, Da Costa was puzzled that no one had described such a disease in any previous war. But telegraphic communications were never before used to such an extent in war. In the British Blue Book of the Crimean War, a conflict which lasted from 1853-56, Da Costa found two references to some troops being admitted to hospitals for "palpitations," and he found possible hints of the same problem reported from India during the Indian Rebellion of 1857-58. These were also the only two conflicts prior to the American

Civil War in which some telegraph lines were erected to connect command headquarters with troop units.[39] Da Costa wrote that he searched through medical documents from many previous conflicts and did not find even a hint of such a disease prior to the Crimean War.

During the next several decades, irritable heart attracted relatively little interest. It was reported among British troops in India and South Africa, and occasionally among soldiers of other nations.[40] But the number of cases was small. Even during the Civil War, what Da Costa considered "common" did not amount to many cases by today's standards. In his day, when heart disease was practically non-existent, the appearance of 1,200 cases of chest pain among two million young soldiers[41] caught his attention like an unfamiliar reef, suddenly materialized in a well-traveled shipping lane across an otherwise calm sea—a sea that was not further disturbed until 1914.

But shortly after the First World War broke out, in a time when heart disease was still rare in the general population and cardiology did not even exist as a separate medical specialty, soldiers began reporting sick with chest pain and shortness of breath, not by the hundreds, but by the tens of thousands. Out of the six and a half million young men that fought in the British Army and Navy, over one hundred thousand were discharged and pensioned with a diagnosis of "heart disease."[42] Most of these men had irritable heart, also called "Da Costa's syndrome," or "effort syndrome." In the United States Army such cases were all listed under "Valvular Disorders of the Heart," and were the third most common medical cause for discharge from the Army.[43] The same disease also occurred in the Air Force, but was almost always diagnosed as "flying sickness," thought to be cause by repeated exposure to reduced oxygen pressure at high altitudes.[44]

Similar reports came from Germany, Austria, Italy, and France.[45]

So enormous was the problem that the United States Surgeon-General ordered four million soldiers training in the Army camps to be given cardiac examinations before being sent overseas. Effort syndrome was "far away the commonest disorder encountered and transcended in interest and importance all the other heart affections combined," said one of the examining physicians, Lewis A. Conner.[46]

Some soldiers in this war developed effort syndrome after shell shock, or exposure to poison gas. Many more had no such history. All, however, had gone into battle using a newfangled form of communication.

The United Kingdom declared war on Germany on August 4, 1914, two days after Germany invaded its ally, France. The British army began embarking for France on August 9, and continued on to Belgium, reaching the city of Mons on August 22, without the aid of the wireless telegraph. While in Mons, a 1500-watt mobile radio set, having a range of 60 to 80 miles, was supplied to the British army signal troops.[47] It was during the retreat from Mons that many British soldiers first became ill with chest pain, shortness of breath, palpitations, and rapid heart beat and were sent back to England to be evaluated for possible heart disease.[48]

Exposure to radio was universal and intense. A knapsack radio with a range of five miles was used by the British army in all trench warfare on the front lines. Every battalion carried two such sets, each having two operators, in the front line with the infantry. One or two hundred yards behind, back with the reserve, were two more sets and two more operators. A mile further behind at Brigade Headquarters was a larger radio set, two miles back at Divisional Headquarters was a 500-watt set, and six miles behind the front lines at Army Headquarters was a 1500-watt radio wagon with a 120-foot steel mast and an umbrella-type aerial. Each operator relayed the telegraph messages received from in front of or behind him.[49]

All cavalry divisions and brigades were assigned radio wagons and knapsack sets. Cavalry scouts carried special sets right on their horses, that were called "whisker wireless" because of the antennae that sprouted from the horses' flanks like the quills of a porcupine.[50]

Most aircraft carried lightweight radio sets, using the metal frame of the airplane as the antenna. German war Zeppelins and French dirigibles carried much more powerful sets, and Japan had wireless sets in its war balloons. Radio sets on ships made it possible for naval battle lines to be spread out in formations 200 or 300 miles long. Even submarines, while cruising below the surface, sent up a short mast, or an insulated jet of water, as an antenna for the coded radio messages they broadcast and received.[51]

In the Second World War irritable heart, now called neurocirculatory asthenia, returned with a vengeance. Radar joined radio for the first time in this war, and it too was universal and intense. Like children with a new toy, every nation devised as many uses for it as possible. Britain, for example, peppered its coastline with hundreds of early warning radars emitting more than half a million watts each, and outfitted all its airplanes with powerful radars that could detect objects as small as a submarine periscope. More than two thousand portable radars, accompanied by 105-foot-tall portable towers, were deployed by the British army. Two thousand more "gun-laying" radars assisted anti-aircraft guns in tracking and shooting down enemy aircraft. The ships of the Royal Navy sported surface radars with a power of up to one million watts, as well as air search radars, and microwave radars that detected submarines and were used for navigation.

The Americans deployed five hundred early-warning radars on board ships, and additional early-warning radars on aircraft, each having a power of one million watts. They used portable radar sets at beachheads and airfields in the South Pacific, and thousands of microwave radars on ships, aircraft, and Navy blimps. From 1941 to 1945 the Radiation Laboratory at the Massachusetts Institute of Technology was kept busy by its military masters developing some one hundred different types of radar for various uses in the war.

The other powers fielded radar installations with equal vigor on land, at sea, and in air. Germany deployed over one thousand ground-based early warning radars in Europe, as well as thousands of ship-borne, airborne, and gun-laying radars. The Soviet Union did likewise, as did Australia, Canada, New Zealand, South Africa, the Netherlands, France, Italy, and Hungary. Wherever a soldier was asked to fight he was bathed in an ever-thickening soup of pulsed radio wave and microwave frequencies. And he succumbed in large numbers, in the armies, navies, and air forces of every nation.[52]

It was during this war that the first rigorous program of medical research was conducted on soldiers with this disease. By this time Freud's proposed term "anxiety neurosis" had taken firm hold among army doctors. Members of the Air Force who had heart symptoms were now receiving a diagnosis of "L.M.F.," standing for "lack of moral fiber." Cohen's team was stacked with psychiatrists. But to their surprise, and guided by cardiologist Paul White, they found objective evidence of a real disease that they concluded was *not* caused by anxiety.

Largely because of the prestige of this team, research into neurocirculatory asthenia continued in the United States throughout the 1950s; in Sweden, Finland, Portugal, and France into the 1970s and 1980s; and even, in Israel and Italy, into the 1990s.[53] But a growing stigma was attached to any doctor who still believed in the physical causation of this disease. Although the dominance of the Freudians had waned, they left an indelible mark not only on psychiatry but on all of medicine. Today, in the West, only the "anxiety" label remains, and people with the symptoms of neurocirculatory asthenia are automatically given a psychiatric diagnosis and, very likely, a paper bag to breathe into. Ironically, Freud himself, although he coined the term "anxiety neurosis," thought that its symptoms were not mentally caused, "nor amenable to psychotherapy."[54]

Meanwhile, an unending stream of patients continued to appear in doctors' offices suffering from unexplained exhaustion, often accompanied by chest pain and shortness of breath, and a few courageous doctors stubbornly continued to insist that psychiatric problems could not explain them all. In 1988, the term "chronic fatigue syndrome" (CFS) was coined by Gary Holmes at the Centers for Disease Control, and it continues to be applied by some doctors to patients whose most prominent symptom is exhaustion. Those doctors are still very much in the minority. Based on their reports, the CDC estimates that the prevalence of CFS is between 0.2 percent and 2.5 percent of the population,[55] while their counterparts in the psychiatric community tell us that as many as one person in six, suffering from the identical symptoms, fits the criteria for "anxiety disorder" or "depression."

To confuse the matter still further, the same set of symptoms was called myalgic encephalomyelitis (ME) in England as early as 1956, a name that focused attention on muscle pains and neurological symptoms rather than fatigue. Finally, in 2011, doctors from thirteen countries got together and adopted a set of "International Consensus Criteria" that recommends abandoning the name "chronic fatigue syndrome" and applying "myalgic encephalomyelitis" to all patients who suffer from "post-exertional exhaustion" plus specific neurological, cardiovascular, respiratory, immune, gastrointestinal, and other impairments.[56]

This international "consensus" effort, however, is doomed to failure. It completely ignores the psychiatric community, which sees far more of these patients. And it pretends that the schism that emerged from World War II never occurred. In the former Soviet Union, Eastern Europe, and most of Asia, the older term "neurasthenia" persists today. That term is still widely applied to the full spectrum of symptoms described by George Beard in 1869. In those parts of the world it is generally recognized that exposure to toxic agents, both chemical and electromagnetic, often causes this disease.

According to published literature, all of these diseases—neurocirculatory asthenia, radio wave sickness, anxiety disorder, chronic fatigue syndrome, and myalgic encephalomyelitis—predispose to elevated levels of blood cholesterol, and all carry an increased risk of death from heart disease.[57] So do porphyria[58] and oxygen deprivation.[59] The fundamental defect in this disease of many names is that although enough oxygen and nutrients reach the cells, the mitochondria—the powerhouses of the cells—cannot efficiently use that oxygen and those nutrients, and not enough energy is produced to satisfy the requirements of heart, brain, muscles, and organs. This effectively starves the entire body, including the heart, of oxygen, and can eventually damage the heart. In addition, neither sugars nor fats are efficiently utilized by the cells, causing unutilized sugar to build up in the blood—leading to diabetes—as well as unutilized fats to be deposited in arteries.

And we have a good idea of precisely where the defect is located. People with this disease have reduced activity of a porphyrin-containing enzyme called cytochrome oxidase, which resides within the mitochondria, and delivers electrons from the food we eat to the oxygen we breathe. Its activity is impaired in all the incarnations of this disease. Mitochondrial dysfunction has been reported in chronic fatigue syndrome[60] and in anxiety disorder.[61] Muscle biopsies in these patients show reduced cytochrome oxidase activity. Impaired glucose metabolism is well known in radio wave sickness, as is an impairment of cytochrome oxidase activity in animals exposed to even extremely low levels of radio waves.[62] And the neurological and cardiac symptoms of porphyria are widely blamed on a deficiency of cytochrome oxidase and cytochrome *c*, the heme-containing enzymes of respiration.[63]

Recently zoologist Neelima Kumar at Panjab University in India proved elegantly that cellular respiration can be brought to a standstill in honey bees merely by exposing them to a cell phone for ten minutes. The concentration of total carbohydrates in their hemolymph, which is what bees' blood is called, rose from 1.29 to 1.5 milligrams per milliliter. After twenty minutes it rose to 1.73 milligrams per milliliter. The glucose content rose from 0.218 to 0.231 to 0.277 milligrams per milliliter. Total lipids rose from 2.06 to 3.03 to 4.50 milligrams per milliliter. Cholesterol rose from 0.230 to 1.381 to 2.565 milligrams per milliliter. Total protein rose from 0.475 to 0.525 to 0.825 milligrams per milliliter. In other words, after just ten minutes of exposure to a cell phone, the bees practically could not metabolize sugars, proteins, or fats. Mitochondria are essentially the same in bees and in humans, but since their metabolism is so much faster, electric fields affect bees much more quickly.

In the twentieth century, particularly after World War II, a barrage of toxic chemicals and electromagnetic fields (EMFs) began to significantly interfere with the breathing of our cells. We know from work at Columbia University that even tiny electric fields alter the speed of electron transport from cytochrome oxidase. Researchers Martin Blank and Reba Goodman thought that the explanation lay in the most basic of physical principles. "EMF," they wrote in 2009, "acts as a force that competes with the chemical forces in a reaction." Scientists at the Environmental Protection Agency—John Allis and William Joines—finding a similar effect from radio waves, developed a variant theory along the same lines. They speculated that the iron atoms in the porphyrin-containing enzymes were set into motion by the oscillating electric fields, interfering with their ability to transport electrons.[64]

It was the English physiologist John Scott Haldane who first suggested, in his classic book, *Respiration*, that "soldier's heart" was caused not by anxiety but by a chronic lack of oxygen.[65] Mandel Cohen later proved that the defect was not in the lungs, but in the cells. These patients continually gulped air not because they were neurotic, but because they really could not get enough of it. You might as well have put them in an atmosphere that contained only 15 percent oxygen instead of 21 percent, or transported them to an altitude of 15,000 feet. Their chests hurt, and their hearts beat fast, not because of panic, but because they craved air. And their hearts craved oxygen, not because their coronary arteries were blocked, but because their cells could not fully utilize the air they were breathing.

These patients were not psychiatric cases; they were warnings for the world. For the same thing was also happening to the civilian population: they too were being slowly asphyxiated, and the pandemic of heart disease that was well underway in the 1950s was one result. Even in people who did not have a porphyrin enzyme deficiency, the mitochondria in their cells were still struggling, to some smaller degree, to metabolize carbohydrates, fats, and proteins. Unburned fats, together with the cholesterol that transported those fats in the blood, were being deposited on the walls of arteries. Humans and animals were not able to push their hearts quite as far as before without showing signs of stress and disease. This takes its clearest toll on the body when it is pushed to its limits, for example in athletes, and in soldiers during war.

The real story is told by the astonishing statistics.

When I began my research, I had only Samuel Milham's data. Since he found such a large difference in rural disease rates in 1940 between the five least and five most electrified states, I wanted to see what would happen if I calculated the rates for all forty-eight states and plotted the numbers on a graph. I looked up rural mortality rates in volumes of the Vital Statistics of the United States. I calculated the percent of electrification for each state by dividing the number of its residential electric customers, as published by the Edison Electric Institute, by the total number of its households, as published by the United States Census.

The results, for 1931 and 1940, are pictured in figures 1 and 2. Not only is there a five- to six-fold difference in mortality from rural heart disease between the most and least electrified states, but all of the data points come very close to lying on the same line. The more a state was electrified—i.e. the more rural households had electricity—the more rural heart disease it had. The amount of rural heart disease was proportional to the number of households that had electricity.[66]

Figure 1 – Rate of Rural Heart Disease in 1931

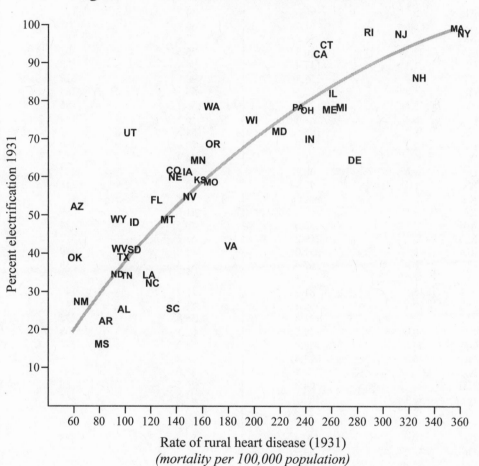

Rate of rural heart disease (1931)
(mortality per 100,000 population)

Table 2

	% electrification (1931)	Rural heart disease 1931 (deaths per 100,000)	% electrification (1940)	Rural heart disease 1940 (deaths per 100,000)
AL	25.7	98.8	34.7	147
AZ	62.5	61.4	56.1	87
AR	22.1	84.6	27.3	109
CA	92.5	250.3	75.6	305
CO	61.5	137.4	56.9	188
CT	94.9	255.7	90.5	328
DE	64.4	277.5	66.1	364
FL	53.8	124.0	50.7	186
GA	28.4	(missing)	36.5	144
ID	48.2	106.5	64.5	187
IL	82.5	259.9	79.4	330
IN	70.0	241.8	74.9	311
IA	61.4	148.3	65.5	234
KS	59.4	157.8	60.2	246
KY	38.0	(missing)	41.6	177
LA	34.1	118.7	41.5	189
ME	77.5	258.5	70.5	344
MD	72.3	219.2	65.2	312
MA	98.5	357.0	91.9	479
MI	78.4	267.4	81.3	339
MN	64.2	156.3	63.4	225
MS	16.5	81.2	22.7	149
MO	59.1	166.3	58.3	241
MT	48.9	131.4	56.8	217
NE	60.0	138.5	62.1	208
NV	54.8	150.0	58.3	370
NH	86.3	327.4	78.7	428
NJ	97.7	313.2	87.0	423
NM	27.3	64.8	26.5	88
NY	98.1	360.3	83.9	465
NC	32.4	120.8	43.7	152
ND	34.5	94.1	40.5	190
OH	77.0	240.1	82.5	323
OK	39.2	59.9	41.3	127
OR	68.8	168.5	67.7	220
PA	78.5	234.2	80.4	331
RI	98.2	289.8	91.0	404
SC	25.6	136.8	32.1	165
SD	41.0	106.0	43.0	188
TN	34.0	100.1	42.1	154
TX	39.5	97.9	43.5	144
UT	71.8	103.9	75.2	198
VT	71.9	(missing)	71.5	367
VA	41.7	181.6	53.1	231
WA	78.7	166.6	73.8	230
WV	41.0	94.7	53.4	146
WI	74.7	198.0	54.2	282
WY	49.5	95.1	50.8	170

Figure 2 – Rate of Rural Heart Disease in 1940

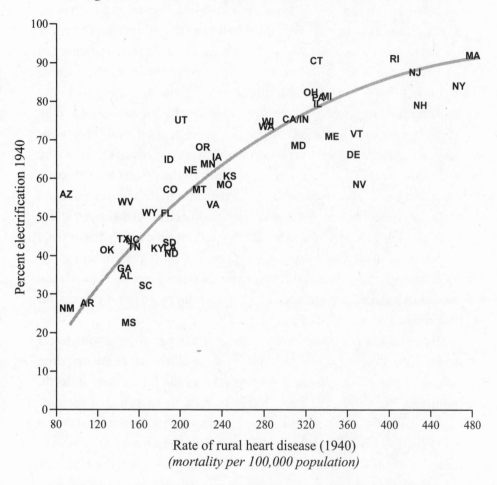

Rate of rural heart disease (1940)
(mortality per 100,000 population)

What is even more remarkable is that the death rates from heart disease in unelectrified rural areas of the United States in 1931, before the Rural Electrification Program got into gear, were still as low as the death rates for the whole United States prior to the beginning of the heart disease epidemic in the nineteenth century.

In 1850, the first census year in which mortality data were collected, a total of 2,527 deaths from heart disease were recorded in the nation. Heart disease ranked twenty-fifth among causes of death in that year. About as many people died from accidental drowning as from heart disease. Heart disease was something that occurred mainly in young children and in old age, and was predominantly a rural rather than an urban disease because farmers lived longer than city-dwellers.

In order to realistically compare nineteenth century statistics with those of today, I had to make some adjustments to the Census figures. The census enumerators for 1850, 1860, and 1870 had only the numbers reported to them from memory by the households they visited as to who had died during the previous year and from what causes. These numbers were estimated by the Census Office to be deficient, on average, by about 40 percent. In the census for 1880, the numbers were supplemented by reports from physicians and averaged only 19 percent short of the truth. By 1890 eight northeastern states plus the District of Columbia had passed laws requiring the official registration of all deaths, and the statistics for those registration states were considered accurate to within two to three percent. By 1910 the registration area had expanded to 23 states, and by 1930 only Texas did not require registration of deaths.

Another complicating factor is that heart failure was sometimes not evident except for the edema it caused, and therefore edema, then called "dropsy,"[67] was sometimes reported as the only cause of death, although the death was most likely to have been caused by either heart or kidney disease. Yet a further complication is the appearance of "Bright's disease" for the first time in the tables for 1870. This was the new term for the type of kidney disease that caused edema. Its prevalence in 1870 was reported to be 4.5 cases per 100,000 population.

With these complexities in mind, I have calculated the approximate rates of death from cardiovascular disease for each decade from 1850 to 2010, adding the figures for "dropsy" when that term was still in use (until 1900), and subtracting 4.5 per 100,000 for the years 1850 and 1860. I added a correction factor of 40 percent for 1850, 1860 and 1870, and 19 percent for 1880. I included reports of deaths from all diseases of the heart, arteries, and blood pressure. Beginning with 1890 I used only the figures for the death registration states, which by 1930 included the entire country except for Texas. The results are as follows:

Death Rates from Cardiovascular Disease
(per 100,000 population)

1850	77
1860	78
1870	78
1880	102
1890	145
1890 (Indians on reservations)	60
1900	154
1910	183
1920	187
1930	235
1940	291
1950	384
1960	396
1970	394
1980	361
1990	310
2000	280
2010	210
2017	214

1910 was the first year in which the mortality in cities surpassed that in the countryside. But the greatest disparities emerged *in* the countryside. In the northeastern states, which in 1910 had the greatest use of telegraphs, telephones, and now electric lights and power, and the densest networks of wires crisscrossing the land, the rural areas had as much mortality from cardiovascular disease, or more, than the cities. The rural mortality rate of Connecticut was then 234, of New York 279, and of Massachusetts 296. By contrast Colorado's rural rate was still 100, and Washington's 92. Kentucky's rural rate, at 88.5, was only 44 percent of its urban rate, which was 202.

Heart disease rose steadily with electrification, as we saw in figures 1 and 2, and reached a peak when rural electrification approached 100 percent during the 1950s. Rates of heart disease then leveled off for three decades and began to drop again—or so it seems at first glance. A closer look, however, shows the true picture. These are just the mortality rates. The number of people walking around with heart disease—the prevalence rate—actually continued to rise, and is still rising today. Mortality stopped rising in the 1950s because of the introduction of anticoagulants like heparin, and later aspirin, both to treat heart attacks and to prevent them.[68] In the succeeding decades the ever more aggressive use of anticoagulants, drugs to lower blood pressure, cardiac bypass surgery, balloon angioplasty, coronary stents, pacemakers, and even heart transplants, has simply allowed an ever growing number of people with heart disease to stay alive. But people are not having fewer heart attacks. They are having more.

The Framingham Heart Study showed that at any given age the chance of having a first heart attack was essentially the same during the 1990s as it was during the 1960s.[69] This came as something of a surprise. By giving people statin drugs to lower their cholesterol, doctors thought they were going to save people from having clogged arteries, which was supposed to automatically mean healthier hearts. It hasn't turned out that way. And in another study, scientists involved in the Minnesota Heart Survey discovered in 2001 that although fewer hospital patients were being diagnosed with coronary heart disease, more patients were being diagnosed with heart-related chest pain. In fact, between 1985 and 1995 the rate of unstable angina had increased by 56 percent in men and by 30 percent in women.[70]

The number of people with congestive heart failure has also continued steadily to rise. Researchers at the Mayo Clinic searched two decades of their records and discovered that the incidence of heart failure was 8.3 percent higher during the period 1996-2000 than it had been during 1979-1984.[71]

The true situation is much worse still. Those numbers reflect only people newly diagnosed with heart failure. The increase in the total number of people walking around with this condition is astonishing, and only a small part of the increase is due to the aging of the population. Doctors from Cook County Hospital, Loyola University Medical School, and the Centers for Disease Control examined patient records from a representative sample of American hospitals and found that the numbers of patients with a diagnosis of heart failure more than doubled between 1973 and 1986.[72] A later, similar study by scientists at the Centers for Disease Control found that this trend had continued. The number of hospitalizations for heart failure tripled between 1979 and 2004, the age-adjusted rate doubled, and the greatest increase occurred in people under 65 years of age.[73] A similar study of patients at Henry Ford Hospital in Detroit showed that the annual prevalence of congestive heart failure had almost quadrupled from 1989 to 1999.[74]

Young people, as the 3,000 alarmed doctors who signed the Freiburger Appeal affirmed, are having heart attacks at an unprecedented rate. In the United States, as great a percentage of forty-year-olds today have cardiovascular disease as the percentage of seventy-year-olds that had cardiovascular disease in 1970. Close to one-quarter of Americans aged forty to forty-four today have some form of cardiovascular disease.[75] And the stress on even younger hearts is not confined to athletes. In 2005, researchers at the Centers for Disease Control, surveying the health of adolescents and young adults, aged 15 to 34, found to their surprise that between 1989 and 1998 rates of sudden cardiac death in young men had risen 11 percent, and in young women had risen 30 percent, and that rates of mortality from enlarged heart, heart rhythm disturbances, pulmonary heart disease, and hypertensive heart disease had also increased in this young population.[76]

In the twenty-first century this trend has continued. The number of heart attacks in Americans in their twenties rose by 20 percent between 1999 and 2006, and the mortality from all types of heart disease in this age group rose by one-third.[77] In 2014, among patients between the ages of 35 and 74 who were hospitalized with heart attacks, one-third were below the age of 54.[78]

Developing countries are no better off. They have already followed the developed countries down the primrose path of electrification, and they are following us even faster to the wholesale embrace of wireless technology. The consequences are inevitable. Heart disease was once unimportant in low-income nations. It is now the number one killer of human beings in every region of the world except one. Only in sub-Saharan Africa, in 2017, was heart disease still outranked by diseases of poverty—AIDS and pneumonia—as a cause of mortality.

In spite of the billions being spent on conquering heart disease, the medical community is still groping in the dark. It will not win this war so long as it fails to recognize that the main factor that has been causing this pandemic for a hundred and fifty years is the electrification of the world.

12. The Transformation of Diabetes

In 1859, AT THE AGE OF TWELVE, the son of a lumber and grain merchant in Port Huron, Michigan strung a telegraph line one mile long between his house and a friend's, placing the two into electrical communication. From that day forward Thomas Alva Edison was intimate with the mysterious forces of electricity. He worked as an itinerant telegraph operator from the age of fifteen until he went into business for himself in Boston at age twenty-one, providing private-line telegraph service for Boston firms, stringing the wires from downtown offices, along the rooftops of houses and buildings, to factories and warehouses on the outskirts of the city. By the time he was twenty-nine, when he moved his laboratory to a small hamlet in New Jersey, he had made improvements to telegraph technology and was engaged in perfecting the newly invented telephone. The "Wizard of Menlo Park" became world famous in 1878 for his invention of the phonograph. He then set himself a much more ambitious task: he dreamed of lighting people's homes with electricity, and displacing the hundred-fifty-million-dollar-a-year gas lighting industry. Before he was done, he had invented the electric light bulb, dynamos that generated electricity at constant voltage, and a system of distributing electricity in parallel circuits. In November 1882, he patented the three-wire distribution system that we all still use today.

At around that time, Edison developed a rare disease known as diabetes.[1]

Another young man, who grew up in Scotland, was teaching elocution at a school in Bath in 1866 when he hooked up a homemade telegraph system between his house and a neighbor's. Five years later he found himself teaching the deaf to speak in Boston, where he was also a professor of elocution at Boston University. But he did not give up his lifelong affair with electricity. One of his deaf students, with whose family he boarded, glanced one day into his bedroom. "I found the floor, the chairs, the table, and even the dresser covered with wires, batteries, coils, cigar boxes, and an indescribable mass of miscellaneous equipment," the man recalled many years later. "The overflow was already in the basement, and it wasn't many months before he had expanded into the carriage house." In 1876, after patenting a number of improvements to the telegraph, Alexander Graham Bell invented the telephone, achieving world renown before the age of thirty. His "endless health complaints"—severe headaches, insomnia, sciatic pain, shortness of breath, chest pains, irregular heartbeat, and an abnormal sensitivity to light—dated from his earliest experiments with electricity in Bath.

In 1915 he, too, was diagnosed with diabetes.[2]

To begin to get a sense of just how rare diabetes once was, I searched the antique books in my medical library. I first looked in the Works of Robert Whytt, a Scottish physician of the early and mid-eighteenth century. I did not find diabetes mentioned in the 750-page volume.

American physician John Brown, at the end of the eighteenth century, devoted two paragraphs to the disorder in his *Elements of Medicine*. In the Works of Thomas Sydenham, who practiced in the seventeenth century and is known as the Father of English Medicine, I found a single page on diabetes. It set forth a sparse description of the disease, recommended an all-meat diet, and prescribed an herbal remedy.

I opened Benjamin Ward Richardson's 500-page work, *Diseases of Modern Life*, published in New York in 1876, a time when Edison and Bell were experimenting intensively with electricity. Four pages were devoted to diabetes. Richardson considered it a modern disease caused by exhaustion from mental overwork or by some shock to the nervous system. But it was still uncommon.

Then I consulted my "bible" of diseases of the nineteenth century, the *Handbook of Geographical and Historical Pathology*, published in stages between 1881 and 1886 in German and English. In this massive three-volume scholarly work, August Hirsch compiled the history of known diseases, along with their prevalence and distribution throughout the world. Hirsch spared six pages for diabetes, noting primarily that it was rare and that little information about it was known. In ancient Greece, he wrote, in the fourth century B.C., Hippocrates never mentioned it. In the second century A.D., Galen, a Greek-born physician practicing in Rome, devoted some passages to diabetes, but stated that he himself had seen only two cases.

The first book on diabetes had actually been written in 1798, but its author, John Rollo of England, had only seen three cases of it himself in his twenty-three years of practicing medicine.

The statistics Hirsch gathered from around the world confirmed to him that the disease "is one of the rarest."[3] About 16 people per year died of it in Philadelphia, 3 in Brussels, 30 in Berlin, and 550 in all of England. Occasional cases were reported in Turkey, Egypt, Morocco, Mexico, Ceylon, and certain parts of India. But an informant in St. Petersburg had not seen a case in six years. Practitioners in Senegambia and the Guinea Coast had never seen a case, nor was there any record of it occurring in China, Japan, Australia, the islands of the Pacific, Central America, the West Indies, Guiana, or Peru. One informant had never seen a case of diabetes during a practice of many years in Rio de Janeiro.

How, then, did diabetes come to be one of the major killers of humanity? In today's world, as we will see, limiting one's intake of sugar plays an important role in the prevention and control of this disease.

But, as we will also see, blaming the rise of diabetes on dietary sugars is as unsatisfactory as blaming the rise of heart disease on dietary fats.

In 1976, I was living in Albuquerque when a friend placed a newly published book in my hands that changed the way I ate and drank. William Dufty, the author of *Sugar Blues*, had done his homework thoroughly. He convinced me that the single most addictive substance that was undermining the health of the masses, and had been doing so for centuries, was not alcohol, tobacco, opium, or marijuana, but sugar. He further blamed four centuries of the African slave trade largely on the need to feed a sugar habit that had been acquired by the Crusaders during the twelfth and thirteenth centuries. Europeans, he said, had wrested control of the world sugar trade from the Arab Empire, and needed a steady supply of labor to tend their sugar plantations. His claim that sugar was "more intoxicating than beer or wine and more potent than many drugs" was supported by an entertaining tale that he spun about his own perplexing illnesses and his heroic efforts to kick the sugar habit, which finally succeeded. Migraine headaches, mysterious fevers, bleeding gums, hemorrhoids, skin eruptions, a tendency to gain weight, chronic fatigue, and an impressive assortment of aches and pains that had tormented him for fifteen years vanished within twenty-four hours, he said, and did not return.

Dufty also explained why sugar causes diabetes. Our cells, especially our brain cells, get their energy from a steady supply of a simple sugar called glucose, which is the end product of digesting the carbohydrates we eat. "The difference between feeling up or down, sane or insane, calm or freaked out, inspired or depressed depends in large measure upon what we put in our mouth," he wrote. He further explained that the difference between life and death depends upon a precise balance between the amount of glucose in our blood and the amount of blood oxygen, insulin being one of the hormones that maintains this balance. If not enough insulin is secreted by the pancreas after a meal, glucose builds up to a toxic level in the blood and we begin excreting it in our urine. If too much insulin is produced, blood glucose levels drop dangerously low.

The problem with eating pure sugar, wrote Dufty, is that it doesn't need to be digested and is absorbed into the blood much too fast. Eating complex carbohydrates, fats, and proteins requires the pancreas to secrete an assortment of digestive enzymes into the small intestine so that these foods can be broken down. This takes time. The glucose level in the blood rises gradually. However, when we eat refined sugar it is turned into glucose almost immediately and passes directly into the blood, Dufty explained, "where the glucose level has already been established in precise balance with oxygen. The glucose level in the blood is thus drastically increased. Balance is destroyed. The body is in crisis."

A year after reading this book I decided to apply to medical school, and had to first take basic courses in biology and chemistry that I did not take in college. My biochemistry professor at the University of California, San Diego essentially confirmed what I had learned from reading *Sugar Blues*. We evolved, said my professor, eating foods like potatoes that have to be digested gradually. The pancreas automatically secretes insulin at a rate that exactly corresponds to the rate at which glucose—over a considerable period of time after a meal—enters the bloodstream. Although this mechanism works perfectly if you eat meat, potatoes, and vegetables, a meal containing refined sugar creates a disturbance. The entire load of sugar enters the bloodstream at once. The pancreas, however, hasn't learned about refined sugar and "thinks" that you have just eaten a meal containing a tremendous amount of potatoes. A lot more glucose should be on its way. The pancreas there-fore manufactures an amount of insulin that can deal with a tremen-dous meal. This overreaction by the pancreas drives the blood glucose level too low, starving the brain and muscles—a condition known as hypoglycemia.[4] After years of such overstimulation the pancreas may become exhausted and stop producing enough insulin or produce none at all. This condition is called diabetes and requires the person to take insulin or other drugs to maintain his or her energy balance and stay alive.

Many besides Dufty have pointed out that an extraordinary rise in sugar consumption has accompanied the equally extraordinary rise in diabetes rates over the past two hundred years. Almost a century ago Dr. Elliott P. Joslin, founder of Boston's Joslin Diabetes Center, published statistics showing that yearly sugar consumption per person in the United States had increased eightfold between 1800 in 1917.[5]

But this model of diabetes is missing an important piece. It teaches us how to avoid getting diabetes in the twenty-first century: don't eat highly refined foods, especially sugar. But it completely fails to explain the terrible prevalence of diabetes in our time. Sugar or no sugar, diabetes was once an impressively rare disease. The vast majority of human beings were once able to digest and metabolize large quantities of pure sugar without eliminating it in their urine and without wearing out their pancreas. Even Joslin, whose clinical experience led him to suspect sugar as a cause of diabetes, pointed out that the consumption of sugar in the United States had increased by only 17 percent between 1900 and 1917, a period during which the death rate from diabetes had nearly doubled. And he underestimated sugar use in the nineteenth century because his statistics were for refined sugar only. They did not include maple syrup, honey, sorghum syrup, cane syrup, and especially molasses. Molasses was cheaper than refined sugar, and until about 1850 Americans consumed more molasses than refined sugar. The following graph[6] shows actual sugar consumption during the past two centuries, including the sugar content of syrups and molasses, and it does not fit the dietary model of this disease. In fact, per capita sugar consumption did not rise at all between 1922 and 1984, yet diabetes rates soared tenfold.

**U.S. Consumption of Sugar
and Other Caloric Sweeteners, 1822-2014**

That diet alone is not responsible for the modern pandemic of diabetes is clearly shown by the histories of three communities at opposite ends of the world from one another. One has the highest rates of diabetes in the world today. The second is the largest consumer of sugar in the world. And the third, which I will examine in some detail, is the most recently electrified country in the world.

American Indians

The poster child for the diabetes story is supposed to be the American Indian. Supposedly—according to the American Diabetes Association—people today are just eating too much food and not getting enough exercise to burn off all the calories. This causes obesity, which, it is believed, is the real cause of most diabetes. Indians, so the story goes, are genetically predisposed to diabetes, and this predisposition has been triggered by the sedentary lifestyle imposed on them when they were confined to reservations, as well as by an unhealthy diet containing large amounts of white flour, fat, and sugar that have replaced traditional foods. And indeed, today, Indians on most reservations in the United States and Canada have rates of diabetes that are the highest in the world.

Yet this does not explain why, since all Indian reservations were created by the end of the nineteenth century, and Indian fry bread, consisting of white flour deep fried in lard and eaten with sugar, became a staple food on most reservations at that time, diabetes nevertheless did not exist among Indians until the latter half of the twentieth century. Before 1940 the Indian Health Service had never listed diabetes as a cause of death for a single Indian. And as late as 1987, surveys done by the Indian Health Service in the United States and the Department of National Health and Welfare in Canada revealed differences in diabetes rates between different populations of Indians that were extreme: 7 cases of diabetes per 1,000 population in the Northwest Territories, 9 in the Yukon, 17 in Alaska, 28 among the Cree/Ojibwa of Ontario and Manitoba, 40 on the Lummi Reservation in Washington, 53 among the Micmac of Nova Scotia and the Makah of Washington, 70 on the Pine Ridge Reservation in South Dakota, 85 on the Crow Reservation in Montana, 125 on the Standing Rock Sioux Reservation in the Dakotas, 148 on the Chippewa Reservation in Minnesota and North Dakota, 218 on the Winnebago/Omaha Reservation in Nebraska, and 380 on the Gila River Reservation in Arizona.[7]

In 1987, neither diet nor lifestyle in the various communities was different enough to account for a fifty-fold difference in diabetes rates. But one environmental factor could account for the disparities. Electrification came to most Indian reservations later than it came to most American farms. Even in the late twentieth century some reservations were still not electrified. This included most Indian reserves in the Canadian Territories and most native villages in Alaska. When the first electric service came to the Standing Rock Reservation in the Dakotas in the 1950s, diabetes came to that reservation at the same time.[8] The Gila River Reservation is located on the outskirts of Phoenix. Not only is it traversed by high voltage power lines serving a metropolis of four million, but the Gila River Indian Community operates its own electric utility and its own telecommunications company. The Pima and Maricopa on this small reservation are exposed to a greater concentration of electromagnetic fields than any other Indian tribes in North America.

Brazil

Brazil, which has grown sugar cane since 1516, has been the largest producer and consumer of that commodity since the seventeenth century. Yet in the 1870s, when diabetes was beginning to be noticed as a disease of civilization in the United States, that disease was completely unknown in the sugar capital of the world, Rio de Janeiro.

Brazil today produces over 30 million metric tons of sugar per year and consumes over 130 pounds of white sugar per person, more than the United States. Analyses of the diets of the two countries—Brazil in 2002–2003, and the United States from 1996–2006—revealed that the average Brazilian obtained 16.7 percent of his or her calories from table sugar or sugar added to processed foods, while Americans consumed only 15.7 percent of their calories from refined sugars. Yet the United States had more than two and a half times the rate of diabetes as Brazil.[9]

Bhutan

Sandwiched between the mountainous borders of India and China, the isolated Himalayan kingdom of Bhutan may be the last country in the world to be electrified. Until the 1960s, Bhutan had no banking system, no national currency, and no roads. In the late 1980s, I learned something about this Buddhist country, thought by some to be the model for James Hilton's Shangri-La, when I made the acquaintance of a Canadian woman who worked for CUSO International, the Canadian version of the United States Peace Corps. She had just returned from a four-year stint in a small Bhutanese village, where she taught English to the local children. Bhutan is somewhat larger, in area, than the Netherlands, and has a population just over 750,000.

The road system at the time was still extremely limited, and most travel outside the immediate vicinity of the small capital, Thimphu, including travel to my friend's village, was by foot or horseback. She felt privileged to be able to live in that country at all, because outside visitors to Bhutan were limited to 1,000 per year. The woven baskets and other handcrafts that she brought back were intricate and beautiful. Technology was unknown, as there was no electricity at all in most of the country. Diabetes was extremely rare, and completely unknown outside the capital.

As recently as 2002, fuel wood provided virtually one hundred percent of all non-commercial energy consumption. Fuel wood consumption, at 1.22 tons per capita, was one of the highest, if not the highest, in the world. Bhutan was an ideal laboratory in which to monitor the effects of electricity, because that country was about to be transformed from near zero percent electrification to one hundred percent electrification in a little over a decade.

In 1998, King Jigme Singye Wangchuk ceded some of his powers to a democratic assembly, which wanted to modernize the country. The Department of Energy and the Bhutan Electricity Authority were created on July 1, 2002. That same day the Bhutan Power Corporation was launched. With 1,193 employees, it immediately became the largest corporation in the kingdom. Its mandate was to generate and distribute electricity throughout the kingdom, with a target of full electrification of the country within ten years. By 2012 the proportion of rural households actually reached by electricity was about 84 percent.

In 2004, 634 new cases of diabetes were reported in Bhutan. The next year, 944. The year after that, 1,470. The following year, 1,732. The next year, 2,541, with 15 deaths.[10] In 2010, there were 91 deaths and diabetes mellitus was already the eighth most common cause of mortality in the kingdom. Coronary heart disease was number one. Only 66.5 percent of the population had normal blood sugar.[11] This sudden change in the health of the population, especially the rural population, was being blamed, incredibly, on the traditional Bhutanese diet which, however, had not changed. "Bhutanese have a penchant for fat-rich foods," reported Jigme Wangchuk in the *Bhutan Observer*. "All Bhutanese delicacies are fat-rich. Salty and fatty foods cause hypertension. Today, one of the main causes of ill-health in Bhutan is hypertension caused by oil-rich and salty traditional Bhutanese diet." Rice, the article continued, which is the staple food of the Bhutanese, is rich in carbohydrates, which turns into fat unless there is physical activity; perhaps the Bhutanese are not getting enough exercise. Two-thirds of the population, the author complained, are not eating enough fruits and vegetables.

But the Bhutanese diet has not altered. The Bhutanese people are poor. Their country is mountainous with few roads. They have not all gone out and suddenly bought automobiles, refrigerators, washing machines, televisions, and computers, and become a lazy, inactive people. Yet rates of diabetes quadruple in four years. Bhutan now ranks eighteenth in the world in its mortality rate from heart disease.

Only one other thing has changed so dramatically in Bhutan in the last decade: electrification, and the resulting exposure of the population to electromagnetic fields.

We recall from the last chapter that exposure to electromagnetic fields interferes with basic metabolism. The power plants of our cells, the mitochondria, become less active, slowing the rate at which our cells can burn glucose, fats, and protein. Instead of being taken up by our cells, excess fats accumulate in our blood and are deposited on the walls of our arteries along with the cholesterol that transports them, forming plaques and causing coronary heart disease. This can be prevented by eating a low-fat diet.

In the same way, excess glucose, instead of being taken up by our cells, also backs up and accumulates in our blood. This increases the secretion of insulin by our pancreas. Normally, insulin lowers blood sugar by increasing its uptake by our muscles. But now our muscle cells can't keep up. They burn glucose as fast as they can after a meal, and it's no longer fast enough. Most of the excess goes into our fat cells, is converted to fat, and makes us obese. If your pancreas becomes worn out and stops producing insulin, you have Type 1 diabetes. If your pancreas is producing enough, or too much insulin, but your muscles are unable to use glucose quickly enough, this is interpreted as "insulin resistance" and you have Type 2 diabetes.

Eating a diet free of highly refined, quickly digested foods, especially sugar, can prevent this. In fact, before the discovery of insulin in 1922, some doctors, including Elliott Joslin, successfully treated severe cases of diabetes with a near-starvation diet.[12] They radically restricted their patients' intake of not just sugar, but all calories, thus ensuring that glucose entered the bloodstream at a rate no faster than the cells could deal with. After a several days' fast normalized the blood glucose, first carbohydrates, then proteins, then fats were gradually reintroduced into the patient's diet. Sugar was eliminated. This saved many people who would have died within a year or two.

But in Joslin's time the very nature of this disease underwent a mysterious transformation.

Insulin resistance—which accounts for the vast majority of diabetes in the world today—did not exist before the late nineteenth century. Neither did obese diabetic patients. Almost all people with diabetes were insulin-deficient, and they were universally thin: since insulin is needed in order for muscle and fat cells to absorb glucose, people with little or no insulin will waste away. They pee away their glucose instead of using it for energy, and survive by burning their stores of body fat.

In fact, overweight diabetics were at first so unusual that late-nineteenth-century doctors couldn't quite believe the change in the disease—and some of them didn't. One of these, John Milner Fothergill, a prominent London physician, wrote a letter to the *Philadelphia Medical Times* in 1884, in which he stated: "When a corpulent, florid-complexioned man, well-fed and vigorous, passes sugar in his urine, only a tyro would conjecture that he was the victim of classical diabetes, a formidable wasting disease."[13] Dr. Fothergill, as it turned out, was in denial. A corpulent, florid-complexioned man himself, Fothergill died of diabetes five years later.

Today the disease has changed entirely. Even children with Type 1, insulin-deficient diabetes tend to be overweight. They are overweight before they become diabetic because of their cells' reduced ability to metabolize fats. They are overweight after they become diabetic because the insulin that they take for the rest of their lives makes their fat cells take up lots of glucose and store it as fat.

Diabetes Is Also a Disorder of Fat Metabolism

Nowadays, all blood that is drawn from a patient is sent right off to a laboratory to be analyzed. The doctor rarely looks at it. But a hundred years ago the quality and consistency of the blood were valuable guides to diagnosis. Doctors knew that diabetes involved an inability to metabolize not just sugar but fat, because blood drawn from a diabetic's vein was milky, and when it was allowed to stand, a thick layer of "cream" invariably floated to the top.

In the early years of the twentieth century, when diabetes had become epidemic and was not yet controllable with any medication, it was not unusual for a diabetic's blood to contain 15 to 20 percent fat. Joslin even found that blood cholesterol was a more reliable measure of the severity of the disease than blood sugar. He disagreed with those of his contemporaries who were treating diabetes with a low-carbohydrate, high-fat diet. "The importance of the modification of the treatment to include control of the fat of the diet is obvious," he wrote. He issued a warning, appropriate not only for his contemporaries but for the future: "When fat ceases to be metabolized in a normal manner no striking evidence of it is afforded, and both patient and doctor continue to journey along in innocent oblivion of its existence, and hence fat is often a greater danger to a diabetic than carbohydrate."[14]

The linked failure of both carbohydrate and fat metabolism is a sign of disordered respiration in the mitochondria, and the mitochondria, we have seen, are disturbed by electromagnetic fields. Under the influence of such fields, respiratory enzyme activity is slower. After a meal, the cells cannot oxidize the breakdown products of the proteins, fats, and sugars that we eat as quickly as they are being supplied by the blood. Supply outstrips demand. Recent research has shown exactly how this happens.

Glucose and fatty acids, proposed University of Cambridge biochemist Philip J. Randle in 1963, compete with each other for energy production. This mutual competition, he said, operates independently of insulin to regulate glucose levels in the blood. In other words, high fatty acid levels in the blood inhibit glucose metabolism, and vice versa. Evidence in support appeared almost immediately. Jean-Pierre Felber and Alfredo Vannotti at the University of Lausanne gave a glucose tolerance test to five healthy volunteers, and then another one a few days later to the same individuals while they were receiving an intravenous infusion of lipids. Every person responded to the second test as though they were insulin resistant. Although their insulin levels remained the same, they were unable to metabolize the glucose as quickly in the presence of high levels of fatty acids in their blood, competing for the same respiratory enzymes. These experiments were easy to repeat, and overwhelming evidence confirmed the concept of the "glucose-fatty acid cycle." Some evidence also supported the idea that not only fats, but amino acids as well, competed with glucose for respiration.

Randle had not been thinking in terms of mitochondria, much less what could happen if an environmental factor restricted the ability of the respiratory enzymes to work at all. But during the last decade and a half, finally some diabetes researchers have begun focusing specifically on mitochondrial function.

Remember that our food contains three main types of nutrients—proteins, fats, and carbohydrates—that are broken down into simpler substances before being absorbed into our blood. Proteins become amino acids. Fats become triglycerides and free fatty acids. Carbohydrates become glucose. Some portion of these is used for growth and repair and becomes part of the structure of our body. The rest is burned by our cells for energy.

Within our cells, inside tiny bodies called mitochondria, amino acids, fatty acids, and glucose are all further transformed into even simpler chemicals that feed into a common cellular laboratory called the

Krebs cycle, which breaks them down the rest of the way so that they can combine with the oxygen we breathe to produce carbon dioxide, water, and energy. The last component in this process of combustion, the electron transport chain, receives electrons from the Krebs cycle and delivers them, one at a time, to molecules of oxygen. If the speed of those electrons is modified by external electromagnetic fields, as suggested by Blank and Goodman, or if the functioning of any of the elements of the electron transport chain is otherwise altered, the final combustion of our food is impaired. Proteins, fats, and carbohydrates begin to compete with each other and back up into the bloodstream. Fats are deposited in arteries. Glucose is excreted in urine. The brain, heart, muscles, and organs become oxygen-deprived. Life slows down and breaks down.

Only recently was it proven that this actually happens in diabetes. For a century, scientists had assumed that because most diabetics were fat, obesity causes diabetes. But in 1994, David E. Kelley at the University of Pittsburgh School of Medicine, in collaboration with Jean-Aimé Simoneau at Laval University in Quebec, decided to find out exactly why diabetics have such high fatty acid levels in their blood. Seventy-two years after insulin was discovered, Kelley and Simoneau were among the first to measure cellular respiration in detail in this disease. To their surprise, the defect turned out not to be in the cells' ability to absorb lipids but in their ability to burn them for energy. Large amounts of fatty acids were being absorbed by the muscles and not metabolized. This led to intensive research into all aspects of respiration at the cellular level in diabetes mellitus. Important work continues to be done at the University of Pittsburgh, as well as at the Joslin Diabetes Center, RMIT University in Victoria, Australia, and other research centers.[15]

What has been discovered is that cellular metabolism is reduced at all levels. The enzymes that break down fats and feed them into the Krebs cycle are impaired. The enzymes of the Krebs cycle itself, which receives the breakdown products of fats, sugars, and proteins, are impaired. The electron transport chain is impaired. The mitochondria are smaller and reduced in number. Consumption of oxygen by the patient during exercise is reduced. The more severe the insulin resistance—i.e., the more severe the diabetes—the greater the reductions in all these measures of cellular respiratory capacity.

In fact, Clinton Bruce and his colleagues in Australia found that the oxidative capacity of the muscles was a better indicator of insulin resistance than their fat content—which threw into question the traditional wisdom that obesity causes diabetes. Perhaps, they speculated, obesity is not a cause but an effect of the same defect in cellular respiration that causes diabetes. A study involving lean, active young African-American women in Pittsburgh, published in 2014, seemed to confirm this. Although the women were somewhat insulin resistant, they were not yet diabetic, and the medical team could find no other physiological abnormalities in the group except two: their oxygen consumption during exercise was reduced, and mitochondrial respiration in their muscle cells was reduced.[16]

In 2009, the Pittsburgh team made an extraordinary finding. If the electrons in the electron transport chain are being disturbed by an environmental factor, then one would expect that diet and exercise might improve all components of metabolism *except* the last, energy-producing step involving oxygen. That is exactly what the Pittsburgh team found. Placing diabetic patients on calorie restriction and a strict exercise regime was beneficial in many respects. It increased the activity of the Krebs cycle enzymes. It reduced the fat content of muscle cells. It increased the number of mitochondria in the cells. These benefits improved insulin sensitivity and helped control blood sugar. But although the number of mitochondria increased, their efficiency did not. The electron transport enzymes in dieted, exercised diabetic patients were still only half as active as the same enzymes in healthy individuals.[17]

In June 2010, Mary-Elizabeth Patti, a professor at Harvard Medical School and researcher at the Joslin Diabetes Center, and Silvia Corvera, a professor at the University of Massachusetts Medical School in Worcester, published a comprehensive review of existing research on the role of mitochondria in diabetes. They were forced to conclude that a defect of cellular respiration may be the basic problem behind the modern epidemic. Due to "failure of mitochondria to adapt to higher cellular oxidative demands," they wrote, "a vicious cycle of insulin resistance and impaired insulin secretion can be initiated."

But they were not willing to take the next step. No diabetes researchers today are looking for an environmental cause of this "failure to adapt" of so many people's mitochondria. They are still, in the face of evidence refuting it, blaming this disease on faulty diet, lack of exercise, and genetics. This in spite of the fact that, as Dan Hurley noted in his 2011 book, *Diabetes Rising*, human genetics has not changed and neither diet, exercise, nor drugs has put a dent in the escalation of this disease during the ninety years since insulin was discovered.

Diabetes in Radio Wave Sickness

In 1917, when Joslin was publishing the second edition of his book on diabetes, radio waves were being massively deployed on and off the battlefield in the service of war. At that point, as we saw in chapter 8, radio waves joined power distribution as a leading source of electromagnetic pollution on this planet. Their contribution has steadily grown until today when radio, television, radar, computers, cell phones, satellites, and millions of transmission towers have made radio waves by far the predominant source of electromagnetic fields bathing living cells.

The effects of radio waves on blood sugar are extremely well documented. However, none of this research has been done in the United States or Europe. It has been possible for western medical authorities to pretend that it doesn't exist because most of it is published in Czech, Polish, Russian, and other Slavic languages in strange alphabets and has not been translated into familiar tongues.

But some of it has, thanks to the United States military, in documents that have not been widely circulated, and thanks to a few international conferences.

During the Cold War, from the late 1950s through the 1980s, the United States Army, Navy, and Air Force were developing and building enormously powerful early warning radar stations to protect against the possibility of nuclear attack. In order to stand sentinel over the air spaces surrounding the United States, these stations were going to monitor the entire coastline and the borders with Mexico and Canada. This meant that a strip of the American border up to hundreds of miles wide—and everyone who lived there—was going to be continuously bombarded with radio waves at power levels that were unprecedented in human history. Military authorities needed to review all ongoing research into the health effects of such radiation. In essence, they wanted to know what were the maximum levels of radiation to which they could get away with exposing the American population. And so one of the functions of the Joint Publications Research Service, a federal agency established during the Cold War to translate foreign documents, was to translate into English some of the Soviet and Eastern European research on radio wave sickness. One of the most consistent laboratory findings in this body of literature is a disturbance of carbohydrate metabolism.

In the late 1950s, in Moscow, Maria Sadchikova gave glucose tolerance tests to 57 workers exposed to UHF radiation. The majority had altered sugar curves: their blood sugar remained abnormally high for over two hours after an oral dose of glucose. And a second dose, given after one hour, caused a second spike in some patients, indicating a deficiency of insulin.[18]

In 1964, V. Bartoníček, in Czechoslovakia, gave glucose tolerance tests to 27 workers exposed to centimeter waves—the type of waves we are all heavily exposed to today from cordless phones, cell phones, and wireless computers. Fourteen of the workers were prediabetic and four had sugar in their urine. This work was summarized by Christopher Dodge in a report he prepared at the United States Naval Observatory and read at a symposium held in Richmond, Virginia in 1969.

In 1973, Sadchikova attended a symposium in Warsaw on the Biologic Effects and Health Hazards of Microwave Radiation. She was able to report on her research team's observations of 1,180 workers exposed to radio waves over a twenty-year period, of whom about 150 had been diagnosed with radio wave sickness. Both prediabetic and diabetic sugar curves, she said, "accompanied all clinical forms of this disease."

Eliska Klimková-Deutschová of Czechoslovakia, at the same symposium, reported finding an elevated fasting blood sugar in fully three-quarters of all individuals exposed to centimeter waves.

Valentina Nikitina, who was involved in some of the Soviet research and was continuing to do such research in modern Russia, attended an international conference in St. Petersburg in 2000. She reported that people who maintained and tested radio communication equipment for the Russian Navy—even people who had ceased such employment five to ten years previously—had, on average, higher blood glucose levels than unexposed individuals.

Attached to the same medical centers at which Soviet doctors were examining patients were laboratories where scientists were exposing animals to the very same types of radio waves. They, too, reported seriously disturbed carbohydrate metabolism. They found that the activity of the enzymes in the electron transport chain, including the last enzyme, cytochrome oxidase, is always inhibited. This interferes with the oxidation of sugars, fats and proteins. To compensate, anaerobic (non-oxygen using) metabolism increases, lactic acid builds up in the tissues, and the liver becomes depleted of its energy-rich stores of glycogen. Oxygen consumption declines. The blood sugar curve is affected, and the fasting glucose level rises. The organism craves carbohydrates, and the cells become oxygen starved.[19]

These changes happen rapidly. As early as 1962, V.A. Syngayevskaya, working in Leningrad, exposed rabbits to low level radio waves and found that the animals' blood sugar rose by one-third in less than an hour. In 1982, Vasily Belokrinitskiy, working in Kiev, reported that the amount of sugar in the urine was in direct proportion to the dose of radiation and the number of times the animal was exposed. Mikhail Navakatikian and Lyudmila Tomashevskaya reported in 1994 that insulin levels decreased by 15 percent in rats exposed for just half an hour, and by 50 percent in rats exposed for twelve hours, to pulsed radiation at a power level of 100 microwatts per square centimeter. This level of exposure is comparable to the radiation a person receives today sitting directly in front of a wireless computer, and considerably less than what a person's brain receives from a cell phone.

If there wasn't a public outcry when most of this information was concealed in foreign alphabets, there should be one now, because it has become possible to confirm directly, in human beings, the degree to which cell phones interfere with glucose metabolism, and the outcomes of such studies are being published in English. Finnish researchers reported their alarming findings in the *Journal of Cerebral Blood Flow and Metabolism* in 2011. Using positron emission tomography (PET) to scan the brain, they found that glucose uptake is considerably reduced in the region of the brain next to a cell phone.[20]

Even more recently, researchers at Kaiser Permanente in Oakland, California, confirmed that electromagnetic fields cause obesity in children. They gave pregnant women meters to wear for 24 hours to measure their exposure to magnetic fields during an average day. The children of those women were more than six times as likely to be obese when they were teenagers if their mothers' average exposure during pregnancy had exceeded 2.5 milligauss. Of course, the children were exposed to the same high fields while growing up, so what the study really proved is that magnetic fields cause obesity in chidren.[21]

Vital Statistics
As with heart disease, rural mortality from diabetes in the 1930s corresponded closely with rates of rural electrification, and varied as much as tenfold between the least and the most electrified states. This is graphically illustrated in figures 3 and 4.

Figure 3 – Rate of Rural Diabetes in 1931

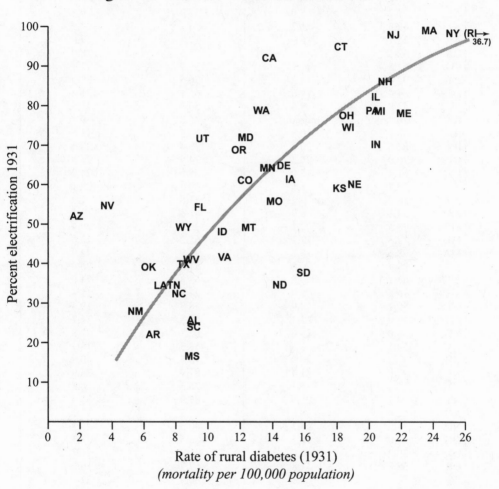

Rate of rural diabetes (1931)
(mortality per 100,000 population)

Table 3

	% electrification (1931)	Rural diabetes 1931 (deaths per 100,000)	% electrification (1940)	Rural diabetes 1940 (deaths per 100,000)
AL	25.7	8.9	34.7	9.8
AZ	62.5	1.7	56.1	4.9
AR	22.1	6.5	27.3	7.8
CA	92.5	13.7	75.6	18.0
CO	61.5	12.2	56.9	11.6
CT	94.9	18.2	90.5	29.0
DE	64.4	14.6	66.1	21.2
FL	53.8	9.4	50.7	11.5
GA	28.4	(missing)	36.5	9.8
ID	48.2	10.8	64.5	13.5
IL	82.5	20.3	79.4	28.4
IN	70.0	20.3	74.9	25.8
IA	61.4	15.0	65.5	24.7
KS	59.4	18.1	60.2	25.1
KY	38.0	(missing)	41.6	11.9
LA	34.1	6.9	41.5	12.1
ME	77.5	22.1	70.5	29.4
MD	72.3	12.2	65.2	23.6
MA	98.5	23.7	91.9	42.9
MI	78.4	20.6	81.3	26.4
MN	64.2	13.6	63.4	24.6
MS	16.5	8.9	22.7	11.3
MO	59.1	14.0	58.3	19.4
MT	48.9	12.4	56.8	16.7
NE	60.0	19.0	62.1	27.8
NV	54.8	3.6	58.3	17.9
NH	86.3	20.9	78.7	40.8
NJ	97.7	21.4	87.0	35.9
NM	27.3	5.3	26.5	4.8
NY	98.1	25.2	83.9	37.4
NC	32.4	8.2	43.7	12.1
ND	34.5	14.3	40.5	23.5
OH	77.0	18.5	82.5	27.3
OK	39.2	6.2	41.3	11.7
OR	68.8	11.8	67.7	16.3
PA	78.5	20.1	80.4	32.2
RI	98.2	36.7	91.0	48.4
SC	25.6	8.9	32.1	9.1
SD	41.0	15.8	43.0	21.4
TN	34.0	7.8	42.1	10.8
TX	39.5	8.4	43.5	10.6
UT	71.8	9.6	75.2	13.9
VT	71.9	(missing)	71.5	32.2
VA	41.7	10.9	53.1	16.6
WA	78.7	13.2	73.8	19.3
WV	41.0	8.8	53.4	12.4
WI	74.7	18.7	54.2	24.4
WY	49.5	8.3	50.8	16.5

Figure 4 – Rate of Rural Diabetes in 1940

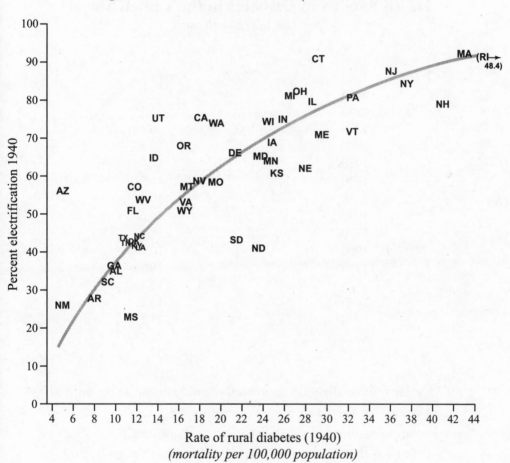

Rate of rural diabetes (1940)
(mortality per 100,000 population)

The overall history of diabetes in the United States is similar to that of heart disease.

Death Rate from Diabetes in the United States
(per 100,000 population)

1850	1.4
1860	1.7
1870	3.0
1880	3.4
1890	6.4
1900	10.6
1910	15.0
1920	16.2
1930	19.0
1940	26.6
1950	16.2
1960	16.7
1970	18.9
1980	15.4
1990	19.2
2000	25.2
2010	22.3
2017	25.7

Mortality from diabetes increased steadily from 1870 until the 1940s—this, despite the discovery of insulin in 1922.

The apparent drop in mortality in 1950 is not real, but is due to a reclassification that occurred in 1949. Previously, if a person had both diabetes and heart disease, the cause of death was reported as diabetes. Beginning in 1949, those deaths were reported as due to heart disease, diminishing the reported mortality from diabetes by about 40 percent. In the late 1950s, Orinase, Diabinase, and Phenformin were brought to market, the first of many oral medications that helped control the blood sugar of people with "insulin-resistant" diabetes for whom insulin was of limited use. These drugs have restrained, but not reduced the mortality from this disease. Meanwhile the number of diagnosed cases of diabetes in the United States has steadily increased:

Year	Cases per 1,000 population
1917	1.9[22]
1944	5.7
1958	9.3
1963	11.5
1968	16.2
1973	20.4
1978	23.7
1983	24.5
1988	25.6
1990	25.2
1992	29.3
1994	29.8
1996	28.9
1997	38.0
1998	39.0
1999	40.0
2000	44.0
2001	47.5
2002	48.4
2003	49.3
2004	52.9
2005	56.1
2006	59.0
2007	58.6
2008	62.9
2009	68.6
2010	69.6
2011	67.8
2012	69.6
2013	71.8
2014	70.1
2015	72.0

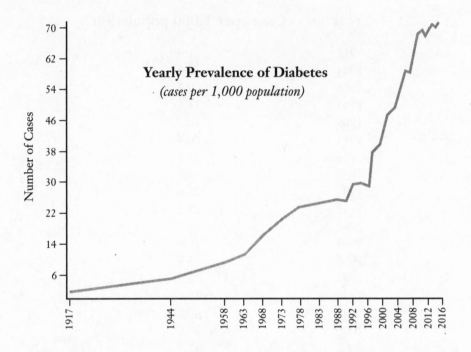

The real change over time may have been even greater because the definition of diabetes, in the United States and worldwide, was relaxed in 1980. A two-hour plasma glucose level of over 130 milligrams per deciliter was formerly taken as an indication of diabetes, but since 1980 diabetes is not diagnosed until the two-hour level exceeds 200 milligrams per deciliter. Levels between 140 and 200, which may not cause sugar in the urine, are now called "prediabetes."

A sudden spike in diabetes cases occurred nationwide in 1997—a 31 percent increase in a single year. No one was able to explain why. But that was the year the telecommunications industry introduced digital cell phones en masse to the United States. The first such phones went on sale in dozens of American cities during the Christmas season of 1996. Construction of cell towers began in those cities during 1996, but 1997 was the year that battalions of towers, previously confined to metropolises, marched out over the rural landscapes to occupy previously virgin territory. That was the year cell phones were transformed from a rich person's luxury to the common person's soon-to-be necessity—the year microwave radiation from towers and antennas became inescapable over large parts of the United States.

The situation today is out of control. The Centers for Disease Control estimates that in addition to the 21 million American adults over the age of twenty who have diagnosed diabetes, 8 million have undiagnosed diabetes, and 86 million have prediabetes. Adding these numbers together gives the shocking statistic that 115 million Americans, or more than half of all adults, have elevated levels of sugar in their blood.

Worldwide, it was estimated that more than 180,000,000 adults had diabetes in 2000. In 2014, the estimate was 387,000,000. In no country on earth is the rate of diabetes, or of obesity, decreasing.

Like diabetes, obesity has tracked exposure to electromagnetic fields. The first official statistics in the United States date from 1960, showing that one-quarter of adults were overweight. That number did not change for twenty years. The fourth survey, however, conducted during 1988–1991, revealed something alarming: fourteen million additional Americans had become fat.

Overweight in the United States
(percent of adults 20 through 74 years of age)

1960-1962	24.3
1971-1974	25.0
1976-1980	25.4
1988-1991	33.3

The authors, writing in the *Journal of the American Medical Association*, commented that studies in Hawaii and England had found similar rises in overweight during the 1980s across the board throughout the population in both sexes and at all ages. They speculated about "dietary knowledge, attitudes, and practices, physical activity levels, and perhaps social, demographic, and health behavior factors" that might have changed, although they did not point to a single piece of evidence that any of those things had changed.[23] In rebuttal, British physician Jeremiah Morris noted in a letter to the *British Medical Journal* that the average lifestyle had improved during this time, not worsened. More people in England were cycling, walking, swimming, and doing aerobics than ever before. Average daily food consumption, even after adjusting for meals eaten outside the home, had declined by 20 percent between 1970 and 1990.

However, in 1977, Apple had marketed its first personal computer, and during the 1980s the majority of people in both the United States and England, either at home or at work or both, were suddenly—and for the first time in history—exposed to high frequency electromagnetic fields continuously for hours everyday.

The problem became so huge that in 1991 the Centers for Disease Control began retroactively tracking not just overweight but obesity. For an American man or woman of average height this is defined as being more than about 30 pounds overweight.

Obesity in the United States[24]
(percent of adults over 20 years of age)

1960-1962	13.4
1971-1974	14.4
1976-1980	14.7
1988-1991	22.3
1999-2000	30.5
2009-2010	35.7
2015-2016	39.6

Grade III obesity, called "morbid obesity," has also been rising since 1980. This is defined as being more than about 100 pounds overweight.

Grade III Obesity in the United States
(percent of adults over 20 years of age)

1960-1962	0.8
1971-1974	1.3
1976-1980	1.3
1988-1991	2.8
1999-2000	4.7
2009-2010	6.3
2015-2016	7.7

More than two-thirds of all adults today—about 150 million Americans—are overweight. Eighty million are obese, as are twelve and a half million children, including one and a half million children aged two to five.[25] Twelve and a half million adults are more than 100 pounds overweight. The experts at the Centers for Disease Control have been able to do little more than shout that similar trends are being reported elsewhere—more than half a billion adults worldwide are obese—and to throw up their hands and say, "We do not know the causes of these increases in overweight and obesity."[26]

Obesity in Wild and Domestic Animals

If obesity is caused by an environmental factor, then it should be occurring in animals too. And it is so.

A few years ago David B. Allison, professor of biostatistics at the University of Alabama School of Public Health, was looking over data on small primates called marmosets from the Wisconsin Non-Human Primate Center, when he noticed that the average weight of the animals had increased remarkably over time. Mystified, he checked with the center, but could find no convincing reason for weight gain in this large population of animals living in a fixed laboratory environment on a controlled diet.

Intrigued, Allison searched online for previous studies of mammals that had lasted at least a decade and contained information about the animals' weight. He involved colleagues at primate centers, toxicology programs, pet food companies, and veterinary programs. The final paper, published in 2010 in the *Proceedings of the Royal Society B*, had eleven coauthors from Alabama, Florida, Puerto Rico, Maryland, Wisconsin, North Carolina, and California, and analyzed data on over 20,000 animals from twenty-four populations representing eight species, including laboratory animals, house pets, and feral rats, both rural and urban. In all twenty-four populations, the average weight of the animals rose over time. The odds of this happening by chance were less than ten billion to one.

Animal population	Average weight gain per decade
macaques, 1971 to 2006 (Wisconsin Primate Center)	5.3%
macaques, 1981 to 1993 (Oregon Primate Center)	9.6%
macaques, 1979 to 1992 (California Primate Center)	11.5%
chimpanzees, 1985 to 2005 (Yerkes Primate Center, Atlanta)	33.6%
vervets, 1990 to 2006 (UCLA Vervet Research Center)	8.8%
marmosets, 1991 to 2006 (Wisconsin Primate Center)	9.3%
laboratory mice, 1982 to 2005	3.4%
domestic dogs, 1989 to 2001	2.9%
domestic cats, 1989 to 2001	9.7%
feral rats, 1948 to 2006 (urban)	6.9%
feral rats, 1948 to 1986 (rural)	4.8%

Chimpanzees gained the most weight: they were twenty-nine times as likely to be obese in 2005 as they were in 1985. But even among country rats there was 15 percent more obesity every decade, consistently for four decades. The authors found similar studies with the same results elsewhere: 19 percent of light breed horses in Virginia were obese in 2006, versus 5 percent in 1998;[27] laboratory rats in France, under identical conditions, had increased in weight between 1979 and 1991.

Because wild and domestic animals were gaining so much weight, and had been doing so since at least the 1940s, Allison and his colleagues challenged the tired old wisdom that the rising tide of human fatness is due to lack of exercise and poor diet. They held up these animals as a warning to us all about an unknown global environmental factor. They titled their report "Canaries in the Coal Mine."[28]

13. Cancer and the Starvation of Life

> At the commencement of the twentieth cen-
> tury the great problem of the causation of
> tumours, like a giant sphinx, looms large on
> the medical horizon.
>
> W. Roger Williams,
> Fellow of the Royal College
> of Surgeons, England, 1908

On February 24, 2011, Italy's Supreme Court upheld the criminal conviction of Cardinal Roberto Tucci, former president of Vatican Radio's
management committee, for creating a public nuisance by polluting
the environment with radio waves. The Vatican's broadcasts to the
world, transmitted in forty languages, emanate from fifty-eight powerful radio towers occupying over one thousand acres of land, surrounded
by suburban communities. And for decades, residents in those communities had been screaming that the transmissions were destroying their
health as well as causing an epidemic of childhood leukemia. At the
request of the Public Prosecutor's office in Rome, which was considering bringing charges against the Vatican for negligent homicide, Judge
Zaira Secchi ordered an official investigation by the National Cancer
Institute of Milan. The results, released November 13, 2010, were
shocking. Between 1997 and 2003, children aged one to fourteen who
lived between six and twelve kilometers (3.7 to 7.5 miles) from Vatican

Radio's antenna farm developed leukemia, lymphoma, or myeloma at eight times the rate of children who lived further away. And adults who lived between six and twelve kilometers from the antennas died of leukemia at almost seven times the rate of adults who lived further away.

Photo by Angelo Franceschi

The third disease of civilization that Samuel Milham related to electrification is cancer. At first blush it is not obvious what the connection is. Impaired metabolism of sugars is surely connected to diabetes, and impaired metabolism of fats to heart disease. But how does cancer fit in? The key was provided by a scientist who studied sea urchin eggs in his laboratory over one hundred years ago. He was a native of the same city where, a century later, three thousand doctors were to sign an appeal to the world stating, among other things, that radio waves cause leukemia.

On October 8, 1883, a son was born to Emil Warburg, a prominent Jewish physicist in Freiburg, Germany. When he was thirteen, the family moved to Berlin, where visitors to his parents' home included some of the giants of the natural sciences—chemist Emil Fischer, physical chemist Walter Nernst, physiologist Theodor Wilhelm Engelmann. Later, when Albert Einstein moved to Berlin, the great scientist used to come over to play chamber music with his father—Einstein on violin and Emil Warburg on piano. No one was surprised when young Otto, growing up in such an atmosphere, enrolled in the University of Freiburg to study chemistry.

But by the time he received his Ph.D. in 1906, a growing disease epidemic had caught the attention of this ambitious young man. His was the first generation seriously to be affected by it. Cancer rates all over Europe had doubled since he was born, and he determined to devote his life to finding the reason and, hopefully, a cure. With this in mind he returned to school, receiving his M.D. from the University of Heidelberg in 1911.

What fundamental changes, he wondered, take place in the tissue when a normal cell becomes cancerous? "Does the metabolism of tumours," he asked, "growing in a disorganized manner, differ from the metabolism of orderly cells growing at the same rate?"[1] Impressed that both tumors and early embryos consist of undifferentiated cells that multiply rapidly, Otto Warburg began his life's work by studying fertilized eggs. Perhaps, he speculated, cancer cells are just normal cells

Otto Heinrich Warburg, M.D., Ph.D.
(1883-1970)

that have reverted to an embryonic pattern of growth. He chose the sea urchin egg to study because its embryo is large and grows particularly fast. His first major work, published while he was still in medical school, showed that on fertilization the rate of oxygen consumption of an egg rises sixfold.[2]

But in 1908, he could pursue his ambition no further because the chemical reactions within cells that involve oxygen were completely unknown. Spectrophotometry—the identification of chemicals from the frequencies of light they absorb—was new, and had not yet been applied to living systems. Existing techniques for culturing cells and measuring gas exchange were primitive. Warburg realized that before any real progress could be made in elucidating the metabolism of cancer, fundamental research on the metabolism of normal cells would have to be done. Cancer research would have to wait.

During the coming years—with a break during which he served in the World War—Warburg, using techniques that he developed, proved that respiration in a cell took place in tiny structures that he called "grana" and that we now call mitochondria. He experimented with the effects of alcohols, cyanide, and other chemicals on respiration and concluded that the enzymes in the "grana" must contain a heavy metal that he suspected, and later proved, was iron. He conducted landmark experiments using spectrophotometry that proved that the portion of the enzyme that reacts with oxygen in a cell is identical with the portion of hemoglobin that binds oxygen in the blood. That chemical, called heme, is a porphyrin bonded to iron, and the enzyme containing it, which exists in every cell and makes breathing possible, is known today as cytochrome oxidase. For this work Warburg was awarded the Nobel Prize in Physiology or Medicine in 1931.

Meanwhile, in 1923, Warburg resumed his research on cancer, picking up where he had left off fifteen years earlier. "The starting point," he wrote, "has been the fact that the respiration of sea urchin eggs increases six-fold at the moment of fertilization," i.e. at the moment that it changes from a state of rest to a state of growth. He expected to find a similar increase of respiration in cancer cells. But to his amazement, he found just the opposite. The rat tumor he was working with used considerably *less* oxygen than normal tissues from healthy rats.

"This result seemed so startling," he wrote, "that the assumption seemed justified that the tumor lacked suitable material for combustion." So Warburg added various nutrients to the culture medium, still expecting to see a dramatic rise in oxygen use. Instead, when he added glucose, the tumor's respiration ceased completely! And in trying to discover why this happened, he found that tremendous amounts of lactic acid were accumulating in the culture medium. The tumor, in fact, was producing fully twelve percent of its weight in lactic acid per hour. Per unit time, it was producing 124 times as much lactic acid as blood, 200 times as much as a frog's muscle at rest, and eight times as much as a frog's muscle working to the limit of its capacity. The tumor was consuming the glucose, all right, but it was not making use of oxygen to do it.[3]

In additional experiments on other types of cancers in animals and humans, Warburg found that this was generally true of all cancer cells, and of no normal cells. This singular fact impressed Warburg as of utmost importance and the key to the causation of this disease.

The extraction of energy from glucose without using oxygen, a type of metabolism called anaerobic glycolysis—also called fermentation— is a highly inefficient process that takes place to a small extent in most living cells but only becomes important when not enough oxygen is available. For example, runners, during a sprint, push their muscles to use energy faster than their lungs can deliver oxygen to them. Their muscles temporarily produce energy anaerobically (without oxygen), incurring an oxygen debt that is repaid when they end their sprint and stop to gulp air. Although capable of supplying energy rapidly in an emergency, anaerobic glycolysis produces much less energy for the same amount of glucose, and deposits lactic acid in the tissues that has to be disposed of.

Fermentation is a very old form of metabolism from which all forms of life obtained their energy for billions of years, before green plants appeared on earth and filled the atmosphere with oxygen. Some primitive forms of life today—many bacteria and yeasts, for example— still rely on it, but all complex organisms have abandoned that way of making a living.

What Warburg discovered in 1923 is that cancer cells differ from normal cells in all higher organisms in this fundamental respect: they maintain high rates of anaerobic glycolysis and produce large amounts of lactic acid even in the presence of oxygen. This discovery, called the Warburg effect, is the basis for the diagnosis and staging of cancer today, using positron emission tomography, or PET scanning. Because anaerobic glycolysis is inefficient and consumes glucose at a tremendous rate, PET scans can easily find tumors in the body by the more rapid uptake of radioactive glucose. And the more malignant the tumor, the more rapidly it takes up glucose.

Warburg reasonably believed he had discovered the cause of cancer. Evidently, in cancer, the respiratory mechanism has been damaged and has lost control over the metabolism of the cell. Unrestrained glycolysis—and unrestrained growth—are the result. In the absence of normal metabolic control the cell reverts to a more primitive state. All complex organisms, proposed Warburg, must have oxygen in order to maintain their highly differentiated forms. Without oxygen, they will revert to a more undifferentiated, simple form of growth, such as existed exclusively on this planet before there was oxygen in the air. "The causative factor in the origin of tumors," proposed Warburg, "is nothing other than oxygen deficiency."[4] When cells are deprived of oxygen only temporarily, glycolysis takes over during the emergency, but ceases again when oxygen is once more available. But when cells are repeatedly or chronically deprived of oxygen, he said, respiratory control is eventually damaged and glycolysis becomes independent. "If respiration of a growing cell is disturbed," wrote Warburg in 1930, "as a rule the cell dies. If it does not die, a tumour cell results."[5]

Warburg's hypothesis was first brought to my attention in the mid-1990s by Dr. John Holt, a colorful figure in Australia who was treating cancer with microwave radiation, and who warned his colleagues that the same radiation could convert normal cells into cancerous ones. I didn't fully understand the connection of Warburg's work on cancer to my work on electricity, so I filed away the research papers Holt sent me for future reference. Today, with so many more pieces of the puzzle in place, the connection is obvious. Electricity, like rain on a campfire, dampens the flames of combustion in living cells. If Warburg was correct, and chronic lack of oxygen causes cancer, then one need look no further than electrification for the origin of the modern pandemic.

Warburg's theory was controversial from the beginning. Hundreds of different kinds of cancers were known in the 1920s, triggered by thousands of kinds of chemical and physical agents. Many scientists were reluctant to believe in a common cause that was so simple. Warburg answered them with a simple explanation: each of those thousands of chemicals and agents, in its own way, starves cells of oxygen. Arsenic, he explained by way of example, is a respiratory poison that causes cancer. Urethane is a narcotic that inhibits respiration and causes cancer. When you implant a foreign object under the skin, it causes cancer because it blocks blood circulation, starving neighboring tissues of oxygen.[6]

Although they didn't necessarily accept Warburg's theory of causation, other researchers lost little time confirming the Warburg effect. Tumors did, universally, have the ability to grow without oxygen. By 1942, Dean Burk at the National Cancer Institute was able to report that this was true of over 95 percent of the cancerous tissues he had examined.

Then, in the early 1950s, Harry Goldblatt and Gladys Cameron, at the Institute for Medical Research at Cedars of Lebanon Hospital in Los Angeles, reported to a skeptical public that they had succeeded in transforming normal cells—cultured fibroblasts from the heart of a five-day-old rat—into cancer cells merely by repeatedly depriving them of oxygen.

In 1959, Paul Goldhaber gave further support to Warburg's hypothesis when he discovered that some types of Millipore diffusion chambers, but not others, when implanted under the skin of mice, caused large tumors to grow around them. Diffusion chambers were used to sample tissue fluid in many kinds of animal experiments. Their ability to cause cancer turned out to depend not on the type of plastic they were made of, but on the size of the pores that allowed fluid to flow through them. Only one animal out of 39 developed a tumor when the pores were 450 millimicrons in diameter. But 9 out of 34 developed tumors when the pore size was 100 millimicrons, and 16 out of 35—close to half—developed tumors when the pore size was only 50 millimicrons. The interference with free fluid circulation when the pore size was too small apparently deprived the tissues next to the chamber walls of oxygen.

In 1967, Burk's team proved that the more malignant a tumor is, the higher its rate of glycolysis, the more glucose it consumes, and the more lactic acid it produces. "The extreme forms of rapidly growing ascites cancer cells," Burk reported, "can produce lactic acid from glucose anaerobically at a sustained rate probably faster than any other living mammalian tissue—up to half the tissue dry weight per hour. Even a hummingbird, whose wings may beat up to at least one hundred times a second, consumes at best only half its dry weight of glucose equivalent per day."

Because he insisted that the origin of cancer was known, Warburg thought that "one could prevent about 80 percent of all cancers if one could keep out the known carcinogens."[7] He therefore advocated, in 1954, for restrictions on cigarette smoking, pesticides, food additives, and air pollution by car exhaust.[8] His incorporation of these attitudes into his personal life earned him a reputation as an eccentric. Long before environmentalism was popular, Warburg had a one-acre organic garden, obtained milk from an organically maintained herd, and purchased French butter because in France the use of herbicides and pesticides was more strictly controlled than in Germany.

Otto Warburg passed away in 1970 at the age of 83—the same year the first oncogene was discovered. An oncogene is an abnormal gene, thought to be caused by mutation, that is associated with the development of cancer. The discovery of oncogenes and tumor suppressor genes promoted a widespread belief that cancer was caused by genetic mutations and not by altered metabolism. Warburg's hypothesis, controversial from the start, was largely abandoned for three decades.

But the widespread use of PET scanning for diagnosing and staging human cancers has catapulted the Warburg effect back onto the main stage of cancer research. No one can now deny that cancers live in anaerobic environments, and that they rely on anaerobic metabolism in order to grow. Even molecular biologists, who once focused exclusively on the oncogene theory, are discovering, after all, that there is a connection between lack of oxygen and cancer. A protein has been discovered that exists in all cells—hypoxia-inducible factor (HIF)—that is activated under conditions of low oxygen, and that in turn activates many of the genes necessary for cancer growth. HIF activity has been found to be elevated in colon, breast, gastric, lung, skin, esophageal, uterine, ovarian, pancreatic, prostate, renal, stomach, and brain cancers.[9]

Cellular changes that indicate damaged respiration—including reductions in the number and size of mitochondria, abnormal structure of mitochondria, lessened activity of Krebs cycle enzymes, lessened activity of the electron transport chain, and mutations of mitochondrial genes—are being routinely found in most types of cancer. Even in tumors caused by viruses, one of the first signs of malignancy is an increase in the rate of anaerobic metabolism.

Experimentally inhibiting the respiration of cancer cells, or simply depriving them of oxygen, has been shown to alter the expression of hundreds of genes that are involved in malignant transformation and cancer growth. Damaging respiration makes cancer cells more invasive; restoring normal respiration makes them less invasive.[10]

A consensus is forming among cancer researchers: tumors can only develop if cellular respiration is diminished.[11] In 2009, a book dedicated to Otto Warburg was published titled "Cellular Respiration and Carcinogenesis." Addressing all aspects of this question, it contains contributions from leading cancer researchers from the United States, Germany, France, Italy, Brazil, Japan, and Poland.[12] In the foreword, Gregg Semenza wrote: "Warburg invented a device, now known as the Warburg manometer, with which he demonstrated that tumor cells consume less oxygen (and produce more lactate) than do normal cells under the same ambient oxygen concentrations. A century later, the struggle to understand how and why metastatic cancer cells manifest the Warburg effect is still ongoing, and 12 rounds of this heavyweight fight await the reader beyond this brief introduction."

The question being asked today by cancer researchers is no longer, "Is the Warburg effect real?" but "Is hypoxia a cause, or an effect, of cancer?"[13] But, as increasingly many scientists are admitting, it really doesn't matter, and may be only a question of semantics. Since cancer cells thrive in the absence of oxygen, oxygen deprivation gives incipient cancer cells a survival advantage.[14] And any environmental factor that damages respiration therefore—whether Warburg was right and it directly causes malignant transformation or whether the skeptics are right and it merely provides an environment in which cancer has an advantage over normal cells—will necessarily increase the cancer rate.

Electricity, as we have seen, is such a factor.

Diabetes and Cancer

If the same cause—a slowing of metabolism by the electromagnetic fields around us—produces both diabetes and cancer, then one might expect diabetics to have a high rate of cancer, and vice versa. And it is so.

The first person to confirm a connection between the two diseases was South African physician George Darell Maynard in 1910. Unlike almost all other diseases, rates of both cancer and diabetes were steadily rising. Thinking that they might have a common cause, he analyzed mortality statistics from the 15 death registration states in the 1900 Census of the United States. And he found, after correcting for population and age, that the two diseases were strongly related. States that had higher incidences of one also had higher incidences of the other. He proposed that electricity might be that common cause:

"Only one cause, it seems to me, will fit the facts as we know them, viz.: the pressure of modern civilisation and the strain of modern competition, or some factor closely associated with these. Radio-activity and various electric phenomenon have from time to time been accused of producing cancer. The increased use of high tension currents is an undoubted fact in modern city life."

A century later, it is an accepted fact that diabetes and cancer occur together. More than 160 epidemiological studies have investigated this question worldwide, and the majority have confirmed a link between the two diseases. Diabetics are more likely than non-diabetics to develop, and to die from, cancers of the liver, pancreas, kidney, endometrium, colon, rectum, bladder, and breast, as well as non-Hodgkin's lymphoma.[15] In December 2009, the American Diabetes Association and American Cancer Society convened a joint conference. The consensus report that resulted concurred: "Cancer and diabetes are diagnosed within the same individual more frequently than would be expected by chance."[16]

Cancer in Animals

We recall from chapter 11 that complete autopsy records of the Philadelphia zoo, kept since 1901, showed an increase in heart disease that accelerated during the 1930s and 1940s, and that affected all species of animals and birds at the zoo. An equivalent increase occurred in rates of cancer. The 1959 report from the Penrose Research Laboratory at the zoo[17] divided the autopsies into two time periods: 1901-1934 and 1935-1955. The rate of malignant tumors among nine families of mammals increased between two- and twenty-fold from the earlier to the later time period. The rate of benign tumors increased even more. Only 3.6 percent of felines, for example, had benign or malignant tumors at autopsy during the earlier period, compared to 18.1 percent during the later period; 7.8 percent of ursines (bears) had tumors during the earlier period, compared to 47 percent during the later period.

The autopsy records of 7,286 birds at the zoo, encompassing four different orders, showed that malignant tumors increased two-and-a-half-fold, and benign tumors eightfold.

Vital Statistics

The real story, again, is revealed by the historical records.

The increase in cancer began slightly before heart disease and diabetes began to rise. Early records from England show that cancer deaths were rising as early as 1850:[18]

Year	Cancer deaths, England (per 100,000 population)
1840	17.7
1850	27.9
1855	31.9
1860	34.3
1865	37.2
1870	42.4
1875	47.1
1880	50.2
1885	57.2
1890	67.6
1895	75.5
1900	82.8
1905	88.5

Cooke and Wheatstone's first telegraph line, running from London to West Drayton, opened for business on July 9, 1839. By 1850, over two thousand miles of wire ran the length and breadth of England. While we don't have earlier statistics from England to prove that cancer rates first began rising between 1840 and 1850, or comparable data from any other national government, we do have them for the parish of Fellingsbro, a small well-to-do rural district 90 miles west of Stockholm, Sweden. We have them because in 1902, Swedish physician Adolf Ekblom, in an effort to discover whether cancer rates had really risen during the previous century, consulted the "death and burial book" kept by the clergy of Fellingsbro parish. These are the numbers that he compiled from that book:

Years	Average yearly cancer mortality (Fellingsbro, per 100,000 population)
1801–1810	2.1
1811–1820	6.5
1821–1830	8.1
1831–1840	3.5
1841–1850	6.6
1851–1860	14.0
***	***
1885–1894	72.5
1895–1900	141.0

The records were incomplete from 1863 to 1884. But the records that survive tell the story that we seek.

The population of Fellingsbro was 4,608 at the beginning of the nineteenth century, and 7,104 at the end of it. One person died of cancer about every three years between 1801 and 1850. Then, in 1853, the first telegraph wire in Sweden was strung between Stockholm, the capital, and Uppsala, a city 37 miles north. The following year a line was run southwestward from Uppsala, via Västerås, to Örebro. This line ran right through the middle of Fellingsbro parish. At that time the cancer rate in Fellingsbro began to rise.[19] By the turn of the twentieth century, the country folk in Fellingsbro were dying of cancer faster than the average residents of London.

In 1900, annual cancer deaths around the world, per 100,000 population, were:

Switzerland	127
Holland	92
Norway	91
England and Wales	83
Scotland	79
Bermuda	75
Germany	72
Austria	71
France	65
USA	64
Australia	63
Ireland	61
New Zealand	56
Belgium	56
Italy	52
Uruguay	50
Japan	46
Spain	39
Hungary	33
Cuba	29
Chile	27
British Guiana	24
Portugal	22
Windward and Leeward Islands	22
Costa Rica	20
British Honduras	19
Jamaica	16
St. Kitts	13
Trinidad	12
Mauritius	12
Serbia	9
Ceylon	5.5
Hong Kong	4.5
Brazil	4.5
Guatemala	4
La Paz, Bolivia	3.4
Bahamas	1.8
Fiji	1.7
New Guinea, Borneo, Java, Sumatra, Philippines, most of Africa, Macao	non-existent

Every historical source shows that cancer always accompanied electricity. In 1914, among about 63,000 American Indians living on reservations, none of which had electricity, there were only two deaths from cancer. The cancer mortality in the United States as a whole was 25 times as high.[20]

An unusual one-year rise in cancer mortality of from 3 to 10 percent occurred in every modernizing country in 1920 or 1921. This corresponded to the beginning of commercial AM radio broadcasting. In 1920, cancer deaths rose 8 percent in Norway, 7 percent in South Africa and France, 5 percent in Sweden, 4 percent in the Netherlands, and 3 percent in the United States. In 1921, cancer deaths rose 10 percent in Portugal, 5 percent in England, Germany, Belgium, and Uruguay, and 4 percent in Australia.

Lung cancer, breast cancer, and prostate cancer rates rose spectacularly throughout the first half of the twentieth century in every country for which we have good data. The number of deaths from breast cancer quintupled in Norway, sextupled in the Netherlands, and increased sixteen-fold in the United States. Lung cancer deaths increase twenty-fold in England. Prostate cancer deaths increased eleven-fold in Switzerland, twelve-fold in Australia, and thirteen-fold in England.

Lung cancer was once so uncommon that it was not even listed separately in most countries until 1929. In the few countries that tracked it, it did not start its dramatic rise until about 1920. Benjamin Ward Richardson, in his 1876 book, *Diseases of Modern Life*, is surprising to a modern reader in this respect. His chapter on "Cancer from Smoking" discusses the controversy over whether tobacco smoking caused cancer of the lip, tongue, or throat, but cancer of the lung is not even mentioned. Lung cancer was still rare in 1913, the year when the American Society for the Control of Cancer was founded. Out of 2,641 cases of cancer reported to the New York State Institute for the Study of Malignant Disease that year, there was only a single case of primary lung cancer. Frederick Hoffman, in his exhaustive 1915 book, *The Mortality From Cancer Throughout the World*, asserted as a proven fact that smoking caused cancer of the lips, mouth, and throat, but like

Richardson four decades previously made no mention of lung cancer in connection with smoking.[21]

Swedish researchers Örjan Hallberg and Olle Johansson have shown that the rates of lung, breast, and prostate cancer continued to rise, just as spectacularly, in the second half of the twentieth century in forty countries, along with malignant melanoma and cancers of the bladder and colon—and that the overall rate of cancer changed precisely with changes in the exposure of the population to radio waves. The rate of increase in cancer deaths in Sweden accelerated in 1920, 1955, and 1969, and took a downturn in 1978. "In 1920 we got AM radio, in 1955 we got FM radio and TV1, in 1969-70 we got TV2 and colour TV and in 1978 several of the old AM broadcasting transmitters were disrupted," they note in their article, "Cancer Trends During the 20th Century." Their data suggest that at least as many cases of lung cancer can be attributed to radio waves as to smoking.

The same authors have focused on FM radio exposure in connection with malignant melanoma, following up on the findings of Helen Dolk at the London School of Hygiene and Tropical Medicine. In 1995, Dolk and her colleagues had shown that the incidence of skin melanoma declined with distance from the powerful television and FM radio transmitters at Sutton Coldfield in The West Midlands, England. Noting that the FM frequency range, 85 to 108 MHz, is close to the resonant frequency of the human body, Hallberg and Johansson decided to compare melanoma incidence with exposure to FM radio waves for all 565 Swedish counties. The results are startling. When melanoma incidence is plotted on a graph against the average number of FM transmitters to which a municipality is exposed, the points fall on a straight line. Counties that get reception from 4.5 FM stations have a rate of malignant melanoma that is eleven times as high as counties that do not get reception from any FM station.

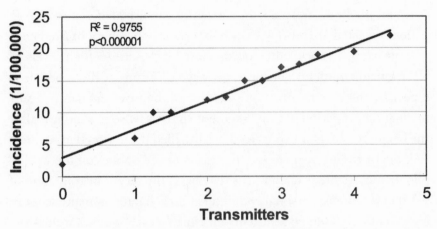

Figure 4, Hallberg & Johansson 2005

In their article, "Malignant Melanoma of the Skin—Not a Sunshine Story," they refute the notion that the tremendous increase in this disease since 1955 is caused primarily by the sun. No increase in ultraviolet radiation due to ozone depletion occurred as early as 1955. Nor, until the 1960s, did Swedes begin to travel to more southerly countries in large numbers to soak up the sun. The embarrassing truth is that rates of melanoma on the head and feet hardly rose at all between 1955 and 2008, while rates for sun-protected areas in the middle of the body increased by a factor of twenty. Most moles and melanomas are now occurring not on the head, arms, and feet, but in areas of the body that are not exposed to sunshine.

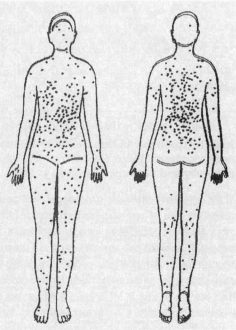

Figure 15, Hallberg & Johansson 2002a

Elihu Richter, in Israel, has recently published a report on 47 patients, treated at Hebrew University-Hadassah School of Medicine, who developed cancer after occupational exposure to high levels of electromagnetic fields and/or radio waves.[22] Many of these people—especially the youngest people—developed their cancers within a surprisingly short period of time—some as short as five or six months after the beginning of their exposure. This dispelled the notion that we must wait ten or twenty years to see the effects of cell phones on the world's population. Richter's team warns that "with the recent introduction of WiFi into schools, personal computers for each pupil in many schools, high frequency voltage transients measured in schools—as well as the population-wide use of cellphones, cordless phones, some exposure to cellphone towers, residential exposure to RF/MW from Smart Meters and other 'smart' electronic equipment at the home and possibly also ELF exposures to high power generators and transformers—young people are no longer free from exposure to EMF."

The range of tumors in Richter's clinic ran the gamut: leukemias, lymphomas, and cancers of the brain, nasopharynx, rectum, colon, testis, bone, parotid gland, breast, skin, vertebral column, lung, liver, kidney, pituitary gland, pineal gland, prostate, and cheek muscle.

United States[23]

Year	Cancer deaths (per 100,000 population)
1850	10.3
1860	14.7
1870	22.5
1880	31.0
1890	46.9
1900	60.0
1910	76.2
1920	83.4
1930	98.9
1940	120.3
1950	139.8
1960	149.2
1970	162.8
1980	183.9
1990	203.2
2000	200.9
2010	185.9
2017	183.9

Figures 5 and 6 show the same linear correspondence between cancer and electrification in the forty-eight United States in 1931 and 1940 that have already been shown for heart disease and diabetes.

Figure 5 – Rate of Rural Cancer in 1931

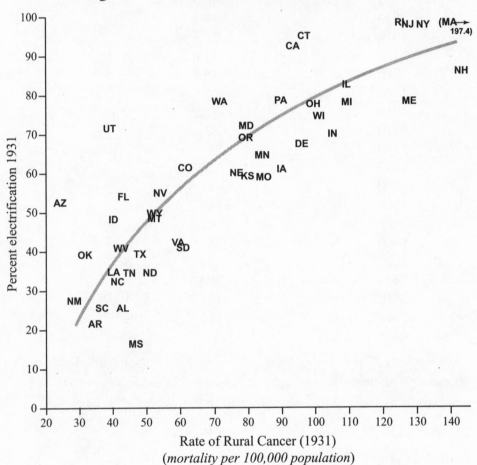

Rate of Rural Cancer (1931)
(*mortality per 100,000 population*)

Table 4

	% electrification (1931)	Rural cancer 1931 *(deaths per 100,000)*	% electrification (1940)	Rural cancer 1940 *(deaths per 100,000)*
AL	25.7	42.5	34.7	55
AZ	62.5	23.4	56.1	43
AR	22.1	34.5	27.3	51
CA	92.5	93.0	75.6	110
CO	61.5	61.5	56.9	80
CT	94.9	96.6	90.5	137
DE	64.4	95.4	66.1	98
FL	53.8	39.6	50.7	68
GA	28.4	(missing)	36.5	47
ID	48.2	39.9	64.5	67
IL	82.5	108.3	79.4	128
IN	70.0	104.3	74.9	121
IA	61.4	89.5	65.5	119
KS	59.4	79.4	60.2	107
KY	38.0	(missing)	41.6	67
LA	34.1	39.2	41.5	61
ME	77.5	127.0	70.5	153
MD	72.3	78.9	65.2	112
MA	98.5	197.4	91.9	177
MI	78.4	108.6	81.3	128
MN	64.2	85.0	63.4	117
MS	16.5	46.6	22.7	61
MO	59.1	83.8	58.3	105
MT	48.9	51.5	56.8	95
NE	60.0	76.5	62.1	110
NV	54.8	63.6	58.3	116
NH	86.3	143.1	78.7	181
NJ	97.7	126.8	87.0	123
NM	27.3	27.7	26.5	43
NY	98.1	131.9	83.9	156
NC	32.4	41.1	43.7	52
ND	34.5	51.4	40.5	91
OH	77.0	98.6	82.5	126
OK	39.2	31.4	41.3	66
OR	68.8	78.3	67.7	85
PA	78.5	88.9	80.4	117
RI	98.2	124.5	91.0	163
SC	25.6	36.6	32.1	46
SD	41.0	60.7	43.0	101
TN	34.0	44.8	42.1	64
TX	39.5	48.1	43.5	62
UT	71.8	37.8	75.2	78
VT	71.9	(missing)	71.5	146
VA	41.7	59.0	53.1	72
WA	78.7	71.3	73.8	110
WV	41.0	41.8	53.4	64
WI	74.7	101.2	54.2	122
WY	49.5	51.7	50.8	66

Figure 6 – Rate of Rural Cancer in 1940

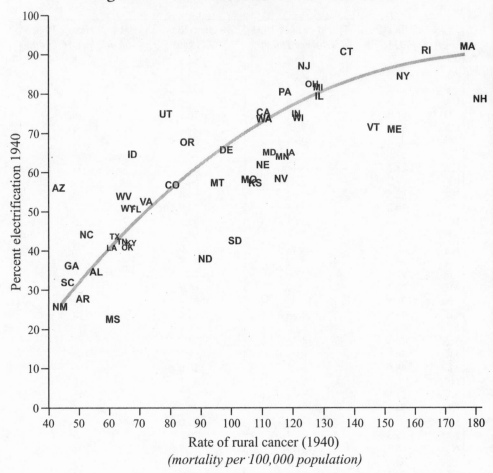

Rate of rural cancer (1940)
(mortality per 100,000 population)

You may notice that the position of Nevada shifted more than any other state between 1931 and 1940. For some reason, deaths from heart disease, diabetes, and cancer rose dramatically in Nevada while the rate of household electrification rose only modestly. I propose that the construction of Hoover Dam, completed in 1936, was that reason. The most powerful hydroelectric plant in the world at that time, its one billion watt capacity supplied Las Vegas, Los Angeles, and most of Southern California via high voltage power lines that coursed through southeastern Nevada on their way to their destinations, exposing the surrounding area—where most of the population of the state lived—to some of the world's highest levels of electromagnetic fields. In June of 1939 the Los Angeles grid was connected to Hoover Dam via a 287,000-volt transmission line, also the most powerful in the world at that time.[24]

Power lines from Hoover Dam carry electricity to the Los Angeles area. This photo by Charles O'Rear is part of the National Archives digital collection.

Two types of cancer deserve additional comment: lung cancer and brain cancer.

As the following graph shows, the percentage of adults who smoke has declined steadily since 1970 among both men and women. Yet lung cancer mortality has almost quadrupled in women, and is virtually the same in men as it was fifty years ago.[25]

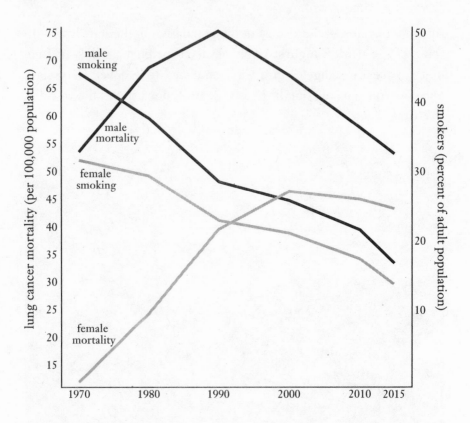

When non-smoker Dana Reeve, the 46-year-old widow of "Superman" actor Christopher Reeve, died of lung cancer in 2006, the public was stunned because we had had it drummed into us for decades that this type of cancer is caused by smoking. Yet lung cancer in people who have never smoked—if you consider it as a separate category—ranks today as the seventh most common cause of cancer deaths worldwide, even before cancer of the cervix, pancreas, and prostate.[26]

Brain tumors deserve mention because, obviously, of cell phones. Several billion people in the world are exposing their brains for up to hours per day to microwave radiation at point blank range—a new situation that began in approximately 1996 or 1997 in most countries. Yet honest data on brain tumors are difficult to obtain because special interests have controlled most of the research funding on brain tumors since the advent of digital cell phones two decades ago. As a result, a media war has pitted the independent scientists, who report a tripling to quintupling of brain cancer rates among those who have used their cell phones for ten years or more, against industry scientists who report no increase in cancer at all.

The problem, as Australian neurosurgeon Charlie Teo tells those who will listen, is that all the data on cell phone usage comes from databanks controlled by cell phone providers, and "no telcos have allowed scientists access to their records for these large studies."

I found out firsthand how closely not only the telecom providers, but the scientists they fund, guard their data, when I requested access to some of it in 2006. Yet another industry-funded study was published, this time in Denmark, purporting to show not only that cell phones did not cause brain cancer, but that cell phone users even had a *lower* rate of brain cancer than everyone else. In other words, those scientists would have the world believe that people might actually protect themselves from brain tumors by holding a cell phone to their heads for hours per day. The study, published in the *Journal of the National Cancer Institute*, was titled "Cellular Telephone Use and Cancer Risks: Update of a Nationwide Danish Cohort."[27] It claimed to come to its conclusions after an examination of the medical records of over 420,000 Danish cell phone users and non-users over a period of two decades. It was clear to me that something was wrong with the statistics.

Although the study found a *lower* rate of brain cancer—in men only—among cell phone users than non-users, it found a *higher* rate of exactly those cancers that Swedish scientists Hallberg and Johansson had reported to be caused by radio waves: bladder cancer, breast cancer, lung cancer, and prostate cancer. The Danish study did not report rates of colon cancer or melanoma, the other two types of cancer that the Swedish researchers had mentioned. However, the Danish study did additionally find that testicular cancer in men was higher and that cervical and kidney cancers in women were significantly higher among the cell phone users. I sensed manipulation of the data, because the only type of cancer for which a "protective" effect was reported was the type of cancer these scientists and their funders were trying to convince the public that cell phones did not cause: brain cancer.

It occurred to me that *all* of the study's subjects had actually been using cell phones for a long time by the year 2004, when the study ended. The only difference between "users" and "non-users" was the date of first subscription: the "users" first bought a cell phone between 1982 and 1995, while the "non-users" didn't buy one until after 1995. And all the "users" were lumped together. The study did not distinguish between people who had used cell phones for 9 years and people who had used them for 22 years. But according to the study, those who subscribed prior to 1994 tended to be wealthier, and drank and smoked much less, than those who first subscribed later. I suspected that controlling for length of use might change the results of the study. So I did the natural, normal, accepted thing that scientists do when they wish to validate a study that is published in a peer-review journal: I requested to look at their data. On December 18, 2006, I sent an email to the lead author, Joachim Schüz, telling him that I had colleagues in Denmark who would like to look at their data. And on January 19, 2007, we were cordially refused permission. The letter of refusal was signed by three of the study's six authors: Schüz, Christoffer Johansen and Jørgen H. Olsen.

Meanwhile, Teo is sounding the alarm. "I see 10 to 20 new patients each week," he says, "and at least one third of those patients' tumors are in the area of the brain around the ear. As a neurosurgeon I cannot ignore this fact."

Many if not most of us have one or more acquaintances or family members who have, or have died from, a brain tumor. My friend Noel Kaufmann, who died in 2012 at the age of 46, never used a cell phone, but he did use a home cordless phone for years, which emits the same type of radiation, and the tumor that killed him was in the part of his brain beneath the ear against which he held that phone. All of us have heard about famous people who have died of brain tumors—Senator Ted Kennedy, attorney Johnnie Cochran, journalist Robert Novak,

Vice President Joe Biden's son Beau. I have in my files, sent to me by the director of the California Brain Tumor Association, a list of over three hundred celebrities who either have a brain tumor or have died from one during the past decade and a half. When I was younger I never heard of any celebrity who had brain cancer.

Yet highly publicized studies assure us that brain tumor rates are not increasing. This is certainly not true, and a little investigation shows why the data cannot be trusted, in the United States or anywhere else. In 2007, researchers at the Swedish National Board of Health and Welfare found out that, for some reason, one-third of the cases of brain cancer diagnosed at university hospitals, and the majority of cases at county hospitals, were not being reported to the Swedish Cancer Registry.[28] All other types of cancer were being routinely reported, but not brain cancer.

A 1994 study revealed that difficulties in brain cancer reporting were already occurring in Finland. Although the Finnish cancer registry was complete for most types of cancer, it seriously underreported brain tumors.[29]

Here in the United States, severe problems have been found with surveillance not just of brain cancer, but across the board. The Surveillance Epidemiology and End Results (SEER) program, run by the National Cancer Institute, depends on state registries to deliver accurate data. But the data are not accurate. American researcher David Harris reported at a conference in Berlin in 2008 that state registries cannot keep up with the increasing load of cancer cases because they are not receiving enough funding to do so. "SEER registries are currently faced with the challenge of collecting more cases in less time with often the same limited resources as the previous year," he said. This means that the greater the rise in cancer, the less it will be reported, barring an improvement in the American economy.

Even worse has been the deliberate refusal by Veterans Administration hospitals and military base medical facilities to report cases to the state cancer registries. A report by Bryant Furlow that appeared in *The Lancet Oncology* in 2007 noted "a precipitous decline in VA reporting of new cases to California cancer registries beginning in late 2004—from 3,000 cases in 2003 to almost none by the end of 2005." After inquiring in other states, Furlow discovered that California was not an exception. The Florida cancer registry had *never* received any VA case reports, and VA facilities in other states were dealing with years of backlogged, unreported cancer cases. "We've been working with the VA for more than 5 years, but it's just got worse," Holly Howe told him. She represents the North American Association of Central Cancer Registries. As many as 70,000 cases of cancer from the VA were not being reported each year. And in 2007, the VA made non-reporting official policy when it issued a directive on cancer nullifying all existing agreements between state registries and VA facilities. Furlow reported that the Department of Defense was also not cooperating with the cancer registries. No cancers diagnosed at military base facilities had been reported to any state registries for several years. As a result of all these failures, Dennis Deapon of the Los Angeles Cancer Surveillance Program warned that studies based on the deficient data may be worthless. "Research from the mid-2000's will forever require an asterisk, or perhaps a sticker on the cover, to remind researchers and the public that they are not correct," he said.

Doctors at the Southern Alberta Cancer Research Institute at the University of Calgary were shocked when records showed a 30 percent increase in malignant brain tumors in Calgary in the single year between 2012 and 2013,[30] despite official government statistics proclaiming no rise in malignant brain tumor rates at all in either the province of Alberta or the nation of Canada. This discrepancy has lit a fire under Faith Davis, professor of epidemiology at the University of Alberta's School of Public Health. As unreliable as official statistics are for malignant tumors, they are even worse for non-malignant tumors: Canada's surveillance system does not record them at all. To remedy this incredible situation, the Brain Tumour Foundation of Canada announced in July 2015 that it is raising money to help Davis create a national brain tumour registry that will finally give clinicians and researchers access to accurate information.

The studies that are assuring us that all is well with cell phones have been funded by the telecommunications industry. But, and in spite of severe underreporting of brain tumors, independent scientists are confirming the impression of brain surgeons and oncologists that their caseloads are increasing, as well as the evident fact that many more people that we all know and hear about are dying of such tumors than ever before. The most prominent of these independent scientists is Lennart Hardell.

Hardell is a professor of oncology and cancer epidemiology at University Hospital in Örebro, Sweden. Although most of his earlier research was on chemicals like dioxins, PCBs, flame retardants, and herbicides, since 1999 he has focused on exposure to cell and cordless telephones. He tells us, based on case control studies involving over 1,250 people with malignant brain tumors, that using both cell phones and cordless phones significantly increases one's risk for brain cancer. The more years you use such a phone, the more cumulative hours you use one, and the younger you are at first exposure, the greater the odds that you will develop a tumor. Two thousand hours of cell phone use, according to Hardell, triples one's risk. Two thousand hours on a cordless phone more than doubles one's risk. First use of a cell phone before the age of twenty increases one's overall risk of brain cancer three-fold, the risk of an astrocytoma—the most common type of malignant brain tumor—five-fold, and the risk of an astrocytoma on the same side of the head as the phone eight-fold. First use of a cordless phone before the age of twenty doubles the risk of any brain tumor, quadruples the risk of an astrocytoma, and increases the risk of an astrocytoma on the same side of the head eight-fold.[31]

The literature on cell towers and radio towers is less compromised. Almost all of the existing studies, until recently, have been funded by independent sources and not by the telecommunications industry, and they have yielded consistent results: living near a transmission tower is carcinogenic.

William Morton, at Oregon Health Sciences University, found that living near VHF-TV broadcast antennas was a significant risk for leukemia and breast cancer in the Portland-Vancouver metropolitan area from 1967 to 1982.

In 1986, the Department of Health of the State of Hawaii found that residents of Honolulu who lived in census tracts that had one or more broadcast towers had a 43 percent increased risk for all types of cancer.[32]

In 1996, Bruce Hocking, an occupational physician in Melbourne, analyzed the childhood cancer incidence for nine Australian municipalities in relation to a group of three high-power television towers. Children who lived closer than four kilometers to the towers were almost two and a half times more likely to die of leukemia as children in more distant cities.

In 1997, Helen Dolk and her colleagues found high rates of adult leukemia, bladder cancer, and skin melanoma near the Sutton Coldfield tower at the northern edge of Birmingham. When Dolk expanded her study to include twenty high power transmission towers throughout Great Britain, she found that, in general, the closer you lived to a tower, the more likely you were to have leukemia.

In 2000, Neil Cherry analyzed the childhood cancer rate in San Francisco as a function of distance from Sutro Tower. Sutro Tower is almost 1,000 feet tall, stands on top of a tall hill, and can be seen from all over San Francisco. At the time of Cherry's study it was broadcasting nearly one million watts of VHF-TV and FM radio signals, plus over 18 million watts of UHF-TV. The rates of brain cancer, lymphoma, leukemia, and all cancers combined, throughout San Francisco, were related to the distance a child lived from that tower. Children who lived on hills and ridgetops had much more cancer than children who lived in valleys and were shielded from the tower. Children who lived less than one kilometer from the tower had 9 times the rate of leukemia, 15 times the rate of lymphoma, 31 times the rate of brain cancer, and 18 times the total cancer rate, as children in the rest of the city.

In 2004, Ronni and Danny Wolf responded to residents in a small neighborhood around a single cell tower in south Netanya, Israel.

During the five years before the tower was erected, two of the 622 residents had developed cancer; during the single year after the tower went up, eight more developed cancer. This turned a neighborhood with one of the lowest cancer rates in the city into a zone where the risk was more than quadruple the average for Netanya.

In the same year, Horst Eger, a physician in Naila, Germany examined 1,000 patient records in his home town. He found that people who lived within 400 meters (1,300 feet) of a cell tower had triple the risk of developing cancer, and developed their cancer, on average, when they were eight years younger, compared to people who lived further away.

In 2011, Adilza Dode headed up a team of university scientists and government officials of a metropolis in southeastern Brazil that confirmed the results of all the previous studies. The risk of cancer for the residents of Belo Horizonte decreased uniformly and steadily with distance from a cell tower.

And on February 24, 2011, the Supreme Court of Italy upheld the 2005 conviction of Cardinal Tucci for polluting Rome with radio waves. A ten-day suspended jail sentence was his only punishment. No one has ever been compensated for their injuries. The Prosecutor's Office has not filed charges of negligent homicide. Vatican Radio's antennas have not been shut down.

14. Suspended Animation

> We admonish mankind to observe and distinguish between what conduces to health, and what to a long life; for some things, though they exhilarate the spirits, strengthen the faculties, and prevent diseases, are yet destructive to life, and, without sickness, bring on a wasting old age; while there are others which prolong life and prevent decay, though not to be used without danger to health.
>
> SIR FRANCIS BACON

> Every animal has allotted to it a constant number of heartbeats per lifetime. If it lives fast and furiously like a shrew or a mouse, it will use up its quota of heartbeats in a much shorter time than if its metabolic personality is a more temperate one.
>
> DONALD R. GRIFFIN
> *Listening in the Dark*

IN 1880, GEORGE MILLER BEARD wrote his classic medical book on neurasthenia, titled *A Practical Treatise on Nervous Exhaustion*. He made an intriguing observation: "Although these difficulties are not directly fatal, and so do not appear in the mortality tables; although, on the contrary, they may tend to prolong life and to protect the system against febrile and inflammatory disease, yet the amount of suffering that they cause is enormous." In *American Nervousness: Its Causes and Consequences*, written a year later for the general public, he reiterated the paradox: "Side by side with this increase of nervousness, and partly as a result of it, longevity has increased." Along with migraine headaches, ringing in the ears, mental irritability, insomnia, fatigue, digestive disorders, dehydration, muscle and joint pains, heart palpitations, allergies, itching, intolerance of foods and medications—in addition to this general degradation in the public health, the world was witnessing an increase in the human lifespan. Those who were suffering the most tended to look young for their age and to live longer than average.

At the end of *American Nervousness* appears a map showing the approximate geographic reach of neurasthenia. It was the same as the reach of railroads and telegraphs, being most prevalent in the northeast where the electric tangle was densest. "The telegraph is a cause of nervousness the potency of which is little understood," wrote Beard. "Within but thirty years the telegraphs of the world have grown to half a million miles of line, and over a million miles of wire—or more than forty times the circuit of the globe." Beard also noticed that a rare disease called diabetes was much more common among neurasthenes than in the general population.[1]

What Beard—an electrotherapist and a friend of Thomas Edison, who was shortly to be diagnosed with diabetes—did not figure out was that the growing cloud of electromagnetic energy, which permeated air, water, and soil wherever telegraph lines advanced, had something to do with the growing numbers of neurasthenes and diabetics that sought his ministrations. He was astute enough, however, to make the connection between longevity and disease, and to understand that the modern expansion of lifespan did not necessarily mean better health or a more excellent life. The mysterious extension of years among individuals who were the sickest was in fact a warning that something was terribly wrong.

Fasting and an austere diet have been recommended since antiquity for the rejuvenation of the body. The prolongation of life, said Francis

Bacon, should be one of the purposes of medicine, along with the preservation of health and the cure of diseases. Sometimes, he added, one must make a choice: "The same things which conduce to health do not always conduce to longevity." But he laid down one sure rule, for those who wished to follow it, that furthered all three goals of the physician: "A spare and almost Pythagorean diet, such as is prescribed by the stricter orders of monastic life or the institution of hermits, which regard want and penury as their rule, produces longevity."

Three hundred years later Bacon's third arm of medicine was still sorely neglected. "What must one do, or rather what must one not do to attain the extreme limits of age?" asked Jean Finot in 1906. "What, after all, are the boundaries of life? These two series of questions together constitute a special science, gerocomy. It exists in name only." Observing the animal world, Finot saw that the length of adolescence had something to do with the length of life. A guinea pig's period of growth endured seven months; that of a rabbit, one year; of a lion, four years, of a camel, eight years, of a man, twenty years. Human initiative was misguided, said Finot. What conduces to health and vigor does not necessarily prolong life. "The education and instruction given to children," he wrote, "are in flagrant contradiction to this law of gerocomy. All of our efforts tend towards the rapid advancement of physical and intellectual maturity." To prolong life, it would be necessary to do just the opposite. And one method, he suggested, was to restrict one's diet.

In the early years of the twentieth century, Russell Chittendon at Yale University, who is often called the father of American biochemistry, experimented on himself and on volunteers at Yale. Over the course of two months he gradually eliminated breakfast, settling into a pattern that consisted of a substantial midday meal and a light supper at night. Although he was eating less than 40 grams of protein daily, one-third the amount then recommended by nutritionists, and only 2,000 calories, he not only suffered no ill effects but the rheumatism in his knee disappeared, as did his migraine headaches and attacks of indigestion. Rowing a boat left him with much less fatigue and muscle soreness than before. His weight dropped to 125 pounds and remained there.

After one year on this diet, with funding from the Carnegie Institution and the National Academy of Sciences, he formally experimented on volunteers. They were: five professors and instructors at Yale; thirteen volunteers from the Hospital Corps of the Army; and eight students, "all thoroughly trained athletes, some with exceptional records in athletic events." He restricted them to about 2,000 calories and no more than fifty grams of protein per day. Without exception his subjects' health was as good as before or better at the end of half a year, with gains in strength, endurance, and well-being.

While Chittendon proved nothing about lifespan, the ancient recommendations have since been thoroughly subjected to the scientific method and, in all species of animals from one-celled organisms on up to primates, proven accurate. Provided an animal receives the minimal nutrients necessary to maintain health, a severe reduction in calories will prolong life. And there is no other method known that will reliably do so.

THESE RATS ARE BOTH 964 DAYS OLD.
From: C.M. McKay et al., "Retarded growth, life span, ultimate body size and age changes in the albino rat after feeding diets restricted in calories." *Journal of Nutrition* 18(1): 1-13 (1939).

A severe restriction in calories will increase the lifespan of rodents by 60 percent, routinely producing four and five year old mice and rats. Calorie restricted rats are not senile. Quite the opposite: they look younger and are more vigorous than other animals their age. If they are female they reach sexual maturity very late and produce litters at impossibly old ages.[2]

The annual fish *Cynolebias adloffi* lived three times as long when restricted in food.[3] A wild population of brook trout doubled their lifespan, some trout living twenty-four years when food was scarce.[4]

Spiders fed three flies a week instead of eight lived an average 139 days instead of 30.[5] Underfed water fleas lived 60 days instead of 46.[6] Nematodes, a type of worm, more than doubled their lifespan.[7] The mollusc *Patella vulgata* lives two and a half years when food is abundant, and up to sixteen years when it is not.[8]

Cows given half the normal amount of feed each winter lived twenty months longer. Their breathing rate was also one-third lower, and their heart rate ten beats per minute less.[9]

During a twenty-five-year-long study at the Wisconsin National Primate Research Center, the death rate of fully-fed adult rhesus monkeys from age-related causes was three times the death rate of calorie-restricted animals. When the study ended in 2013, twice as many diet-restricted monkeys as fully fed monkeys were still alive.[10]

Calorie restriction works whether it is lifelong or only during a portion of life, and whether it is begun early, during adulthood, or relatively late in life. The longer the period of restriction, the longer the prolongation of life.

Calorie restriction prevents age-related diseases. It delays or prevents heart disease and kidney disease, and drastically decreases the cancer rate: in one study, rats that were fed one-fifth as much food had only seven percent as many tumors.[11] In rhesus monkeys it reduces the cancer rate by half, heart disease by half, prevents diabetes, prevents atrophy of the brain, and reduces the incidence of endometriosis, fibrosis, amyloidosis, ulcers, cataracts, and kidney failure.[12] Older diet-restricted monkeys have less wrinkled skin and fewer age spots, and their hair is less gray.

A natural human experiment exists. In 1977, there lived 888 people over one hundred years old in Japan, the greatest concentration of whom lived on the southwestern coast and a few islands. The percentage of centenarians on Okinawa was the highest in Japan, forty times higher than in the northeastern prefectures. Yasuo Kagawa, professor of biochemistry at Jichi Medical School, explained: "People in areas of longevity have a lower caloric intake and smaller physique than those in the rest of Japan." The daily diet of school boys and girls in Okinawa was about 60 percent of recommended caloric intake.

The reason calorie restriction works is controversial, but the simplest explanation is that it slows metabolism. While the aging process is not fully understood, anything that slows the metabolism of cells must slow the aging process.

The idea that we are each allotted a fixed number of heartbeats is ancient. In modern times, Max Rubner at the University of Berlin, in 1908, proposed a variation on this idea: instead of a fixed number of heartbeats, our cells are allotted a fixed amount of energy. The slower an animal's metabolism, the longer it will live. Most mammals, Rubner calculated, use about 200 kilocalories per gram of body weight during their lifetime. For humans, assuming a lifespan of ninety years, the value is about 800. If an individual is able to delay the use of that amount of energy, his or her life will be correspondingly longer. Raymond Pearl, at The Johns Hopkins University, published a book along these lines in 1928 titled *The Rate of Living*.

During 1916 and 1917 Jacques Loeb and John Northrop, at The Rockefeller Institute, experimented on fruit flies. Since flies are cold-blooded, their metabolism can be slowed merely by lowering the ambient temperature. The average duration of life, from egg to death, was 21 days at a temperature of 30° C; 39 days at 25° C; 54 days at 20° C; 124 days at 15° C; and 178 days at 10° C. The rule that low temperatures prolong life applies to all cold-blood animals.

Another common way animals reduce their metabolism is by hibernating. Hibernating species of bats, for example, live on average six years longer than species that don't. And bats live far longer than other animals their size because, in effect, they hibernate on a daily basis. Bats are active, on the wing hunting for dinner, for only a few hours each night. They sleep the rest of the time, and sleeping bats are not warm-blooded. "It is sometimes possible in the laboratory to keep a rectal thermocouple in place while a bat settles down for a nap," wrote bat expert Donald Griffin, "and in one such case the body temperature fell in an hour from 40° when the bat was active to 1°, which was almost exactly the temperature of the air in which it was resting."[13] This explains why bats weighing only a quarter of an ounce can live more than thirty years, while no laboratory mouse has ever lived more than five.

Calorie restriction, the only method of prolonging life that works for all animals—warm-blooded, cold-blooded, hibernators, and non-hibernators—obviously slows metabolism, as measured by how much oxygen an animal consumes. Food-restricted animals always use less oxygen. A controversy arose among gerontologists because food-restricted animals also lose weight, and oxygen use *per unit weight* does not necessarily decline. But it declines where it counts. In humans, the internal organs, despite comprising less than 10 percent of our weight, are responsible for about 70 percent of our resting energy use. And it is our internal organs, not our fat or muscle tissues, that determine how long we will live.[14]

As researchers into the aging process have emphasized, the engine of our lives is the electron transport system in the mitochondria of our cells.[15] It is there that the oxygen we breathe and the food we eat combine, at a speed that determines our rate of living and our lifespan. That speed is in turn determined by our body temperature, and by the amount of food we digest.

But there is a third way to slow our rate of living: by poisoning the electron transport chain. One way to do this is to expose it to an electromagnetic field. And since the 1840s, at a gradual but accelerating rate, we have immersed our world, and all biology, in a thickening fog of such fields, that exert forces on the electrons in our mitochondria and slow them down. Unlike calorie restriction, this does not promote health. It starves our cells not of calories, but of oxygen. Resting metabolic rate does not change, but maximum metabolism does. No cell—no brain cell, no heart cell, no muscle cell—can work to its capacity. Where calorie restriction prevents cancer, diabetes, and heart disease, electromagnetic fields promote cancer, diabetes, and heart disease. Where calorie restriction promotes well-being, oxygen deprivation promotes headaches, fatigue, heart palpitation, "brain fog," and muscular aches and pains. But both will slow overall metabolism and prolong life.

Industrial electricity in any of its forms always injures. If the injury is not too severe, it also prolongs life.

In an experiment funded by the Atomic Energy Commission, exposure to simple electric shock for one hour each day throughout adulthood increase the average lifespan of mice by 62 days.[16]

Radio waves also increase lifespan.

In the late 1960s, a proton accelerator was being built at Los Alamos National Laboratory that was going to use radio waves at a frequency of 800 MHz. As a precaution, forty-eight mice were enrolled in an experiment to see if this radiation might be dangerous for workers in the facility. Twenty-four of the mice were irradiated at a power level of 43 milliwatts per square centimeter for two hours a day, five days a week, for three years. This is a huge exposure that is powerful enough to produce internal burns. And indeed four of the mice died from burn injuries. A fifth mouse became so obese that it could not be extracted from the exposure compartment and it died there. But the mice that weren't directly killed by the experiment lived a long time—on average, 19 days longer than the unexposed mice.[17]

In the late 1950s, Charles Süsskind at the University of California, Berkeley received funding from the Air Force to determine the lethal dose of microwave radiation in mice, and to investigate its effects on growth and longevity. At that time, the Air Force thought that 100 milliwatts per square centimeter was a safe dose; Süsskind soon found out that it was not. It killed most mice within nine minutes. So after that, Süsskind only exposed mice for four and a half minutes at a time. He irradiated one hundred mice for 59 weeks, five days per week for four and a half minutes a day at a power density of 109 milliwatts per square centimeter. Some of the irradiated mice, which subsequently died, developed extraordinarily high white blood cell counts, and had enlarge lymphoid tissue and enormous liver abscesses. Testicular degeneration occurred in 40 percent of the irradiated mice, and 35 percent developed leukemia. However, the unirradiated mice, although they were much healthier, did not live as long. After 15 months, half the control mice were dead, and only 36 percent of the irradiated ones.

From 1980 to 1982, Chung-Kwang Chou and Arthur William Guy led a famous experiment at the University of Washington. They had a contract with the United States Air Force to investigate the safety of the early warning radar stations recently installed at Beale Air Force Base in California, and on Cape Cod in Massachusetts. Known as PAVE PAWS, these were the most powerful radar stations in the world, emitting a peak effective radiated power of about three billion watts and irradiating millions of Americans. The University of Washington team approximated the PAVE PAWS signals at a "very low" level, irradiating one hundred rats 21.5 hours a day, 7 days a week, for 25 months. The Specific Absorption Rate—approximately that of the average cell phone today—was 0.4 watts per kilogram. During the two years of the experiment the exposed animals developed four times as many malignant tumors as the control animals. But they lived, on average, 25 days longer.

Recently gerontologists at the University of Illinois exposed cell cultures of mouse fibroblasts to radio waves (50 MHz, 0.5 watts) for either 0, 5, 15, or 30 minutes at a time, twice a week. The treatments lowered the mortality rate of the cells. The greater the exposure time, the lower the mortality, so that the 30-minute exposure reduced cell death by one-third after seven days, and increased their average lifespan from 118 days to 138 days.[18]

Even ionizing radiation—X-rays and gamma rays—will prolong life if not too intense. Everything from Paramecia to coddling moths to rats and mice to human embryo cells have had their average and/or maximal life spans increased by exposure to ionizing radiation. Even wild chipmunks have been captured, irradiated, and released—and had their average lifespans thereby extended.[19] Rajindar Sohal and Robert Allen, who irradiated house flies at Southern Methodist University, discovered that at moderate doses, an increase in lifespan occurred only if the flies were placed in compartments small enough so that they could not fly. They concluded that radiation always produces two opposite kinds of effects: injurious effects that shorten the lifespan, and a reduction in basal metabolic rate that lengthens the lifespan. If the dose of radiation is low enough, the net effect is a lengthening of life despite obvious injuries.

Loren Carlson and Betty Jackson at the University of Washington School of Medicine reported that rats exposed daily to moderate doses of gamma rays for a year had their lives extended, on average, by 50 percent, but suffered a significant increase in tumors. Their oxygen consumption was reduced by one-third.

Egon Lorenz, at the National Cancer Institute, exposed mice to gamma rays—one-tenth of a roentgen per eight-hour day—beginning at one month of age and for the rest of their lives. The irradiated females lived just as long, and the irradiated males one hundred days longer, than the unirradiated animals. But the irradiated mice developed many more lymphomas, leukemias, and lung, breast, ovarian, and other types of cancers.

Even extremely low doses of radiation will both injure and extend lifespan. Mice exposed to only 7 centigrays per year of gamma radiation—only 20 times higher than background radiation—had their lives extended by an average of 125 days.[20] Human fibroblasts, exposed in cell culture once for only six hours to the same level of gamma rays that is received by astronauts in space, or during certain medical exams, lived longer than unexposed cells.[21] Human embryo cells exposed to very low dose X-rays for ten hours a day had their lifespans increased by 14 to 35 percent, although most of the cells also suffered several kinds of damage to their chromosomes.[22]

Modern medicine can take some but not all of the credit for the modern increase in the average human lifespan. For that increase began a century before the discovery of antibiotics, in a time when doctors still bled their patients and dosed them with medicines containing lead, mercury, and arsenic. But medicine can take none of the credit for the modern extension of the *maximum* human life span. For medicine still does not pretend to understand the aging process, and only a tiny minority of doctors are even beginning to try to do anything to reverse aging. Yet the maximum age at death, worldwide, has been steadily rising.

Sweden has the most accurate and longest continuous records on the extreme limits of human age of any country, dating back to 1861. They reveal that the recorded maximum age at death was 100.5 years in 1861, that it rose gradually but steadily until 1969, when it was 105.5 years, and that it has risen more than twice as fast since then, reaching 109 years by the turn of the twenty-first century.

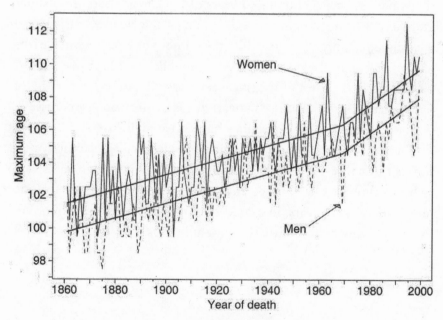

Figure 1, Wilmoth et al. 2000

In 1969, the trends in both Swedish longevity and Swedish cancer accelerated. It was the year color TV and UHF-TV were introduced into the country (see chapter 13).

In 1994, Väinö Kannisto, former United Nations advisor on demographic and social statistics, showed that the number of people living more than one hundred years was increasing spectacularly in the twenty-eight countries for which good data existed. The number of centenarians in Sweden had risen from 46 in 1950 to 579 in 1990. During the same period, the number of centenarians had risen from 17 to 325 in Denmark; from 4 to 141 in Finland; from 265 to 4,042 in England and Wales; from 198 to 3,853 in France; from 53 to 2,528 in West Germany; from 104 to 2,047 in Italy; from 126 to 3,126 in Japan; from 14 to 196 in New Zealand. The number of centenarians in all these countries, roughly doubling every ten years, had far outraced the increase in population.

Even in Okinawa, long known for its longevity, there lived only a single person over one hundred years old as late as 1960. In Japan as a whole, noted Kagawa in 1978, the number of male centenarians had quadrupled in only 25 years, while the number of female centenarians had sextupled. And yet he observed, in middle-aged Japanese, almost a doubling in the rates of breast cancer and colon cancer, a tripling of lung cancer, a 40 percent rise in heart disease, and an 80 percent rise in diabetes: "extended life expectancy but increased diseases."

The explanation for both phenomena is electricity—electricity that travels through wires as well as earth, that radiates through air as well as bones. We are all, to an extent that has been intensifying for one hundred and sixty years, in a mild state of suspended animation. We live longer, but are less alive, than our ancestors.

15. You mean you can hear electricity?

In 1962, A LOCAL WOMAN contacted the University of California, Santa Barbara for help tracking down a mysterious noise. She had moved into a newly-built home in a quiet neighborhood and this noise, whose location she could not find, was accompanying her wherever she went like an unwanted ghost. It was impairing her health, keeping her awake, and forcing her, in desperation, to abandon her home for large periods of time just to get relief. In response to her plea for help, an engineer showed up at her house with a load of electronic equipment.

Clarence Wieske, who was with the Laboratory for the Study of Sensory Systems in Tucson, a military contractor that was working on the interface between man and machine, happened to be involved with a project at the University at Santa Barbara when the woman's call came. His initial intention was to look for electric fields on her property that might be setting some metal object into vibration, creating the noise that was bothering her. He was startled by what he found.

His search coil, as he anticipated, did pick up unusually strong harmonic frequencies. They emanated not only from her electric wires, but also from her telephone wires, gas pipes, water pipes, and even the metal in her heating system. But his stethoscope could find no audible noise being emitted by any of these items. He therefore tried what he thought was a far-fetched experiment: he attached a tape recorder to his search coil, which recorded the electric frequency patterns and translated them into sounds, and then played the recording for the woman. When she put on the headphones and listened to the tape, she recognized the sounds as identical to the noise that was tormenting her. Wieske then took the experiment one step farther. He disconnected the headphones and played the tape directly back into his search coil. The woman said instantly, "You mean you cannot hear that?" She was hearing the same thing again directly from the search coil although it emitted only an electromagnetic field and no actual noise.

In a further experiment Wieske, without telling the woman, connected a low-power frequency generator to the water pipe about one hundred feet from her house. She remarked that there was peculiar noise "like a barking dog." When Wieske turned on the pickup equipment in her house and put on the headphones, he found that she was correct. He heard a sound like a barking dog!

These experiments and others done at her home and at the university left no doubt that the woman was hearing electricity—and that the noise was not coming from her dental fillings. Wieske then set about to try to alleviate her problem. Electrically grounding her refrigerator, freezer case, door chimes, and other appliances reduced the noise level a bit but did not get rid of it. One day, during a power outage, she telephoned Wieske, ecstatic. The noise had stopped! But it returned as soon as the power came back on. Therefore Wieske contacted all the utility companies. With their cooperation, he put filters on her phone line, an isolation transformer on her electric line, and sections of non-conducting pipe into her water line and her gas line. These time-consuming, expensive measures prevented unwanted electric frequencies originating elsewhere in the neighborhood from being conducted over these paths. Finally, the noise was reduced to an endurable level and the woman could inhabit her home.

After investigating a number of similar cases, Wieske predicted that with the continued electrification of society, complaints like hers would one day be common. His article about his experiences, published in *Biomedical Sciences Instrumentation* in 1963, concluded with a technical description of human hearing, including all of the places within the ear where electromagnetic fields might cause electric currents to flow. He speculated as to the reasons some people can hear them and not others: "If the nerve for some reason in some individuals is not as well insulated from these currents as in the normal individual, or if the cochlea is not as well insulated from these currents in some individuals, perhaps this could make them sensitive to these electrical fields."

Wieske's prediction has come to pass. Today companies serving the population that can feel and hear electromagnetic fields form a significant cottage industry in every part of the United States. One organization, the International Institute for Building Biology and Ecology, lists sixty consultants, scattered throughout the United States and Canada, that it has trained in the methods of detecting and mitigating residential electromagnetic pollution.

About eighty million Americans today have "ringing in the ears" to some degree. Some hear their sounds intermittently. Some hear them only when everything else is quiet. But for increasingly many, the sounds are so loud all of the time that they cannot sleep or function. Most of these people do not have tinnitus, which is an internally generated sound, often in one ear, usually accompanied by some degree of hearing loss. Most people today who have "ringing in the ears" hear it equally in both ears, have perfect hearing, and are hearing a pitch at the very top of their hearing range. They are hearing the electricity around them, and it is getting louder all the time. The clues to what is happening were planted over two centuries ago.

French electrotherapist Jean Baptiste Le Roy, in 1755, was apparently the first to elicit an auditory response to static electricity. He was treating a man blind from cataracts by winding wire around the man's head and giving him twelve shocks from a Leyden jar. The man reported hearing the explosion of "twelve pieces of cannon."

Experimentation began in earnest when Alessandro Volta invented the electric battery in 1800. The metals he first used, silver and zinc, with salt water for an electrolyte, generated about a volt per pair—less when he stacked them up in his original "pile." Applying a single pair of metals to his own tongue produced either a sour or sharp taste, depending on the direction of the current. Applying a piece of silver to his eye, and touching it with a piece of zinc held in his moistened hand, produced a flash of light—a flash, he said, that was "much more beautiful" if he placed the second piece of metal, or both pieces, inside his mouth.

Stimulating the sense of hearing proved more difficult. Volta tried in vain to elicit a noise with only a single pair of metal plates. But with thirty pairs, roughly equivalent to a twenty-volt battery, he succeeded. "I introduced, a considerable way into both ears," he wrote, "two probes or metallic rods with their ends rounded, and I made them to communicate immediately with both extremities of the apparatus. At the moment when the circle was thus completed I received a shock in the head, and some moments after (the communication continuing without any interruption) I began to hear a sound, or rather noise, in the ear, which I cannot well define: it was a kind of crackling with shocks, as if some paste or tenacious matter had been boiling." Being afraid of permanent injury to his brain, Volta did not repeat the attempt.

But hundreds of other people did. After this report by one of the most famous men in the world, everyone wanted to see if they could hear electricity. Carl Johann Grapengiesser, a physician, was careful to use only small currents on his patients, and he was a much more careful observer than Volta. His subjects varied widely in their sensitivity and in the sounds they heard. "The noises, in respect of their quality and strength, are very variable," he wrote. "Most often, it seems to the patient that he hears the hissing of a boiling teakettle; another hears ringing and bell-pealing, a third thinks that outside a storm wind blows; to a fourth it seems that in each ear a nightingale sings most lustily."[1] A few of his patients heard the electricity generated by only a single pair of metals applied to blister plaster wounds underneath their ears.

Physicist Johann Ritter was not afraid of currents much greater than those risked by Volta. Using batteries containing 100, 200, and more pairs of metals, he was able to hear a pure musical tone that was approximately g above middle c, and that persisted as long as the current flowed through his ears.

Many were the doctors and scientists who, in the heady years following Volta's gift to the world of its first reliable source of steady electricity, stimulated the acoustic nerve with greater or lesser quantities of current. The following list, limited to German scientists who published their research, was compiled by Rudolf Brenner in 1868:

> Carl Johann Christian Grapengiesser (*Attempts to Use Galvanism in the Healing of Some Diseases*, 1801)
>
> Johann Wilhelm Ritter (*Contributions to the Recent Knowledge of Galvanism and the Results of Research*, 1802)
>
> Friedrich Ludwig Augustin (*Attempt at a Complete Systematic History of Galvanic Electricity and its Medical Use*, 1801; *On Galvanism and its Medical Use*, 1801)
>
> Johann Friedrich Alexander Merzdorff (*Treatment of Tinnitus with the Galvanic Current*, 1801)
>
> Carl Eduard Flies (*Experiments of Dr. Flies*, 1801)
>
> Christoph Friedrich Hellwag (*Experiments on the Healing Powers of Galvanism, and Observations on its Chemical and Physiological Effects*, 1802)
>
> Christian August Struve (*System of Medical Electricity with Attention to Galvanism*, 1802)
>
> Christian Heinrich Wolke (*Report on Deaf and Dumb Blessed by the Galvani-Voltaic Hearing-Giving Art at Jever and on Sprenger's Method of Treating Them with Voltaic Electricity*, 1802)
>
> Johann Justus Anton Sprenger (*Method of Using Galvani-Voltaic Metal-Electricity as a Remedy for Deafness and Hearing Loss*, 1802)

Franz Heinrich Martens (*Complete Instructions on the Therapeutic Use of Galvanism; Together with a History of This Remedy*, 1803)

Ironically, the man who laid the foundation for such research—Alessandro Volta—was also the man whose mechanistic world view has so dominated scientific thinking for more than two centuries that it has not been possible to understand the results of these experiments. They have been regarded as little more than parlor tricks, when they have been remembered at all. For Volta, we recall, pronounced that electricity and life are distinct and that there are no electric currents flowing in the body. To this day, as a result, in the teaching of biology, including the biology of the ear, chemistry is king and electricity is omitted.

By Brenner's time the work of all these early scientists had already been forgotten. A physician who specialized in diseases of the ear, he described this state of affairs in terms that could just as easily apply to today: "Nothing can be more instructive for the history of scientific development than the fate of the old experiments on galvanic stimulation of the acoustic nerve. Among contemporary researchers who deny the possibility of such stimulation are names of the very best repute. One must therefore ask: do these men really believe that Volta, Ritter, and the other old galvanists only imagined the tones and noises they heard?" Brenner's goal was to establish, once and for all, not only that electicity could be heard, but exactly how, why, and to what degree this occurs. "It is not established *if*, and it is unknown *how* the acoustic nerve reacts to the influence of electrical current," he wrote.[2] The results of his experiments filled a 264-page book. His apparatus contained 20 zinc-copper Daniell cells, each producing a maximum of about one volt, connected to a rheostat that could be adjusted to any of 120 positions. Any desired number of cells could be inserted into the circuit at the turn of a dial. He performed 47 different kinds of experiments on a large number of individuals.

The average person, with 7 volts of direct current coursing through his or her ear canal, heard a clear metallic sound resembling a small dinner bell. The range of sensitivity of normal human beings, however, was enormous. Some heard nothing at all, even when all twenty Daniell cells were in the circuit. For others, who were deemed to have "acoustic nerve hyperaesthesia," the sound from only one cell was intense. Some heard nothing unless their ear canal was filled with salt water, which helped conduct the electricity. Others, their ear canals dry, heard the ringing bell when the knob-shaped electrode was simply placed on the cheek in front of the ear, or on the mastoid process, the bony protrusion behind the ear.

The direction of the current was crucial. The sound—unless the person had "hyperaesthesia"—was heard only when the negative, not the positive, electrode was in the ear. With minimal current the sound typically resembled the "buzzing of a fly." This was elevated to "distant wagon roll," then to "cannon roll," "striking of a metal plate," and finally the "ringing of a silver dinner bell," as the current was gradually raised. The greater the current, the purer the tone, and the greater the resemblance to a bell. When Brenner asked his subjects to sing the tone they heard, some, agreeing with Ritter's report of 1802, heard a g above middle c. Others disagreed. But although the threshold of perception varied enormously, and the quality and exact pitch were different for everyone, each individual always heard the same thing. They always heard the identical sound and pitch, and had the same threshold, whenever they were tested, even at intervals years apart.

After experimenting with different placements of the second, non-ear electrode on the skull, neck, torso, arms, and legs, Brenner became convinced that a sound was heard only when the inner ear was in the path of the current, and that direct stimulation of the acoustic nerve was the cause of the sensation of sound.

American physician Sinclair Tousey, one of the last electrotherapists of the old school, wrote about electricity and the ear in the third edition of his textbook on *Medical Electricity*, published in 1921. Brenner's results with direct current, completely forgotten today, were still at that time taught, accepted, and verified by every electrical practitioner. Sounds were normally caused by cathodic (negative) stimulation of the auditory nerve. The range of sensitivity was extraordinary. "Many individuals," wrote Tousey, closely echoing Brenner's words, "give no reactions whatever." In others, the sound was so loud that the person was deemed to have "a distinct hyperesthesia of the auditory nerve."[3]

With the disappearance of the electrotherapist's art and the dwindling of opportunities for the average physician to become familiar with the auditory response to electricity, the old knowledge was again almost forgotten.

Then, around 1925, amateur radio enthusiasts thought they found a way to listen to the radio without a loudspeaker, by directly stimulating the acoustic nerve. "Thus, even deaf persons whose eardrums no longer function properly, but whose nerve centers are intact, can hear radio," wrote Gustav Eichhorn. The device he patented, however—a kind of flat electrode held against the ear—was soon dismissed as being nothing more than a "condenser receiver." Apparently, the surfaces of the skin and the electrode, vibrating, took the place of a loudspeaker, creating an ordinary sound that reached the inner ear by bone conduction.[4]

Nevertheless, the experiments of the radio engineers spawned a spate of genuine efforts by biologists to stimulate the inner ear with alternating current. This was typically done after the manner of Brenner—by inserting one electrode in the ear canal, which was first filled with salt water, and completing the circuit with a second electrode on the back of the forearm or hand. The subjects most often heard a tone that corresponded in pitch to the frequency of the applied current. The sensitivity of the subjects, as before, varied tremendously. In experiments done in Leningrad, the most sensitive individual, when tested with a current of 1,000 cycles per second, heard a sound as soon as the voltage exceeded a fifth of a volt; the least sensitive subject required six volts—a thirty-fold difference in sensitivity. There was nothing wrong with the hearing of any of these people. The variations in their ability to hear electricity bore no relation to the subjects' ability to hear ordinary sound.[5]

In 1936, Stanley Smith Stevens, an experimental psychologist at Harvard University, gave the hearing phenomenon a new name: "electrophonic hearing." Four years later, at his newly-created Psycho-Acoustics Laboratory, he proposed three different mechanisms of hearing by electrical stimulation. Most people with normal hearing, when stimulated by an electrode in their ear, heard a pitch that was exactly one octave higher than the frequency of the applied current. However, if a negative DC voltage was applied at the same time, they heard the fundamental frequency as well. His knowledge of physics led Stevens to conclude that the ear was responding like a condenser receiver, with the ear drum and the opposite wall of the middle ear being the vibrating "plates" of that condenser.

People without ear drums, however, heard either the fundamental frequency or a "buzzing" noise, or both. None heard the higher octave. And as Brenner had also reported, eardrumless ears were much more sensitive to electricity than normal ears. One of Stevens' subjects heard a pure tone when stimulated with only one-twentieth of a volt. Stevens proposed that the hearing of the fundamental frequency was caused by direct stimulation of the hair cells of the inner ear. For those that heard a buzzing sound, he proposed that the auditory nerve was being stimulated directly.

Thus, by 1940, three different parts of the ear were being proposed as capable of turning electricity into sound: the middle ear, the hair cells of the inner ear, and the auditory nerve. All three mechanisms appeared to operate throughout the normal hearing range of human beings.

Stevens tried one additional experiment, whose significance he failed to appreciate, and which was not repeated by anyone else for two decades: he exposed subjects to a low frequency, 100 kHz radio wave that was modulated at 400 Hz. Somehow the ear demodulated this signal and the person heard a 400-cycle pure tone, close to a g above middle c.[6]

In 1960, biologist Allan Frey introduced yet another method of hearing electromagnetic energy, this time without placing electrodes on the body. A radar technician at Syracuse, New York swore to him that he could "hear" radar. Taking him at his word, Frey accompanied the man back to the Syracuse facility and found that he could hear it too. Frey was soon publishing papers about the effect, proving that even animals, and people with conduction deafness—but not nerve deafness—could hear brief pulses of microwave radiation at extremely low levels of average power. This phenomenon, known as "microwave hearing," attracted a fair amount of publicity, but is probably not responsible for most of the sounds that torment so many people today.

However, the 1960s would bring still more surprises. Renewed research into electrophonic hearing had both civilian and military goals. The medical community wanted to see if the deaf could be made to hear. The military community wanted to see if a new method of communication could be devised for soldiers or astronauts.

In 1963, Gerhard Salomon and Arnold Starr, in Copenhagen, proved that the inner ear was far more sensitive to electrical energy than anyone had previously suspected. They placed electrodes directly adjacent to the cochlea in two patients who had had surgical reconstruction of their middle ear. One patient heard "clicks" or "cracklings" when stimulated by only three microamperes (millionths of an ampere) of direct current. The second patient required 35 microamperes to hear the same sound. As the current was gradually increased, the clicks changed to "walking on dry snow" or the rush of "blowing air." Alternating current elicited pure tones whose pitch matched the applied frequency, but this required about a thousand times more current.

Then the Electromagnetic Warfare and Communication Laboratory at Wright-Patterson Air Force Base in Ohio published a report written by Alan Bredon of Spacelabs, Inc., investigating both electrophonic hearing and microwave hearing for their potential use in space. The goal was to develop "an efficient, dual-purpose transducer which can be worn with an absolute minimum of discomfort during long missions in the confines of pressure clothing and aerospace environments." Bredon found that electrophonic devices were unsuitable because the sound they produced was too faint to be useful in the noisy environment of aircraft or space vehicles. And microwave hearing was judged useless because it appeared to depend on short pulses of energy and did not produce continuous sound. But Patrick Flanagan's Neurophone, which had been recently publicized in *Life* magazine,[7] caught Bredon's interest. This device, which Flanagan claimed to have invented at the age of 15, was a radio wave device almost identical to the one Eichhorn had patented in 1927, and appeared to work by skin vibration. It differed, however, in one crucial respect: Flanagan used a carrier frequency in the ultrasonic range, specified as being between 20,000 and 200,000 Hz. He had rediscovered the phenomenon that Stevens had briefly described back in 1937 and never followed up on.

As a further result of the publicity surrounding Flanagan's invention, Henry Puharich, a physician, and Joseph Lawrence, a dentist, under contract with the Air Force, investigated what they called "transdermal electrostimulation." They delivered electromagnetic energy at ultrasonic frequencies via electrodes placed next to the ear. The audio signal, added to the ultrasonic carrier, was somehow demodulated by the body and heard like any other sound. Like Flanagan's device, it appeared at first glance to work by skin vibration. However, several astonishing results were reported.

First, most people's hearing range was significantly extended. Say the upper limit of a person's hearing was normally 13,000 or 14,000 cycles per second. By using this device, they typically heard sounds as high in pitch as 18,000 cycles per second. Some even heard a true pitch as high as 25,000 cycles per second—5,000 cycles higher than most human beings are supposed to be able to hear.

Second, the use of an ultrasonic carrier wave eliminated distortion. When the audio signal was fed directly into the electrodes without the carrier wave, speech could not be understood and music was unrecognizable. But when the speech or music was delivered only as a modulation to a high frequency carrier wave—in the same way that AM radio broadcasts deliver speech and music—the body, like a radio receiver, somehow decoded the signal and the person heard the speech or music perfectly without any distortion. The optimal carrier frequency, delivering the purest sound, was found to be between 30,000 and 40,000 Hz.

Third, and most surprising, nine out of nine deaf people—even those with profound sensorineural deafness from birth—could hear sound in this way by transdermal stimulation. But the electrodes had to be pressed more firmly on the skin, and the deaf subject had to move the electrode around beneath or in front of the ear until he or she located the exact spot that stimulated hearing—as though the signal had to be focused on a target inside the head. The four subjects with residual hearing described the sensation as "sound," not "vibration." The two who were deaf from birth described it as something "new and intense." The three who had acquired total deafness described it as hearing as they remembered it.

When insulated electrodes were used, people with normal hearing responded to power levels as low as 100 microwatts (millionths of a watt). When bare metal electrodes were pressed directly against the skin, more current was required, but the deaf could hear as well or better, with this method, than hearing people. Once the proper skin pressure and location were found, the threshold electromagnetic stimulus was between one and ten milliwatts (thousandths of a watt) for both hearing and deaf people, while only the slightest increase in power brought the sound, as described by one of the deaf subjects, "from a comfortable level to one of great force."

Even more amazingly, ten out of ten profoundly deaf subjects, who had never heard speech before, were able to understand words, after very brief training, when delivered in this manner. And patients who had lesser sensorineural hearing loss, who could identify only 40 to 50 percent of words spoken through the air, scored 90 percent or better by transdermal stimulation, without training.

For the first time in fifty years, there was evidence that an electrode carrying radio waves to the skin might be doing something more than just causing the skin to vibrate. These researchers speculated, based on measurements of cochlear microphonics (electrical signals generated by the inner ear), that transdermal stimulation produced a sound by a combination of acoustic and electrical effects—by both vibrating the skin and directly stimulating the hair cells in the inner ear. "However," they wrote, "these two effects do not give a satisfactory explanation of word recognition response in those patients whose cochlea is non-functional."

The results of animal experiments were just as astonishing. Two dogs were rendered deaf—one through injections of streptomycin, which destroyed the cochlear hair cells, and one by surgical removal of the ear drums, middle ear bones, and cochleas. Both dogs had previously been conditioned to respond to transdermal stimulation by jumping over a divider in a box, and both had learned to respond correctly better than 90 percent of the time. Incredibly, both dogs continued to respond correctly 90 percent of the time to the high frequency stimulus when it was modulated with the audio signal, but only 1 percent of the time to the unmodulated high frequency signal alone.

The implications of this research are profound. Since people and animals without any cochlear function at all, or even without any cochlea, can apparently hear this type of stimulation, either the brain is being stimulated directly—which is unlikely since the source of the sound always appears to the person to be coming from the direction of the electrode producing it—or there is another part of the inner ear besides the cochlea that responds to ultrasound, or to electromagnetic waves at ultrasonic frequencies. Since most hearing subjects were able to hear much higher frequencies than they could hear in the normal way, this is the most likely explanation. And we will see that there are good reasons to believe that most people who are bothered by electrical "tinnitus" are hearing electrically-delivered ultrasound.

Puharich and Lawrence patented their device, and the Army acquired two prototype units for testing aboard Chinook helicopters and airboats used in Vietnam. The news editor for *Electronic Design* reported, after trying out one of the devices, that "the signals were almost, but not quite, like airborne sounds."[8]

In 1968, Garland Frederick Skinner repeated some of Puharich and Lawrence's experiments at higher power, using a carrier frequency of 100 kHz, for his master's thesis at the Naval Postgraduate School. He did not test his "Trans-Derma-Phone" on any deaf people, but like Puharich and Lawrence, he concluded that "be it the ear, the nerves, or the brain, an AM detection mechanism exists."

In 1970, Michael S. Hoshiko, under a post-doctoral fellowship from the National Institutes of Health, tested the device of Puharich and Lawrence at the Neurocommunications Laboratory at The Johns Hopkins University's School of Medicine. Subjects not only heard pure tones from 30 Hz up to the remarkable frequency of 20,000 Hz equally well at low sound levels, but scored 94 percent in speech discrimination. The twenty-nine college students who were tested performed equally well whether the words were delivered through the air as ordinary sounds, or whether they were delivered electronically as modulations to a radio wave in the ultrasonic range.

Two more efforts to make people hear modulated radio waves were made by members of the military, but probably because they did not use ultrasonic frequencies they were unable to identify any cause of hearing besides the vibrating skin. One of the reports, a master's thesis submitted by Lieutenants William Harvey and James Hamilton to the Air Force Institute of Technology at Wright-Patterson Air Force Base, specified a carrier frequency of 3.5 MHz. The other project was undertaken by M. Salmansohn, Command and Control Division at the Naval Air Development Center in Johnsville, Pennsylvania. He also did not use an ultrasonic carrier, in fact he later dispensed with the carrier wave altogether and used direct audio-frequency current.

Finally, in 1971, Patrick Woodruff Johnson, for his master's thesis at the Naval Postgraduate School, decided to revisit "ordinary" electrophonic hearing. He wanted to see how little electricity it took to make people hear a sound. Most previous researchers had exposed their subjects' heads to up to one watt of power, resulting in large and potentially dangerous levels of AC current. Johnson found that by using a silver disc plated with silver chloride as one of the electrodes, and simultaneously applying a positive direct current, an alternating current of as little as 2 microamperes (millionths of an ampere), delivered with only 2 microwatts (millionths of a watt) of power, could be heard. Johnson proposed that "an extremely small low cost hearing aid" could be developed using this system.

In June 1971, at M.I.T., Edwin Charles Moxon reviewed the entire field for his Ph.D. dissertation and added the results of his own experiments on cats. By recording the activity of the cats' auditory nerves while their cochleas were electrically stimulated, he proved definitely that two distinct phenomena were occurring at the same time. The electrical signal was somehow being converted into ordinary sound, which was being processed by the cochlea in the normal way. And in addition, the current itself was stimulating the auditory nerve directly, producing a second, abnormal component of the discharge pattern of the nerve.

At this point efforts at understanding how electricity affects the normal ear ceased, as practically all funding was diverted to the development of cochlear implants for the deaf. This was a natural outcome of the development of computers, which were beginning to transform our world. The brain was being modeled as a fantastically elaborate digital computer. Hearing researchers thought that if they separated sounds into their different frequency components, they could feed those components in digital pulses to the appropriate fibers of the auditory nerve for direct processing by the brain. And, considering they are stimulating thirty thousand nerve fibers with only eight to twenty electrodes, they have been remarkably successful. By 2017, the number of cochlear implants worldwide exceeded five hundred thousand. But the results are robotic and do not duplicate normal sound. Most patients can learn to understand carefully articulated speech well enough to use the telephone in a quiet room. But they cannot distinguish voices, recognize music, or converse in average noisy environments.

Meanwhile, progress in understanding electrophonic hearing came to a complete halt. Some research into microwave hearing continued for another decade or so, and then ceased as well. The peak power levels that appear to be required for microwave hearing make it unlikely to be the source of sounds that bother most people today. The phenomenon discovered by Puharich and Lawrence is a much more likely candidate. To understand why requires an excursion into the anatomy of one of the most complex and least understood parts of the body.

The Electromodel of the Ear

In the normal ear, the ear drum receives sound and passes the vibrations on to three tiny bones in the middle ear. They are the malleus, incus, and stapes (hammer, anvil, and stirrup), named after the implements they resemble. The stapes, the last bone in the chain, although only half the size of a grain of rice, funnels the world of vibrational sound to the bony cochlea, a snail-shaped structure which itself is a marvel of miniaturization. No bigger than a hazelnut, the cochlea is able to take the roar of a lion, the song of a nightingale, and the squeak of a mouse, and reproduce them all with perfect fidelity in the form of electrical signals sent to the brain. To this day no one knows exactly how this is accomplished. And what little is known is probably wrong.

"It is unfortunate," wrote Augustus Pohlman, an anatomy professor and dean of the school of medicine at the University of South Dakota, "that no machinery is available for deleting from the literature those interpretations which have proven to be incorrect." Pohlman stood, in 1933, looking back on seventy years of research that had failed to eradicate what he regarded as a fundamentally flawed assumption about the operation of the liquid-filled cochlea. Another eighty years have still failed to eradicate it.

The tiny cochlear spiral is divided along its length into an upper and lower chamber by a partition called the basilar membrane. Upon this membrane sits the organ of Corti, containing thousands of hair cells with their attached nerve fibers. And in 1863, the great German physicist Hermann Helmholtz had proposed that the cochlea was a sort of an underwater piano, and suggested that the ear's resonant "strings" were the different length fibers of the basilar membrane. The membrane increases in width as it winds round the cochlea. The longest fibers at the apex, he suggested, like the long bass strings of a piano, resonate with the deepest tones, while the shortest fibers at the base are set into vibration by the highest tones.

Helmholtz assumed that the transmission of sound was a simple matter of mechanics and levers, and subsequent research, for one and a half centuries, has simply built upon his original theory with remarkably little change. According to this model, the motion of the stapes, like a tiny piston, pumps the fluid in the two compartments of the cochlea back and forth, causing the membrane separating them to flex up and down, thereby stimulating the hair cells on top of it to send nerve impulses to the brain. Only those parts of the membrane that are tuned to the incoming sounds flex, and only those hair cells sitting on those parts send signals to the brain.

But this model does not explain the hearing of electricity. It also fails to explain some of the most obvious features of the inner ear. Why, for example, is the cochlea shaped like a snail shell? Why are the thousands of hair cells lined up in four perfectly spaced rows, one behind the other like the keyboards of a pipe organ? Why is the cochlea encased in the hardest bone of the human body, the otic capsule? Why is the cochlea the precise size that it is, fully formed in the womb at six months of gestation, never to grow any larger? Why is the cochlea only marginally bigger in a whale than in a mouse? How is it possible to fit a full set of resonators, vibrating over a greater musical range than the largest pipe organ, into a space no bigger than the tip of your little finger?

Pohlman thought that the standard model of the ear was contradicted by modern physics, and a number of courageous scientists after him have agreed. By including electricity in their model of hearing, they have made progress in explaining the basic features of the ear. But they are up against a cultural barrier, which still does not permit electricity to play a fundamental role in biology.

The ear is much too sensitive to work by a system of mechanics and levers, and Pohlman was the first to point out this obvious fact. The real resonators in the ear—the "piano strings"—had to be the thousands of hair cells, lined up in rows and graded in size from bottom to top of the cochlea, and not the fibers of the membrane they were sitting on. And the hair cells had to be pressure sensors, not motion detectors. The extreme sensitivity of the ear made that evident. This also explained why the cochlea is embedded in the densest bone in the human body. It is a soundproof chamber, and the function of the ear is to transmit sound, not motion, to the delicate hair cells.

The next scientist to add pieces to the puzzle was an English physician and biochemist, Lionel Naftalin, who passed away in March 2011 at the age of 96 after working on the problem for half a century. He began by making precise calculations that proved conclusively that the ear is much too sensitive to work in the accepted fashion. It is a known fact that the quietest sound that a person can hear has an energy of less than 10^{-16} watts (one ten-thousandth of one trillionth of a watt) per square centimeter, which, calculated Naftalin, produces a pressure on the eardrum that is only slightly greater than the pressure exerted by randomly moving air molecules. Naftalin stated flatly that the accepted theory of hearing was impossible. Such tiny energies could not move the basilar membrane. They could not even move the bones of the middle ear by the assumed lever mechanism.

The absurdity of the standard theory was obvious. At the threshold of hearing the eardrum is said to vibrate through a distance (0.1 ångstrom) that is only one-tenth the diameter of a hydrogen atom. And the motion of the basilar membrane is calculated to be as small as ten trillionths of a centimeter—only slightly larger than the diameter of an atomic nucleus, and much smaller than the random motions of the molecules that make up the membrane. This "movement" of subatomic dimensions supposedly causes the hairs on the hair cells to "bend," triggering an electric depolarization of the hair cells and the firing of the attached nerve fibers.

Recently some scientists, realizing the foolishness of such a notion, have introduced various ad hoc assumptions that increase the distance the basilar membrane must move from subatomic to only atomic dimensions—which still doesn't overcome the fundamental problem. Naftalin pointed out that the contents of the cochlea are not solid metal objects but liquids, gels, and flexible membranes, and that such infinitessimal distances could have no basis in physical reality. He then calculated that to move a resonant portion of the basilar membrane only one ångstrom—about the distance now claimed necessary to trigger a response from the hair cells[9]—would require over ten thousand times more energy than is contained in a threshold sound wave that hits the ear drum.

During his fifty years of work on hearing, Naftalin thoroughly demolished the prevailing mechanical theory and created a model in which electrical forces are central. Instead of focusing on the basilar membrane, on which the hair cells sit, he drew his attention to a much more unusual membrane—the one that covers the tops of the hair cells. It has a jelly-like consistency and composition that occurs nowhere else in the human body. It also has unusual electrical properties, and a large voltage is always present across it. Elsewhere in the body, voltages of this magnitude—about 100 to 120 millivolts—are usually found only across cell membranes.

In 1965, Naftalin, thinking in terms of solid state physics, postulated that this membrane—called the tectorial membrane—is a semiconductor that is also piezoelectric. Piezoelectric substances, we recall, are those that convert mechanical pressure into electrical voltages, and vice versa. Quartz crystals are the most familiar example. Often used in radio receivers, they convert electrical vibrations into sound vibrations. Judging by its structure and chemical composition, Naftalin suggested that the tectorial membrane ought to have this property. He proposed that it is a piezoelectric liquid crystal that converts sound waves into electrical signals, which it communicates to the hair cell resonators embedded in it. He suggested that the large voltage across the membrane causes great amplification of these signals.

Naftalin then built scale models of both the cochlea and the tectorial membrane, and began to find answers to some of the outstanding mysteries of the ear. He discovered that the snail-like shape of the cochlea is important to its function as a precision musical instrument. He also discovered that the makeup of the tectorial membrane has something to do with the instrument's small size. While the speed of sound in air is 330 meters per second, and in water is 1500 meters per second, in ten percent gelatin it is only 5 meters per second, and in the tectorial membrane it is likely to be considerably less. By slowing the speed of sound, the jelly-like substance of the membrane contracts the wavelengths of sounds from meters to millimeters, allowing a millimeter-sized instrument like the cochlea to receive and play the world of sound we live in for our brain.

George Offutt came to this problem as a marine biologist, and reached similar conclusions from an evolutionary perspective. His doctoral dissertation at the University of Rhode Island's School of Oceanography dealt with codfish hearing. His theory of human hearing, first published in 1970, was later expanded into a book, *The Electromodel of the Auditory System*. I interviewed him in early 2013, shortly before his death.

Like Naftalin, Offutt concluded that the tectorial membrane is a piezoelectric pressure sensor. And because of his background, he argued that human hair cells, by evolution and by function, are electroreceptors.

The mammalian cochlea, after all, evolved from a fish organ called the lagena, which has hair cells not too unlike ours, covered by a gelatinous membrane, also similar to ours. But the fish's membrane is in turn topped by structures called otoliths ("ear stones"), which are crystals of calcium carbonate and are known to be about one hundred times more piezoelectric than quartz. Offutt said that this is not accidental. The hair cells in the ears of fish, he said, are sensitive to the voltages generated by the otoliths in response to sound pressure.[10] This, he said, explains why sharks can hear. Fish, being composed largely of water, are supposed to be transparent to water-borne sounds unless they have a swim bladder containing air. Therefore sharks, which have no swim bladder, ought to be deaf according to standard theories, but they aren't. In 1974, Offutt elegantly solved this contradiction by introducing electricity into his model for how fish hear. And by extension, he said, there is no reason why human hearing should not still work in the same basic fashion. If the cochlea evolved from the lagena, then the tectorial membrane evolved from the otolithic membrane and ought still to be piezoelectric. And the hair cells, which are substantially the same, should still function as electroreceptors.

In fact, fish have other, related hair cells that are *known* to be electroreceptors. Lateral line organs, for example, arranged in lines along the sides of every fish's body in order to sense water currents, actually respond not only to water currents but also to low frequency sounds and to electric currents.[11] These organs' hair cells, too, are covered by a jellylike substance, called the cupula, and they, too, are supplied by a branch of the acoustic nerve. In fact, the lateral line and the inner ear are so closely related functionally, evolutionarily, and embryonically, that all such organs in all types of animals are referred to as the acoustico-lateralis system.

Some fish have other organs, which evolved from this system, that are exquisitely and primarily sensitive to electrical currents. With these organs, sharks can detect the electric fields of other fish or animals, and can locate them in darkness, in murky water, or even when hidden in the sand or mud at the bottom. The hair cells of these electric organs lie beneath the surface of the body in sacs called ampullae of Lorenzini and are covered, again, with a gelatinous substance.

All such fish organs, no matter their specialization, have proven to be sensitive to both pressure and electricity. Lateral line organs that primarily sense water currents also react to electrical stimuli, and ampullae of Lorenzini that primarily sense electric currents also react to mechanical pressure. Therefore marine biologists were once of the opinion that piezoelectricity was at play in both the lateral line and the ear.[12] Hans Lissman, once the world's foremost authority on electric fishes, thought that this was so. Later, anatomist Muriel Ross, who had a grant from NASA to study the effects of weightlessness on the ear, emphasized that the otoliths of fish, and the related otoconia ("hair sand") of our own ears' gravity sensors, are known to be piezoelectric.

Mechanical and electrical energy, she said, are interchangeable, and feedback between hair cells and piezoelectric membranes will transform one form of energy into the other.

In a related study in 1970, Dennis O'Leary exposed the gelatinous cupulas of frogs' semicircular canals—the organs of balance in the inner ear—to infrared radiation. The response of the canals' hair cells was consistent with the electrical and not the mechanical model of such organs.

Recently the outer hair cells of the cochlea have themselves proven to be piezoelectric. They acquire a voltage in response to pressure, and they lengthen or shorten in response to an electric current. Their sensitivity is extreme: one picoamp (one trillionth of an amp) of current is enough to cause a measurable change in a hair cell's length.[13] Electric currents, traveling in complex paths, have also been found traversing the tectorial membrane and coursing through the organ of Corti.[14] And pulsating waves have been discovered, in the thin space between the tops of the hair cells and the bottom of the tectorial membrane, that reverberate between the outer hair cells, the tectorial membrane, and the inner hair cells.[15] Australian biologist Andrew Bell has calculated that in the human cochlea these fluid waves should have wavelengths roughly between 15 and 150 microns (millionths of a meter)—just the right size to put hair cells 20 to 80 microns in length into musical resonance. Bell has compared these waves to surface acoustic waves, and the organ of Corti to a surface acoustic wave resonator, a common electronic device that has replaced quartz crystals in a wide variety of industries.

In the electromodel of hearing that these scientists have constructed, there are several places where electricity can act directly on the ear. The inner hair cells are electroreceptors. The outer hair cells are piezoelectric. The tectorial membrane is piezoelectric. And since both direct and alternating current can act on any of these structures, many of the early reports of the hearing of electricity, said Offutt, including reports that were dismissed as being due to "skin vibration," should be reevaluated.

The exquisite sensitivity of the organ of Corti to electricity explains the nineteenth century reports of the hearing of direct current and the twentieth century reports of the hearing of alternating current. And it forms a basis for understanding the torment suffered half a century ago by Clarence Wieske's client in Santa Barbara, and the suffering of so many millions today. But a piece of the auditory puzzle is still missing.

Direct or alternating current applied to the ear canal requires about one milliampere (one thousandth of an ampere) to stimulate hearing.[16] If an electrode is placed directly in the cochlear fluid, about one microampere (millionth of an ampere) suffices.[17] If current is applied directly to a hair cell, one picoampere (trillionth of an ampere) is all that is necessary to cause a mechanical reaction.[18] Clearly, sticking electrodes in your outer ear is an inefficient way to stimulate the hair cells. Very little of the applied current ever reaches those cells. But in today's world, electrical energy is reaching the hair cells directly in the form of radio waves, to which bones and membranes are transparent. The hair cells are also bathed in electric and magnetic fields originating in the electric power grid and all the electronic appliances that are plugged into it. All those fields and radio waves penetrate the inner ear and induce electric currents to flow inside the cochlea itself. The question then becomes, why do we all *not* hear a constant cacophony of noise drowning out all conversation and music? Why is most electrical noise confined to either very low or very high frequencies? The answer very likely has to do with a part of the ear that is not ordinarily associated with hearing at all.

Hearing Ultrasound

Human ultrasonic hearing has been rediscovered more than a dozen times since the 1940s, most recently by Professor Martin Lenhardt at Virginia Commonwealth University. "So outlandish is the concept that humans can have the hearing range of specialized mammals, such as bats and toothed whales," he writes, "that ultrasonic hearing has generally been relegated to the realm of parlor tricks rather than being considered the subject of scientific inquiry."[19] At the present time, apparently, ultrasonic hearing is being intensively investigated only by Lenhardt and by a small group of researchers in Japan.

Yet it is a fact that most human beings—even many profoundly deaf human beings—can hear ultrasound by bone conduction, and that this ability encompasses the entire hearing range of bats and whales. It extends well beyond 100 kHz. Dr. Roger Maass reported to British Intelligence in 1945 that young people can hear up to 150 kHz,[20] and one group in Russia reported in 1976 that the upper limit for ultrasonic hearing is 225 kHz.[21]

Bruce Deatherage, while doing shipboard research for the Department of Defense in the summer of 1952, rediscovered the ability to hear ultrasound by accident when he swam into a sonar beam broadcasting at 50 kHz. Repeating the experiment with volunteers, he reported that each subject heard a very high-pitched sound that was the same as the highest pitch that person could ordinarily hear. Recently scientists at the Naval Submarine Base in New London, Connecticut verified the hearing of underwater ultrasound up to a frequency of 200 kHz.[22]

What is known today is this:

Virtually everyone with normal hearing can hear ultrasound. Elderly people who have lost their high frequency hearing can still hear ultrasound. Many profoundly deaf people with little or no functioning cochlea can hear ultrasound. The perceived pitch varies from person to person but is usually between 8 and 17 kHz. Pitch discrimination does occur, but requires a greater change in frequency in the ultrasonic range than in the normal auditory range. And, most surprisingly, when speech is transposed into the ultrasonic range and spread out over that range, it can be heard and understood. Somehow the brain recondenses the signal, and instead of high pitched "tinnitus," the person hears the speech as though it were normal sound. Speech can also be modulated onto an ultrasonic carrier, and it is demodulated by the brain and heard as normal sound. Lenhardt, who has built and patented bone conduction ultrasonic hearing aids based on these principles, reports that word comprehension is around 80 percent in normal hearing individuals, even in a noisy environment, and 50 percent in the profoundly deaf.

Since even many deaf people can hear ultrasound, several investigators over the years, including Lenhardt and the Japanese group, have suggested that the ultrasound receptor lies not in the cochlea but in an older part of the ear, one which functions as a primary hearing organ in fish, amphibians and reptiles: the saccule. It still exists in humans, containing hair cells capped by a gelatinous membrane covered with piezoelectric calcium carbonate crystals.

Although it is adjacent to the cochlea, and although its nerve fibers connect to both the vestibular and auditory cortex of the brain, the human saccule has usually been thought to be an exclusively vestibular, or balance organ, and to play no part in hearing. This dogma, however, has come under challenge periodically for the past eighty years. In 1932, Canadian physician John Tait presented a provocative paper, titled "Is all hearing cochlear?" at the 65th Annual Meeting of the American Otological Society in Atlantic City. He said that he and other investigators had failed to find any connection between the saccule and posture in fish, frogs, or rabbits, and proposed that even in humans the saccule is part of the hearing apparatus. Its construction, he said, indicates that the saccule is designed to detect vibrations of the head, including the vibrations that occur in speaking. The saccule in air-breathing animals, he proposed, "is a proprioceptor involved in the emission and regulation of voice. This would mean that we hear our own voice with the help of two kinds of receptors, while we hear the voice of our neighbors with only one." In other words, Tait suggested that the cochlea is an innovation that allows air-breathing animals to hear airborne sounds, while the saccule retains its ancient function as a sensitive receptor for bone-conducted sounds.

Since that time, saccular hearing has been proven to exist in a variety of mammals and birds, including guinea pigs, pigeons, cats, and squirrel monkeys. Elephants may use their saccule to hear low frequency vibrations received through the earth by bone conduction. Even in human beings, audiologists have developed a hearing test involving the electrical response of neck muscles to sound—called "vestibular evoked myogenic potentials" (VEMP)—to evaluate the functioning of the saccule. This test is often normal in people with profound sensorineural hearing loss.

Lenhardt believes that ultrasonic hearing may be both cochlear and saccular in normally hearing people, while it is strictly saccular in the deaf.

Many pieces of evidence indicate that what is tormenting people around the world today is electromagnetic energy in the ultrasonic range—from about 20 kHz to about 225 kHz—which is converted to sound in the cochlea and/or the saccule:

1. Most frequently people are complaining of "loud tinnitus" at the highest pitch they can hear.

2. Although airborne ultrasound is not audible, Puharich and Lawrence showed that electromagnetic energy at ultrasonic frequencies *is* audible, to both hearing and deaf people.

3. The otoconia (calcium carbonate crystals) in the saccule, and the outer hair cells in the cochlea, are known to be piezoelectric, i.e. they will convert electric currents to sound.

4. Electric and magnetic fields induce electric currents in the body whose strength is proportional to frequency. The higher the frequency, the greater the induced current. These principles of physics mean that the same field strength will produce 1,000 times more current at 50,000 Hz, in the ultrasonic range, than at 50 Hz, in the audible range.

5. The measured threshold for hearing in the ultrasonic range is as low, or lower, than the threshold at 50 or 60 Hz. An exact comparison cannot be made because ultrasound is only audible by bone conduction, and extremely low frequencies are better heard by air conduction. But superimposing typical air, bone, and ultrasonic hearing threshold curves gives an overall hearing curve that looks something like this:[23]

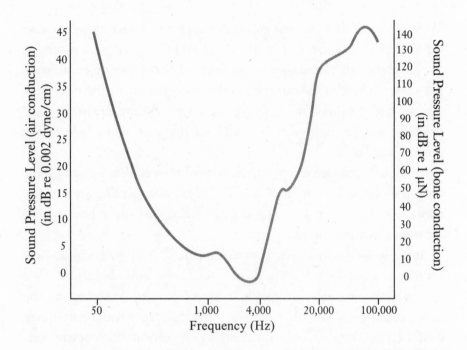

The inner ear looks to be about 5 to 10 times more sensitive to sound at 50 kHz than at 50 Hz. Therefore the ear may be 5,000 to 10,000 times more sensitive to electric and magnetic fields at ultrasonic frequencies than at power lines frequencies. The ear's much greater sensitivity to sound in the middle of the hearing range is largely due to the resonant properties of the outer, middle and inner ear before they are transformed into electrical impulses.[24] This means that the ear is much more sensitive to electric currents at ultrasonic frequencies than to currents in either the middle or low parts of its range. The ear's insensitivity to electromagnetic fields at 50 or 60 Hz explains why, thankfully, we do not hear a 60-cycle buzz from the power grid at all times.

By consulting charts published by the World Health Organization,[25] it is possible to estimate the approximate minimum frequency at which we might expect to begin hearing an electromagnetic field. Since 1 picoampere of current is enough to stimulate one hair cell, and 50 picoamperes to trigger 50 hair cells—about the number required to stimulate hearing—one can look this amount of current up on the

WHO's charts. It turns out to be the amount of current per square centimeter that is induced in the ear at 20 kHz by either a magnetic field of about one microgauss or an electric field of about ten millivolts per meter. These are about the magnitudes of some of the ultrasonic electric and magnetic fields that pollute our modern environment.[26] And one square centimeter is just about the area of the base of the human cochlea.

In other words, given the dimensions of the cochlea, we can expect to hear electromagnetic fields in today's environment that are roughly above 20 kHz and below 225 kHz, which is the upper limit of our ultrasonic hearing range.

If the saccule is more sensitive to ultrasound than the cochlea, these estimates could be too conservative. As I was reminded some years ago by Canadian acoustic physicist Marek Roland-Mieszkowski, the ear is sensitive to sound energies of less than 10^{-16} watts per square centimeter. Assuming, as he did, only a one percent efficiency in converting electrical energy into sound energy, the ear could be sensitive to magnetic fields of a hundredth of a microgauss, or to electric fields of 100 microvolts per meter. The ability of some people to hear the northern lights—said to resemble the sound of rustling silk[27]—indicates a potential sensitivity of about that level.[28]

SOURCES OF ELECTRICAL SOUND

Electronic consumer devices

On April 2, 2000, Dave Stetzer, a former electronics technician for the Air Force, testified about "nonlinear loads" before the Michigan Public Service Commission. By this, he explained, he meant "computers, fax machines, copiers, and many other electronic devices, as well as various utility equipment including capacitors, solid state monitoring and switching devices, and transformers." All these devices—in other words, virtually all modern electronic equipment—were putting tremendous amounts of high frequencies onto the power grid, and the grid, which was designed to transmit only 60 Hz, could no longer contain what was on it. The electrons in the wires, he explained, once they pass through a computerized device, vibrate not only at 60 Hz, but at frequencies extending throughout the ultrasonic range and well into the radio frequency spectrum. Since as much as seventy percent of all electric power flowing on the wires at any given time had passed through one or more computerized devices, the entire grid was being massively polluted.

Stetzer first described some of the technical problems this was causing. The high frequencies increased the temperature of the wires, shortened their life span, degraded their performance, and forced substantial amounts of electric current to return to the power plant through the earth instead of through a return wire. And the high frequencies and "transients" (spikes of high current) emanating from everyone's electronic equipment were causing interference and damage to everyone else's electronic equipment. This was becoming expensive for homeowners, businesses, and utility companies.

Even worse, all of the high frequency currents that were coursing through the earth, and the high frequency electromagnetic fields vibrating through the air, were making millions of people sick. Society was, and is, in denial about that, and that was not of great interest to the Michigan Public Service Commission. However, these fields and earth currents were also making dairy cows sick, all over Michigan, which was a threat to the state's economy. So the commissioners listened attentively while Stetzer spoke.

"In my visits to the various farms," he said, "I have observed over 6,000 dairy cows and some horses. I have observed damaged cows with swollen joints, open sores, and other maladies, as well as aborted and deformed calves. I have even observed aborted twin calves, one of which was fully developed while its twin was grossly deformed. Ironically, the grossly deformed twin was the one directly in the current flow pathway between the cow's back legs."

"In addition," Stetzer told the stunned commissioners, "I have also observed stressed cows, cows reluctant to enter certain spaces, including barns and milking parlors, and even cows reluctant to drink water, such that they lap at the water instead of sucking it up as they normally do. I have seen numerous cows fall over dead for no apparent reason. I have observed cows whose entire sides and muscles spasm uncontrollably. The articles from the Wisconsin *La Crosse Tribune* accurately highlight and describe a few of the conditions that I have personally observed on farms in Wisconsin, Minnesota, and Michigan. These symptoms and impacts are not limited to Wisconsin; they appear everywhere I have found dirty power."

My first experience of a health nature with modern electronics occurred back in the mid-1960s, when my family junked its old vacuum tube television set and acquired a transistorized model. As soon as it was plugged in, I heard an awful high-pitched sound—even though I was in the other end of the house with walls and doors in between—that apparently no one else could hear. Such was my introduction to the electronics age. I took care of myself by not watching television, which is one of the reasons why, from the day I moved out on my own to the present, I have never owned one.

Auditory unpleasantness of that sort was not a widespread problem—at least not for me—until the 1990s. As long as I avoided televisions and computers, the world, in the places I chose to live, contained mostly natural sounds, and complete silence was easy to find.

But at some point in the 1990s—the change was so gradual that I can't pinpoint when—I realized that I could not find silence any more. It happened after 1992, when I rented a cabin in northern Ontario—which was still silent—and before 1996, when I fled the new crop of digital cell towers in my native New York to save my life. Since at least 1996, I have found no escape, anywhere in North America, from the awful high-pitched sound that I first heard when I was about fifteen. In 1997, I sought silence in an underground cave in Clarksville, New York—and did not find it. The sound was greatly diminished underground but did not vanish. In 1998, I sought silence in Green Bank, West Virginia, the only place on earth that is legally protected from radio waves—and did not find it. The sound did not even diminish. I can make it louder by plugging in electronic devices, and softer again by unplugging them, but I cannot make it go away, not even by turning the power off in my house. I can hear appliances being turned on in a neighbor's house. Without warning or explanation the sound sometimes becomes suddenly much louder all over my neighborhood. It becomes quieter when there is a power outage. But it never disappears. It matches 17,000 Hz, which is the highest pitch that I can hear.

Low frequency sounds

The low frequency Hum is heard by between two and eleven percent of the population.[29] This is fewer than hear the high frequency sound, but the effects of the Hum can be far more disturbing. At its best it sounds like a diesel engine idling somewhere in the distance. At its worst it vibrates one's whole body, causes intense dizziness, nausea and vomiting, prevents sleep, and is completely incapacitating. It has driven people to suicide.

The probable sources of the Hum are powerful ultrasonic radio broadcasts modulated at extremely low frequencies to communicate with submarines. To penetrate the oceans requires radio signals of immense power and long wavelengths, and the frequencies called VLF (very low frequency) and LF (low frequency)—corresponding to the ultrasonic range—fit the bill. The American military systems currently in use for this purpose include enormous antennas located in Maine, Washington, Hawaii, California, North Dakota, Puerto Rico, Iceland, Australia, Japan, and Italy, in addition to sixteen mobile antennas flown on aircraft whose locations at any given time are kept secret. Land stations of this type are also operated by Russia, China, India, England, France, Sweden, Japan, Turkey, Greece, Brazil, and Chile, and by NATO in Norway, Italy, France, the United Kingdom, and Germany.

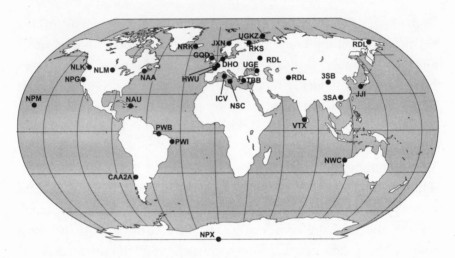

Since the wavelengths are so long, every VLF antenna is tremendous. The antenna array at Cutler, Maine, which has been operating since 1961, is in the form of two giant six-pointed stars, covering a peninsula of nearly five square miles and supported by 26 towers up to 1,000 feet tall. It broadcasts with a maximum power of 1.8 million watts. The facility at Jim Creek, Washington, built in 1953, has a 1.2-million-watt transmitter. Its broadcast antenna is strung between two mountaintops.

The low frequencies that are required in order to penetrate the oceans limit the speed at which messages can be transmitted. The American stations send a binary code at 50 pulses per second, which is consistent with the frequency of the Hum that most people hear today. The enhanced system recently adopted by the Navy uses multiple channels to transmit more data, but each channel still pulses at 50 Hz. In addition, the binary code itself is created by two ultrasonic frequencies spaced 50 Hz apart. These signals are therefore doubly modulated at approximately the frequency that is tormenting people worldwide.

Geology Professor David Deming at the University of Oklahoma, who was driven to investigate the Hum that he hears, has focused his attention on the mobile TACAMO ("Take Charge and Move Out") system. TACAMO planes, which trail long antennas behind them, have been flying out of Tinker Air Force Base in Oklahoma City since

1963, and the maximum power of each airborne transmitter is 200,000 watts. They use a variety of frequencies between 17.9 and 29.6 kHz, which are doubly modulated at 50 Hz like all other VLF stations that communicate with submarines. Navy TACAMO planes are always in the sky whenever there is a Hum in Oklahoma. The aircraft head out from Oklahoma to Travis Air Force Base in California and Naval Air Station Patuxent River in Maryland. From there the planes fly six to ten hour missions in predetermined orbits over the Atlantic and Pacific Oceans.

One other ultrasonic, pulsed communication network deserves mention here—one which ceased broadcasting in North America in 2010, but which is still functioning in some parts of the world and may yet be fully resurrected here: the LORAN-C system. LORAN, which stands for LOng RAnge Navigation, is an old network of extremely powerful land-based navigation beacons whose function is now duplicated by Global Positioning Satellites. LORAN may have been responsible for the earliest reports of a Hum in England as well as the famous Hum in Taos, New Mexico that was the subject of a government investigation launched in 1992.

LORAN-C operates at 100 kHz and is pulsed at multiples of 10 to 17 Hz, depending on location. Placed under the control of the Coast Guard, the first LORAN-C stations were built along the east coast in 1957—in Martha's Vineyard, Massachusetts; Jupiter, Florida; and Cape Fear, North Carolina. In the late 1950s, chains of LORAN-C stations were also built around the Mediterranean Sea and the Norwegian Sea, and by 1961 others were on the air in the Bering Sea, and in the Pacific Ocean centered on Hawaii. Although it was not the first long range navigation system, its predecessor, LORAN-A, operated at frequencies between 1850 and 1950 kHz and was not in the ultrasonic range.

My own encounters with the Hum date from 1983, when I first moved to the remote, otherwise-quiet sanctuary in the redwoods that is Mendocino, California. Although Cornell University is quite near the 800,000-watt Seneca LORAN station that began operating in 1978, I had graduated from college there in 1971 and never heard a

Hum. But in Mendocino, I was kept awake by it. Like so many other people, I first thought I was hearing a distant motor or generator—until I realized that this noise followed me even on camping trips deep into roadless areas of wild, far-northern California. Its pitch was about a low E-flat—roughly 80 Hz—and I discovered that I could bring the Hum into my head, even on days when it was otherwise not there, by playing my piano in the key of E-flat—as though there were an E-flat piano string inside my head vibrating in sympathetic resonance.

When, some years later, a Coast Guard official told me there was a LORAN antenna over in Middletown, I wondered if there was a connection to the annoying and puzzling Hum. The official had casually mentioned that the signal was so powerful that the people who worked at the facility could hear it. So I got in my car one morning and made the three-hour drive. As I approached within a half mile of the 63-story tower, my ears began to hurt. And I began to hear not only my accustomed 80 Hz pulsating Hum, but also a purer tone one octave lower. I obtained a copy of the LORAN-C User Handbook from the Coast Guard, and learned that the repetition rate for LORAN-C transmissions on the west coast was almost exactly 10 Hz. Apparently I was hearing the fourth and eighth harmonics. Further consultation with the Handbook provided an explanation. The West Coast Chain consisted of four stations—the one in Middletown, one in George, Washington, and two in Nevada—that transmitted once every tenth of a second in a precisely timed sequence.

Fallon...George...Middletown...Searchlight..............Fallon... George...Middletown...Searchlight.............. It took exactly one twentieth of a second to transmit the sequence of signals from the four beacons—corresponding to a repetition rate of 80 Hz, and reinforcing the eighth harmonic of the fundamental frequency. Taking the signals two at a time—Fallon-George and Middletown-Searchlight—gives a repetition rate of 40 Hz, reinforcing the fourth harmonic. The predominance of the Middletown signal, when one was close enough to the Middletown tower, apparently made the fourth harmonic audible.

By this time the Taos Hum was well known, and I wondered if it, too, was caused by LORAN. It had been investigated by a team of scientists from Los Alamos and Sandia National Laboratories, the Air Force's Phillips Laboratory, and the University of New Mexico—who predictably didn't find anything. But three items in their report stood out. First, 161 of the 1,440 residents of the Taos area who responded to their survey heard the Hum. Second, the team heard back not only from Taos-area residents, but from people throughout the northern hemisphere who had heard about the investigation and contacted the team to report being tormented by the same sound. Third, the frequencies that hearers said matched the Hum ranged from 32 Hz to 80 Hz, and several trained musicians identified it as a tone near 41 Hz. The South Central LORAN Chain had a repetition rate of 10.4 Hz, and the fourth harmonic was 41.6 Hz. The third harmonic was 31.2 Hz. Apparently many people were hearing the eighth harmonic as well.

The evidence that LORAN-C caused the Taos Hum is abundant. The South Central Chain was the only LORAN chain that had six transmitting beacons, and Taos was near the geographic center of them. The South Central Chain was built from 1989 to 1991 and fully commissioned in April 1991, precisely when residents of Taos began complaining. The combined electric field strength at Taos, from the six stations, was about 30 millivolts per meter, more than enough to trigger a hearing sensation.[30]

Some of the other Hums around the world seem also to have been caused by LORAN-C. The LORAN-C chain in the Norwegian Sea, with stations in Norway, Jan Mayen Island, Iceland, and the Faeroe Islands, provided coverage to England since 1959. The British Hum, which has been reported for about that long, suddenly decreased in loudness around 1994—the same year Iceland turned off the most powerful LORAN station in that chain. It increased in loudness again in 1996—at the same time that a new station in Værlandet in southern Norway was put into operation to again give better coverage to the British Isles. The new station also provided coverage for the first time to the area around Vanern Lake, Sweden—where the Hum was first reported in 1996.

I can also add another piece of my own experience. I now live in Santa Fe, New Mexico—not too far from Taos—where I hear the Hum only infrequently. It is not audible to me most of the time, and it is no longer a low E-flat. It is now closer to an A or A-flat, which corresponds to the frequencies used by the Navy in communicating with submarines.

At this writing, an Enhanced LORAN-C, or eLORAN, network is being built in several areas of the world to ensure the operation of a backup navigation and timing system in case the GPS satellites fail or their broadcasts are jammed. eLORAN relies on the same immensely powerful long-wave radio transmissions as before, but the addition of a data channel provides much greater position accuracy. To achieve position accuracies to within 10 meters, networks of receiving stations, called differential-LORAN, or DLoran, are also being built. They monitor the powerful eLORAN signals and broadcast correction factors over the data channel, or over a cell tower network, to local mariners. South Korea is currently operating three eLORAN stations and plans to achieve full nationwide coverage in 2020. Iran has built an eLORAN system, and India, Russia, China, and Saudi Arabia are upgrading their existing LORAN-C stations to eLORAN. France, Norway, Denmark, and Germany ceased their LORAN-C transmissions at the end of 2015 and have dismantled their towers. The situation in the United States is less certain. The 625-foot LORAN-C tower at Wildwood, New Jersey went back on the air temporarily in 2015 under the aegis of the Department of Homeland Security. And in December 2018, President Trump signed into law the National Timing Resilience and Security Act, which mandates the establishment of a terrestrial backup system for the Global Positioning Satellites that will be able to penetrate underground and inside buildings throughout the United States. It authorizes the acquisition of the mothballed LORAN facilities for this purpose.

To see if the shutoff of most of the European LORAN-C stations had any effect on the Hum in that part of the world, I consulted a worldwide database of Hum reports kept by Glen MacPherson, an instructor at the University of British Columbia. On January 1, 2016, the day after the planned LORAN-C shutoff, reports came in from Scotland and Northern Ireland saying that the Hum had suddenly stopped between 2:00 a.m. and 3:00 a.m. that morning.

OTHER SOURCES OF ULTRASONIC RADIATION

Time broadcasts

The National Institute of Standards and Technology broadcasts a time-of-day signal that synchronizes "atomic" clocks and watches through-out North America. Transmitting from Fort Collins, Colorado, the 60-kHz signal of station WWVB is even usable in parts of South America and Africa at night. Time stations using ultrasonic frequencies also broadcast from Anthorn, England; Mount Hagane and Mount Ootakadoya, Japan; Mainflingen, Germany; and Lintong, China.

Energy efficient light bulbs

In a contagious fit of insanity, countries are falling like dominoes for the myth that fluorescent lighting is good for the environment. Cuba, in 2007, was the first to ban outright all sales of ordinary incandescent bulbs—bulbs that have shed soft light into our dark evenings for a hundred and thirty-five years. Australia banned imports of incandescents in November 2008, and sales a year later. The European Union completed a three-year phase-out on September 1, 2012, and China banned 100-watt bulbs one month later, with total prohibition scheduled for 2016. Brazilians can no longer buy bulbs of 60 watts or greater as of July 1, 2015. Canada and the United States, which had planned to ban 100-watt bulbs in 2012, temporarily relented in the face of strong public opposition.

And the public are right. Fluorescents give off harsh light, and they contain mercury vapor, which gives off ultraviolet radiation when it is energized by high voltage. The inside of the glass is coated with a chemical that emits visible light when it is hit by the ultraviolet radiation. All fluorescents, without exception, work this way. All homes and business that use fluorescents long enough will eventually break one and be contaminated with mercury dust and vapor. And landfills throughout the world are being heavily polluted with mercury from the disposal of billions of broken and used up fluorescent light bulbs. Not to mention the inconvenient fact that little, if any, energy is being saved if you live anywhere but the tropics. In summer, the heat given off by light bulbs is wasted and increases the demand for air conditioning. But in winter, we gain that cost back because the heat from light bulbs then warms our homes. When we lose that extra source of heat, we have to make up the difference by burning more oil and gas. In the United States, we have probably neither gained nor lost, environmentally. But in Canada, for example, which gets virtually all its electricity from hydro power, banning incandescent bulbs has been an unqualified mistake. It has done nothing but increase the consumption of fossil fuels, putting more carbon dioxide into the atmosphere and worsening global warming.

And that mistake is being compounded. All manufacturers of fluorescent bulbs, under pressure from government regulators, are making a bad situation worse by attaching a miniature radio transmitter to each and every light bulb under the theory that this makes them even more energy efficient. The radio waves energize the mercury vapor without having to subject it to a high voltage. All compact fluorescent bulbs, and a large percentage of long fluorescent bulbs today are energized with these radio transmitters, which are called "electronic ballasts." The frequencies used, between 20 and 60 kHz, are in the ultrasonic hearing range. The ubiquity of this type of lighting, and the growing difficulty of obtaining ordinary incandescents, even where they are still legal, means that these bulbs are a predominant source of ultrasonic radiation in homes and businesses, and on power lines throughout the world. Virtually all electricity that flows on the power grid and in the earth is contaminated to some extent with 20 to 60 kHz, having passed through hundreds or thousands of these radio transmitters on its way to the next consumer, or back to the utility's generating plant. And because the electronic ballasts put out so much electrical distortion, today's fluorescent bulbs also emit measurable energy far into the microwave range. The FCC's rules allow each and every energy efficient bulb to emit microwave radiation, at frequencies up to 1,000 MHz, at a field strength of up to 20 microvolts per meter, as measured at a distance of 100 feet from the bulb.

LED bulbs, which are being offered as another substitute for incandescents, are no better. They too give off harsh light, and they contain a variety of toxic metals and require special electronic components that convert the alternating current in our homes to low-voltage direct current. Most often, these components are switch mode power supplies which operate at ultrasonic frequencies and are discussed below in connection with computers.

Sadly, the North American reprieve was only temporary. Canada officially banished most incandescent bulbs as of January 1, 2015, and the U.S. effort to further postpone its death knell ended at the same time. The last examples of Edison's enduring invention vanished from the shelves of my local hardware stores a couple of months later. The gentle incandescent has disappeared from much of the world. Only specialty bulbs and halogen lamps are left, and many countries are prohibiting those also. Incandescents are still completely legal, however, in most of Africa, most of the Middle East, much of southeast Asia, and all the island nations of the Pacific.[31]

Cell phones and cell towers

Although cell phones and cell towers are best known as emitters of microwave radiation, that radiation is modulated at a bewildering array of much lower frequencies that the human body, as a radio receiver, perceives. For example, GSM (Global System for Mobile) is a telecommunications system long used by AT&T and T-Mobile in the United States, and by most companies in the rest of the world. The radiation from GSM cell phones and cell towers has components at 0.16, 4.25, 8, 217, 1733, 33,850 and 270,833 Hz. In addition, the microwave carrier is divided into 124 subcarriers, each 200 kHz wide, all of which can broadcast simultaneously, in order to accommodate up to about a thousand cell phone users at once in any given area. This generates many harmonics of 200,000 Hz.

Although GSM is a "2G" technology, it has not gone away. Layered over it are "3G" and "4G" networks that smart phones of more recent vintage use. The 3G system, called Universal Mobile Telecommunications System, or UMTS, is completely different, containing modulation components at 100, 1500, 15,000, and 3,840,000 Hz. The 4G system, called Long-Term Evolution, or LTE, is modulated at yet another set of lower frequencies, including 100, 200, 1000, 2,000, and 15,000 Hz. In 4G, the carrier frequency is divided into hundreds of 15-kHz wide subcarriers, adding yet another set of harmonics. And since smart phones and flip phones of different vintages presently coexist, every cell tower has to emit all of the different modulation frequencies, old and new. Otherwise older phones would not continue to work. AT&T towers, for example, are therefore presently emitting modulation frequencies of 0.16, 4.25, 8.33, 100, 200, 217, 1000, 1500, 1733, 2,000, 15,000, 33,850, 270,833 and 3,840,000 Hz, plus harmonics of these frequencies and additional harmonics of 15,000 Hz and 200,000 Hz, not to mention the microwave carrier frequencies of 700 MHz, 850 MHz, 1700 MHz, 1900 MHz, and 2100 MHz. Like the proverbial boiled frog, we are all immersed in a giant pot of radiation, whose intensity is increasing, and whose effect, though unperceived, is nevertheless certain.[32]

Cell phones spend a higher percentage of their energy on their low frequency components than do cell towers[33]—which may explain the high prevalence of "tinnitus" among cell phone users with otherwise normal hearing. In 2003, at a time when cell phone use was not as universal as it is today, it was still possible to do epidemiological studies of users and non-users. A team of scientists led by Michael Kundi at the Medical University of Vienna, comparing people with and without tinnitus at an ear, nose, and throat clinic, found a greater prevalence of tinnitus—often in both ears—among cell phone users than among non-users, and a clear trend of more tinnitus with increasing intensity of cell phone use.[34] The more minutes, the more tinnitus.

Remote control devices

Most remote control devices—the gadgets that open garages and car doors, and operate television sets—communicate using infrared radiation. But the infrared signals are pulsed between 30 and 60 thousand times per second, in the middle of the ultrasonic range. The most common frequency chosen by manufacturers is 36 kHz.

The problem with computers

In 1977, Apple gave the world a revolutionary new device. The personal computer, as it came to be known, was powered by a new type of gadget called a switch mode power supply. If you have a laptop, it's the little transformer/charger that you plug into the wall. This gift from Apple was much lighter in weight, more efficient, and more versatile than previous methods of supplying low-voltage DC power to electrical equipment. It had only one glaring fault: instead of delivering only pure DC, it also polluted the electric power grid, the earth, the atmosphere, and even outer space with a broad range of frequencies. But its usefulness made it rapidly indispensible to the mushrooming electronics industry. Today computers, televisions, fax machines, cell phone chargers, and most other electronic equipment used in home and industry depend on it.

Its method of operation makes it obvious why it causes such a huge amount of electrical pollution. Instead of regulating voltage in the traditional way with variable resistors, a switch mode power supply interrupts the current flow tens of thousands to hundreds of thousands of times per second. By chopping up the current into slightly more or fewer pieces, these little devices can regulate voltage very precisely. But they change 50- or 60-cycle current into something very different. The typical switch mode power supply operates at a frequency between 30 and 60 kHz.

Computers, and all other electronic equipment that contains digital circuitry, also emit ultrasonic radiation from other components, as anyone can verify using an ordinary (non-digital) AM radio. Simply tune the radio to the beginning of the dial (about 530 kHz), bring it near a computer—or a cell phone, television, fax machine, or even a handheld calculator—and you will hear a variety of loud screaming noises coming from the radio.

What you are hearing is called "radio frequency interference," and much of that is harmonics of emissions that are in the ultrasonic range. A laptop computer produces such noise even when it is running off the battery. When it is plugged in, the switch mode power supply not only intensifies the noise, but communicates it to your house wiring. From your house wiring it travels onto the distribution line in your neighborhood and into everyone else's homes, and down the ground wire attached to your electric meter into the earth. And the electric power grid, and the earth itself, contaminated with ultrasonic frequencies from billions of computers, becomes an antenna that radiates ultrasonic energy throughout the atmosphere and beyond.

Dimmer switches

Another device that chops up 50- or 60-cycle current is the ubiquitous dimmer switch. Here, too, the traditional variable resistor has been replaced with something else. The strategy is different than in your computer's transformer—the modern dimmer switch interrupts the current only twice in each cycle—but the result is similar: the sudden starting and stopping of the current produces dirty power. Instead of a smooth flow of 50- or 60-cycle electricity, you get a tumultuous mixture of higher harmonics that flows through the light bulb, pollutes house wiring, and irritates the nervous system. A large portion of these unwanted frequencies are in the ultrasonic range.

Power lines

As early as the 1970s, Hiroshi Kikuchi, at Nihon University in Tokyo, reported that significant amounts of high frequency currents were occurring on the power grid due to transformers, motors, generators, and electronic equipment. And some of it was radiating into space. On the ground, radiation in a continuous spectrum from 50 Hz to as high as 100 MHz was being measured at distances as far as one kilometer from both low and high power lines. Frequencies up to about 10 kHz, originating from power lines, were being measured by satellites.

In 1997, Maurizio Vignati and Livio Giuliani, at the National Institute for Occupational Health and Prevention in Rome, reported that they were detecting radio frequency emissions as far as 50 meters (165 feet) from power lines, at frequencies ranging from 112 to 370 kHz, that were amplitude modulated and seemed to be carrying data. These frequencies, they discovered, were deliberately put on the electric power grid by Italian utility companies. And the same technology is being used worldwide. It is called Power Line Communications. The technology is not new but its use has exploded.

Electric companies have been sending radio signals over power lines since about 1922, using frequencies ranging from 15 to 500 kHz, for monitoring and control of their substations and distribution lines. The signals, as powerful as 1,000 watts or more, travel hundreds of miles.

In 1978, small devices appeared in Radio Shack stores that transmitted at 120 kHz. Consumers could plug them in and use the wiring in their walls to carry signals that enabled them to control lamps and other appliances remotely from command consoles. Later the HomePlug Alliance developed devices that use home wiring to connect computers. HomePlug devices work at 2 to 86 MHz, but have modulation components at 24.4 kHz and 27.9 kHz, in the ultrasonic range.

Smart Meters

The use of the power grid to deliver Internet to homes and businesses—called Broadband over Power Lines—has not been commercially successful. But the use of the power grid to transmit data between homes, businesses and power plants is now being implemented for something called the Smart Grid, presently under construction all over the world.

When the Smart Grid is fully implemented, electricity will be automatically sent where it is needed, when it is needed—even rerouted from one region to another to satisfy instantaneous demand. Utilities will continuously monitor every major appliance in every home and business, and will have the ability to automatically regulate thermostats and turn their customers' air conditioners and washing machines on and off during times of greater or lesser demand for electricity. In order to accomplish this, radio transmitters are being installed on everyone's electric meters and appliances, which communicate not only with each other, but with the utility company, either wirelessly, or via fiber optic cable, or by radio signals sent over the power lines. The FCC has allocated frequencies from 10 to 490 kHz for this latter purpose, but utility companies most often use frequencies below 90 kHz, in the ultrasonic range, for long-distance communication over the power lines.

The wireless version of smart meters, especially the variety called a "mesh network," has spread around the world like technological wildfire in the past few years, rapidly becoming the single most intrusive source of electronic noise in modern life. The meters in a mesh network communicate not only with the utility company but with each other, each meter chattering loudly to its neighbors as frequently as two hundred and forty thousand times a day. And the chattering is not silent. Shrill, high-pitched ringing and a variety of hissing and clicking noises are so consistently reported by utility customers following the installation of these smart meters that cause and effect can no longer be denied. The symbol transmission frequency of 50 kHz for many of these systems, and the sheer power of the signal, outclassing other sources of radiation in the modern home, are likely responsible—that, and the pulsatile nature of the signal, like a woodpecker beating incessantly at all hours of the day and night.

Tinnitus today
Tinnitus rates have been rising for at least the last thirty years, and dramatically so for the last twenty.

From 1982 to 1996, the National Health Interview Survey conducted by the United States Public Health Service included questions about both hearing impairment and tinnitus. Although the prevalence of hearing loss declined during those years, the rate of tinnitus climbed by one-third.[35] Later, the National Health and Nutrition Examination Surveys (NHANES), conducted by the Centers for Disease Control, found that the rate continued to climb. In 1982, about 17 percent of the adult population complained of tinnitus; in 1996, about 22 percent; between 1999 and 2004, about 25 percent. The authors of the NHANES study estimated that by 2004, 50 million adults suffered from tinnitus.[36]

In 2011, Sergei Kochkin, the Executive Director of the Better Hearing Institute in Washington, D.C., reported the very surprising result of a nationwide survey, conducted in 2010. What was so surprising was that 44 percent of Americans who complained of ringing in their ears said they had normal hearing. Kochkin simply did not believe it. "It is widely acknowledged that people with tinnitus almost always have hearing loss," he said. He therefore assumed that millions of Americans who complain of ringing in their ears must have hearing loss but don't know it. But his assumption is no longer valid.

Researchers who wish to study real tinnitus have to be careful. If you put the average human being in a soundproof room for several minutes, he or she will begin to hear sounds that are not there. Veterans Administration doctors Morris Heller and Moe Bergman demonstrated this in 1953, and a research team at the University of Milan repeated the experiment fifty years later with the same result: over 90 percent of their subjects heard sounds.[37] Therefore the results of tinnitus surveys may depend on the way the data are gathered as well as the way the questions are worded and even on the definition of "tinnitus." To really find out if tinnitus is increasing, we need virtually identical studies done a number of years apart by the same researchers in the same place on the same population. And we have just such a series of studies.

During the years 1993 to 1995, 3,753 residents of Beaver Dam, Wisconsin, aged 48 to 92, were enrolled in a hearing study at the University of Wisconsin, Madison. Follow-up examinations were done on these subjects at five, ten, and fifteen year intervals. In addition, the children of the original subjects were enrolled in a similar study between 2005 and 2008. As a result, data on the prevalence of tinnitus in this population are available almost continuously from 1993 to 2010.

Since hearing disorders among older adults declined during this period, the researchers expected to see a corresponding decline in tinnitus. They found just the opposite: a steady increase in tinnitus in all age groups during the 1990s and 2000s. For example, the rate of tinnitus among people aged 55 to 59 increased from 7.6 percent (at the beginning of the study) to 11.0 percent, to 13.6 percent, to 17.5 percent (at the end of the study). Overall, the rate of tinnitus in this population increased by about 50 percent.[38]

We also have a series of studies, conducted during these same years, on young children, who have long been assumed to have almost no tinnitus.

Kajsa-Mia Holgers is a professor of audiology at the University of Jonkoping in Sweden. She conducted her first study in 1997 on 964 seven-year-old school children in Göteborg who were undergoing routine audiometry testing—470 girls and 494 boys. Twelve percent of the children said they had experienced ringing in their ears, the vast majority of whom had perfect hearing. Nine years later Holgers, using the same study design and the same tinnitus questions, conducted an identical study on another large group of seven-year-old school children in Göteborg who were undergoing audiometry testing. This time an astonishing 42 percent of the children reported ringing in their ears. "We face a several fold increase in the problem in just a few years," an alarmed Holgers told the national daily newspaper, *Dagens Nyheter*.

To further explore the problem, Holgers gave a detailed questionnaire to middle and high school students aged 13 to 16 during the 2003-2004 school year. More than half of these older students reported tinnitus in some form. Some experienced only "noise-induced tinnitus"

(tinnitus after being exposed to loud noise), but almost one-third of the students had "spontaneous tinnitus" with some frequency.

And in 2004, Holgers studied another group of school children aged 9 to 16, almost half of whom had spontaneous tinnitus. Even more alarming was the fact that 23 percent reported their tinnitus to be annoying, that 14 percent heard it everyday, and that hundreds of children were showing up at Holgers' audiology clinic seeking help for their tinnitus.

If what is occurring in Wisconsin and Sweden is also occurring in the rest of the world—and there is no reason to think otherwise—then in less than two decades, as computers, cell phones, fluorescent lights, and a crescendo of digital and wireless communication signals have penetrated every recess of our environment, at least a quarter of all adults and half of all children have entered a new world in which they must live, learn, and function while attempting to ignore an inescapable presence of intrusive electronic noise.

16. Bees, Birds, Trees, and Humans

ALFONSO BALMORI MARTÍNEZ is a wildlife biologist who lives in Valladolid, Spain. In his official capacity he works in wildlife management for the Environment Department of his region, Castilla y León. But for over a decade he has also labored for a cause that he considers at least as important. "It was in about the year 2000," he says, "that I began to be aware of serious health problems that were being provoked by cell phone antennas in certain individuals who were my neighbors and acquaintances, including a serious situation in the school

Alfonso Balmori Martínez

which my two oldest sons were attending at that time." The problem at the school, the Colegio García Quintana, was not easy to ignore, since he was confronted with it every time he dropped off his sons there. For looming over the playground, like a giant pincushion, a neighboring building's rooftop harbored about sixty transmission antennas of all shapes and sizes.

The antenna farm sprouted its communication crop rapidly, and during the first year of its growth, between December 2000 and January 2002, five cases of leukemia and lymphoma were diagnosed in succession at the school—four in children aged four to nine, and the fifth in a seventeen-year-old young woman who did cleaning. Considering that only four cases of leukemia and lymphoma in children under twelve had been diagnosed during the previous year in all of the province of Valladolid, the community was frightened. The school was closed by the Health Department on January 10, 2002, and was reopened several weeks later after inspectors could find no dangerous conditions within. The antennas, however, were removed by court order of December 2001, and a new organization, AVAATE—Asociación Vallisoletana de Afectadas por Antenas de TElefonía (Valladolid Association of People Affected by Telecommunication Antennas)—arose from their ashes, nurtured in part by a newly motivated Balmori, who was disturbed by what he was learning. People exposed to antennas were not just getting cancer, but in much greater numbers they were getting headaches, insomnia, memory loss, heart arrhythmias, and acute, even life-threatening neurological reactions. "After educating myself over a period of several months," he recalls, "and discovering that something so evident was considered by the authorities to be groundless fear and little more than a 'social psychosis' without scientific basis, I decided to study the effects on fauna and flora. I thought that a 'collective psychosis' or 'groundless fear' could not be attributed to non-human organisms. And so I began to study storks, pigeons, trees, insects, tadpoles... and to publish the results that I was obtaining."

The effects Balmori found were dramatic and universal. Radiation from cell phone antennas affected every species he looked at. Storks, for example. White storks (*Ciconia ciconia*) are a common urban bird in many Spanish cities, inhabiting buildings and church steeples alongside sparrows and pigeons. Selecting 60 rooftop nests scattered throughout Valladolid—30 that were within 200 meters of one or more cell sites, and 30 that were further than 300 meters (about 1,000 feet) from any cell sites—Balmori observed the storks with telescopes during the spring of 2003 to determine their breeding success. By measuring the electric field at each location, he verified that the radiation, on average, was four and a half times more intense at the closer locations. Between

February 2003 and June 2004 he also paid several hundred visits to 20 nests that were within only 100 meters of a cell site in order to observe the birds during all phases of breeding.

The results, for a wildlife biologist, were profoundly disturbing. The nests that were closer than 200 meters from the nearest cell tower fledged half the number of baby storks as the nests that were further away. Of the 30 highly exposed nests, 12 fledged no chicks at all, while only one of the lesser exposed nests was barren. Of the 12 highly exposed nests where no young were flown, some had hatched out no chicks, and others had produced chicks that died soon after hatching. The behavior of the birds that nested within 100 meters of a tower was just as troubling. Stork couples fought over nest construction. Sticks fell to the ground while the couple tried to build the nest. "Some nests were never completed and the storks remained passively in front of the cellsite antennae."

In light of the plummeting numbers of house sparrows in Europe, Balmori also undertook to monitor the number of sparrows at thirty parks and park-like locations in Vallodolid between 2002 and 2006. He visited each of these points on Sunday mornings, once a month for four years, counting birds and measuring radiation. He found not only that sparrows were becoming generally much fewer over time, but that they were incredibly more numerous in less irradiated areas—42 sparrows per hectare where the electric field was 0.1 volts per meter, down to only one or two sparrows per hectare where the electric field was over 3 volts per meter. It was clear to Balmori why the species was disappearing. The United Kingdom had even added the house sparrow to its Red List of threatened and endangered species after the bird's population in British cities fell by 75 percent between 1994 and 2002. "This coincides with the rollout of mobile telephony," he wrote. If the declining trend that he observed in his home town continued, he said, the house sparrow would be extinct in Valladolid by 2020.[1]

And the apparent effects of the radiation were not confined to storks and sparrows. Antennas had been installed in the "Campo Grande" urban park in Valladolid during the 1990s, and Balmori monitored the avian population there for the next decade. These are some of Balmori's observations from 2003:

Kestrel: "A general disappearance of the kestrels that had bred every year on nearby roofs, after antennas for mobile telecommunications were installed."

White Stork: "Although this species is quite opposed to abandon its nest, even under adverse conditions, the nests placed near phone masts' radiation beams gradually disappeared."

Rock Dove (domestic): "Many dead specimens appeared near phone mast areas."

Magpie: "Anomalies were detected in a great number of specimens at points highly contaminated with microwave radiation; such as, plumage deterioration, especially in head and neck, locomotive problems (limps and difficulties in flying), partial albinism and melanism, especially in flanks, and a tendency to stay long in low parts of trees and on the ground."

Green woodpeckers, short toed treecreepers, and **Bonnelli's warblers,** all previously common, disappeared sometime between 1999 and 2001 and were not seen again.

Half of the park's 14 resident bird species had either seriously declined or vanished despite the fact, as Balmori points out, that air pollution improved.

The decline of the house sparrow is a worldwide tragedy. "Twenty, even 10 years ago, it was unimaginable that the house sparrow would be the focus for discussion at an international ornithological or environmental conference," wrote Jenny De Laet and James Denis Summers-Smith. Their 2007 study found spectacular declines of over 90 percent in house sparrow populations in London, Glasgow, Edinburgh, Dublin, Hamburg, Ghent, Antwerp, and Brussels. Scattered throughout Princes Street Gardens, a 50-acre park in central Edinburgh, at least 250 sparrows had resided as recently as 1984. In 1997, only 15 to 30 birds were left, in only a single location. The population of sparrows in Kensington Gardens, a 275-acre park gracing central London, declined from 2,603 in 1925 to just four in 2002. This bird, which has associated with human beings for at least ten thousand years, is vanishing even where there are plenty of seeds and insects, where ornithologists can find no obvious cause for its decline. But there is a cause, and it is hidden in plain sight. Today, twenty-six antenna installations are lined up on the northern, western, and southern borders of Kensington Gardens, operated by Vodafone, T-Mobile, Orange, O2, 3, and Airwave. They are saturating this beautiful park with microwaves so that human visitors can use their cell phones and the police can use their radios. The situation in Edinburgh's Princes Street Gardens is even worse. Thirty-four cell sites surround this much smaller park, most of them less than five meters above the ground. The only location where sparrows still nested in 1997—the Gatekeeper's cottage—is nestled against the bottom of an artificial hill called The Mound, and is the only spot in the entire park that is not in the direct beam of multiple microwave antennas. The irradiation of these parks that began in 1992 parallels the catastrophic collapse of their house sparrow communities.

The situation in Switzerland has become so alarming that the Swiss Association for the Protection of Birds declared the house sparrow "bird of the year" for 2015. A study conducted by zoologist Sainudeen Pattazhy in Kerala, India during 2008 and 2009 found that house sparrows were virtually extinct there. In Delhi, ornithologist Mohammed Dilawar reminisces that "till March 2001, they were in and out of our home. We left for a while to return to see, the commonest bird had flown the nest."[2] Pattazhy's conclusion is the same as Balmori's: cell towers are leaving sparrows no place to live. "Continuous penetration of electromagnetic radiation through the body of birds affects their nervous system and their navigational skills. They become incapable of navigation and foraging. The birds which nest near towers are found to leave the nest within one week," he says. "One to eight eggs can be present in a clutch. The incubation lasts for 10 to 14 days. But the eggs which are laid in nests near towers failed to hatch even after 30 days."[3]

It may seem surprising that sparrows, of all birds, seem to be among the most sensitive to electricity. But we recall from chapter 7 that sparrows were noted to suffer the most among all birds during the influenza pandemic of 1732-1733, following upon the return of sunspots to the sun, and the celestial aurora to polar skies.

The impact of radio waves on bird reproduction is no longer a matter of conjecture. While Balmori was doing his field study on storks, scientists in Greece were proving the effects in their laboratory. Ioannis Magras and Thomas Xenos at Aristotle University of Thessaloniki first exposed 240 newly laid quail eggs in an incubator to the type of radiation emitted by FM radio transmitters. The levels of radiation were about the same as if the birds had built nests one to three hundred yards away from a 50,000-watt tower. But these eggs were exposed for only three days, and for only one hour a day: thirty minutes in the morning and thirty minutes in the afternoon. Forty-five of the embryos died. None of 60 quail eggs, nearby in an unirradiated incubator, died.

Then the same researchers exposed 60 more quail eggs to pulsed microwaves—the type of radiation emitted by cell towers—continuously for three days, this time at only 5 microwatts per square centimeter, a level of exposure commonly found in cities today. Under these conditions 65 percent of the embryos were killed.

In a third experiment 380 chicken eggs were exposed to microwave radiation at a power level of 8.8 microwatts per square centimeter. Instead of irradiating them as soon as they were laid, the researchers exposed the eggs between the third and tenth days of their development. Under these conditions most of the embryos lived but developed abnormally. Under continuous-wave radiation 86 percent of the eggs hatched, but 14 percent of the chicks died soon after birth. Almost half of the remaining chicks were developmentally retarded and 3 percent had severe birth defects. Pulsed radiation produced a similar number of deaths, about half the number of retarded chicks, and twice the number of birth defects. Of 116 unexposed eggs, only two failed to hatch, none had birth defects, and only two were retarded in development.

Disastrous effects of radio waves on birds were first noticed during the 1930s by those who were most intimately connected with them: homing pigeon racers, and divisions of the military that were still using carrier pigeons for communication. Charles Heitzman, a father of the pigeon-racing sport in the United States, and Major Otto Meyer, former head of the United States Army's Pigeon Corps, were both alarmed by the large numbers of pigeons losing their way during the heyday years of the expansion of radio broadcasting.[4]

Apparently, after many pigeon generations, the birds learned to adjust to the new conditions and the problem was largely, though not entirely, forgotten.

Then in the late 1960s a team of Canadian researchers shed some new light on the problem. They were J. Alan Tanner, at Control Systems Laboratory, National Research Council, Canada; César Romero-Sierra, professor of neuroanatomy at Queens University; and Jaime Bigu del Blanco, biophysicist and research associate in the Queens University Department of Anatomy. They began by exposing young chickens to microwave radiation at relatively high power levels, between 10 and 30 milliwatts per square centimeter. The birds usually collapsed to the floor of their cage within 5 to 20 seconds. Even if only their tail feathers were exposed they would scream, defecate, and try to escape. Experiments using pigeons and seagulls gave similar results. But not if the birds were defeathered. Chickens that had been plucked showed no evident reaction to being irradiated until about the twelfth day when their regrowing feathers were about one centimeter long.

The researchers then measured radiation patterns in the laboratory using both individual feathers and arrays of feathers spaced varying distances apart, and proved that bird feathers make fine receiving aerials for microwaves. If this takes place while the bird is flying, they said, "an increase in the microwave field strength should be 'sensed' by the bird."[5]

In the 1970s, Professor William Keeton at Cornell University proved that pigeons are so sensitive to magnetic disturbances that a change in the earth's magnetic field amounting to less than one ten-thousandth of its average value significantly altered the takeoff direction of a bird's homeward flight.

In the 1990s and early 2000s, when cell phone towers proliferated, raising the ambient levels of microwave radiation tens to hundreds of times higher everywhere in the world, when white storks had trouble reproducing near antennas, and when house sparrows made it onto the endangered species list in the United Kingdom, membership in pigeon-racing clubs plummeted and pigeon fanciers were forced to pay renewed attention to a problem they had laid aside in the 1950s. The secretary of the New Ross and District Pigeon Club in Ireland, Jim Power, blamed the new problem of lost birds, which had begun in about 1995, on "satellite television and the mobile telecommunications network." The story made the front page of the *Irish Times*.[6] Both events—the explosion of cell towers and severe pigeon losses—came to America in 1997.[7]

In early October 1998 the story made headlines all over the United States as, during a two-week period, pigeon races far and wide ended in disaster, with up to ninety percent of birds going missing. "They're turning up in barns. Under bird feeders. On window ledges. And sometimes just standing out in the rain," read the first paragraph of an article in the *Washington Post*. Out of 1,800 birds competing in a race from New Market, Virginia to Allentown, Pennsylvania, about 1,500 vanished. In a race from western Pennsylvania to suburban Philadelphia, 700 out of 900 pigeons failed to return. In a 350-mile race from Pittsburgh to Brooklyn, 1,000 out of 1,200 birds never showed up. Very few wild birds were out flying. Hawks were not out hunting.[8] Geese were scattered all over the sky, instead of in normal "V" formations.[9] The trigger for the two weeks of sudden bird disorientation was apparently the commencement of microwave rain falling from satellites. On September 23, 1998, Motorola's 66 newly-launched

Iridium satellites had begun providing the first-ever cell phone service from space, everywhere on earth, to its first 2,000 trial subscribers.

Many members of the British Royal Pigeon Racing Association changed the route their birds flew so as to avoid cell towers and lose fewer pigeons.[10] In 2004, the Association called for more research into the impact of microwave radiation on birds. And as old-time pigeon racers gradually left the sport in discouragement, they were replaced by young enthusiasts who do not remember what it was like when almost all released pigeons would fly directly back to their roosts. The kinds of extraordinary losses Larry Lucero of New Mexico complained about in 1997—an 80 percent loss of birds in eight weeks of racing—are no longer considered unusual. Sankaralingam, the president of the Chennai Homer Pigeons Association in India, reminisces. "Earlier," he says, "before the advent of cell phones, if I liberated 100 pigeons in my Kodungaiyur neighborhood, all would return home in a couple of minutes."[11] Texas pigeon racer Robert Benson states that today, "under the best of conditions, a 25% loss before the race can be expected. It is not surprising to see a 75% loss." "The number of losses occurring each year," says Kevin Murphy at Scotland's Angus College, "is showing no signs of improvement and whenever you speak to pigeon fanciers it's the same old story; high losses in young birds and very few fanciers that are able to build up an established team of 3, 4 and 5 year old experienced birds."

Radio-tagging animals

In an exercise in scientific folly, Murphy is proposing to solve the problem by developing a GSM/GPS device that will be fitted onto pigeons' legs to keep track of wayward birds. Initially this is a research project—designed, he says, to see if solar flares and magnetic storms affect the birds' homing ability. But the devices will track birds by satellites and cell phone towers—the very things that are now responsible for far more pigeon losses than solar flares. Worse, the devices, being radio transmitters themselves, will expose the birds at point blank range to far more radiation than distant cell towers.

Microchipping pigeons to keep track of them is not yet standard practice in this sport. But in recent years pigeon racers are already making a bad situation worse by attaching radio frequency identification (RFID) "chip rings" to each bird's foot during every race, so that when the bird arrives home and crosses the finish line, an RFID scanner automatically records the time of arrival. These are passive devices containing no batteries and rely on external sources of energy to activate them. But sudden deaths of exotic birds immediately after being microchipped are not unusual.[12] And as so many electrically sensitive people are discovering—people who can't handle their own chip-embedded drivers' licenses and passports—the radio frequency oscillators inside even passive devices pollute their immediate environment enough to affect the nervous systems even of organisms without any homing ability.

Attaching a radio tracking device to a wild animal is like giving the animal a cell phone to wear. Land-based wildlife tracking systems use frequencies between 148 and 220 MHz and emit 10 milliwatts of power, day and night. Satellite tracking systems, such as are used to track dolphins and whales, require the animal to wear a much stronger transmitter, radiating from 250 milliwatts up to 2 watts of power— equivalent to giving the animal a satellite phone to wear. These are also used to track turtles, sharks, polar bears, musk oxen, camels, wolves, elephants, and other animals that roam or swim very long distances. They are also used on long-migrating or elusive birds, such as albatrosses, bald eagles, penguins, and swans.

Snakes, amphibians, and bats are being radio tagged. Even butterflies, and fish in lakes and rivers are being outfitted with transmitters. If a creature exists today that is large enough to fashion antennas for, you may be assured

that resourceful wildlife biologists have devised ways to affix them onto members of its species, be it by means of collars, harnesses, or surgical implants. In a misguided effort to discover why honey bees are disappearing, Australia's leading scientific research agency, the Commonwealth Scientific and Industrial Research Organization, is in process of attaching RFID tags with superglue to the backs of two and a half million bees and placing RFID readers inside one thousand hives.

On February 6, 2002, the U.S. National Park Service issued a report warning wildlife biologists that radio tracking devices could radically alter the very behaviors they are using the devices to study, and that not only the physical dimensions of the devices, but the radio waves they emit could be detrimental to the animals' health.[13] Effects of radio-tagging birds, according to this and other reports, have included increased preening, weight loss, abandonment of brood, reduced time spent in flight, increased metabolism, avoidance of water, decreased courtship activity, decreased feeding activity, decreased clutch survival, reduced wing growth, greater susceptibility to predation, lowered reproductive success, and increased mortality.[14]

Radio collared mammals, including rabbits, voles, lemmings, badgers, foxes, deer, moose, armadillos, river otters, sea otters, and wild dogs in the Serengeti[15] have suffered increased mortality, impaired digging ability, weight loss, reduced activity levels, increased self-grooming, altered social interactions, reproductive failure, and profoundly altered sex ratios of offspring. In one study of moose, calves with plain ear tags and calves without any ear tags had equal mortality rates—about 10 percent—while 68 percent of calves with ear tags that contained transmitters died. This had the researchers scratching their heads because they could find no difference between the plain tags and the ones that killed the calves except the presence of radio waves.[16] In another study, involving water voles at England's Bure Marshes National Nature Reserve, colonies that contained radio tagged females gave birth to more than four times as many males as females. The

·researchers concluded that likely none of the radio tagged female voles gave birth to any female offspring.[17]

In some cases radio tagging endangered species may drive them further toward extinction. In 1998, the first Siberian snow tiger ever to go through her pregnancy and give birth while wearing a radio collar delivered a litter of four, of which two died from genetic abnormalities.[18]

The results of an extensive review of the literature, published in 2003, examining 836 scientific studies on radio tagged animals, found that 90 percent of them ignored the effects of the radio tags on the animals, making a tacit assumption that they had no significant impact. But of those studies that asked the question, the majority found one or more detrimental effects of these devices on their bearers.[19]

Migratory Birds

Professor Keeton's work has widespread importance for bird conservation. Even in captivity, when the migratory season is upon them, songbirds will face the direction in which they have an urge to fly. Therefore, scientists at the University of Oldenburg in Germany were shocked to find, beginning in 2004, that the migratory songbirds they had been studying were no longer able to orient themselves toward the north in spring and toward the southwest in autumn. Suspecting that electromagnetic pollution might be responsible, they surrounded the aviaries in which they kept European robins with grounded aluminum sheeting beginning in the winter of 2006-2007. "The effect on the birds' orientation capabilities was profound," wrote the authors of the study, which they published in 2014. Only when the aluminum sheeting was grounded did the birds orient normally in springtime. And since the enclosure, when not grounded, only admitted frequencies below 20 MHz, the birds were evidently being disoriented not by cell towers, but by radiation originating from AM radio towers, as well as from ordinary household electronic equipment. In a rural area outside Oldenburg, the robins were still able to orient themselves without the aluminum screening. But the scientists issued a warning:

"If anthropogenic electromagnetic fields prevent migratory songbirds from using their magnetic compass, their chances of surviving the migratory journey might be significantly reduced, in particular during periods of overcast weather when sun and star compass information is unavailable. Night-migratory songbird populations are declining rapidly."[20]

Amphibians

In 1996, when I was writing my first book, *Microwaving Our Planet: The Environmental Impact of the Wireless Revolution*, the decline of frogs, toads, salamanders, and other amphibians the world over caught my attention like an alarm bell. Why weren't people more concerned, I wondered? Like the debris of recently wrecked craft, this catastrophe should provide the ship of humanity with urgent cause to shift direction. "An Amphibian Horror Story," screamed a headline from *New York Newsday*.[21] "Trouble in the Lily Pads," announced *Time Magazine*.[22] "Space Aliens Stealing Our Frogs," read a supermarket tabloid.[23] It seemed that mutant frogs were turning up by the thousands in pristine lakes, streams, and forests all across the American midwest. Their deformed legs, extra legs, missing eyes, misplaced eyes, and other genetic mistakes were frightening school children out on field trips.[24] Every species of frog and toad in Yosemite National Park, I learned, was disappearing. The boreal toad, which used to be so abundant near Boulder, Colorado that drivers would squish large numbers on mountain roads, had dwindled to about five percent of its former population.[25] When I dug deeper I learned that frogs were falling silent in other countries too, and had been doing so for over a decade. In the Monteverde Cloud Forest Preserve of Costa Rica, the famous and highly protected golden toad, named for its brightly colored skin, had gone extinct. Eight of thirteen frog species in a Brazilian rainforest preserve had vanished. The gastric-brooding frog of Australia, I read, named for its habit of incubating its young in its stomach, "broods no longer."[26] Seventy-five species of the colorful harlequin frogs that once lived near streams in the tropics of the Western Hemisphere had not been seen since the 1980s.[27]

What so puzzled scientists was not just that an entire very ancient class of animals—the Amphibia—were disappearing, but that they were vanishing in so many pristine, remote environments that were thought to be unpolluted. Which is one of the aspects of the story that so grabbed my attention. Environmentalists, for the most part, like the rest of modern humanity, have one terrific blind spot: they don't acknowledge electromagnetic radiation as an environmental factor, and they are comfortable with placing power lines, telephone relay towers, cell towers, and radar stations in the middle of the most remote, pristine mountainous locations, never realizing that they are intensely polluting those environments. I was only speculating, at that time, that the discovery of grossly deformed frogs in the midwest was related to the increasingly frequent reports from farmers in the midwest of cows and horses born with webbed necks and legs on backwards after cell towers were built on or next to their farms.[28] It seemed more than coincidental that the reports of misshapen amphibia were coming from popular lake vacation districts, which were almost certain to have had cell towers built during 1996.

Balmori's curiosity paralleled mine, and in 2009, he put his speculations to the test. During a two-month period he took care of two almost identical tanks of tadpoles of the common frog that he set out on the fifth floor terrace of an apartment in Valladolid. One hundred forty meters (450 feet) away, on the roof of an eight-story building, stood four cellular phone base stations, which were irradiating the neighborhood. The only difference between the two tanks of tadpoles was that a layer of thin fabric was draped over one. The fabric, woven with metallic fibers, admitted air and light but kept out radio waves. The results were a shocking confirmation of what was occurring out in the rest of the world: in a period of two months, the mortality rate was 90 percent in the exposed tank, and only 4 percent in the shielded tank. Almost all of the exposed tadpoles—exposed only to what the residents of the apartment building were also exposed to—swam in an uncoordinated fashion, showed little interest in food, and died after six weeks. Balmori titled his 2010 article, "Mobile Phone Mast Effects on Common Frog (*Rana temporaria*) Tadpoles: The City Turned Into a Laboratory."

In the late 1990s, researchers in Moscow had put these kinds of effects to the test in another urban laboratory, using another device that we all take for granted. They exposed developing frog embryos and tadpoles to an ordinary personal computer. The resulting frogs had severe malformations that included anencephaly (absence of a brain), absence of a heart, absence of limbs, tail necrosis, and other deformities that were "incompatible with survival."[29]

Insects

The insect world is as susceptible to electromagnetic pollution as the amphibian world. In fact, as Alexander Chan discovered in 2004, it is so easy to demonstrate the effects of computers and cell phones on diminutive creatures that even a sophomore in high school can do it for a science fair project. Then fifteen years old and a student at Benjamin Cardozo High School in Queens, New York, Chan exposed fruit fly larvae daily to a loudspeaker, a computer monitor, and a cell phone and observed their development. The flies that were exposed to the cell phone failed to develop wings. "Radiation and electromagnetic emissions are really more harmful than anyone realizes," the stunned teenager concluded.[30]

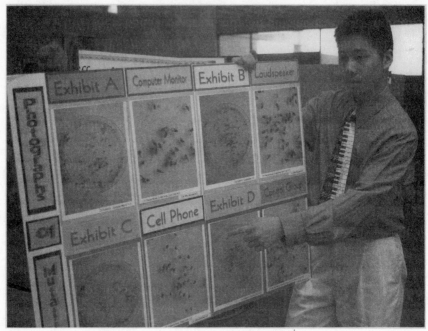

Photo by Alan Raia, *New York Newsday*

At the University of Athens, Dimitris Panagopoulous has been doing similar work with fruit flies for a decade and a half, and producing results that are just as alarming. Like Chan—and unlike most other scientists doing research on electromagnetic radiation—he and his colleagues in the Department of Cell Biology and Biophysics decided to expose their flies not to specialized equipment, but to an ordinary cell phone in use. In their first experiments, in 2000, they found that a few minutes' exposure was enough to radically interfere with fly reproduction. Exposing adult flies to the antenna of a working cell phone for just six minutes a day for five consecutive days reduced the number of eggs they laid by 50 to 60 percent. When the insects were exposed for only two days, i.e. a total of twelve minutes of radiation, the number of eggs was reduced by an average of 42 percent. Even flies that were exposed for only one minute a day for five days produced 36 percent fewer offspring than their unexposed cousins. Regardless of whether just male flies, just female flies, or both were exposed, the number of offspring was greatly reduced. Their experiments cried out for an explanation, because such rapid sterilization was an effect scientists were used to seeing from X-rays, not from an ordinary cell phone.[31] So in follow-up experiments, after zapping the flies with a cell phone for five days—again for six minutes a day—the researchers killed the flies and used a standard technique—the TUNEL assay—to look for fragmented DNA in the ovaries and egg chambers of the female flies. Using this technique they proved that the brief exposure to a cell phone was causing the death and degeneration of 50 to 60 percent of both eggs and their supporting cells at all stages of development.[32]

In later experiments these scientists have found "intensity windows" of maximal effect—a not uncommon finding in electromagnetic research. In other words, the greatest damage is not always done by the greatest levels of radiation. Holding your cell phone away from your head may actually worsen the damage. Using a 900 MHz phone, Panagopoulos' flies produced even fewer offspring when the antenna was held a foot away—reducing the exposure level by a factor of almost 40—than when the antenna was actually touching the vial of flies. With an 1800 MHz phone, maximum mortality occurred at a distance of eight inches.[33] In a large series of further experiments, exposure to a cordless phone base station, a cordless phone handset, a WiFi router, a baby monitor, a microwave oven, and several different kinds of bluetooth devices each lowered the numbers of offspring of two different species of fruit flies by up to 30 percent. Exposure time varied from 6 minutes, just once, up to thirty minutes a day for nine days. Every experiment, regardless of exposure time, produced cell death in the developing eggs and at least a ten percent reduction in the number of offspring.[34]

And in Belgium, entomologist Marie-Claire Cammaerts has shown, in experiments that any high school student could duplicate, that a cell phone is clearly and obviously dangerous even when it is turned off, as long as the battery remains in it. She brought thousands of ants into her laboratory at the Free University of Brussels, placed an older model flip phone under their colonies where they could neither see nor smell it, and simply watched them walk. When the phone contained no battery it affected the ants not at all. Neither did the battery alone. But as soon as the battery was placed in the phone—even though it was still turned off—the ants' helter-skelter movements became radically disturbed. The little creatures darted back and forth with increased vigor, as if trying to escape an enemy they could not see. The rate at which they changed directions—their angular speed—increased by 80 percent. When the phone was then put into standby mode, they changed directions even more. Finally, Cammaerts turned the phone on. Within two to three seconds, the insects visibly slowed down.

Cammaerts next exposed a fresh ant colony to a smartphone and then a "DECT" cordless phone. In each case the creatures' angular speed doubled or tripled, while their actual walking speed drastically slowed. This happend within one to three seconds. When the DECT phone was on, the ants were "nearly paralyzed." After being thus exposed for three minutes to each of the two pieces of equipment, they required two to four hours before they appeared normal again. Cammaerts then repeated the experiment with a fresh colony, this time placing a flip phone in standby mode under the ants' nest instead of under its foraging area. Immediately all the ants left their nest, taking their eggs, larvae, and nymphs with them. "It looked spectacular," she said. "They relocated their nest far from the place under which the mobile phone was located. After the experimentation, when the mobile phone has been removed, the ants returned to their initial nest, transporting back their brood into the nest. This relocation lasted about one hour."

Finally, Cammaerts tested a WiFi router, placed between two colonies of ants, about one foot away from each colony. While the router was still switched off, nothing unusual happened. But "after a few seconds of exposure, the ants clearly presented signs of bad health and, consequently, a disturbed behavior." After being exposed to the router for thirty minutes, the ants had to recover for six to eight hours before foraging as usual again. "Unfortunately," wrote Cammaerts, "several ants never recovered and were found dead a few days later."

For his part, Panagopoulous, in a chapter of a 2012 book about *Drosophila melanogaster*, has issued a severe and unusual warning to the world: "The experimental results of ours as well as of other experimenters show that microwave exposure even for a few minutes per day and for only a few days, at exposure levels encountered in our everyday environment, is maybe the most intense modern environmental stress factor compared to other environmental stress factors tested so far, like starvation, heat, chemicals, electric or magnetic fields." He warned that DNA damage to the developing egg may "result in inherited mutations transferred to subsequent generations. For this reason the biological changes due to microwave radiation may be far more dangerous as they may not be restricted only to changes in reproductive capacity."

Colony Collapse Disorder

In recent years an apocryphal story has circulated about Albert Einstein. "If the bee disappears from the surface of the earth," he is supposed to have said, "man would have no more than four years to live."

Perishing honey bees do constitute a warning to the world, but the real story is not circulating because it is not yet acceptable to remove the cultural blinders regarding electricity. Beekeepers the world over are still poisoning their bees against parasites that are not killing them, instead of paying attention to the influence that is.

"I observed a pronounced restlessness in my bee colonies," wrote Ferdinand Ruzicka to the Austrian beekeeping community in 2002, "and a greatly increased urge to swarm." Ruzicka, a medical physicst retired from the University of Vienna, is also an amateur beekeeper. He observed the strange behavior after telecommunications antennas appeared in a field near his hives. "I am a frame-hive beekeeper," he wrote. "The bees now built their honeycombs not in the manner prescribed by the frames, but in a helter skelter fashion. In the summer, the colonies collapsed without any obvious cause. In the winter, despite snow and below zero temperatures, the bees would fly out and freeze to death next to the hive. Colonies that exhibited this behavior collapsed, even though they were strong, healthy colonies with active queens before the winter. They were provided with adequate additional food and the fall pollen supply had been more than sufficient."

Ruzicka told his story in *Bienenwelt* ("Bee World") and published a survey form in *Bienenvater* ("Beekeeper"),[35] requesting to be contacted by others with antennas near their hives. The majority of *Bienenvater* readers who filled out his form corroborated what he had written: their bees had become suddenly aggressive when the antennas appeared, and had begun to swarm; their healthy colonies had vanished for no other reason.[36]

As we saw in chapter 9, bee colonies have been disappearing near communication towers for over a century. On the small island lying off England's southern coast where Marconi sent the world's first long-distance radio transmission in 1901, the bees began to vanish. By 1906, the island, then host to the greatest density of radio transmissions in the world, was almost empty of bees. Thousands, unable to fly, were found crawling and dying on the ground outside their hives. Healthy bees imported from the mainland began dying within a week of arrival.

During the next few decades "Isle of Wight disease" was reported throughout Great Britain and in Italy, France, Switzerland, Germany, Brazil, Australia, Canada, South Africa, and the United States.[37] Almost everyone assumed it was infectious, and in 1912, when Graham Smith at Cambridge University found a parasite called *Nosema apis* in the stomachs of some diseased bees, most people thought the mystery had been solved. However, this theory was soon disproven by John Anderson and John Rennie in Scotland; swarms of bees that were "crawling" with Isle of Wight disease were free of *Nosema*, while healthy stocks were

found teeming with the parasite. Finally the two researchers deliberately infected a colony with *Nosema*. It did not produce disease.

So the search went on for a different parasite, and in 1919 Rennie
presented *Acarapis woodi*, which inhabited bees' breathing passages.
His article in the *Transactions of the Royal Society of Edinburgh* had such
wide influence that the tracheal mite is today regarded as one of the
two major parasitic infections of bees that are responsible for colony
collapse disorder. It supposedly kills bees by sucking their blood and
clogging their breathing tubes. In fact, this is so widely accepted that
it is standard practice for commercial beekeepers to treat all their bees
with miticides to kill both tracheal mites and a second kind of mite, the
Varroa mite. However, in the late 1950s the tracheal mite theory was
disproven, too, by the eminent British bee pathologist, Leslie Bailey.
Not only did he show that mite-infested bees did not die at greater
rates than non-infested bees, but he deliberately infected healthy bees
with the parasite and proved that it did not cause disease. The only
effect of infestations, wrote Bailey in 1991, is to "shorten very slightly
the life of bees, but usually causing no obvious sickness in spite of the
abnormal appearance of infested tracheae."

Bailey also warned against attaching too much importance to the
Varroa mite, which, he said, had achieved its notoriety partly because
of its size: it is the only common parasite of honey bees that can be
seen with the naked eye and identified with a hand lens.[38] *Varroa* mites,
after all, while not harmless, have coexisted with wild populations of
honey bees for a century in Japan[39] and Russia,[40] and more recently
in Serbia,[41] Tunisia,[42] Sweden,[43] Brazil,[44] Uruguay,[45] and even parts of
California[46] and New York.[47] Other environmental factors, said Bailey,
determine the amount of damage done by this parasite.

The problem of Isle of Wight disease smoldered for decades, not
often making the news. But the number of managed honey bee colonies in the United States has been quietly declining since the 1940s.[48]
During the 1960s and 1970s unexplained large losses acquired a
new name— "disappearing disease"—and was reported in Montana,
Nebraska, Louisiana, California, Texas, Europe, Mexico, Argentina,

and Australia. Beekeepers would open their hives in autumn or winter to find ample supplies of stored pollen and honey but no bees. Where some dead or living bees remained, they were not malnourished and had no mites or other parasites, bacteria, viruses, or poisons. Attempts to transmit the condition by introducing bees from "diseased" hives into healthy ones failed. When a survey was conducted by the United States Department of Agriculture in 1975, the problem turned up in 33 states, with beekeepers often volunteering that it had been prevalent in their colonies for ten or fifteen years, and that it was getting worse with each passing year.[49]

Then, during the last half of the 1990s, when the telecommunications industry was beginning to weave its web of antennas over cities, farmland, and wildland, American farmers reawoke to a crisis. The smoldering, half-forgotten problem about disappearing bees was erupting in flames. "Farmers Stung By Bee Shortages," warned a headline in the June 15, 1996 edition of the *Washington Post*. During the previous winter, beekeepers had lost 45 percent of their hives in Kentucky, 60 percent in Michigan, 80 percent in Maine.[50] Farmers were also waking up to the fact that wild bees weren't going to be there to take over the job of pollinating their crops, because 90 percent of all feral honey bee colonies nationwide had already disappeared.[51] All this havoc—at least in the United States—was thought to have been caused by two bee parasites, the tracheal mite and the even more voracious *Varroa* mite, assumed to have hitchhiked into the United States in shipments of infected bees from Europe and Asia during the 1980s.

But the alarm spread to Europe during the winter of 2002-2003. Officially there was no panic: colony losses were "only" 20 percent in Sweden and 29 percent in Germany. Swedish beekeeper Börje Svensson, who published an article titled "Silent Spring in northern Europe?", begged to differ. When he opened up his hives that winter, 50 out of 70 colonies were devoid of life. A neighbor had lost 95 of 120 colonies, and another neighbor lost 24 of 25. Fellow beekeepers in Austria, Germany, Belgium, Denmark, and Finland were reporting similar huge losses, although many could find no *Varroa* mites, and no sign of foulbrood, sacbrood, chalkbrood, *Nosema*, or other bee diseases.

Finally, during the winter of 2006-2007, what was once known as Isle of Wight disease became a worldwide panzootic, frightening farmers and the public everywhere, and was given yet another name: colony collapse disorder.[52] The United States lost one-third of its honey bees in just a few months, with many beekeepers experiencing a total loss of their bees.[53] First thought to be confined to Europe, North America and Brazil,[54] colony collapse disorder soon spread to China, India, Japan, and Africa.[55] Farmers in many countries are pollinating growing acreages of crops with half as many bees, and replenishing their losses with greater difficulty and expense with each succeeding year.

And the culprit, according to a study conducted by a joint team of American and Belgian researchers, does not seem to be tracheal mites, *Varroa* mites, *Nosema*, or any other particular infectious disease vector. During the disastrous winter of 2006-2007, this team, headed by Jeffery Pettis of the United States Department of Agriculture's Bee Research Laboratory, examined thirteen large apiaries owned by eleven different commercial beekeepers in Florida and California, and to their amazement were unable to find any specific nutritional, toxic, or infectious factor that differentiated bees or colonies with and without colony collapse disorder. Tracheal mites were actually more than three times as prevalent in the *healthy* colonies as in the decimated colonies. Even the supposedly devastating *Varroa* mite was not more prevalent in collapsed or collapsing colonies. The only helpful conclusion that these scientists were able to come to was that "some other factor" must be responsible for the bees' weakened state, and that the "other factor" seemed to be location-specific: colonies with this disorder tended to cluster together.

The picture of this disease that has beekeepers so thoroughly baffled resembles nothing so much as the scene of an apparent mass murder where there is not even any real evidence of a crime. A million colonies a year in the United States disappear overnight without leaving a trace. The queen bee and mother of the hive is simply abandoned by the workers and left to starve and die. What has scientists even more stumped is that the dead colonies tend to be left alone even by the parasites that normally infest dead honey bee colonies. It is as though there were a large "KEEP OUT" sign at the entrance to these hives that is respected by friend and foe alike.

The international beekeeping community is extremely resistant to giving up its long-standing belief in the infectious nature of bee losses, and so, in the absence of evidence, most beekeepers are falling back on the only thing they know: more toxic pesticides to kill mites.[56]

But the decimation of so many other insect species that are not subject to the same parasites is a strong hint that a non-infectious agent is at work. The Franklin bumble bee, once prevalent in southwestern Oregon, has not been seen in a decade. Until the mid-1990s, the western bumblebee was abundant in forests, fields, and urban backyards throughout western North America, from New Mexico to Saskatchewan to Alaska. It has vanished except for small pockets in the Colorado Rockies. The rusty-patched bumble bee, a familiar visitor to flowers on the Cornell University campus when I was a student there, has not been seen in New York State since 2004. Once common in 26 states and two Canadian provinces, this insect has disappeared from the eastern United States and Canada and has drastically declined in the American midwest. The Xerces Society for Invertebrate Conservation lists 57 species of bees and 49 species of butterflies and moths native to North America and Hawaii as vulnerable, imperiled, or extinct in their entire range.[57] The Massachusetts Division of Fisheries and Wildlife lists 46 species of butterflies and moths that are threatened and endangered in Massachusetts.

Exquisite sensitivity to electromagnetic fields has been demonstrated in a variety of insects. Termites, for example, will avoid building their galleries near other groups of termites, so as not to compete for food. In 1977, Günther Becker proved that the signal that enables groups of termites to avoid competing with each other passes through walls and can be blocked by aluminum, but not by thick polystyrene and not by solid glass. The signal blocked by the aluminum had to be alternating electric fields emitted by the insects.

It must not be forgotten, warns German biologist Ulrich Warnke, that every insect is equipped with a pair of antennas, which are demonstrably electromagnetic sensors.[58] In fact, the signals communicated between honey bees when they meet and touch antennas can be recorded by an oscilloscope and appear to be frequency modulated between 180 Hz and 250 Hz.[59]

And the famous waggle dance, Warnke reminds us, by means of which honey bees tell each other the precise direction of food sources with respect to the sun, depends on their knowing the exact position of the sun, even on cloudy days, and within the darkness of the hive. Bees accomplish this feat by sensing minute variations in the earth's magnetic field—a sense, he says, that can be rendered useless under the assault of wireless transmissions with their constantly changing magnetic fields.[60]

The quickest way to destroy a bee hive, investigators have found, is to place a wireless telephone inside it. The results of such experiments, considering the complete denial by our society that wireless technology has any environmental effects at all, have been almost unbelievable.

In 2009, environmental scientist Ved Parkash Sharma and zoologist Neelima Kumar, at Panjab University in India, placed two cell phones each—one in talk mode and one in listening mode in order to maintain the connection—in two of four hives. They turned them on at 11:00 in the morning for 15 minutes, and at 3:00 in the afternoon for another 15 minutes. They did this twice a week between February and April. As soon as the phones were turned on the bees would become quiet and still "as if unable to decide what to do." During the course of three months fewer and fewer bees flew in and out of those two hives. The number of eggs laid by the queen declined from 546 to 145 per day. The area under brood declined from 2,866 to 760 square centimeters. Honey stores declined from 3,200 to 400 square centimeters. "At the end of the experiment there was neither honey, nor pollen nor brood nor bees in the colony resulting in complete loss of the colony," wrote the authors.

The following year Kumar performed a landmark experiment, described in more detail in chapter 11, that showed dramatically and simply how electromagnetic fields interfere with cellular metabolism. She repeated the exposure of the previous year and then analyzed the bees' blood, or hemolymph, as it is called. After the cell phones had been on for only ten minutes, the concentration of glucose, cholesterol, total carbohydrates, total lipids, and total protein rose tremendously. In other words, after just ten minutes of exposure to cell phones, the bees practically could not metabolize sugars, proteins, or fats. As in humans (see chapters 11, 12, 13, and 14), their cells were becoming oxygen starved. But it happens much faster in bees. When the phones were left on longer than 20 minutes, the bees, at first quiet, became aggressive and started beating their wings in agitation.

Daniel Favre, at the Apiary School of the City of Lausanne, Switzerland, repeated the experiment and took it yet another step farther: he made a detailed analysis of the sounds made by the suddenly aggressive bees. He confirmed that bees exposed to a cell phone would become quiet and still when first exposed to a cell phone, and that within 30 minutes they would start to produce loud, high frequency sounds. When the phones had been on for 20 hours, the bees were still buzzing like mad 12 hours later. When Favre analyzed the sounds, he determined that they were the so-called "worker piping," which is usually produced by bees only when they are preparing to swarm, shortly before takeoff.

Favre's bees did not actually leave their hive after a single 20-hour exposure, but Sainudeen Pattazhy's bees did, after a much shorter total exposure. A professor at Sree Narayana College, Pattazhy basically repeated Kumar's initial experiment, except that instead of exposing his bees only twice a week he exposed them briefly everyday. He placed one cell phone inside each of six bee hives and turned the phone on for just ten minutes, once a day for ten days. While the phone was on, the bees became still. An average of 18 bees left the hive per minute while the phone was on, compared to 38 per minute at other times. The egg-laying rate of the queen declined from 355 to 100 per day. And after ten days no bees were left in any of the hives.[61]

Europe's first UMTS network, which is now known as "3G," short for "third generation," and which turned every cell phone into a computer, and every cell tower into a transmitter of broadband radiation, went into service in the fall of 2002—just before the disastrous winter during which so many of Europe's honey bees vanished.

Warnke believes that HAARP—the High-frequency Active Auroral Research Project—is responsible for the worldwide outbreak of Colony Collapse Disorder that began in the winter of 2006-2007.[62]

An "ionospheric heater" owned until recently by the United States Air Force and operated jointly with the Navy and the University of Alaska, HAARP is only the most powerful radio transmitter on earth. Capable of emitting a peak effective radiated power of four billion watts, its purpose is to set the biosphere to ringing. HAARP, whose 180 antenna towers sit on the northwest tip of Alaska's Wrangell-St. Elias National Park, has turned the ionosphere itself—the life-giving layer of sky to which every creature is tuned (see chapter 9)—into a gigantic radio transmitter useful for military communications, including communication with submarines. By aiming a narrow beam of pulsating energy upwards, there near the North Pole where the aurora meets the earth, Project HAARP can force rivers of sky to broadcast radio transmissions at the frequency of the pulsations, and to send those signals to almost everywhere on earth. In 1988, when planning for HAARP was still in its early stages, physicist Richard Williams, a consultant to Princeton University's David Sarnoff Laboratory, called the project "an irresponsible act of global vandalism." "Look at the power levels that will be used!" he wrote in *Physics and Society*, the newsletter of the American Physical Society. "This is equivalent to the output of ten to 100 large power-generating stations." In 1994, when HAARP's first 18 antennas were about to be put into service, Williams was interviewed by *Earth Island Journal*. "A ten-billion-watt generator," he said, "running continuously for one hour, would deliver a quantity of energy equal to that of a Hiroshima-sized atomic bomb."

In March 1999, HAARP expanded to 48 antennas and an effective radiated power of almost one billion watts. The rest of its complement of 180 antennas were delivered between 2004 and 2006, enabling the facility to reach its full intended power during the winter of 2006-2007. Although the Air Force shut HAARP down in 2014 and proposed to dismantle the facility, it instead was acquired by the University of Alaska Fairbanks, which reopened the facility in February 2017 and has made it available to the scientific community for research. The university is operating the facility at a loss, and it announced in 2019 that if it does not get sufficient funding, it will shut down HAARP permanently.

The frequencies of HAARP, says Warnke, superimpose unnatural magnetic fields on the natural resonant frequencies of the sky, whose daily variations have not changed since life appeared on earth. This is disastrous for bees. They "lose an orientation," he says, "that served them for millions of years as a reliable indicator of the time of day."

The Path Into the Dying Forest

Around 1980 the world awoke to a new, seemingly random environmental problem: forest die-off. Large swaths of trees would grow up stunted, age prematurely, drop their leaves, and perish without visible cause. Other stands, tall and vigorous, would suddenly lose all their upper leaves and die from the top down. In the Great Smoky Mountains of Tennessee, in Canada's Bay of Fundy, and in Central Europe such tragedies were blamed on acid rain, contaminated by the sulphuric effluent of industrial civilization. But on remote mountain ridges, forests breathing unpolluted air were suffering from a similar infirmity. Wolfgang Volkrodt, retired physicist and electrical engineer, thought he knew why.

Volkrodt, who formerly worked for Siemens, the multinational technology giant, had become interested in trees because of the strange behavior of the forests in the wooded development at Bad Neustadt, Germany, where he lived. On the north side of his home the fir trees had been sickly for years, while on the south side all the trees were strong and robust. How, he mused, could acid rain fall only on one side of his house? This astute observation led him to investigate not only the trees but the soil. "It seems clear that soil acidification in Central Europe has increased significantly during the past decades," he wrote later. "Paradoxically this is true even in clean air regions that receive only traces of 'acid rain.' This poses the puzzling question of how soil can become acidic in the absence of chemical precipitation from the air. There must be additional culprits."

The existence of a military installation twelve miles to the north of his home made an impression on Volkrodt as an electrical engineer, and when he took measurements on his property he found that the dying trees north of his house were not only being exposed to distant military radar, but happened to be in the direct beam of a nearby transmitter used for postal communications. The healthy trees south of his home were situated where they were not exposed to either. He then set out to determine whether this was just coincidental.

"I traveled through the mountains of the Fichtelgebirge, the Black Forest, the Bavarian Forest and the Salzburger Land," he wrote. "And in every location where military radar stations or postal, telephone and telegraph relay towers are subjecting the forest to radiation, the tree damage cannot be overlooked. I also traveled around Switzerland. The situation is exactly the same." And wherever he saw damaged forests near radar stations, there the soil was dead and acidic.

At the International Congress on Forest Decline Research in Lake Constance in 1989, Volkrodt displayed hundreds of photographs of dead forests, all of which were in line of sight of a radar installation, and he presented his theory. "Needles and leaf-ribs of trees are resonant absorbers like antennas," he said. "And it may be that microwave energy will be changed into an electrical current. The electrons migrate as ionic bonds from the leaves, the trunk and then through the roots into the soil. In the soil a kind of electrolytic deposition happens, making aluminum, among other things, soluble and generally making the soil acidic similar to the effect of acid rain." Of course, no formal studies had been done on the magnitude of induced currents in trees caused by radar stations, but his theory generated interest among the forest biologists at the conference and elsewhere. He soon was receiving reports from observers in Canada confirming his prediction that the line of early warning radar stations lining the Canadian far north from Atlantic to Pacific were killing the trees in front of them.

Forest damage in West Germany during the Cold War.

From *Forest Decline*, Jülich, Germany, 1988, published by Jülich Nuclear Research Center for the U.S. Environmental Protection Agency and German Ministry of Research and Technology.

Following up on experiments by forest biologist Aloys Hüttermann, who had measured microwave-induced current flow in tree needles and leaves, Volkrodt did some elementary calculations. He assumed that a tiny amount of energy—a tenth of a watt—was being absorbed by a section of forest standing before a directional radio antenna transmitting long distance telephone service at a few watts of power from one point to another. He further assumed that the stand contained 100 trees, each having 100 square meters of leaf surface, which was capable of converting the microwave energy into an electrical current. Intuitively, the total of only a tenth of a watt of microwave radiation, spread out over an acre of soil, seemed insignificant, but when Volkrodt took into consideration the factor of time, he came to an astonishing conclusion. "Within 10 years of exposure to the directional energy," he wrote, "the seemingly minute 0.1 watts received by the group of trees adds up to 8.8 kilowatt hours." 8.8 kilowatt hours of electricity, he calculated, is sufficient to create 2,000 liters of hydrogen gas within the soil by the electrolytic splitting of water. This would acidify the soil, even without a trace of acid rain. And when Volkrodt considered that radar installations sometimes broadcast not a few watts but a few million watts, he realized that such an installation could acidify a phenomenal amount of soil.

Partial confirmation of Volkrodt's theory came from unpublished field experiments in Switzerland. Young fir trees were irradiated with microwaves at a power density below 10 milliwatts per square centimeter. After four months the trees had lost nearly all their needles, and the soil in which they were growing was dead and acid.

Meanwhile, foresters in Central Europe were observing a very rapid deterioration in forest health. In West Germany, where the alarm was first sounded, white fir trees began mysteriously to decline around 1970. Spruce caught the affliction in about 1979, Scots pine in about 1980, and European beech in about 1981. Before long, symptoms of ill health and abnormal growth afflicted almost every species of forest tree and several herbs and shrubs. The area of forest affected rose from about 8 percent in 1982, to about 34 percent in 1983, to about half the forests in 1984.[63] Die-off was most severe at high elevations. To Volkrodt, a simple explanation was at hand: a large number of powerful radar stations, built or upgraded during the 1970s and 1980s, were irradiating the mountain ranges on both sides of the border between East and West Germany.

When Germany was reunited, and the radars protecting its former parts were scrapped, Volkrodt made another prediction: "The forest, with parts of it having been irradiated by these installations for two to three decades, now has a chance to regenerate." And this prediction also came true. In 2002, the United Nations Economic Commission for Europe, in cooperation with the European Commission, surveyed the conditions of all of the forests of Europe. The resulting report painted a telling portrait: during the mid-1990s, following the end of the Cold War, the forests not only in Germany, but throughout Europe, had recovered their vitality.

During those years of the 1990s, famous experiments were done in Switzerland, Poland, and Latvia, sponsored by the governments of those countries, proving the effects of radio transmissions on people, farm animals, wildlife, and forests—experiments that it would shortly not be possible to do any more.

***

The small town of Skrunda, 150 kilometers from Latvia's capital, Riga, was once just a few kilometers away from a Russian early warning radar station that scanned the northwestern sky. Its two units went into operation in 1967 and 1971. From the very beginning these radars, situated in a green valley surrounded by farms, were the subject of vigorous complaints from local residents—complaints that the radiation was destroying their health, their crops, their animals, and their forests. Finally, in 1989, as the Berlin Wall fell and the Cold War was ending, the government put out a call for scientists to submit proposals for studies that would put these claims to the test. Physicians, epidemiologists, cell biologists, botanists, ornithologists, and physicists from throughout Latvia converged on the region to do field studies.

And to the surprise of the organizers the researchers, almost without exception, found evidence of biological damage. The findings were presented at a conference held June 17 to 21, 1994, called *The Effect of Radio Frequency Electromagnetic Radiation on Organisms.*

School children in the area—even children who lived twenty kilometers away from the radar—had impaired motor function, memory, and attention. When asked to press two keys with their right and left hands as fast as they could for thirty seconds, children from Skrunda could not do it as fast as children from Preiļi, an agricultural community similar in every respect except that no radar station stood nearby. Asked to press a button when they heard a tone or saw a flash of light, they could not react as fast. The Preiļi children could remember longer and more complex numbers than the Skrunda children. And within Skrunda, children who lived on the western slope of the valley, directly exposed to the radar, had worse memories than children who lived further away. Standard psychological tests evaluated their ability to focus attention on a task, and to switch attention between tasks. Again the Preiļi children did better than the lesser exposed Skrunda children, who did better than the children living on the western slope.

The directly exposed children also had lower lung capacity and higher white blood cell counts than other children. In fact, the entire population of Skrunda had higher white cell counts and suffered from more headaches and sleep disturbances than a more distant community.[64] The radiation even appeared to have impacted human reproduction, affecting the sex ratio of the community. Fewer boys than girls had been born during the early years of the radar. There were 16 percent fewer grade 9 boys in Skrunda as a whole, and 25 percent fewer in the directly exposed area.[65]

The effects on farm animals and wildlife were just as obvious. Blood samples were drawn from sixty-seven Latvian Brown cows that grazed on land in front of the radar station. Chromosome damage was found in more than half.[66]

Six hundred nest-boxes were provided for birds, placed at distances of up to nineteen kilometers from the radar station. Only 14 percent of the nest-boxes were occupied by pied flycatchers, an extremely low number for Latvia. The numbers of great and blue tits that took up residence in the nest-boxes increased steadily with distance from the radars.[67]

The effects on the area's forests were equally profound. Stands of Scots pines were sampled at twenty-nine locations at various distances in front of the radars. The trees in all of the stands, without exception, had laid down much thinner growth rings, beginning precisely in 1971 and continuing throughout the period of operation of the radars. The average growth rings were half as wide as before the radars were constructed.[68]

Pine cones were collected from the tops of fifty- or sixty-year-old trees. All of the seeds from trees that were less exposed to the radars germinated, while only a quarter to a half of the seeds from highly exposed locations did. Abundant secretion of resin from pine needles indicated that the exposed trees were aging prematurely.[69]

In yet another experiment, newly germinated duckweed plants were exposed to the radars from two kilometers away for just 88 hours and then moved to a distant location. Duckweed is a tiny floating plant that lives on the surfaces of ponds and reproduces by budding. For the first twenty days after exposure, the plants reproduced at nearly double the normal rate. Reproduction then dropped precipitously. Ten days later, many of the plants began to grow abnormally. They became misshapen, sprouted roots that grew upwards, budded from the wrong side, and produced deformed daughter plants. Exposure of additional plants to the radar for just 120 hours reduced their average lifespan from 86 days to 67 days, and lowered their reproductive capacity by 20 percent.[70]

The Skrunda Radio Location Station was shut down permanently on August 31, 1998.

———•◦•———

Konstantynów is a country crossroads near the river Vistula in the center of Poland, about 60 miles northwest of Warsaw. Extensive pine forests grow to the west. For seventeen years, from 1974 until 1991, it was also the Voice of Poland, for next to the village stood the long wave radio antenna that broadcast Polish language programming throughout Europe. More than 2,100 feet high, it was the tallest manmade structure in the world, and at two million watts, Warsaw Central Radio was also one of the most powerful radio stations in the world. And for seventeen years, the people in the surrounding villages complained that their health was being destroyed.

In 1991, a government study proved them right. The research, overseen by Dr. Wiesław Flakiewicz, who worked in the Radiation Protection Department in the County of Płock, was simple and inexpensive: it consisted of analyzing blood samples drawn from 99 randomly selected residents of two communities, Sanniki and Gabin, each six kilometers from the tower. The first results indicated that something was indeed affecting the residents' health. For 68 percent of the people in Gabin had abnormally high levels of cortisol, a stress hormone. Forty-two percent had hypoglycemia, 30 percent had elevated thyroid hormones, 32 percent had high cholesterol, and 32 percent had abnormally high red blood cell counts. Fifty-eight percent had disturbed electrolytes: they tended to have high calcium, sodium, and potassium levels, and low phosphorus. The pattern in Sanniki was similar, except that thyroid and electrolyte disturbances were even more common and serious, and 41 percent of the population also had elevated platelets, indicating overstimulation of their bone marrow.

Then, on August 8, 1991, a serendipitous event took place: the tallest structure in the world fell down. Flakiewicz took full advantage of the opportunity, and in October he recalled the 50 subjects from Gabin into his laboratory to draw a fresh set of blood samples. The new results were startling. A handful of the youngest subjects, who had been the most severely affected by the radiation, still had abnormal glucose levels and red blood cell counts, and the older subjects still had elevated cholesterol. But all of the electrolyte levels, all of the thyroid levels, and all of the cortisol levels, without exception, were now completely normal.

Experiments on plants exposed to the radio station produced equally stunning results. Dr. Antonina Cebulska-Wasilewska, who worked at the Institute of Nuclear Physics in Kraków, directed this phase of the research. As subjects she selected spiderwort (*Tradescantia*) plants, with which she was very familiar in her work on nuclear radiation, and which are used as standard assays for ionizing radiation throughout the world. When exposed to X-rays or gamma rays, the stamen hairs of spiderwort flowers mutate, changing from blue to pink. The more ionizing radiation they are exposed to, the greater the number of pink hair cells.

Here, too, there was a before and after study. Potted plants containing at least 30 spiderwort blossoms were placed at each of four locations in Gabin and Sanniki from June 10-20, 1991, while the radio station was still operating, and then taken to a laboratory in Kraków where, between 11 and 25 days after the exposure, their stamen hairs were examined. The flowers that had been at three of the sites had approximately double the number of pink mutations as flowers that had never been near the radio station. Flowers that had been at the fourth site, which was inside a schoolroom near a telephone stand—whose wires acted as an antenna that amplified the radiation—had nearly nine times as many pink mutations. The plants near the telephone stand also had 100 times as many lethal mutations, and only three of their thirty blossoms ever opened.

After the tower fell down the experiment was repeated, with a ten-day exposure period from August 14-23, 1991. This time there was no increase in mutations at the first three locations. The plants near the telephone stand still had double the normal number of pink mutations, but all of their blossoms opened this time. Dr. Cebulska-Wasilewska, who usually used these plants to assess levels of ionizing radiation, stated that exposing the plants to the radio tower for only eleven days, at a distance of six kilometers, had been the equivalent to exposing them to a 3 centigray dose of X-rays or gamma rays. That is roughly 1,000 times more radiation than a chest X-ray, 10 times more than a

CT-scan, and about as much radiation as the average survivor of the atomic bomb received in Hiroshima.

In January 1995, the Polish parliament passed, and the President signed, an act authorizing the reconstruction of the long wave radio station at Konstantynów. Fierce local protests followed. The Society for the Protection of People Living near the Highest Mast in Europe formed in the village of Topólno. Fifteen people participated in a month-long hunger strike.

The tower was not rebuilt.

———— ·•· ————

Schwarzenburg is a small rural community on the river Sense, surrounded by lush green fields, nestled in the northern foothills of the Swiss Alps. In 1939, a short wave radio station was constructed about three kilometers east of town in order to broadcast Radio Swiss International to Swiss emigrants living abroad. The station broadcast to every continent, shifting the direction of its transmissions every two to four hours, so as to reach a different part of the world.

At first the town got along well with its neighbor. But after a new antenna was added in 1954, boosting the station's power to 450,000 watts, the surrounding residents began to complain that it was damaging the health of themselves, their farm animals, and the surrounding forests. Almost four decades later the Federal Department of Transport and Energy finally launched an investigation. The Swiss Federal Office of Environment, Forests and Landscape was involved, and Professor Theodor Abelin, Head of the Department of Social and Preventive Medicine at the University of Berne, was placed in charge.

In the summer of 1992 an extensive health survey was conducted. Measurements of the magnetic field strength were taken at numerous outdoor locations, and in the bedrooms of participants. Residents were given diaries in which to record symptoms and complaints at one-hour intervals during four ten-day periods, spread out over two summers. Blood pressure was monitored, school records examined, and urine samples collected to measure melatonin levels. Saliva, collected from area cows, measured their melatonin levels as well. During the second summer, at an unannounced time, the transmitter was turned off for three days.

The results confirmed the long-standing complaints. Of the people who lived within 900 meters (about half a mile) from the antennas, one-third complained of difficulty sleeping—three and a half times as frequently as people who lived four kilometers away. They complained of limb and joint pains four times as often, and of weakness and tiredness three and a half times as often. They woke up at night three times as often. They were more constipated, had more trouble concentrating, and had more stomach pains, heart palpitations, shortness of breath, headaches, vertigo, and "cough and sputum." One-third had abnormal blood pressure. Forty-two percent spent their leisure time away from home, compared with only six percent of the people who lived four kilometers away.

The second year's diaries showed the dramatic effect of turning off the transmitter. Even the people who lived four kilometers away woke up only about half as often during the nights when the transmitter was off. Melatonin levels did not change significantly in humans, but cows' melatonin levels rose two- to seven-fold during the three days the transmitter was off, and were suppressed again when the transmitter was turned back on.

School records from two schools showed that between 1954 and 1993, children at the school nearer to the antennas had a significantly smaller chance of being promoted from primary to secondary school.

It was left to the citizens of Schwarzenburg, however, to document the damage to their forests. Ulrich Hertel published photographs of the stumps of trees that died, showing decades of compression of their growth rings, but only on the side of the trees facing the antennas, as though, he wrote, the trees had tried "to get out of the path of a threat to their lives." His 1991 article in *Raum & Zeit*, published two months before Volkrodt's article, is strewn with photographs of forests in the Schwarzenburg area that were sick and dying.

On May 29, 1996, Phillippe Roch, the Director of the Federal Bureau for Environment, Forests, and Landscapes, stated that "a connection between the established sleep disturbances and the transmitting operation is proven." The Federal Bureau of Health agreed. On March 28, 1998, the short wave transmitter station of Schwarzenburg was shut down forever.

Hans-Ulrich Jakob, a long time resident, wrote: "The most surprising thing for me is the fact that the people have got back their joyfulness, their frankness, which I never saw before. And I have been living here for more than 40 years, in this region. The depressive, sometimes also aggressive behavior of many of my acquaintances has completely disappeared. A farmer, about 50 years old, told me that two weeks after the transmitter was switched off, he slept through the whole night for the first time in his life."

And Jakob had a story to tell about the trees. "It is wonderful to see," he remarked, "how quickly the forests, which were treated with radiation, are recovering now. The rate of growth, I think, is twice the growth of years past. The young trees are also growing up straight as a dart and don't try to flee in a direction away from the transmitter."

Dr. Abelin's team took advantage of the planned termination to conduct a before-and-after sleep study on 54 of their original subjects. It lasted from March 23 until April 3, 1998. Not only did sleep quality improve after the shutdown on March 28, but melatonin levels rebounded just as they had in the cows. During the week after the shutdown, melatonin levels in the people who lived closest to the antennas rose between one-and-a-half- and six-fold.

The recovery of Europe's forests at the end of the Cold War lasted only a decade. In 2002, almost one-quarter of the trees visited by a United Nations team again showed signs of damage, with one out of every five trees in Europe suffering from defoliation.[71] Acid rain, meanwhile, had been transferred along with heavy industry to China and India. Many foresters revised their textbooks to attribute forest dieback to global warming instead. But that is not the real culprit either.

Cedar trees, some of which are three thousand years old, having outlasted the Medieval Warm Period, the Little Ice Age, and innumerable droughts and floods, are disappearing from the face of the earth.

The venerable Cedars of Lebanon, whose twelve remaining stands cover about 5,000 acres, are in visible decay.

The cedars of Algeria's Atlas Mountains began to decline about 1982, and the cedars of Morocco have been dying rapidly since 2000.[72]

More than 600,000 acres of yellow cedars in remote areas of southeast Alaska and British Columbia are vanishing. Approximately 70 percent of the mature trees are dead, with some areas now completely devoid of cedars. Foresters are left thunderstruck by massive mortality on wet soils where yellow cedars have always thrived, and where no disease organisms can be isolated, on which to pin the blame.

In 1990, Paul Hennon, a United States Forest Service scientist stationed in Juneau, made a startling discovery: old aerial photographs showed that some of the stands of yellow cedars that are damaged today were already damaged in 1927, 1948, 1965, and 1976. And to his further amazement, the areas of decline in 1990 were only slightly larger than they had been in 1927. He then scoured the old forestry literature. Reports from expeditions throughout the 1800s had all included observations of yellow cedar near Sitka and elsewhere in southeast Alaska, and none had mentioned dying trees. Charles Sheldon, the first to report dead yellow cedar anywhere in Alaska, had seen them on Admiralty Island near Pybus Bay in the Sitka region in 1909, stating that "vast areas are rolling swamp, with yellow cedar, mostly dead." Harold E. Anderson, in 1916, also saw dying cedars near Sitka.[73]

Hennon concluded that no human factor could have caused cedar decline in the Alaska panhandle so long ago, but he was wrong. NPB Sitka, a 20-kilowatt long wave radio station operated by the Navy, was installed west of Pybus Bay in 1907. Army radio stations were installed at Petersburg and Wrangell in 1908. Private radio stations were also operating. A 1913 list of the radio stations of the United States includes five operated by the Marconi Company in southeast Alaska, including one at Kake, on Kupreanof Island, directly across Frederick Sound from Pybus Bay.[74]

That trees are dying without obvious cause throughout the Amazon rainforest was first noticed in 2005 and is being blamed, again, on global warming, which caused an unusual drought in that year.[75] Researchers connected with the worldwide RAINFOR network went back to the plots of forest, scattered through Brazil and seven neighboring countries, that they had been monitoring every three to five years, in some cases since the 1970s. To their surprise the intensity of the drought in individual locations was only weakly related to the health of the forest. Some areas had tree mortality but no drought, and some had drought but no mortality. Pockets of high mortality were surrounded by trees with little or no decline in growth. But overall, only half the plots gained biomass during 2005, an unprecedented circumstance. The Amazon, they feared, was changing from a net carbon sink to a net carbon source, with grave implications for our atmosphere. They blamed the change on global warming since they could find no other reason for a shift. But like Hennon and his team in Alaska, they were wrong.

On July 27, 2002, the environment everywhere in the Amazon was suddenly, drastically altered. For on that day, an American-financed, Raytheon-built, 1.4-billion-dollar system of radars and sensors called SIVAM (System for Vigilance of the Amazon) began its monitoring activities in a two-million-square-mile area of remote and inaccessible wilderness. The primary purpose of the new system was to deprive drug traffickers and guerrillas of the protection that the trackless jungle had always offered. But this required pretending that blasting the rainforest with radiation at levels that were unprecedented in the history of the world was of no consequence to the forest's precious inhabitants, human or otherwise. Since 2002, the system's 25 enormously powerful surveillance radars, 10 Doppler weather radars, 200 floating water-monitoring stations, 900 radio-equipped "listening posts," 32 radio stations, 8 airborne state-of-the-art surveillance jets equipped with fog-penetrating radar, and 99 "attack/trainer" support aircraft have enabled Brazil to track images as small as human beings anywhere. The system is so pervasive that Brazilian officials boast that they can hear a twig snap anywhere in the Amazon.[76] But it comes at the expense of the greatest diversity of animals and plants on earth, of the people who depend on them, and of our atmosphere.

In a small backyard laboratory in the foothills of Colorado's Rocky Mountains, Katie Haggerty performed the simplest, most elegant experiment of all: she hung aluminum window screening around nine potted trembling aspen seedlings to keep out the radio waves, and watched them grow. The screens didn't keep out much light, but to make sure the experiment was well controlled, she bought twenty-seven aspen trees and grew them side by side. Nine grew without any enclosure, nine were surrounded by aluminum screening, and nine were surrounded by fiberglass screening, which kept out just as much light but let in all the radio waves. She began the experiment on June 6, 2007. After just two months, the new shoots of the radio-shielded aspens were 74 percent longer, and their leaves 60 percent larger in area, than those of either the mock-shielded or the unshielded aspens.

On October 5-6, she evaluated the conditions of the three groups of plants. The mock-shielded and unshielded plants looked just like what most aspens in Colorado now look like every fall, their leaves and leaf veins yellow to green, their leafstalks light red to pink, and all their leaves covered to some degree with gray and brown areas of decay.

The shielded aspens looked like what aspens used to look like not long ago. Their leaves were much bigger, largely free of spots and decay, and displayed a wide palette of brilliant fall colors: bright orange, yellow, green, dark red, and black. Their leaf veins were dark to bright red, and their leafstalks were bright red as well.

The suddenness and simultaneity of aspen decline throughout Colorado, which began precisely in 2004, has been a source of wonder and despair to all who love and miss the vivid fall colors of these striking trees. In just three years, from 2003 to 2006, the area of aspen damage increased from twelve thousand acres to one hundred and forty thousand acres. Aspen mortality in the national forests rose three- to sevenfold, with some stands losing 60 percent of these trees.[77] There is a reason.

The State of Colorado operates a sophisticated public safety communications network, called the Digital Trunked Radio System, consisting of 203 tall radio towers whose transmissions cover every square inch of the state. They are heavily used by police, firefighters, park rangers, emergency medical service providers, schools, hospitals, and a wide variety of other municipal, state, federal, and tribal officials. Between 1998 and 2000 the pilot phase of the system, covering the Denver metropolitan area, was built and tested. In 2001 and 2002, radio towers were built throughout northeastern and southeastern Colorado and the eastern plains. And in 2003, 2004, and 2005, the system invaded the western, mountainous part of the state: aspen territory.

"At times," says Alfonso Balmori, "I compare what is occurring to a collective ritual of suicide in slow motion." But he does not think it can continue indefinitely. "I don't know when," he continues, "but there will come a day of realization, when society will awaken to the serious problem of electromagnetic contamination and its dangerous effects on sparrows, frogs, bees, trees, and all other living beings, including ourselves."

17. In the Land of the Blind

WHAT IF, ON ANOTHER PLANET, in a distant universe, the sun was dark. God never said, "Let there be light," and there was none. But people invented it anyway and lit up the world, lit it with light so bright that it burned all it touched. What if you were the only person who could see it. What if there were a thousand, a million, ten million others? How many aware people would it take to make the destruction stop?

How many will it take before people no longer feel too alone to say, "Your cell phone is killing me," instead of "I'm electrically sensitive"?

A tremendous number of people get headaches from their cell phone. Almost one-quarter of Norwegians who would now be considered moderate cell phone users (more than one hour per day) admitted it to the scientists who asked the question in 1996.[1] Almost two-thirds of Ukrainian university students who were heavy cell phone users (more than three hours per day) admitted it to the scientists who asked the question in 2010.[2] Perhaps there are some who really don't get headaches, but few people are asking

Gro Harlem Brundtland,
M.D., M.P.H

the question, and to publicly admit to the true answer is not socially acceptable.

Gro Harlem Brundtland got headaches from cell phones. And since she was the Director-General of the World Health Organization and the former Prime Minister of Norway, she did not feel the need to apologize for it, and simply ordered that no one was to enter her office in Geneva carrying a cell phone on their person. She even gave an interview about it in 2002 to a Norwegian national newspaper.[3] The following year she was no longer Director-General of the World Health Organization. No other public officials have repeated her mistake.

Even for those who really don't get headaches, their cell phones affect their sleep and their memory. Folk singer Pete Seeger wrote to me twenty years ago. "At age 81," he said, "it's normal for me to start losing my memory. But everybody I tell this to, says, 'Well, I seem to be losing my memory, too.'"

Those of us whose injuries are so severe, so devastating that we can no longer ignore them, and who are lucky enough to figure out what has happened to us and why, have here and there formed tiny, isolated groups, and for lack of a more acceptable term we call our injury "electrical sensitivity," or worse, "electromagnetic hypersensitivity" (EHS), a travesty of a name for a disease that affects the whole world and everyone in it, a name as absurd as "cyanide sensitivity" would be if anyone were foolish enough to apply such a name to those poisoned. The problem is that we are all being electrocuted to a greater or lesser extent, and because society has been in denial about that for more than two hundred years, we invent terms that hide the truth instead of speaking in plain language and admitting what is happening.

After pulsed microwave radiation came to my hometown for the first time, all over the city at once, on November 14, 1996, I was so sure it had killed masses of people that I telephoned epidemiologist John Goldsmith to ask for advice on how to prove it. Formerly with the California Department of Health Services, Goldsmith was then at Ben Gurion University of the Negev in Israel. He directed me to the weekly mortality statistics published online by the Centers for Disease

Control for 122 cities, and advised me to find out exactly when, for each city, digital cell phone service had begun. Here, for nine large cities in different parts of the country, whose digital service began at different times, are the results:

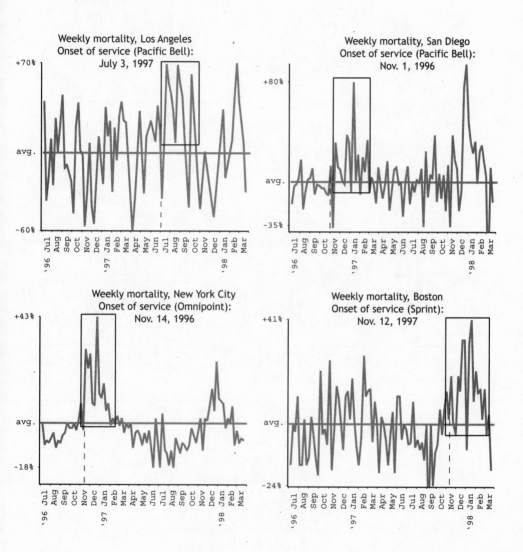

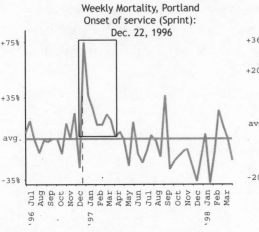

Weekly Mortality, Portland
Onset of service (Sprint):
Dec. 22, 1996

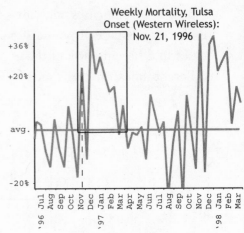

Weekly Mortality, Tulsa
Onset (Western Wireless):
Nov. 21, 1996

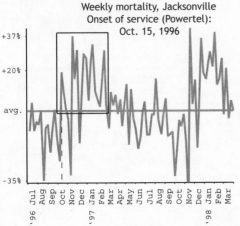

Weekly mortality, Jacksonville
Onset of service (Powertel):
Oct. 15, 1996

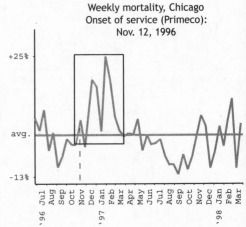

Weekly mortality, Chicago
Onset of service (Primeco):
Nov. 12, 1996

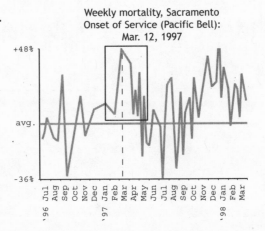

Weekly mortality, Sacramento
Onset of Service (Pacific Bell):
Mar. 12, 1997

I had been sure, because the sudden irradiation of my city had almost killed me, and because I knew people who had died from it.

On November 14, I had traveled to Killington, Vermont to attend "Unplugged: Health and Policy Implications of the Wireless Revolution," a conference sponsored by the Vermont Law School. When I returned home on November 16, I became dizzy. I assumed one of my neighbors had sprayed something toxic; perhaps the exterminator had been in the building. This would pass, I thought. But within a few days I became nauseated, and I had uncontrollable tremors. I had the first asthma attack of my life. My eyeballs felt like they were bulging out, my throat swelled, my lips felt dry, fat, and puffy, I felt pressure in my chest, and the bottoms of my feet hurt. I became so weak I couldn't lift a book. My skin became so sensitive I couldn't bear to be touched. My head was roaring like a freight train. After November 20 I did not sleep, and could not eat. During the night of November 22, my larynx went into spasm and I couldn't draw a breath in or out. In the morning I grabbed my sleeping bag, got on the Long Island Railroad, and left town.

My relief was unbelievable.

I learned that on November 14, while I was in Vermont, Omnipoint Communications, New York's first digital cell phone company, had begun selling its service to the public. Thousands of rooftop antennas at six hundred locations were operational: New Yorkers were now living inside a computer.

I compared notes with a few friends. Together we compiled a list of symptoms and placed the following classified ad in a local newspaper: "If you have been ill since 11/15/96 with any of the following: eye pain, insomnia, dry lips, swollen throat, pressure or pain in the chest, headaches, dizziness, nausea, shakiness, other aches and pains, or flu that won't go away, you may be a victim of a new microwave system blanketing the city. We need to hear from you."

And we did hear from them, by the hundreds—men and women, whites, blacks, Hispanics, and Asians, office workers, computer operators, stockbrokers, teachers, doctors, nurses, and lawyers, all of whom had woken up suddenly sometime between mid-November and Thanksgiving, their hearts racing, their heads pounding, thinking they were having a heart attack, a stroke, or a nervous breakdown—now relieved to find out they were not alone. The very first person to answer the ad was a forty-one-year-old airline employee who lived in the Bronx. Joe Sanchez's head suddenly began to hurt on about November 15, so badly that he was afraid he was having a stroke. Five and a half months later, on May 8, 1997, he died—of a hemorrhagic stroke.

For the next two years, without letup, Janet Ostrowski, a nurse who worked in a family practice office in Manhattan, and then on Long Island, saw a constant stream of patients with "viral syndrome," typically with excruciating headache, ear pain, swollen gland deep in the neck, nasal congestion they could not get rid of, facial pain, sore throat, fatigue, and sometimes profound dehydration. "No flu lasts an entire year," Ostrowski told us. She also noticed that the majority of her patients were suddenly not responding to medication. "I have done triage in various emergency rooms throughout the Tri-State area over the course of twenty-five years of nursing," she said. "Whatever used to be stabilized on routine medication, be it hypertension, diabetes, whatever, now seems to become unstabilized easily and not responding to current meds." She also saw a tremendous increase in the number of people complaining of stress and anxiety, many of whom, in their thirties and forties, were found, on routine EKG, to have cardiac changes.

Officially, this North American "influenza" epidemic began in October 1996 and lasted through May 1997.

The organization I started in 1996, called the Cellular Phone Task Force, is struggling to serve a growing population of injured. And the title of the magazine I published for five years, *No Place To Hide*, has come true. *Say To Countryside Goodbye, When Even Healthy People Die,*[4] wrote Olle Johansson, the guru of electrical sensitivity in Sweden and one of the world's foremost authorities on electrical illness and injury. The old wisdom, that if you wish to escape civilization you can do so if you go far enough away, is no longer true, because secondhand radiation no longer comes only from cell phones, WiFi, and other personal devices. The invisible tentacles of civilization, in the form of cell towers, radar installations, and two-way satellite dishes, have made radiation ubiquitous, impossible to escape no matter how far away you go and how much land you buy. And even if you find one of the last hidden sanctuaries, it can be destroyed in an instant, invisibly and without warning. There is no protection. Quite the opposite—laws have been passed preventing citizens from protecting themselves, or elected officials from doing anything about the radiation. But no one is immune.

"Recently I celebrated my forty-first birthday," said Dafna Tachover in 2013, "and I am not sure that the word *celebration* is appropriate." An attractive young attorney with an MBA, Tachover was licensed in New York and Israel, and just a few years previously had been working for an investment company in Manhattan as advisor to the chairman. She had been married to a doctor who was also a research scientist at Princeton University. They had decided to have a baby, and she had decided to open up a private law practice. All of life, seemingly, was hers for the taking.

When I interviewed her in 2013, she was divorced, unemployed, still childless, and struggling just to survive in a remote farmhouse in upstate New York. "My life is pretty much impossible," she said, "as I am a prisoner in my own house. I cannot go anywhere, I cannot even walk on the street and drive into town. I cannot work and be in the presence of other people. I cannot fly, travel, go to a restaurant, or sleep in a hotel. I cannot access a doctor, a hospital, or even go to court to enforce my rights which are being crushed. When I needed to move, I could not look for a house by myself, as driving on roads saturated with antennas and cars with wireless systems has become impossible. My father had to come from Israel to help me and after two months of searching, and five hundred houses, I found just one house which I could tolerate. The closest neighbor is 300 yards away (such distance is required in order to not be affected by a neighbor's WiFi, cordless phones, and other gadgets), there is only spotty cell phone reception, and radiation from only one radio station. I live in an isolated cabin in the woods and my only 'outing' to civilization is a once-a-month trip to buy groceries. Many times I am not well enough to even do that and I depend on friends to buy me food. As I cannot work and my money is almost exhausted, I don't know how I will survive financially, and with the spread of 'smart' meters, soon there will not be even one house I would be able to live in. It is very frustrating knowing that without this radiation I can live a normal and full life, but because of it I am forced into an absurd existence."

Tachover was a confirmed cell phone user who had no landline and spent hours on her cell phone and in front of her wireless computer. "My laptop was my best friend," she says. "I was one of the first to purchase a cellular wireless Internet connection to my laptop, to ensure that I had Internet access wherever I went." Finally, like so many other people, she was injured—injured by a new laptop computer she had bought for the law practice she was starting. "Every time I used the computer I felt pressure in my chest, the rapid pounding of my heart, difficulty breathing, dizziness, pressure in my head, my face would become red and hot, and I was nauseous. I had weird cognitive problems—I could not find words and when my husband spoke to me, five minutes later I would not remember that he did. I suddenly was unable to touch my cell phone and if I put it near my head it felt as if someone were drilling into my brain."

The first action she took was to go home to Israel to recover her health. "It was an unfortunate choice," she said. "On my first day there my body collapsed. While I was driving I felt excruciating pain. I looked up and saw 'white stripes' on the roof of the mall, and when I asked my mother what they were, she told me that they were cell phone antennas. Until that moment I did not know I felt antennas. I had tears in my eyes and all I could say was 'For God's sake, there are children growing up here!' From that moment on my condition quickly went downhill and my life became a nightmare. I could not sleep any more and the pain was unbearable."

Back in New York, Tachover spent months living in her car. "I could not be in my apartment, could not find a house, and I spent my days

Unshielded seedling
October 6, 2007

Mock-shielded seedling
October 6, 2007

Shielded seedlings. October 6, 2007

Photos by Katie Haggerty 2008

Effect of radar on city landscaping plant in Valladolid, Spain
(24 GHz speed detector). *Photo by Alfonso Balmori.*

desperately trying to find a place without radiation in which to park my car. At nights I parked my car in parking lots and would cover the windows with dark cloths and sheets so people would not see me."

Unfortunately Tachover's experience is very common, and becoming more so. Although she is now focusing her efforts as a lawyer to try to win "basic human and civil rights" for those who are called electrically sensitive, Tachover knows that the real problem is much bigger. "Humans are electric beings," she says, "and there is no mechanism in the human body that protects it from the radiation. Therefore, to claim that this radiation is not affecting us is ignorant and absurd. EHS is not a disease, it is an environmentally induced condition to which no one is immune. I want to believe that the day in which the extent of this disaster will be exposed is not far. Ignoring the facts and reality do not change them and ignoring a problem is guaranteed to worsen its scale."

Olle Johansson, who for decades was on the faculty of the world-famous Karolinska Institute—the institute that awards the Nobel Prize in Medicine every year—first became interested in the effects of microwave radiation in 1977 when he heard a presentation about leakage of the blood-brain barrier at a conference in Finland. He began to study the problem of skin rashes in computer operators in the early

Olle Johansson, Ph.D.

1980s after hearing a radio program by Kajsa Vedin. Vedin, who later wrote "In the Shadow of a Microchip," an analysis of the occupational risks of computer work, asked for expertise in neurology. "As a neuroscientist," says Johansson, "I thought I was close enough, and I strongly believed that the issues she wanted to highlight, using the conventional repertoire of scientific 'tools,' ought to be easily investigated. I did not realize at all that there were other forces not wanting to see such studies initiated, but very soon I understood that these very clear-cut and simple and obvious investigations proposed by Kajsa Vedin would be very, very hard to start.

"For me," he recalls, "it was immediately clear that persons claiming skin reactions after having been exposed to computer screens very well could be reacting in a highly specific way and with a completely correct avoidance reaction, especially if the provocative agent was radiation and/or chemical emissions—just as you would do if you had been exposed to, for example, sun rays, X-rays, radioactivity, or chemical odors. Very soon, however, from different clinical colleagues a large number of other 'explanations' became fashionable—that the persons claiming screen dermatitis were only imagining this, or they were suffering from post-menopausal psychological aberrations, or they were old, or had a short school education, or were the victims of classical Pavlovian conditioning. Strangely enough, most of the, often self-made, 'experts' who proposed these explanations had themselves never met anyone with screen dermatitis and had never done any investigations of their proposed explanatory models."

When he first contacted Vedin, Johansson did not personally know anyone with screen dermatitis either, but he quickly learned that they were hidden all around him in plain view. He learned that skin rashes were only the most visible manifestations of a devastating impairment, and that exposure not only to computer screens but other sources of radiation, and even ordinary electricity, could seriously damage the heart, nervous system, and other systems of the body. "After all these years," he says, "today I now regularly communicate with many thousands of such people, spread all around the world, and coming from all aspects of life. Nothing protects you from this functional impairment, not political stance, not your income, not sex, skin color, age, where you live or what you do for a living. Anyone can be affected. These people suffer radiation damage from gadgets that have been very rapidly introduced without ever having been formally tested for potential toxic environmental exposures or any other types of health hazards."

Johansson has not only seen his research funding disappear, and has lost his position at the Karolinska Institute, but he has had death threats, and on one occasion an attempt on his life. He went riding on his motorcycle with his wife one day, and while still going slow, he suddenly lost control of the vehicle. Twenty-seven spokes of the rear wheel had been cleanly sawed through, so professionally that it had been impossible to see. I asked Johansson what keeps him going. He began by telling me about the lives of the people who are called electrically sensitive.

"The lives of EHS persons most often are a living hell," he said. "I very soon realized that the very famous Swedish social security net did not catch them in its arms, but allowed them to fall and crash. That disturbed me a lot. The EHS had become a model of the democratic world, or rather a model of how democracies fail to protect their citizens. It was, and is, not hard to imagine yourself in such a situation. Today the EHS person, but what about tomorrow? Who will then be an outsider? Myself even? You? Who? The EHS became a kind of medical outcast, facing difficulties not shared by the rest of society. A very scary panorama. Anyone, as a fellow human being, would have been equally affected by what I witnessed over and over again.

"At the same time, another side also grew on me. The EHS persons, most of them, actually are very strong. They have to endure harassments of various sorts from the society, from physicians, scientists, experts, politicians, civil servants, their kin, and so forth, and all this makes their mental 'skin' very tough. I admire them a lot! I know I never would be able to constantly take such immense beatings.

"What keeps me going? One must stick to the task; to give in and move to another field would leave these persons very much without hope. As a government scientist I am supposed to work for people in need, not for my own personal career. When I grew up in the 1950s and 1960s, in Sweden, my family were very poor. I learned then the value of a hand stretched out, willing to support and help you. Such a lesson you never forget."

Dr. Erica Mallery-Blythe is an engaging physician, born in England, who has dual British and American citizenship, and who has also dedicated her life to this problem, having experienced it firsthand. After graduating from medical school in 1998, she worked at hospitals all over England, becoming an instructor in trauma medicine. In 2007, she moved to the United States with her husband, who was an F-16 pilot with the British Royal Air Force, working as an exchange officer with NATO. She became injured while she was pregnant. Like so many other young professionals, Mallery-Blythe had become dependent on technology. In fact, she was one of the earliest cell phone users, her father having bought her one when she was ten years old, in the mid-1980s. She had always noticed that she got a headache if she used her cell phone too long, but like most people, she had not paid too much attention.

Now, however, the pain became intense after every phone call, and the right side of her face would become bright red as if she were sunburned. She had also just acquired her first WiFi-enabled laptop computer, which she used a great deal for medical research, and which she rested on her legs—but not for long, because every time she did that she would get severe, deep aching pain inside her legs. "It felt like my legs were cooking from the inside," she recalls. Soon she could no longer use her computer at all, even at a distance. "As a doctor," she says, "I knew that when there's pain there's something wrong." Eventually she had to give up using both the computer and the telephone. By this time she was not sleeping, and had acquired a heart arrhythmia and severe tremors, in addition to the dizziness and headaches that were tormenting her. But everything she read on the Internet reassured her that she was not going to get cancer from her cell phone, and she could not put her experience into any medical context that she had ever been taught. She finally heard the term "electromagnetic hypersensitivity" after her daughter was born, but still did not grasp the seriousness of it. "How could there be a condition that was so profound that I'd never heard of it?" she wondered. It was not until she underwent an MRI to rule out a brain tumor that she finally realized that her life had been permanently, utterly altered. For when the high frequency pulse of the MRI was turned on she saw "a million grains of golden sand exploding outwards," and had "a feeling of impending doom." The final piece of the puzzle fell into place when she and her husband visited an isolated campground on the edge of Death Valley where there was no WiFi and no cell phone reception. "The relief was unbelievable," she says. For the first time in a long time, she felt completely well and completely normal.

But, like Tachover, and like so many other people throughout the world, life was now impossible. Mallery-Blythe and her husband moved out of their home and began camping in tents or sleeping in the back of their car. She describes it as "living like war refugees." She could not enter a market or a gas station without becoming crippled. "You can't do the basic things you need to live. You almost feel like you're going to wake up, like it's some kind of bizarre dream." Almost worse than the physical hardship was the fact that they had to hide the truth of what was happening from everybody they knew and met. They lived like that for more than half a year, until they found a log cabin by a lake in South Carolina, where they were forced to live without electricity so that she could recover her health. She was living there when I first met her. Eventually she moved back to England, but before she did she had met many other people who were injured by electricity, especially by wireless technology, and had attended a medical conference on the subject in Dallas. And she decided that she had no choice but to devote the rest of her life to the needs of this population, including the most urgent need for a sanctuary where people can save their lives, recover their health, and become productive individuals again. "The first and foremost need," says Mallery-Blythe, "is a safe refuge for those who need urgent care, with supportive medical staff. What makes me sad is to see all the people who can't escape and get to a pure environment, because if you can't get to a pure environment, it will destroy you." Considering that an estimated five percent of the population know they have been injured,[5] and that perhaps one out of four of them have had to leave their homes, the need for refugee aid is enormous.

Yury Grigoriev, known affectionately as the grandfather of EMF research in Russia, has been working on radiation since 1949. After graduating from the Military Medical Academy he was assigned to research the biological effects of atomic weapons at the Institute of Biophysics at the U.S.S.R. Ministry of Health. Since 1977 he has been the head of research on non-ionizing radiation (i.e., radio waves) at the same institute, since renamed the A. I. Burnazyan Federal Medical and Biophysical Center. He is also the Honorary Chairman of the

Russian National Committee on Non-Ionizing Radiation Protection. His most recent book, *Mobile Communication and Children's Health*, was published in 2014, a year before his ninetieth birthday. His greatest fear is for the children. "For the first time in history," he says, "human beings are exposing their own brains to an open, unprotected source of micro-wave radiation. From my viewpoint as a radiobiologist, the brain is a critical organ and children have become the group at greatest peril."

Yury Grigorievich Grigoriev, M.D.

"In the early period," says Grigoriev, "the government deliberately underestimated the risk of nuclear radiation, before the accident at Chernobyl. This accident caused fear among the population, and as a result the Russian government agreed to provide full information to the public about the dangers of ionizing radiation. Now we are dealing with similar issues surrounding mobile communications. I believe that the time has arrived, here too, to provide full information to the general public."

Scarcely a day goes by when I don't receive terrifying new information that is being tragically ignored.

"Children's Cell Phone Use May Increase Their Risk of ADHD," reads a recent news headline about a Korean study. The more calls made by a child, the more time spent on the phone, and the more time playing games on the phone, the greater the risk of ADHD.[6]

"Computer Screens Can Make You Blind," screams another headline. This research, out of Japan, found that spending more than four hours per day on a computer for ten years more than doubles one's risk of glaucoma.[7]

"Are Mobiles Bad for Your Skin?" Also out of Japan, this research found that mobile phones worsen eczema.[8]

"Mobiles Can Make You Blind." This study in China found that microwave radiation at levels emitted by cell phones caused cataracts to form on the eyes of rabbits.[9]

"Could Microwaves Be Associated with Children's Asthma?" This investigation was done at Kaiser Permanente in Oakland, California. Women who were exposed to higher magnetic fields while pregnant gave birth to children who were at greater risk for asthma.[10]

"Talking on the Phone Makes You Deaf." I have received a number of studies saying this. Teams of researchers at Dicle University in Turkey,[11] at a hospital in Chandigarh, India,[12] and at the University of Malaysia in Kuala Lumpur[13] found that heavy cell phone use is associated with permanent hearing loss. Scientists at King Edward Memorial Hospital in Mumbai, India found that chronic use of a cell phone for ten minutes a day causes hearing loss.[14] Research at the University of Southampton, England showed that even a single short exposure to a cell phone causes temporary hearing loss.[15]

"Cell Phones Now Tied to Alzheimer's." A team of Swedish scientists, led by neurosurgeon Leif Salford, proved in the late 1990s that a cell phone disrupts the blood-brain barrier of laboratory rats within two minutes of exposure. When they reduced the power of the phone a thousandfold—the equivalent of a person keeping a phone several feet away from his or her head—the damage *increased*. In 2003, they proved that a *single two-hour exposure* causes permanent brain damage. They exposed 12- to 26-week-old rats to an ordinary cell phone, just once for two hours, and waited eight weeks before sacrificing them and examining their brains. Like human teenagers, these rats had brains that were still developing. In those animals that had been exposed once to a cell phone, up to two percent of the neurons in all areas of the brain were shrunken and degenerated.[16] Salford called the potential implications "terrifying." In 2007, they exposed rats chronically, for two hours once a week for 55 weeks, beginning in their "teenage years." At the end of the experiment, the exposed rats, by now in middle age, had memory deficits.[17] To mimic cell phone use by very young children, scientists in Turkey experimented on 8-week-old rats. In their study, published in 2015, they exposed the animals to cell phone-like radiation for one hour a day for a month, and then examined a particular area of the brain called the hippocampus, which is involved in learning and memory. The exposed rats had 10 percent fewer brain cells in the hippocampus than the unexposed rats. And a large number of brain cells in the exposed rats were abnormal, dark, and shrunken, just like the brain cells in Salford's rats.[18] In another large set of experiments, the Turkish team exposed pregnant female rats to cell phone-like radiation at low power for one hour a day for nine days. The exposed rats' offspring had degenerative changes in their brains, spinal cords, hearts, kidneys, livers, spleens, thymuses, and testes.[19] In yet a further experiment, the same scientists exposed young rats to cell phone-like radiation for one hour a day during their early and mid-adolescence, which for a rat is from 21 to 46 days of age. The exposed rats' spinal cords were atrophied and had significant losses of myelin, similar to what occurs in multiple sclerosis.[20]

Since the first edition of this book was written, the mountain of truth confronting every cell phone user has only grown larger. Millennials—the generation born between 1981 and 1996 and the first to grow up using cell phones—are experiencing an unprecedented decline in their health when they reach their late twenties. On April 24, 2019, the American health insurance company Blue Cross Blue Shield released a report titled "The Health of Millennials." It showed not only that the health of this generation takes a sharp decline beginning at age 27, but also that the prevalence of many medical conditions had risen precipitously among millennials in just three years.

The prevalence of eight of the top ten conditions among all millennials showed a double-digit increase in 2017 as compared with 2014. Major depression increased 31 percent. Hyperactivity increased 29 percent. Type 2 diabetes increased 22 percent. Hypertension increased

16 percent. Psychoses increased 15 percent. High cholesterol increased 12 percent. Crohn's disease and ulcerative colitis increased 10 percent. Substance use disorder increased 10 percent.

The decline in millennials' health from 2014 to 2017 was not due to their being three years older. The report also compared the health of millennials who were 34 to 36 years old in 2017 to the health of Gen Xers who were 34 to 36 years old in 2014. At the same age, millenials in 2017 had 37 percent more hyperactivity, 19 percent more diabetes, 18 percent more major depression, 15 percent more Crohn's disease and ulcerative colitis, 12 percent more substance use disorder, 10 percent more hypertension, and 7 percent more high cholesterol than Gen Xers had in 2014.

When the researchers looked at all health conditions, they found that 34- to 36-year-olds in 2017 had a 21 percent increase in cardiovascular conditions, a 15 percent increase in endocrine conditions, and an 8 percent increase in other physical conditions compared to 34- to 36-year-olds in 2014.

The only reasonable explanation for the alarming decline in health of the millennial generation is the life-long irradiation of their brains and bodies from their cell phones. Cell phones did not work in most of the United States until 1997, and their use was not prevalent among teenagers until 2000. Millennials are the first generation that began using cell phones in their teenage years or earlier, when their brains and bodies were still developing. People who were 34 to 36 years old in 2017 were 17 to 19 years old in 2000. People who were 34 to 36 years old in 2014 were 20 to 22 years old in 2000. No other environmental factor changed so radically in just three years. Microwave radiation is responsible for the tragic state of the millennial generation's health compared to the health of every other generation that preceded them.[21]

The incidence of stroke overall is steady or declining but it is rising in adults younger than 50, and shockingly so in very young adults, who are the heaviest users of cell phones. Studies out of France,[22] Sweden,[23] and Finland[24] all say the same thing. A Danish study published in 2016 examined the rate of strokes in people aged 15 to 30—a population that never used to have strokes at all. The annual number of strokes in that age group in Denmark rose 50 percent between 1994 and 2012, and the annual number of transient ischemic attacks (mini-strokes) in that age group tripled.[25] Cell phones were marketed in Europe three years earlier than in America.

Women in their twenties and thirties who keep their cell phones in their bras are getting a distinctive type of breast cancer directly underneath where they keep their phones.[26] Rates of total hip replacements have skyrocketed since cell phones began living in hip pockets. Between 2000 and 2010 the number of annual hip replacements in the United States more than doubled, and the rate of hip replacements among people aged 45 to 54 more than tripled.[27] Rates of colon cancer among Americans aged 20 to 54, which had been declining for decades, began to rise suddenly in 1997. The rise has been steepest and began earliest in people aged 20 to 29; the rate of colon cancer in young men and women aged 20 to 29 doubled between 1995 and 2013.[28] Rates of prostate cancer—the prostate is also in the same part of the body—have been rising worldwide since 1997.[29] The number of cases of prostate cancer among Swedish men aged 50 to 59 was stable for decades until 1996 and rose nine-fold between 1997 and 2004.[30] The incidence of metastatic prostate cancer among American men under 55 increased 62 percent between 2004 and 2013, and nearly doubled for men aged 55 to 69 during the same period.[31] An American study conducted from 2003 to 2013 found that young men had lower sperm counts than their elders for the first time in human history, and that men born between 1990 and 1995 had on average 40 percent lower sperm counts than men born earlier.[32]

And the kind of brain damage that occurred in a Swedish laboratory in teenaged rats, and in a Turkish laboratory in preteen rats, is now being found in preschool children in America. Not only did the scientists at Cincinnati Children's Hospital Medical Center find that children who spent more time per day on a wireless device have poorer language and literacy skills, but MRIs of the children showed structural damage to the white matter of their brains.[33]

The damage to the natural world is mounting up just as high. In 2017, Mark Broomhall presented his report to the United Nations

Educational, Scientific and Cultural Organization (UNESCO) on the exodus of so many species of wildlife from the Nightcap National Park World Heritage area surrounding Mount Nardi in Australia. Broomhall has lived on Mount Nardi for more than forty years. After antennas for 3G cell phones were installed on the Mount Nardi communications tower in 2002, he saw an immediate decline in insect populations. In 2009, when "enhanced 3G" was added to the tower, along with channels for 150 television stations, 27 bird species left the mountain. In early 2013, when 4G was installed on Mount Nardi, a further 49 bird species left, all bat species became scarce, four common species of cicada almost disappeared, frog populations were drastically reduced, and the massive and diverse populations of moths, butterflies, and ants became uncommon to rare.[34]

At about the same time that Broomhall presented his report, people all over the world woke up to the fact that their cars' windshields were not being splattered with tiny life, and that insects of all kinds were disappearing from the earth. In 2017, scientists reported a 75 to 80 percent decline in total flying insects in 63 nature protection areas in Germany.[35] In 2018, another group of scientists reported a 97 to 98 percent decline in total insects caught in sticky traps in a Puerto Rican rainforest.[36] In 2019, scientists from Australia, Vietnam, and China reviewed 73 reports of insect declines from across the globe, and concluded that 40 percent of all insect species on earth are threatened with extinction.[37]

We are living in a world where information does not increase knowledge, nor open eyes. The cultural barriers are too great. Society has been in denial for too long. And yet it is impossible to continue on the present path any longer. Decisions are being made to intensify the global microwave rain, before 2020, from a steady drizzle to a downpour.

Instead of cell towers every few miles, there are going to be cell towers every few houses. This is already being implemented throughout China and South Korea and is spreading like wildfire to every city in the world. Although the new antennas are small—little boxes on top of telephone poles—they expose the population to tens or hundreds of times more radiation than the tall structures they are replacing.

Dense rows of similar antennas are being sown like so much rice along the sides of highways and beneath the pavement, and the electric fields that sprout from their seeds to cover the adjacent countrysides will guide cars and trucks that are outfitted with their own antennas and driven by robots instead of human beings.

These are the structures that are replacing men and women with machines within cities and along highways. It is called "5G" because it is the 5th generation of wireless technology. 5G will enable the creation of the "Internet of Things": not only cars, trucks, and home appliances, but virtually everything we buy is being outfitted with antennas and microchips in order to be connected to the wireless cloud that will take over the business of the world from human beings. Cars will drive themselves, milk cartons will instruct refrigerators to order milk, and your baby's diaper will tell your phone when it needs to be changed. By some estimates, as many as one trillion antennas will soon be talking to one another, outnumbering people on the earth by a hundred to one.

Not just people, but all of nature is being replaced by electrical pulsations, and not just in cities and suburbs. Radio waves are replacing eagles and hawks in national parks and wilderness areas, fish and whales in the earth's oceans, and penguins and auks in Antarctica and Greenland, where ice is melting into electric fog.

Four billion people, you see, still have little or no access to the Internet. And the remedy for that deficiency is now at hand, via balloons, drones, or satellites from space. Humankind is now willing and able to finally fulfill the original promise of the telegraph, put into words for the first time a century and a half ago. Space and time are poised to be thoroughly annihilated. That promise, however, is the ultimate Trojan horse, containing within it an unsuspected threat: the annihilation or severe impoverishment of life itself. Unsuspected, that is, by those who cannot yet see what is happening. Those of us with EHS who remember the beginning of satellite phone service foresee catastrophe.

In 1998, the launch of the 66-satellite constellation called Iridium brought cell phone service for the first time to the vast unserved regions of the earth, previously owned by penguins and whales. As we saw in the last chapter, however, it also unleashed a new kind of rain that emptied the skies of birds for a couple of weeks. The loss of thousands of racing pigeons during the two weeks following September 23, 1998, made headline news. The fact that wild birds were also not flying received only brief mention. The human toll was not mentioned at all.

On about October 1, 1998, I contacted fifty-seven electrically sensitive people in six countries. I also surveyed two support groups, and interviewed two nurses and one physician who served this population. My survey[38] found that eighty-six percent of the electrically sensitive people interviewed, and a majority of patients and support group members, had become ill on Wednesday, September 23 exactly, with typical symptoms of electrical illness such as headaches, dizziness, nausea, insomnia, nosebleeds, heart palpitations, asthma attacks, ringing in the ears, and so forth. One person said it felt like a knife went through the back of her head early Wednesday morning. Another had stabbing pains in his chest. A number of people, including me, were so sick we weren't sure we were going to live. Follow-ups revealed that some of these people were acutely ill for up to three weeks. I suddenly lost my sense of smell on September 23, 1998, and it still today has not returned to normal.

Mortality statistics obtained from the Centers for Disease Control reveal the following numbers for 1998:

Week	Deaths
Sept. 6	11,351
Sept. 13	11,601
Sept. 20	11,223
Sept. 27	**11,939**
Oct. 4	**11,921**
Oct. 11	11,497
Oct. 18	11,387

As recommended by the CDC, the above numbers are based on an average three-week delay between the time of death and the filing of a death certificate, and have been adjusted to account for missing data for some cities. A four to five percent rise in the national death rate occurred during those two weeks in which electrically sensitive people were the sickest and birds were not flying in the sky.

The commencement of service by the second satellite cell phone company, Globalstar, was again accompanied by widespread sudden illness. Globalstar announced the beginning of full commercial service in the United States and Canada from its 48 satellites on Monday, February 28, 2000. Widespread reports of nausea, headaches, leg pain, respiratory problems, depression, and lack of energy began on Friday, February 25, the previous business day, and came from people both with and without EHS.[39]

Iridium, which had gone bankrupt in the summer of 1999, was resurrected on December 5, 2000, when it signed a contract to provide satellite phones to the United States Armed Forces. On March 30, 2001, commercial service was resumed, and on June 5, Iridium added mobile satellite data services, including the ability to connect to the Internet. Nausea, flu-like symptoms, and feelings of oppression accompanied both events. Hoarseness was a prominent complaint of many who contacted me in early June. But the reports that grabbed headlines had nothing to do with human beings.

The March 30 event was unusual in several respects. First, it was the night of a rare red aurora that was visible in the northern hemisphere as far south as Mexico, as well as in the southern hemisphere. It was a time of intense solar activity, so I was tempted to attribute this to pure coincidence, except that I was reminded of the reddish sky that some reported the night of September 23, 1998, when Iridium was turned on the first time. No one understands all the interactions of these satellite operations with the earth's magnetic field and atmosphere.

But the second item that attracted notice was a catastrophic loss of Kentucky race horse foals in late April and early May.[40] Since mares, according to the Merck Veterinary Manual, abort several weeks to a month after, for example, a viral infection, this would put the triggering event at the end of March. Except that no such virus was ever found. In the United States, unusual foaling problems were reported simultaneously not only from Kentucky and nearby states like Ohio, Tennesee, Pennsylvania, and Illinois, but also from Maryland, Texas, and northern Michigan. Lenn Harrison, director of the University of Kentucky's Livestock Disease Diagnostic Center, said he had received similar reports from as far away as Peru.[41]

Between 2001 and now, our skies have not essentially changed. The number of satellites in low orbit has gradually increased, but Iridium and Globalstar are still the only providers of satellite phones, and the amount of data raining on us all from space is still dominated by those two fleets. That, however, is poised to change in a grand way. In 2017, we had a total of some 1,100 functioning artificial satellites of all types circling the earth. By the end of 2019, the number had already doubled. In 2020, several companies are competing to launch new fleets of 500 to 42,000 satellites *each*, for the sole purpose of bringing high-speed wireless Internet to the furthest reaches of the earth, and recruiting billions of untapped consumers into the ranks of social media. These plans call for the satellites to fly in orbits as low as 210 miles in altitude, and to aim highly focused beams at the earth with an effective radiated power per beam of up to twenty million watts.[42] The names of some of these companies are familiar to everyone: Google, Facebook, and Amazon. Others, as yet, are less well known. SpaceX is the space transport company created by billionaire Elon Musk, the man who wants to put a colony on Mars—and to provide high speed Internet to both planets. OneWeb, based in the United Kingdom, has attracted major investments by Qualcomm and Virgin Galactic, and has signed up Honeywell International as its first large customer. Google, in addition to investing one billion dollars in Musk's satellite project, has a contract to supply Internet from high-flying balloons to remote parts of the Amazon rainforest in Peru.

As this book goes to press, SpaceX has submitted applications for 42,000 satellites to the U.S. Federal Communications Commission and the International Telecommunication Union and is already is in process of launching them, 60 at a time. SpaceX has announced that as soon as 420 satellites are in place, which could be as early as February 2020, it will turn them on and begin providing service to some areas of the earth. OneWeb has submitted applications for 5,260 satellites, plans to begin launching 30 at a time in January 2020, and has projected the beginning of service to the Arctic and Antarctic in late 2020 and full global service from 650 satellites in 2021. Telesat, based in Canada, expects to begin launching a fleet of up to 512 satellites in 2021 and to provide global service in 2022. Amazon projects that its 3,236 satellites will serve the entire world *except* the Arctic and Antarctic. Facebook, thus far, has an experimental satellite license from the FCC under which it is not required to disclose its plans to the public. A new company called Lynk also has an experimental license; it plans to deploy "several thousand" satellites by 2023 and boasts that "we're going to turn all mobile phones into satellites phones."

These plans must not happen. The roots of our life-support system are firmly anchored in the pillars of the earth's magnetic field, far above our heads, where the pulsations of the universe, nourished and watered by the sun, are absorbed, animating all living things below. The engineers, who believe that all these satellites will be too far away to affect life, miss the mark. Even the first small fleet of 28 military satellites, launched into orbit in 1968, ushered in a worldwide pandemic of influenza. The direct radiation is only part of the problem. Satellites have a profound effect, as we learned in chapter 9, because they are already *in* the earth's magnetosphere. Unlike radiation from earthly towers, which is greatly attenuated by the time it reaches outer space, radiation from satellites works its full force on the magnetosphere, and is demodulated and amplified there by mechanisms that are poorly understood.

Not only will all these satellites be located in the magnetosphere, but most will be located in the ionosphere, which is the lower part of the magnetosphere. The ionosphere, as we learned in chapter 9, is charged to an average of 300,000 volts and provides the power for the global electrical circuit. The global electrical circuit provides the energy for all living things: it is why we are alive, and it is the source of all health and healing. All doctors of oriental medicine know this, except they call that energy "qi" or "chi." It flows from the sky to the earth, and it circulates through our meridians and gives us life. It is electricity. You cannot contaminate the global electrical circuit with millions of pulsed, modulated electronic signals without destroying all of life.

The reason the engineering perspective fails is fundamental: it perpetuates the error that our ancestors made in 1800, the terrible decision to treat electricity as a foreign element, a strange beast that operates outside the laws of nature. We acknowledge the existence of electricity only to the extent that it does work for us; otherwise we pretend it is not there. We ignore the warning, issued in 1748 by Jean Morin, that harnessing electricity is tampering with life. We pretend, contrary to all scientific evidence, that there is a safe level of exposure, and that if the authorities only set the safety standards low enough, we can have our radar stations and computer screens and cell phones and not suffer the consequences. We forget the admonitions of Ross Adey, the grandfather of bioelectromagnetics, and of atmospheric physicist Neil Cherry, that we are electrically tuned to the world around us and that the safe level of exposure to radio waves is zero.

The satellite projects have made the growing efforts to educate the world much more urgent. In 2009, an international coalition formed whose mission is to bring the matters addressed in this book to global awareness. At this writing, the International EMF Alliance (IEMFA) collaborates with one hundred and twenty-one organizations from twenty-four countries. The Global Union Against Radiation Deployment from Space (GUARDS) formed in 2015; its mission is to prevent the planned rain of wireless Internet from satellites, drones, and balloons. And in 2019, an International Appeal to Stop 5G on Earth and in Space has gathered the signatures of thousands of organizations and hundreds of thousands of individuals from two hundred and two countries and territories. Scientists, doctors, engineers, nurses, psychologists, architects, builders, veterinarians, beekeepers, and other individuals from almost every nation have signed this appeal, and preparations are underway to deliver it to all of the world's governments.

In 2014, Japanese physician Tetsuharu Shinjyo published a before-and-after study that is a harbinger of the direction in which the world needs to go. He evaluated the health of the residents of an apartment building in Okinawa, upon whose roof cell phone antennas had been operating for a number of years. One hundred and twenty-two individuals, representing 39 of the 47 apartments, were interviewed and examined. Prior to the removal of the antennas, 21 people suffered from chronic fatigue; 14 from dizziness, vertigo, or Ménière's disease; 14 from headaches; 17 from eye pain, dry eyes, or repeated eye infections; 14 from insomnia; 10 from chronic nosebleeds. Five months after the antennas were removed, no one in the building had chronic fatigue. No one had nosebleeds any more. No one had eye problems. Only two people still had insomnia. One still had dizziness. One still had headaches. Cases of gastritis and glaucoma resolved. Like the residents of that building before the study, the majority of the people in the world today do not know that their acute and chronic illnesses are in large part caused by electromagnetic pollution. They do not talk to each other about their health problems, and are unaware that they are shared by many of their neighbors.

As awareness spreads, it will become acceptable to turn to your neighbor and ask them to turn off their cell phone, or unplug their WiFi. And that will be the beginning of recognition that we have a problem, one that is more than two centuries old. It is a problem that pits the apparent ease of living, the limitless power at our fingertips brought to us by electrical technology, against the unavoidable, irreversible effects of that same technology on the natural world of which we are part. The unfolding human rights emergency, already affecting perhaps one hundred million people worldwide, and the environmental emergency threatening so many plant and animal species with extinction, must be faced with open eyes.

Notes

Chapter 1. Captured in a Bottle

1. Musschenbroek 1746.
2. Letter from Allamand to Jean Antoine Nollet, partially quoted in Nollet 1746b, pp. 3-4; summarized in Trembley 1746.
3. Priestley 1767, pp. 82-84.
4. Mangin 1874, p. 50.
5. *Ibid.*
6. Franklin 1774, pp. 176-77.
7. Wesley 1760, pp. 42-43.
8. Graham 1779, p. 185.
9. Lowndes 1787, pp. 39-40. See discussion in Schiffer 2003, pp. 155-56.
10. Heilbron 1979, pp. 490-91.

Chapter 2. The Deaf to Hear, and the Lame to Walk

1. La Beaume 1820, p. 25.
2. Duchenne (de Boulogne) 1861, pp. 988-1030.
3. Humboldt 1799, pp. 304-5, 313-16.
4. Volta 1800, p. 308.
5. Humboldt 1799, pp. 333, 342-46.
6. Kratzenstein 1745, p. 11.
7. Gerhard 1779, p. 148.
8. Steiglehner 1784, pp. 118-19.
9. Jallabert 1749, p. 83.
10. Sauvages de la Croix 1749, pp. 372-73.
11. Mauduyt de la Varenne 1779, p. 511.
12. Bonnefoy 1782, p. 90.
13. Sigaud de la Fond 1781, pp. 591-92.
14. Sguario 1756, pp. 384-85.
15. Veratti 1750, pp. 112, 118-19.
16. van Barneveld 1787, pp. 46-55.
17. Sguario 1756, p. 384.
18. Humboldt 1799, p. 318.
19. Gerhard 1779, p. 147.
20. Thillaye-Platel 1803, p. 75.

21. Humboldt 1799, p. 310.

22. Donovan 1847, p. 107.

23. Nollet 1753, pp. 390-99.

24. Steiglehner 1784, p. 123.

Chapter 3. Electrical Sensitivity

1. Wilson 1752, p. 207.

2. Reported in Gralath 1756, p. 544, and in *Nouvelle Bibliothèque Germanique* 1746, p. 439.

3. Letter of March 5, 1756 to Elizabeth Hubbart; letters of March 30, 1756, January 14, 1758, September 21, 1758, February 21, 1760, February 27, 1760, March 18, 1760, December 27, 1764, and August 5, 1767 to Deborah Franklin; letter of January 22, 1770 to Mary Stevenson; letter of March 23, 1774 to Jane Mecom.

4. Morin 1748, pp. 171-73.

5. Bertholon 1780, pp. 53-54.

6. Sigaud de la Fond 1781, pp. 572-3.

7. Mauduyt 1777, p. 511.

8. Nollet 1746a, p. 134; 1753, pp. 39-40.

9. Stukeley 1749, p. 534.

10. Humboldt 1799, p. 154.

11. Brydone 1773, vol. 1, pp. 219-20.

12. Humboldt 1799, pp. 151-52.

13. Martin 1746, p. 20.

14. Musschenbroek 1769, vol. 1, p. 343.

15. Bertholon 1786, vol. 1, p. 303.

16. Louis 1747, p. 32.

17. Sguario 1756, p. 288.

18. Morin 1748, p. 192.

19. Wilson 1752, p. 208.

20. Morin 1748, pp. 170-71, 192-97.

21. Nollet 1748, p. 197.

22. Morin 1748, pp. 183-86.

23. Nollet 1753, pp. 90-91.

24. Heilbron 1979, p. 288.

25. Beard and Rockwell 1883, pp. 248-56.

26. Sulman 1980.

27. Michael Persinger, personal communication.

28. Sulman, pp. 11-12.

29. *ICB 2008*. Proceedings of the 18th International Congress of Biometeorology, 22-26 Sept. 2008, Tokyo, p. 128.

30. Michael Persinger, personal communication.

31. Mauduyt 1777, p. 509.

32. Bertholon 1786, vol. 1, p. 61.

33. Priestley 1775, pp. 429-30.

34. *Yellow Emperor's Classic of Internal Medicine*, chap. 5. Translation by Zhang Wenzhi, Center for Zhouyi and Ancient Chinese Philosophy, Shanding University, Jinan, China.

35. Faust 1978, p. 326; Mygge 1919.

Chapter 4. The Road Not Taken

1. Newton 1713, p. 547.

2. Nollet 1746, p. 33.

3. Marcelin Du Carla-Bonifas, *Cosmogonie*, quoted in Bertholon 1786, vol. 1, p. 86.

4. Voltaire 1772, pp. 90-91.

5. Marat 1782, p. 362.

6. Wesley 1760, p. 1.

Chapter 5. Chronic Electrical Illness

1. Charles Dickens, "House-Top Telegraphs," *All the Year Round*, Nov. 26, 1859.

2. Highton 1851, pp. 151-52.

3. Dana 1923, p. 429.

4. Beard 1875.

5. Prescott 1860, pp. 84, 270, 274.

6. Morse 1870, p. 613.

7. London District Telegraph Company used a single-needle apparatus and an alphabet code that required an average 2.9 needle positions per letter.

8. Gosling 1987; Lutz 1991; Shorter 1992; Winter 2004.

9. Flint 1866, pp. 640-41.

10. Tourette 1889, p. 61.

11. Cleaves 1910, pp. 9, 80, 96, 168-69.

12. Anonymous 1905.

13. Letter to W. Wilkie Collins, Jan. 17, 1858.

14. Gellé 1889; Castex 1897a, b; Politzer 1901; Tommasi 1904; Blegvad 1907; Department of Labour, Canada 1907; Heijermans 1908; Julliard 1910; Thébault 1910; Butler 1911; Capart 1911; Fontègne 1918; Picaud 1949; Le Guillant 1956; Yassi 1989.

15. Desrosiers 1879, citing Jaccoud.

16. Arndt 1885, pp. 102-4.

17. Kleinman 1988, p. 103; World Psychiatric Association 2002, p. 9. Flaskerud 2007, p. 658 reports that neurasthenia is the second most common psychiatric diagnosis in China.

18. World Psychiatric Association 2002, p. 10.

19. Tsung-Yi Lin 1989b, p. 112.

20. Goering 2003, p. 35.

Chapter 6. The Behavior of Plants

1. Nollet 1753, pp. 356-61.

2. Jallabert 1749, pp. 91-92.

3. Bose 1747, p. 20.

4. Bertholon 1783, p. 154.

5. Marat 1782, pp. 359-60.

6. Quotation in Hull 1898, pp. 4-5.

7. Stone 1911, p. 30.

8. Paulin 1890; Crépeaux 1892; Hull 1898, pp. 9-10.

9. Bose 1907, pp. 578-86, "Inadequacy of Pflüger's Law."

10. Bose 1915.

11. Bose 1919, pp. 416-24, "Response of Plants to Wireless Stimulation."

12. Bose 1923, pp. 106-7.

13. Bose 1927, p. 94.

Chapter 7. Acute Electrical Illness

1. *Scientific American* 1889d.

2. Stuart-Harris 1965, fig. 54, p. 87.

3. Hope-Simpson 1992, p. 59.

4. Mygge 1930, p. 10.

5. Mygge 1919, p. 1255.

6. Hogan 1995, p. 122.

7. Here is a sampling of opinion as to the time span of this pandemic: 1727-34 (Gordon 1884); 1729-38 (Taubenberger 2009); 1729-33 (Vaughan 1921; van Tam and Sellwood 2010). Some authors divide it into two separate pandemic periods: 1725-30 and 1732-33 (Harries 1892); 1727-29 and 1732-33 (Creighton 1894); 1728-30 and 1732-33 (Arbuthnot 1751 and Thompson 1852); 1729-30 and 1731-35 (Schweich 1836); 1729-30 and 1732-37 (Bosser 1894, Leledy 1894, and Ozanam 1835); 1729-30 and 1732-33 (Webster 1799; Hirsch 1883; Beveridge 1978; Patterson 1986).

8. Thompson 1852, pp. 28-38.

9. *Ibid.*, p. 43.

10. Marian and Mihăescu 2009.

11. Parsons 1891, pp. 9, 14.

12. Lee 1891, p. 367.

13. Parsons 1891, p. 43.

14. *Journal of the American Medical Association* 1890a.

15. Parsons 1891, p. 33.

16. Brakenridge 1890, pp. 997, 1007.

17. Parsons 1891, p. 11 note.

18. Clemow 1903, p. 198.

19. Parsons 1891, p. 20.

20. *Ibid.*, p. 16.

21. *Ibid.*, p. 24.

22. Clemow 1903, p. 200.

23. Parsons 1891, p. 15.

24. *Ibid.*, p. 24.

25. *Ibid.*, p. 22.

26. *Ibid.*, p. 22.

27. *Ibid.*, p. 19.

28. Bowie 1891, p. 66.

29. Lee 1891, p. 367.

30. Creighton 1894, p. 430. See also Webster 1799, vol. 1, p. 289; Hirsch 1883, pp. 19-21; Beveridge 1978, p. 47.

31. Beveridge 1978, p. 35.

32. Ricketson 1808, p. 4.

33. Jones 1827, p. 5.

34. Thompson 1852, p. ix.

35. Mackenzie 1891, p. 884.

36. Birkeland 1949, pp. 231-32.

37. Bordley and Harvey 1976, p. 214.

38. McGrew 1985, p. 151.

39. Beveridge 1978, 15-16.

40. Parsons 1891, pp. 54, 60.

41. Lee 1891, p. 367.

42. Mackenzie 1891, pp. 299-300.

43. Beveridge 1978, p. 11.

44. Schnurrer 1823, p. 182.

45. Webster 1799, vol. 1, p. 98; Jones 1827, p. 3; *Journal of the Statistical Society of London* 1848, p. 173; Thompson 1852, pp. 42, 57, 213-15, 285-86, 291-92, 366, 374-75; Gordon 1884, p. 363-64; Creighton 1894, p. 343; Beveridge 1978, pp. 54-67; Taubenberger 2009, p. 6.

46. Beveridge 1978, p. 56.

47. E.g., *Lancet* 1919; Beveridge 1978, p. 57.

48. Hope-Simpson 1979, p. 18.

49. Kilbourne 1975, p. 1; Beveridge 1978, p. 38.

50. Jefferson 2006, 2009. See also Glezen and Simonsen 2006; Cannell 2008.

Chapter 8. Mystery on the Isle of Wight

1. d'Arsonval 1892a.
2. d'Arsonval 1893a.
3. *Ibid.*
4. Underwood and van Engelsdorp 2007.
5. Carr 1918.
6. Baker 1971, p. 160.
7. Nimitz 1963, p. 239.
8. *Annual Report of the Surgeon General* 1919, p. 367.
9. Berman 1918.
10. *Annual Report of the Surgeon General* 1919, pp. 411-12.
11. Nuzum 1918.
12. *Journal of the American Medical Association* 1918e, p. 1576.
13. Pflomm 1931; Schliephake 1935, p. 120; Kyuntsel' and Karmilov 1947; Richardson 1959; Schliephake 1960, p. 88; Rusyaev and Kuksinskiy 1973; Kuksinskiy 1978. See also Person 1997; Firstenberg 2001.
14. Jordan 1918.
15. Berman 1918, p. 1935.
16. Bircher 1918.
17. *Journal of the American Medical Association* 1918g.
18. Armstrong 1919, p. 65; Sierra 1921.
19. *Journal of the American Medical Association* 1919b.
20. Firstenberg 1997, p. 29.
21. *Annual Report of the Surgeon General* 1919, p. 408.
22. *Ibid.*, pp. 409-10.
23. Menninger 1919a.
24. *Annual Report of the Surgeon General* 1919, pp. 426-35.
25. Erlendsson 1919.
26. Soper 1918, p. 1901.
27. Rosenau 1919. See also Leake 1919; *Public Health Reports* 1919.

Chapter 9. Earth's Electric Envelope

1. *The Immense Journey.* NY: Random House, 1957, p. 14.
2. Burbank 1905, p. 27.
3. Rheinberger and Jasper 1937, p. 190; Ruckebusch 1963; Klemm 1969; Pellegrino 2004, pp. 481-82.
4. König 1974b; König 1975, pp. 77-81.
5. Helliwell 1965, p. 1.

6. Reiter 1954, p. 481.

7. Lyman and O'Brien 1977, pp. 1-27.

8. Brewitt 1996; Larsen 2004.

9. Xiang et al. 1984; Hu et al. 1993; Huang et al. 1993; Wu et al. 1993; Zhang et al. 1999; Starwynn 2002.

10. Wei et al. 2012.

11. de Vernejoul et al. 1985.

12. Jiang et al. 2002; Baik, Park, et al. 2004; Baik, Sung, et al. 2004; Cho et al. 2004; Johng et al. 2004; Kim et al. 2004; Lee 2004; Park et al. 2004; Shin et al. 2005; Johng et al. 2006; Lee et al. 2008; Lee et al. 2010; Soh et al. 2012; Avijgan and Avijgan 2013; Park et al. 2013; Soh et al. 2013.

13. Lee et al. 2009.

14. Fujiwara and Yu 2012.

15. Lim et al. 2015.

16. Helliwell 1977.

17. Davis 1974; Fraser-Smith et al. 1977.

18. Park and Chang 1978.

19. Bullough 1995.

20. Fraser-Smith 1979, 1981; Villante et al. 2004; Guglielmi and Zotov 2007.

21. Fraser-Smith 1979.

22. Guglielmi and Zotov 2007.

23. Bullough et al. 1976; Tatnall et al. 1983; Bullough 1995.

24. Boerner et al. 1983.

25. Bullough 1985.

26. Cannon and Rycroft 1982.

27. Bullough et al. 1976; Luette et al. 1977, 1979; Park et al. 1983; Imhof et al. 1986.

28. Kornilov 2000.

Chapter 10. Porphyrins and the Basis of Life

1. Randolph 1987, chap. 4.

2. Leech 1888; Matthes 1888; Hay 1889; Ireland 1889; Marandon de Montyel 1889; *Revue des Sciences Médicales* 1889; Rexford 1889; Bresslauer 1891; Fehr 1891; Geill 1891; Hammond 1891; Lepine 1893; With 1980.

3. Morton 2000.

4. Morton 1995, 1998, 2000, 2001, personal communication.

5. Morton 1995, p. 6.

6. Hoffer and Osmond 1963; Huszák et al. 1972; Irvine and Wetterberg 1972; Pfeiffer 1975; McCabe 1983; Durkó et al. 1984; McGinnis et al. 2008a, 2008b; Mikirova 2015.

7. Moore et al. 1987, pp. 42-43.

8. Gibney et al. 1972; Petrova and Kuznetsova 1972; Holtmann and Xenakis 1978, 1978; Pierach 1979; Hengstman et al. 2009;.

9. Quoted in Mason et al. 1933.

10. Athenstaedt 1974; Fukuda 1974.

11. Adler 1975.

12. Kim et al. 2001; Zhou 2009; Hagemann et al. 2013.

13. Aramaki et al. 2005.

14. Szent-Györgyi 1957, p. 19.

15. Becker and Selden 1985, p. 30.

16. Burr 1945b, 1950, 1956.

17. Ravitz 1953.

18. Becker 1960; Becker and Marino 1982, p. 37; Becker and Selden 1985, p. 116.

19. Gilyarovskiy et al. 1958.

20. Becker 1985, pp. 238-39.

21. Rose 1970, pp. 172-73, 214-15; Lund 1947 (comprehensive review and bibliography).

22. Becker and Selden 1985, p. 237.

23. Becker 1961a; Becker and Marino 1982, pp. 35-36.

24. Klüver 1944a, 1944b; Harvey and Figge 1958; Peters et al. 1974; Becker and Wolfgram 1978; Chung et al. 1997; Kulvietis et al. 2007; Felitsyn et al. 2008.

25. Peters 1993.

26. Felitsyn et al. 2008.

27. Soldán and Pirko 2012.

28. Hargittai and Lieberman 1991; Ravera et al. 2009; Morelli et al. 2011; Morelli et al. 2012; Ravera, Bartolucci, et al. 2013; Rivera, Nobbio, et al. 2013; Ravera et al. 2015; Ravera and Panfoli 2015.

29. Peters 1961.

30. Peters et al. 1957; Peters et al. 1958; Peters 1961; see also Painter and Morrow 1959; Donald et al. 1965.

31. Lagerwerff and Specht 1970; Wong 1996; Wong and Mak 1997; Apeagyei et al. 2011; Tamrakar and Shakya 2011; Darus et al. 2012; Elbagermi et al. 2013; Li et al. 2014; Nazzal et al. 2014.

32. Flinn et al. 2005.

33. Hamadani et al. 2002.

34. Hamadani et al. 2001.

35. Buh et al. 1994.

36. McLachlan et al. 1991; Cuajungco et al. 2000; Regland et al. 2001; Ritchie et al. 2003; Frederickson et al. 2004; Religa et al. 2006; Bush and Tanzi 2008.

37. Religa et al. 2006.

38. Hashim et al. 1996.

39. Cuajungco et al. 2000; Que et al. 2008; Baum et al. 2010; Cristóvão et al. 2016.

40. Voyatzoglou et al. 1982; Xu et al. 2013.

41. Milne et al. 1983; Taylor et al. 1991; Johnson et al. 1993; King et al. 2000.

42. Johnson et al. 1993; King et al. 2000.

43. Andant et al. 1998. See also Kauppinen and Mustajoki 1988.

44. Linet et al. 1999.

45. Halpern and Copsey 1946; Markovitz 1954; Saint et al. 1954; Goldberg 1959; Eilenberg and Scobie 1960; Ridley 1969; Stein and Tschudy 1970; Beattie et al. 1973; Menawat et al. 1979; Leonhardt 1981; Laiwah et al. 1983; Laiwah et al. 1985; Kordač et al. 1989.

46. Ridley 1975.

47. I. P. Bakšiš, A. I. Lubosevičute, and P. A. Lopateve, "Acute Intermittent Porphyria and Necrotic Myocardial Changes," *Terapevticheskiĭ arkhiv* 8: 145-46 (1984), cited in Kordač et al. 1989.

48. Sterling et al. 1949; Rook and Champion 1960; Waxman et al. 1967; Stein and Tschudy 1970; Herrick et al. 1990.

49. Berman and Bielicky 1956.

50. Labbé 1967; Laiwah et al. 1983; Laiwah et al. 1985; Herrick et al. 1990; Kordač et al. 1989; Moore et al. 1987; Moore 1990.

Chapter 11. Irritable Heart

1. Maron et al. 2009.

2. Milham 2010a, p. 345.

3. White 1938, pp. 171-72, 586; White 1971; Flint 1866, p. 303.

4. Chadha et al. 1997.

5. Milham 2010b.

6. Dawber et al. 1957; Doyle et al. 1957; Kannel 1974; Hatano and Matsuzaki 1977; Rhoads et al. 1978; Feinleib et al. 1979; Okumiya et al. 1985; Solberg et al. 1985; Stamler et al. 1986; Reed et al. 1989; Tuomilehto and Kuulasmaa 1989; Neaton et al. 1992; Verschuren et al. 1995; Njølstad et al. 1996; Wilson et al. 1998; Stamler et al. 2000; Navas-Nacher et al. 2001; Sharrett et al. 2001; Zhang et al. 2003.

7. Phillips et al. 1978; Burr and Sweetnam 1982; Frentzel-Beyme et al. 1988; Snowdon 1988; Thorogood et al. 1994; Appleby et al. 1999; Key et al. 1999; Fraser 1999, 2009.

8. Phillips et al. 1978; Snowden 1988; Fraser 1999; Key et al. 1999.

9. Sijbrands et al. 2001.

10. Dawber et al. 1957.

11. Doyle et al. 1957.

12. Fox 1923, p. 71.

13. Ratcliffe et al. 1960, p. 737.

14. Rigg et al. 1960.

15. Vastesaeger and Delcourt 1962.

16. Daily 1943; Barron et al. 1955; McLaughlin 1962.

17. Barron et al. 1955; Brodeur 1977, pp. 29-30.

18. Sadchikova 1960, 1974; Klimková-Deutschová 1974.

19. See Pervushin 1957; Drogichina 1960; Letavet and Gordon 1960; Orlova 1960; Gordon 1966; Dodge 1970 (review); Healer 1970 (review); Marha 1970; Gembitskiy 1970; Subbota 1970; Marha et al. 1971; Tyagin 1971; Barański and Czerski 1976; Bachurin 1979; Jerabek 1979; Silverman 1979 (review); McRee 1979, 1980 (reviews); Sadchikova et al. 1980; McRee et al. 1988 (review); Afrikanova and Grigoriev 1996. For bibliographies, see Kholodov 1966; Novitskiy et al. 1970; Presman 1970; Petrov 1970a; Glaser 1971-1976, 1977; Moore 1984; Grigoriev and Grigoriev 2013.

20. Personal communication, Oleg Grigoriev and Yury Grigoriev, Russian National Committee on Non-Ionizing Radiation Protection. Russian textbooks include Izmerov and Denizov 2001; Suvorov and Izmerov 2003; Krutikov et al. 2003; Krutikov et al. 2004; Izmerov 2005, 2011a, 2011b; Izmerov and Kirillova 2008; Kudryashov et al. 2008.

21. Tyagin 1971, p. 101.

22. Frey 1988, p. 787.

23. Brodeur 1977, p. 51.

24. Presman and Levitina 1962a, 1962b; Levitina 1966.

25. Frey and Seifert 1968; Frey and Eichert 1986.

26. Cohen, Johnson, Chapman, et al. 1946.

27. Cohen 2003.

28. Haldane 1922, p. 56; Jones and Mellersh 1946; Jones and Scarisbrick 1946; Jones 1948.

29. Cohen, Johnson, Chapman, et al. 1946, p. 121.

30. See also Jones and Scarisbrick 1943; Jones 1948; Gorman et al. 1988; Holt and Andrews 1989; Hibbert and Pilsbury 1989; Spinhoven et al. 1992; Garssen et al. 1996; Barlow 2002, p. 162.

31. Cohen and White 1951, p. 355; Wheeler et al. 1950, pp. 887-88.

32. Craig and White 1934; Graybiel and White 1935; Dry 1938. See also Master 1943; Logue et al. 1944; Wendkos 1944; Friedman 1947, p. 23; Blom 1951; Holmgren et al. 1959; Lary and Goldschlager 1974.

33. Orlova 1960; Bachurin 1979.

34. Dumanskiy and Shandala 1973; Dumanskiy and Rudichenko 1976; Zalyubovskaya et al. 1977; Zalyubovskaia and Kiselev 1978; Dumanskiy and

Tomashevskaya 1978; Shutenko et al. 1981; Dumanskiy and Tomashevskaya 1982; Tomashevskaya and Soleny 1986; Tomashevskaya and Dumanskiy 1989; Tomashevskaya and Dumanskiy 1988.

35. Chernysheva and Kolodub 1976; Kolodub and Chernysheva 1980.

36. Da Costa 1871, p. 19.

37. Plum 1882.

38. Johnston 1880, pp. 76-77.

39. Plum 1882, vol. 1, pp. 26-27.

40. Oglesby 1887; MacLeod 1898.

41. Smart 1888, p. 834.

42. Howell 1985, p. 45; International Labour Office 1921, Appendix V, p. 50.

43. Lewis 1918b, p. 1; Cohn 1919, p. 457.

44. Munro 1919, p. 895.

45. Aschenheim 1915; Brasch 1915; Braun 1915; Devoto 1915; Ehret 1915; Merkel 1915; Schott 1915; Treupel 1915; von Dziembowski 1915; von Romberg 1915; Aubertin 1916; Galli 1916; Korach 1916; Lian 1916; Cohn 1919.

46. Conner 1919, p. 777.

47. Scriven 1915; Corcoran 1917.

48. Howell 1985, p. 37.

49. Corcoran 1917.

50. Worts 1915.

51. Scriven 1915; *Popular Science Monthly* 1918.

52. Lewis 1940; Master 1943; Stephenson and Cameron 1943; Jones and Mellersh 1946; Jones 1948.

53. Mäntysaari et al. 1988; Fava et al. 1994; Sonimo et al. 1998.

54. Freud 1895, pp. 97, 107; Cohen and White 1972.

55. Reyes et al. 2003, Reeves et al. 2007.

56. Caruthers and van de Sande 2011.

57. Cholesterol in anxiety disorder: Lazarev et al. 1989; Bajwa et al. 1992; Freedman et al. 1995; Peter et al. 1999. Heart disease in anxiety disorder: Coryell et al. 1982; Coryell et al. 1986; Coryell 1988; Hayward et al. 1989; Weissman et al. 1990; Eaker et al. 1992; Nutzinger 1992; Kawachi et al. 1994; Rozanski et al. 1999; Bowen et al. 2000; Paterniti et àl. 2001; Huffman et al. 2002; Grace et al. 2004; Katerndahl 2004; Eaker et al. 2005; Csaba 2006; Rothenbacher et al. 2007; Shibeshi et al. 2007; Vural and Başar 2007; Frasure-Smith et al. 2008; Phillips et al. 2009; Scherrer et al. 2010; Martens et al. 2010; Seldenrijk et al. 2010; Vogelzangs et al. 2010; Olafiranye et al. 2011; Soares-Filho et al. 2014. Cholesterol in chronic fatigue syndrome: van Rensburg et al. 2001; Peckerman et al. 2003; Jason et al. 2006. Heart disease in chronic fatigue syndrome: Lerner et al. 1993; Bates et al. 1995; Miwa

and Fujita 2009. Heart disease in myalgic encephalomyelitis: Caruthers and van de Sande 2011. Cholesterol in radio wave sickness: Klimkova-Deutschova 1974; Sadchikova 1981.

58. Heart disease in porphyria: Saint et al. 1954; Goldberg 1959; Eilenberg and Scobie 1960; Ridley 1969, 1975; Stein and Tschudy 1970; Beattie et al. 1973; Bonkowsky et al. 1975; Menawat et al. 1979; Leonhardt 1981; Kordač et al. 1989; Crimlisk 1997. Cholesterol in porphyria: Taddeini et al. 1964; Lees et al. 1970; Stein and Tschudy 1970; York 1972, pp. 61-62; Whitelaw 1974; Kaplan and Lewis 1986; Shiue et al. 1989; Fernández-Miranda et al. 2000; Blom 2011; Park et al. 2011.

59. Chin et al. 1999; Newman et al. 2001; Coughlin et al. 2004; Robinson et al. 2004; Li et al. 2005; McArdle et al. 2006; Li et al. 2007; Savransky et al. 2007; Steiropoulous et al. 2007; Gozal et al. 2008; Dorkova et al. 2008; Lefebvre et al. 2008; Çuhadaroğlu et al. 2009; Drager et al. 2010; Nadeem et al. 2014.

60. Behan et al. 1991; Wong et al. 1992; McCully et al. 1996; Myhill et al. 2009.

61. Marazziti et al. 2001; Gardner et al. 2003; Fattal et al. 2007; Gardner and Boles 2008, 2011; Hroudová and Fišar 2011.

62. See note 34. Also Ammari et al. 2008.

63. Goldberg et al. 1985; Kordač et al. 1989; Herrick et al. 1990; Moore 1990; Thunell 2000.

64. Sanders et al. 1984.

65. Haldane 1922, pp. 56-57; Haldane and Priestley 1935, pp. 139-41.

66. Numbers of residential electric customers for 1930-1931 were obtained from National Electric Light Association, *Statistical Bulletin* nos. 7 and 8, and for 1939-1940 from Edison Electric Institute, *Statistical Bulletin* nos. 7 and 8. For states east of the 100th meridian, "Farm Service" customers (1930-1931) or "Rural Rate" customers (1939-1940) were added to "Residential or Domestic" customers to get the true residential count, as recommended in the *Statistical Bulletins*. "Farm" and "Rural Rate" service in the west referred mainly to commercial customers, usually large irrigation systems. The same terms, east of the 100th meridian, were used for residential service on distinct rural rates. A discrepancy in the number of farm households in Utah was resolved by consulting *Rural Electrification in Utah*, published in 1940 by the Rural Electrification Administration.

67. Johnson 1868.

68. Koller 1962.

69. Parikh et al. 2009.

70. McGovern et al. 2001.

71. Roger et al. 2004.

72. Ghali et al. 1990.

73. Fang et al. 2008.

74. McCullough et al. 2002.

75. Cutler et al. 1997; Martin et al. 2009.

76. Zheng et al. 2005.

77. National Center for Health Statistics 1999, 2006.

78. Arora et al. 2019.

Chapter 12. The Transformation of Diabetes

1. *The Sun* 1891; Howe 1931; Joslin Diabetes Clinic 1990.

2. Gray 2006, pp. 46, 261, 414.

3. Hirsch 1885, p. 645.

4. Harris 1924; Brun et al. 2000.

5. Joslin 1917, p. 59.

6. Annual consumption of sugar and other sweeteners from 1822 to 2014 was obtained from tables published in *Annual Report of the Commissioner of Agriculture for the Year 1878*; *American Almanac and Treasury of Facts* (New York: American News Company, 1888); *Proceedings of the Interstate Sugar Cane Growers First Annual Convention* (Macon, GA: Smith and Watson, 1903); A. Bouchereau, *Statement of the Sugar Crop Made in Louisiana in 1905-'06* (New Orleans, 1909); *Statistical Abstracts of the United States* for 1904-1910; *Ninth Census of the United States*, vol. 3, *The Statistics of Wealth and Industry of the United States* (1872); *Twelfth Census of the United States*, vol. 5, *Agriculture* (1902); *Thirteenth Census of the United States*, vol. 5, *Agriculture* (1914); *United States Census of Agriculture*, vol. 2 (1950); *Statistical Bulletin* No. 3646 (U.S. Dept. of Agriculture, 1965); Supplement to *Agricultural Economic Report* No. 138 (U.S. Dept. of Agriculture, 1975); and *Sugar and Sweeteners Outlook*, Table 50 – U.S. per capita caloric sweeteners estimated deliveries for domestic food and beverage use, by calendar year (U.S. Dept. of Agriculture, 2003). Honey was estimated to contain 81 percent sugar; molasses, 52 percent sugar; cane syrup, 56.3 percent sugar; maple syrup, 66.5 percent sugar; and sorghum syrup, 68 percent sugar.

7. Gohdes 1995.

8. Black Eagle, personal communication.

9. Levy et al. 2012; Welsh et al. 2010.

10. Pelden 2009.

11. Giri et al. 2013.

12. Joslin 1917, 1924, 1927, 1943, 1950; Woodyatt 1921; Allen 1914, 1915, 1916, 1922; Mazur 2011.

13. Fothergill 1884.

14. Joslin 1917, pp. 100, 102, 106, 107.

15. Simoneau et al. 1995; Gerbiz et al. 1996; Kelley et al. 1999; Simoneau and Kelley 1997; Kelley and Mandarino 2000; Kelley et al. 2002; Bruce et

al. 2003; Morino et al. 2006; Toledo et al. 2008; Ritov et al. 2010; Patti and Corvera 2010; DeLany et al. 2014; Antoun et al. 2015.

16. DeLany et al. 2014.

17. Ritov et al. 2010.

18. Gel'fon and Sadchikova 1960.

19. Gel'fon and Sadchikova 1960; Syngayevskaya 1962; Bartoníček and Klimková-Deutschová 1964; Petrov 1970a, p. 164; Sadchikova 1974; Klimková-Deutschová 1974; Dumanskiy and Rudichenko 1976; Dumanskiy and Shandala 1974; Dumanskiy and Tomashevskaya 1978; Gabovich et al. 1979; Kolodub and Chemysheva 1980; Belokrinitskiy 1981; Shutenko et al. 1981; Dumanskiy et al. 1982; Dumanskiy and Tomashevskaya 1982; Tomashevskaya and Soleny 1986; Tomashevskaya and Dumanskiy 1988; Navakatikian and Tomashevskaya 1994.

20. Kwon et al. 2011.

21. Li et al. 2012.

22. 1917 figure from Joslin 1917, p. 25.

23. Kuczmarski et al. 1994. See also Prentice and Jebb 1995.

24. Flegal et al. 1998, 2002, 2010; Ogden et al. 2012.

25. Kim et al. 2006.

26. Flegal 1998, p. 45.

27. Thatcher et al. 2009.

28. Klimentidis et al. 2011.

Chapter 13. Cancer and the Starvation of Life

1. Warburg 1925, p. 148.

2. Warburg 1908.

3. Warburg et al. 1924; Warburg 1925.

4. Warburg 1925, p. 162.

5. Warburg 1930, p. x.

6. Warburg 1956.

7. Warburg 1966b.

8. Krebs 1981, pp. 23-24, 74.

9. Harris 2002; Ferreira and Campos 2009.

10. Ristow and Cuezva 2006; van Waveren et al. 2006; Srivastava 2009; Sánchez-Aragó et al. 2010.

11. Kondoh 2009, p. 101; Sánchez-Aragó et al. 2010.

12. Apte and Sarangarajan 2009a.

13. Ferreira and Campos 2009, p. 81.

14. Vaupel et al. 1998; Gatenby and Gillis 2004; McFate et al. 2008; Gonzáles-Cuyar et al. 2009, pp. 134-36; Semenza 2009; Werner 2009, pp. 171-72; Sánchez-Aragó et al. 2010.

15. Vigneri et al. 2009.
16. Giovannuca et al. 2010.
17. Lombard et al. 1959.
18. From Williams 1908, p. 53.
19. Guinchard 1914.
20. Hoffman 1915, p. 151.
21. *Ibid.*, pp. 185-186.
22. Stein et al. 2011.
23. From volumes of *Vital Statistics of the United States* (United States Bureau of the Census) and *National Vital Statistics Reports* (Centers for Disease Control and Prevention).
24. Moffat 1988.
25. Data on smoking rates from National Center for Health Statistics. Data on lung cancer from *Vital Statistics of the United States* (1970, 1980, 1990) and *National Vital Statistics Reports* (2000, 2010, 2015).
26. National Cancer Institute 2009.
27. Schüz et al. 2006.
28. Barlow et al. 2009.
29. Teppo et al. 1994.
30. Jacob Easaw, Southern Alberta Cancer Research Institute, personal communication.
31. Hardell and Carlberg 2009; Hardell et al. 2011a.
32. Anderson and Henderson 1986.

Chapter 14. Suspended Animation

1. Beard 1980, pp. 2-3; Beard 1881a, pp. viii, ix, 105.
2. Weindruch and Walford 1988.
3. Walford 1982.
4. Riemers 1979.
5. Austad 1988.
6. Dunham 1938.
7. Johnson et al. 1984.
8. Fischer-Piette 1989.
9. Hansson et al. 1953.
10. Colman et al. 2013.
11. Ross and Bras 1965; for other studies of tumors in rats, see Weindruch and Walford, pp. 76-84.
12. Colman et al. 2009; Mattison et al. 2003.
13. Griffin 1958, p. 35.
14. Ramsey et al. 2000; Lynn and Wallwork 1992.
15. Ramsey et al. 2000.

16. Ordy et al. 1967.

17. Spalding et al. 1971.

18. Perez et al. 2008.

19. Tryon and Snyder 1971.

20. Caratero et al. 1998.

21. Okada et al. 2007.

22. Suzuki et al. 1998.

Chapter 15. You mean you can hear electricity?

1. Grapengiesser 1801, p. 133. Quoted in Brenner 1868, p. 38.

2. Brenner 1868, pp. 41, 45.

3. Tousey 1921, p. 469.

4. Meyer 1931.

5. Gersuni and Volokhov 1936.

6. Stevens and Hunt 1937, unpublished, described in Stevens and Davis 1938, pp. 354-55.

7. Moeser, W. "Whiz Kid, Hands Down," *Life*, September 14, 1962.

8. Einhorn 1967.

9. Russell et al. 1986.

10. See also Degens et al. 1969.

11. Lissman, p. 184; Offutt 1984, pp. 19-20.

12. de Vries 1948a, 1948b.

13. Honrubia 1976; Mountain 1986; Ashmore 1987.

14. Zwislocki 1992; Gordon, Smith and Chamberlain 1982, cited in Zwislocki.

15. Nowotny and Gummer 2006.

16. Brenner 1868.

17. Mountain 1986.

18. Mountain et al. 1986; Ashmore 1987; Honrubia and Sitko 1976.

19. Lenhardt 2003.

20. Combridge and Ackroyd 1945, Item No. 7, p. 49.

21. Gavrilov et al. 1980.

22. Qin et al. 2011.

23. Stevens 1938, p. 50, fig. 17; Corso 1963; Moller and Pederson 2004, figs. 1-3; Stanley and Walker 2005.

24. Stevens 1937.

25. *Environmental Health Criteria 137*, 1993 edition, pp. 160 and 161, figs. 23 and 24.

26. Duane Dahlberg, Ph.D., personal communication.

27. Petrie 1963, pp. 89-92.

28. Maggs 1976.

29. Reported by the Low Frequency Noise Sufferer's Association of

England, Jean Skinner, personal communication; by Sara Allen of Taos, New Mexico, personal communication; and by Mullins and Kelley 1995.

30. Calculation based on Jansky and Bailey 1962, fig. 35, Ground Wave Field Intensity; and Garufi 1989, fig. 6, U.S. Coast Guard Conductivity Map.

31. In Africa, only Egypt, Tunisia, Ghana, Senegal, Ethiopia, Zambia, Zimbabwe, and South Africa currently have bans in place or in progress. In the Middle East, only Israel, Lebanon, Kuwait, Bahrain, Qatar, and the United Arab Emirates currently have bans. Other countries where prohibition is neither in place nor in progress include Haiti, Jamaica, St. Kitts and Nevis, Granada, Antigua and Barbuda, St. Vincent and the Grenadines, St. Lucia, Trinidad and Tobago, Dominica, Venezuela, Bolivia, Paraguay, Uruguay, Suriname, Albania, Moldova, Belarus, Uzbekistan, Kyrgyzstan, Turkmenistan, Mongolia, Turkey, Afghanistan, Pakistan, Nepal, Bhutan, India, Bangladesh, Myanmar, Singapore, Cambodia, Laos, Indonesia, East Timor, Papua New Guinea, New Zealand, Bosnia and Herzegovinia, Kosovo, and North Macedonia.

32. Signal structure for GSM: superframe (6.12 sec), control multiframe (235.4 msec), traffic multiframe (120 msec), frame (4.615 msec), time slot (0.577 msec), symbol (270,833 per second per channel, 33,850 per second per user). Signal structure for UMTS: frame (10 msec), time slot (0.667 msec), symbol (66.7 µsec), chip (0.26 µsec). Signal structure for LTE: frame (10 msec), half-frame (5 msec), subframe (1 msec), slot (0.5 msec), symbol (0.667 msec).

33. Mild and Wilén 2009.

34. Hutter et al. 2010.

35. National Center for Health Statistics 1982-1996.

36. Shargorodsky et al. 2010.

37. Del Bo et al. 2008.

38. Nondahl et al. 2012.

Chapter 16. Bees, Birds, Trees, and Humans

1. Balmori and Hallberg 2007.

2. Sen 2012.

3. *Deccan Herald* 2012.

4. Personal communication from New Mexico pigeon racer Larry Lucero, 1999.

5. Bigu del Blanco et al. 1973.

6. Haughey 1997.

7. Larry Lucero, personal communication.

8. Robert Costagliola, of Fogelsville, Pennsylvania, personal communication.

9. Gary Moore, the "liberator" for the western Pennsylvania-to-Philadelphia race, personal communication.

10. Elston 2004.

11. *Indian Express* 2010.

12. Roberts 2000.

13. Mech and Barber 2002, p. 29.

14. Withey et al. 2001, pp. 47-49; Mech and Barber 2002, p. 30.

15. Burrows et al. 1994, 1995 on wild dogs; Mech and Barber 2002, pp. 50-51.

16. Swenson et al. 1999.

17. Moorhouse and Macdonald 2005.

18. *Reader's Digest* 1998.

19. Godfrey and Bryant 2003.

20. Engels et al. 2014.

21. Souder 1996.

22. Hallowell 1996.

23. Stern 1990.

24. Hallowell 1996; Souder 1996.

25. Watson 1998.

26. *Ibid.*

27. Revkin 2006.

28. Hawk 1996.

29. Hoperskaya et al., p. 254 .

30. Serant 2004.

31. Panagopoulos et al. 2004.

32. Panagopoulous, Chavdoula, Nezis, and Margaritis 2007; Panagopoulos 2012a.

33. Panagopoulos and Margaritis 2008, 2010; Panagopoulos, Chavdoula, and Margaritis 2010; Panagopoulos 2011.

34. Margaritis et al. 2014.

35. *Bienenvater*, issue no. 9, 2003.

36. Ruzicka 2006.

37. Phillips 1925; Bailey 1964; Underwood and vanEngelsdorp 2007.

38. Bailey 1991, pp. 97-101.

39. *Ibid.*, p. 101.

40. Rinderer et al. 2001.

41. Sanford 2004.

42. Boecking and Ritter 1993.

43. Fries et al. 2006.

44. Page 1998; Rinderer et al. 2001.

45. Rinderer et al. 2001.

46. Kraus and Page 1995.

47. Seeley 2007.

48. National Research Council 2007; Kluser and Peduzzi 2007; vanEngelsdorp 2009.

49. Wilson and Menapace 1979; Underwood and vanEngelsdorp 2007; McCarthy 2011.

50. Also Finley et al. 1996.

51. O'Hanlon 1997.

52. Hamzelou 2007.

53. Kluser and Peduzzi 2007.

54. Borenstein 2007.

55. McCarthy 2011; Pattazhy 2012.

56. Le Conte et al. 2010.

57. Evans et al. 2008.

58. Warnke 1976; Becker 1977.

59. Warnke 1989.

60. Lindauer and Martin 1972; Warnke 2009.

61. Pattazhy 2011a, 2011b, 2012, and personal communication.

62. Warnke 2009.

63. Schütt and Cowling 1985.

64. *Microwave News* 1994.

65. Kolodynski and Kolodynska 1996.

66. Balode 1996.

67. Liepa and Balodis 1994.

68. Balodis et al. 1996.

69. Selga and Selga 1996.

70. Magone 1996.

71. Lorenz et al. 2003.

72. Bentouati and Bariteau 2006.

73. Hennon et al. 1990; Hennon and Shaw 1994; Hennon et al. 2012.

74. Navy Department, Bureau of Equipment 1907, 1908; United States Department of Commerce, Bureau of Navigation 1913.

75. Phillips et al. 2009.

76. Rohter 2002.

77. Worrall et al. 2008.

Chapter 17. In the Land of the Blind

1. Mild et al. 1998.

2. Yakymenko et al. 2011.

3. Dalsegg 2002.

4. Johansson 2004.

5. Hallberg and Oberfeld 2006.

6. Byun et al. 2013.

7. Tatemichi et al. 2004.

8. Kimata 2002.

9. Ye et al. 2001.

10. Li et al. 2011.

11. Oktay and Dasdag 2006.

12. Panda et al. 2011.

13. Velayutham et al. 2014.

14. Mishra 2011.

15. Mishra 2010, p. 51.

16. Salford et al. 2003.

17. Nittby et al. 2008.

18. Şahin et al. 2015.

19. Baş et al. 2013; Hancı et al. 2013; İkinci et al. 2013; Odacı et al. 2013; Hancı et al. 2015; Odacı, Hancı, İkinci et al. 2015; Odacı and Özyılmaz 2015; Odacı, Ünal, et al. 2015; Topal et al. 2015; Türedi et al. 2015; Odacı, Hancı, Yuluğ et al. 2016.

20. İkinci et al. 2015.

21. Blue Cross Blue Shield 2019.

22. Bejot et al. 2014.

23. Rosengren et al. 2013.

24. Putaala et al. 2009.

25. Tibæk et al. 2016.

26. West et al. 2013.

27. Wolford et al. 2015.

28. Siegel et al. 2017.

29. Wong et al. 2016

30. Hallberg and Johansson 2009.

31. Weiner et al. 2016.

32. Centola et al. 2016.

33. Hutton et al. 2019.

34. Broomhall 2017.

35. Hallman et al. 2017

36. Lister and Garcia 2018.

37. Sánchez-Bayo and Wyckhuys 2019.

38. "Satellites Begin Worldwide Service," *No Place To Hide* 2(1): 3 (1999).

39. "Satellites: An Urgent Situation," *No Place To Hide* 2(3): 18 (2000).

40. "Update on Satellites," *No Place To Hide* 3(2): 15 (2001).

41. Janet Patton, "Foal deaths remain a mystery," *Lexington Herald-Leader*, May 9, 2001; Lenn Harrison, personal communication.

42. The actual power in each beam will be 100 watts or less, but since all of that power will be focused in a laser-like beam, the *effective* radiated power (EIRP) is reported to the FCC. The EIRP is the amount of power the satellite would have to emit in order to have the same strength in all directions as it has in the focused beam.

Bibliography

Note: *JPRS* = *Joint Publications Research Service*.

Chapters 1-4

Adams, George. 1787, 1799. *An Essay on Electricity*, 3rd ed. London: R. Hindmarsh; 5th ed. London: J. Dillon.

Aldini, Jean. 1804. *Essai Théorique et Expérimental sur le Galvanisme*. Paris: Fournier Fils.

Baker, Henry. 1748. "A Letter from Mr. Henry Baker, F.R.S. to the President, concerning several Medical Experiments of Electricity." *Philosophical Transactions* 45: 370-75.

Beard, George Miller and Alphonso David Rockwell. 1883. *A Practical Treatise on the Medical and Surgical Uses of Electricity*, 4th ed. New York: William Wood.

Beaudreau, Sherry Ann and Stanley Finger. 2006. "Medical Electricity and Madness in the 18th Century: The Legacies of Benjamin Franklin and Jan Ingenhousz." *Perspectives in Biology and Medicine* 49(3): 330-45.

Beccaria, Giambatista. 1753. *Dell'Elettricismo Artificiale e Naturale*. Torino: Filippo Antonio Campana.

Becket, John Brice. 1773. *An Essay on Electricity*. Bristol.

Bell, Whitfield Jenks, Jr. 1962. "Benjamin Franklin and the Practice of Medicine." *Bulletin of the Cleveland Medical Library* 9: 51-62.

Berdoe, Marmaduke. 1771. *An Enquiry Into the Influence of the Electric Fluid in the Structure and Formation of Animated Beings*. Bath: S. Hazard.

Bertholon, Pierre Nicholas. 1780. *De l'Électricité du Corps Humain dans l'État de Santé et de Maladie*. Lyon: Bernusset.

———. 1783. *De l'Électricité des Végétaux*. Paris: P. F. Didot Jeune.

———. 1786. *De l'Électricité du Corps Humain dans l'État de Santè et de Maladie*, 2 vols. Paris: Didot le jeune.

Bertucci, Paola. 2007. "Sparks in the Dark: the Attraction of Electricity in the Eighteenth Century." *Endeavor* 31(3): 88-93.

Bonnefoy, Jean-Baptiste. 1782. *De l'Application de l'Éléctricité a l'Art de Guérir*. Lyon: Aimé de la Roche.

Bose, Georg Matthias. 1744a. *Tentamina electrica in Academiis Regiis Londinensi et Parisina*. Wittenberg: Johann Joachim Ahlfeld.

———. 1744b. *Die Electricität nach ihrer Entdeckung und Fortgang, mit poetischer Feder entworffen*. Wittenberg: Johann Joachim Ahlfeld.

Bresadola, Marco. 1998. "Medicine and Science in the Life of Luigi Galvani." *Brain Research Bulletin* 46(5): 367-80.

Bryant, William. 1786. "Account of an Electric Eel, or the Torpedo of Surinam." *Transactions of the American Philosophical Society* 2: 166-69.

Brydone, Patrick. 1773. *A Tour Through Sicily and Malta*, 2 vols. London: W. Strahan and T. Cadell.

Cavallo, Tiberius. 1786. *Complete Treatise on Electricity in Theory and Practice*. London: C. Dilly.

Chaplin, Joyce E. 2006. *The First Scientific American: Benjamin Franklin and the Pursuit of Genius*. New York: Basic Books.

Delbourgo, James. 2006. *A Most Amazing Scene of Wonders: Electricity and Enlightenment in Early America*. Cambridge, MA: Harvard University Press.

Donndorf, Johann August. 1784. *Die Lehre von der Elektricität theoretisch und praktisch aus einander gesetzt*, 2 vols. Erfurt: Georg Adam Kayser.

Donovan, Michael. 1846, 1847. "On the Efficiency of Electricity, Galvanism, Electro-Magnetism, and Magneto-Electricity, in the Cure of Disease; and on the Best Methods of Application." *Dublin Quarterly Journal of Medical Science* 2: 388-414, 3: 102-28.

Dorsman, C. and C. A. Crommelin. 1951. *The Invention of the Leyden Jar*. Leyden: National Museum of the History of Science. Communication no. 97.

Duchenne (de Boulogne), Guillaume Benjamin Amand. 1861. *De l'Électrisation Localisée*, 2nd ed. Paris: J.-B. Baillière et Fils.

Duhamel du Monceau, Henri Louis. 1758. *La Physique des Arbres*. Paris: H. L. Guérin & L. F. Delatour.

Elliott, Paul. 2008. "More Subtle than the Electric Aura: Georgian Medical Electricity, the Spirit of Animation and the Development of Erasmus Darwin's Psychophysiology." *Medical History* 52(2): 195-220.

Flagg, Henry Collins. 1786. "Observations on the Numb Fish, or Torporific Eel." *Transactions of the American Philosophical Society* 2: 170-73.

Franklin, Benjamin. 1758. "An Account of the Effects of Electricity in Paralytic Cases. In a Letter to John Pringle, M.D. F.R.S." *Philosophical Transactions* 50: 481-83.

———. 1774. *Experiments and Observations on Electricity*, 5th ed. London: F. Newbery.

———. *Benjamin Franklin Papers*, <http://franklinpapers.org>.

Gale, T. 1802. *Electricity, or Ethereal Fire*. Troy: Moffitt & Lyon.

Galvani, Luigi. 1791. *De viribus electricitatis in motu musculari. Commentarius*. Bologna: Istituto delle Scienze. Translation by Robert Montraville Green, *Commentary on the Effect of Electricity on Muscular Motion* (Cambridge: Elizabeth Licht), 1953.

Gerhard, Carl Abraham. 1779. "De l'Action de l'Électricité Sur le Corps humain, et de son usage dans les Paralysies." *Observations Sur la Physique, Sur l'Histoire Naturelle, et Sur les Arts* 14: 145-53.

Graham, James. 1779. *The General State of Medical and Chirurgical Practice, Exhibited; Showing Them to be Inadequate, Ineffectual, Absurd, and Ridiculous*. London.

Gralath, Daniel. 1747, 1754, 1756. "Geschichte der Electricität." *Versuche und Abhandlungen der Naturforschenden Gesellschaft in Danzig* 1: 175-304, 2: 355-460, 3: 492-556.

Haller, Albrecht von. 1745. "An historical account of the wonderful discoveries, made in Germany, etc. concerning Electricity." *The Gentleman's Magazine* 15: 193-97.

Hart, Cheney. 1754. "Part of a Letter from Cheney Hart, M.D. to William Watson, F.R.S. giving some Account of the Effects of Electricity in the County Hospital at Shrewsbury." *Philosophical Transactions* 48: 786-88.

Heilbron, John L. 1979. *Electricity in the 17th and 18th Centuries: A Study of Early Modern Physics.* University of California Press: Berkeley.

Histoire de l'Academie Royale des Sciences. 1746. "Sur l' Électricité," pp. 1-17.

———. 1747. "Sur l' Électricité," pp. 1-32.

———. 1748. "Des Effets de l'Électricité sur les Corps Organisés," pp. 1-13.

Houston, Edwin J. 1905. *Electricity in Every-Day Life*, 3 vols. New York: P. F. Collier & Son.

Humboldt, Friedrich Wilhelm Heinrich Alexander von. 1799. *Expériences sur le Galvanisme.* Paris: Didot Jeune.

Jallabert, Jean. 1749. *Expériences sur l'Électricité.* Paris: Durand & Pissot.

Janin, Jean. 1772. *Mémoires et Observations Anatomiques, Physiologiques, et Physiques sur l'Œil.* Lyon: Perisse.

Kratzenstein, Christian Gottlieb. 1745. *Abhandlung von dem Nutzen der Electricität in der Arzneywissenschaft.* Halle: Carl Hermann Hemmerde.

La Beaume, Michael. 1820. *Remarks on the History and Philosophy, But Particularly on the Medical Efficacy of Electricity in the Cure of Nervous and Chronic Disorders.* London: F. Warr.

———. 1842. *On Galvanism.* London: Highley.

Ladame, Paul-Louis. 1885. "Notice historique sur l'Électrothérapie a son origine." *Revue Médicale de la Suisse Romande* 5: 553-72, 625-56, 697-717.

Laennec, René. 1819. *Traité de l'Auscultation Médiate*, 2 vols. Paris: Brosson & Chaudé.

Lindhult, Johann. 1755. "Kurzer Auszug aus des Doctors der Arztneykunst, Johann Lindhults, täglichem Verzeichnisse wegen der Krankheiten, die durch die Electricität sind gelindert oder glücklich geheilet worden. In Stockholm im November und December 1752 gehalten." *Abhandlungen aus der Naturlehre* 14: 312-15.

Louis, Antoine. 1747. *Observations sur l'Électricité.* Paris: Osmont & Delaguette.

Lovett, Richard. 1756. *The Subtil Medium Prov'd.* London: Hinton, Sandby and Lovett.

Lowndes, Frances. 1787. *Observations on Medical Electricity.* London: D. Stuart.

Mangin, Arthur. 1874. *Le Feu du Ciel: Histoire de l'Électricité*, 6th ed. Tours: Alfred Mame et Fils.

Marat, Jean-Paul. 1782. *Recherches Physiques sur l'Électricité.* Paris: Clousier.

———. 1784. *Mémoire sur l'électricité médicale.* Paris: L. Jorry.

Martin, Benjamin. 1746. *An Essay on Electricity: being an Enquiry into the Nature, Cause and Properties thereof, on the Principles of Sir Isaac Newton's Theory of Vibrating Motion, Light and Fire.* Bath.

Mauduyt de la Varenne, Pierre-Jean-Claude. 1777. "Premier Mémoire sur l'électricité, considérée relativement à l'économie animale et à l'utilité dont elle peut être en Médecine." *Mémoires de la Société Royale de Médecine*, Année 1776, pp. 461-513.

———. 1778. "Lettre sur les précautions nécessaires relativement aux malades qu'on traite par l'électricité." *Journal de Médecine, Chirurgie, Pharmacie, &c.* 49: 323-32.

———. 1780. "Mémoire sur le traitement électrique, administré à quatre-vingt-deux malades." *Mémoires de la Société Royale de Médecine*, Années 1777 et 1778, pp. 199-455.

———. 1782. "Nouvelles observations sur l'Électricité médicale." *Histoire de la Société Royale de Médecine*, Année 1779, pp. 187-201.

———. 1785. "Mémoire sur les différentes manières d'administrer l'Électricité." *Mémoires de la Société Royale de Médecine*, Année 1783, pp. 264-413.

Mazéas, Guillaume, Abbé. 1753-54. "Observations Upon the Electricity of the Air, made at the Chateau de Maintenon, during the Months of June, July, and October, 1753." *Philosophical Transactions* 48(1): 377-84.

Morel, Auguste Désiré Cornil. 1892. *Étude historique, critique et expérimentale de l'action des courants continus sur le nerf acoustique à l'état sain et à l'état pathologique*. Bordeaux: E. Dupuch.

Morin, Jean. 1748. *Nouvelle Dissertation sur l'Électricité des Corps*. Chartres: J. Roux.

Musschenbroek, Pieter van. 1746. Letter to René de Réaumur. *Procès-verbaux de l'Académie Royale des Sciences* 65: 6.

———. 1748. *Institutiones Physicæ*. Leyden: Samuel Luchtman and Son.

———. 1769. *Cours de Physique Expérimentale et Mathématique*, 3 vols. Paris: Bailly.

Mygge, Johannes. 1919. "Om Saakaldte Barometermennesker: Bidrag til Belysning af Vejrneurosens Patogenese." *Ugeskrift for Læger* 81(31): 1239-59.

Nairne, Edward. 1784. *Description de la machine électrique*. Paris: P. Fr. Didot le jeune.

Newton, Isaac. 1713. *Philosophiæ Naturalis Principia Mathematica*, 2nd ed. Cambridge. English translation by Andrew Motte, *Newton's Principia. The Mathematical Principles of Natural Philosophy* (New York: Daniel Adee), 1846.

Nollet, Jean Antoine (Abbé). 1746a. *Essai sur l'Électricité des Corps*. Paris: Guérin.

———. 1746b. "Observations sur quelques nouveaux phénomènes d'Électricité." *Mémoires de l'Académie Royale des Sciences* 1746: 1-23.

———. 1747. "Éclaircissemens sur plusieurs faits concernant l'Électricité." *Mémoires de l'Académie Royale des Sciences* 1747: 102-131.

———. 1748. "Éclaircissemens sur plusieurs faits concernant l'Électricité. Quatrième Mémoire. Des effets de la vertu électrique sur les corps organisés." *Mémoires de l'Académie Royale des Sciences* 1748: 164-99.

———. 1753. *Recherches sur les Causes Particulières des Phénomènes Électriques*. Paris: Guérin.

Nouvelle Bibliothèque Germanique. 1746. "Nouvelles Littéraires, Allemagne, de Greifswald." 2 (part 1): 438-40.

Paulian, Aimé-Henri. 1790. *La Physique à la Portée de Tout le Monde*. Nisme: J. Gaude.

Pera, Marcello. 1992. *The Ambiguous Frog: The Galvani-Volta Controversy on Animal Electricity*. Princeton University Press. Translation of *La rana ambigua* (Torino: Giulio Einaudi), 1986.

Plique, A. F. 1894. "L'électricité en otologie," *Annales des Maladies de l'Oreille, du Larynx, du Nez et du Pharynx* 20: 894-910.

Priestley, Joseph. 1767. *The History and Present State of Electricity*. London: J. Dodsley, J. Johnson, B. Davenport, and T. Cadell.

———. 1775. *The History and Present State of Electricity*, 3rd ed. London: C. Bathurst and T. Lowndes.

Recueil sur l'Électricité Médicale. 1761. Second ed., 2 vols. Paris: P. G. Le Mercier.

Rowbottom, Margaret and Charles Susskind. 1984. *Electricity and Medicine: History of Their Interaction*. San Francisco Press.

Sauvages de la Croix, François Boissier de. 1749. "Lettre de M. de Sauvages." In: Jean Jallabert, *Expériences sur l'Électricité* (Paris: Durand & Pissot), pp. 363-79.

Schiffer, Michael Brian. 2003. *Draw the Lightning Down: Benjamin Franklin and Electrical Technology in the Age of Enlightenment*. Berkeley: University of California Press.

Sguario, Eusebio. 1746. *Dell'elettricismo*. Venezia: Giovanni Battista Recurti.

Sigaud de la Fond, Joseph. 1771. *Lettre sur l'Électricité Médicale*. Amsterdam.

———. 1781. *Précis Historique et Expérimental des Phénomènes Électriques*. Paris: Rue et Hôtel Serpente.

———. 1803. *De l'Électricité Médicale*. Paris: Delaplace et Goujon.

Sparks, Jared. 1836-40. *The Works of Benjamin Franklin*, 10 vols. Boston: Hilliard, Gray.

Sprenger, Johann Justus Anton. 1802. "Anwendungsart der Galvani-Voltaischen Metall-Electricität zur Abhelfung der Taubheit und Harthörigkeit." *Annalen der Physik* 11(7): 354-66.

Steiglehner, Celestin. 1784. "Réponse à la Question sur l'Analogie de l'Électricité et du Magnétisme." In: Jan Hendrik van Swinden, *Recueil de Mémoires sur l'Analogie de l'Électricité et du Magnétisme* (The Hague: Libraires Associés), vol. 2, pp. 1-214.

Stukeley, William. 1749. "On the Causes of Earthquakes." *Philosophical Transactions Abridged* 10: 526-41.

Sue, Pierre, aîné. 1802-1805. *Histoire du Galvanisme*, 4 vols. Paris: Bernard.

Symmer, Robert. 1759. "New Experiments and Observations concerning Electricity." *Philosophical Transctions* 51: 340-93.

Thillaye-Platel, Antoine. 1803. *Essai sur l'Emploi Médical de l'Électricité et du Galvanisme*. Paris: André Sartiaux.

Torlais, Jean. 1954. *L'Abbé Nollet*. Paris: Sipuco.

Trembley, Abraham. 1746. "Part of a Letter concerning the Light caused by Quicksilver shaken in a Glass Tube, proceeding from Electricity." *Philosophical Transactions* 44: 58-60.

van Barneveld, Willem. 1787. *Medizinische Elektricität*. Leipzig: Schwickert.

van Swinden, Jan Hendrik. 1784. *Recueil de Mémoires sur l'Analogie de l'Électricité et du Magnétisme*, 3 vols. The Hague: Libraires Associés.

Veratti, Giovan Giuseppi. 1750. *Observations Physico-Médicales sur l'Électricité*. Geneva: Henri-Albert Gosse.

Volta, Alessandro. 1800. "On the Electricity excited by the mere Contact of conducting Substances of different Kinds." *The Philosophical Magazine* 7 (September): 289-311.

———. 1802. "Lettera del Professore Alessandro Volta al Professore Luigi Brugnatelli sopra l'applicazione dell'elettricità ai sordomuti dalla nascita." *Annali di Chimica e Storia Naturale* 21: 100-5.

Voltaire (François-Marie Arouet). 1772. *Des Singularités de la Nature*. London.

Wesley, John. 1760. *The Desideratum: Or, Electricity Made Plain and Useful*. London: W. Flexney.

Whytt, Robert. 1768. *The Works of Robert Whytt, M.D.* Edinburgh: Balfour, Auld, and Smellie. Reprinted by The Classics of Neurology and Neurosurgery Library, Birmingham, AL, 1984.

Wilkinson, Charles Hunnings. 1799. *The Effects of Electricity*. London: M. Allen.

Wilson, Benjamin. 1752. *A Treatise on Electricity*. London: C. Davis and R. Dodsley.

Winkler, John Henry. 1746. "An Extract of a Letter from Mr. John Henry Winkler, Græc. & Lat. Litt. Prof. publ. Ordin. at Leipsick, to a Friend in London; concerning the Effects of Electricity upon Himself and his Wife." *Philosophical Transactions* 44: 211-12.

Wosk, Julie. 2003. *Women and the Machine*. Baltimore: Johns Hopkins University Press.

Zetzell, Pierre. 1761. "Thèses sur la médecine électrique." In: *Recueil sur l'Électricité Médicale*, 2nd ed. (Paris: P. G. Le Mercier), vol. 1, pp. 283-300.

Weather Sensitivity

Buzorini, Ludwig. 1841. *Luftelectricität, Erdmagnetismus und Krankheitsconstitution*. Constanz: Belle-Vue.

Craig, William. 1859. *On the Influence of Variations of Electric Tension as the Remote Cause of Epidemic and Other Diseases*. London: John Churchill.

Faust, Volker. 1978. *Biometeorologie: Der Einfluss von Wetter und Klima auf Gesunde und Kranke*. Stuttgart: Hippokrates.

Hippocrates. *The Genuine Works of Hippocrates*. Translation by Francis Adams (Baltimore: Wilkins & Williams), 1939.

Höppe, Peter. 1997. "Aspects of Human Biometeorology in Past, Present and Future." *International Journal of Biometeorology* 40(1): 19-23.

International Journal of Biometeorology. 1973. "Symposium on Biological Effects of Natural Electric, Magnetic and Electromagnetic Fields. Held During the 6th International Biometeorological Congress at Noordwijk, The Netherlands, 3-9 September 1972." 17(3): 205-309.

———. 1985. Issue on air ions and atmospheric electricity. 29(3).

Kevan, Simon M. 1993. "Quests for Cures: a History of Tourism for Climate and Health." *International Journal of Biometeorology* 37(3): 113-24.

König, Herbert L. 1975. *Unsichtbare Umwelt: Der Mensch im Spielfeld Elektromagnetischer Kräfte*. München: Heinz Moos Verlag.

Peterson, William F. 1935-1937. *The Patient and the Weather*, 4 vols. Ann Arbor, MI: Edwards Brothers.

———. 1947. *Man, Weather and Sun*. Chicago: Thomas.

Sulman, Felix Gad. 1976. *Health, Weather and Climate*. Basel: Karger.

———. 1980. *The Effect of Air Ionization, Electric Fields, Atmospherics and Other Electric Phenomena on Man and Animal*. Charles C. Thomas: Springfield, IL.

———. 1982. *Short- and Long-Term Changes in Climate*, 2 vols. Boca Raton, FL: CRC Press.

Sulman, Felix Gad, D. Levy, Y. Pfeifer, E. Superstine, and E. Tal. 1975. "Effects of the Sharav and Bora on Urinary Neurohormone Excretion in 500 Weather-Sensitive Females." *International Journal of Biometeorology* 19(3): 202-209.

Tromp, Solco W. 1983. *Medical Biometeorology: Weather, Climate and the Living Organism*. Amsterdam: Elsevier.

Chapter 5

American Psychiatric Association. 2013. *DSM-V, Diagnostic and Statistical Manual for Mental Disorders*. Washington, DC.

Anonymous. 1905. "Die Nervosität der Beamten." *Zeitschrift für Eisenbahn-Telegraphen-Beamte* 23: 179-81.

Aronowitz, Jesse N., Shoshana V. Aronowitz, and Roger F. Robison. 2007. "Classics in Brachytherapy: Margaret Cleaves Introduces Gynecologic Brachytherapy." *Brachytherapy* 6: 293-97.

Arndt, Rudolf. 1885. *Die Neurasthenie (Nervenschwäche).* Wien: Urban & Schwarzenberg.

Bartholow, Roberts. 1884. "What is Meant by Nervous Prostration?" *Boston Medical and Surgical Journal* 110(3): 53-56, 63-64.

Beard, George Miller. 1869. "Neurasthenia, or Nervous Exhaustion." *Boston Medical and Surgical Journal,* new ser., 3(13): 217-21.

———. 1874. "Cases of Hysteria, Neurasthenia, Spinal Irritation and Allied Affections, with Remarks." *Chicago Journal of Nervous and Mental Disease* 1: 438-51.

———. 1875. "The Newly-Discovered Force." *Archives of Electrology and Neurology* 2(2): 256-82.

———. 1876. *Hay-Fever; Or, Summer Catarrh: Its Nature and Treatment.* New York: Harper.

———. 1877. "The Nature and Treatment of Neurasthenia (Nervous Exhaustion), Hysteria, Spinal Irritation, and Allied Neuroses." *The Medical Record* 12: 579-85, 658-62.

———. 1878. "Certain Symptoms of Nervous Exhaustion." *Virginia Medical Monthly* 5(3): 161-85.

———. 1879a. "The Nature and Diagnosis of Neurasthenia (Nervous Exhaustion)." *New York Medical Journal* 29(3): 225-51.

———. 1879b. "The Differential Diagnosis of Neurasthenia – Nervous Exhaustion." *Medical Record* 15(8): 184-85.

———. 1880. *A Practical Treatise on Nervous Exhaustion (Neurasthenia).* New York: William Wood.

———. 1881a. *American Nervousness: Its Causes and Consequences.* New York: G. P. Putnam's Sons.

———. 1881b. *A Practical Treatise on Sea-Sickness: Its Symptoms, Nature and Treatment.* New York: Treat.

Berger, Molly W. 1995. "The Old High-Tech Hotel." *Invention and Technology Magazine* 11(2): 46-52.

Bernhardt, P. 1906. *Die Betriebsunfälle der Telephonistinnen.* Berlin: Hirschwald.

Beyer, Ernst. 1911. "Prognose und Therapie bei den Unfallneurosen der Telephonistinnen." *Medizinische Klinik,* no. 51, pp. 1975-78.

Blegvad, Niels Reinhold. 1907. "Über die Einwirkung des berufsmässigen Telephonierens auf den Organismus mit besonderer Rücksicht auf das Gehörorgan." *Archiv für Ohrenheilkunde* 71: 111-16, 205-36; 72: 30-49. Original in Swedish in *Nordiskt Medicinskt Arkiv (Kirurgi)* 39(3): 1-109.

Böhmig, H. 1905. "Hysterische Unfallerkrankungen bei Telephonistinnen." *Münchener medizinische Wochenschrift* 52(16): 760-62.

Bouchut, Eugène. 1860. *De l'État Nerveux Aigu et Chronique, ou Nervosisme.* Paris: J. B. Baillière et Fils.

Bracket, Cyrus F., Franklin Leonard Pope, Joseph Wetzler, Henry Morton, Charles L. Buckingham, Herbert Laws Webb, W. S. Hughes, John Millis, A. E. Kennelly, and M. Allen Starr. 1890. *Electricity in Daily Life*. New York: Charles Scribner's Sons.

Butler, Elizabeth Beardsley. 1909. "Telephone and Telegraph Operators." In: Butler, *Women and the Trades, Pittsburgh, NY, 1907-1908* (New York: Charities Publication Committee), pp. 282-94.

Calvert, J. B. 2000. *District Telegraphs*. University of Denver.

Campbell, Hugh. 1874. *A Treatise on Nervous Exhaustion*. London: Longmans, Green, Reader, and Dyer.

Capart, fils (de Bruxelles). 1911. "Maladies et accidents professionnels des téléphonistes." *Archives Internationales de Laryngologie, d'Otologie et de Rhinologie* 31: 748-64.

Castex, André. 1897a. "La médecine légale dans les affections de l'oreille, du nez, du larynx et des organes connexes: L'oreille dans le service des téléphones." *Bulletins et Mémoires de la Société Française d'Otologie, de Laryngologie et de Rhinologie* 13 (part 1): 86-87.

———. 1897b. *La médecine légale dans les affections de l'oreille, du nez, du larynx et des organes connexes*. Bordeaux: Férét et Fils.

Cerise, Laurent. 1842. *Des fonctions et des maladies nerveuses dans leur rapports avec l'éducation sociale et privée, morale et physique*. Paris: Germer-Baillière.

Chatel, John C. and Roger Peele. 1970. "A Centennial Review of Neurasthenia." *American Journal of Psychiatry* 126(10): 1404-13.

Cherry, Neil. 2002. "Schumann Resonances, a Plausible Biophysical Mechanism for the Human Health Effects of Solar/Geomagnetic Activity." *Natural Hazards Journal* 26(3): 279-331.

Cheyne, George. 1733. *The English Malady: Or, a Treatise of Nervous Diseases of all Kinds*. London: G. Strahan.

Cleaves, Margaret Abigail. 1899. *Report of the New York Electro-therapeutic Clinic and Laboratory. For the Period Ending June 1, 1899*.

———. 1904. *Light Energy: Its Physics, Physiological Action and Therapeutic Applications*. New York: Rebman.

———. 1910. *Autobiography of a Neurasthene*. Boston: Richard G. Badger.

Cronbach, E. 1903. "Die Beschäftigungsneurose der Telegraphisten." *Archiv für Psychiatrie und Nervenkrankheiten* 37: 243-93.

Dana, Charles Loomis. 1921. *Text-book of Nervous Diseases*, 9th ed. Bristol: John Wright and Sons. Chapter 24, "Neurasthenia," pp. 536-56.

———. 1923. "Dr. George M. Beard: A Sketch of His Life and Character, with Some Personal Reminiscences." *Archives of Neurology and Psychiatry* 10: 427-35.

Department of Labour, Canada. 1907. *Report of the Royal Commission on a Dispute Respecting Hours of Employment between The Bell Telephone Company of Canada, Ltd. and Operators at Toronto, Ont.* Ottawa: Government Printing Bureau.

Desrosiers, H. E. 1879. "De la neurasthénie." *L'Union Médicale du Canada* 8: 145-54, 201-11.

D'Hercourt, Gillebert. 1855. "De l'hydrothérapie dans le traitement de la surexcitabilité nerveuse." *Bulletin de l'Académie Impériale de Médecine* 21: 172-76.

———. 1867. *Plan d'études simultanées de Nosologie et de Météorologie, ayant pour but de rechercher le rôle des agents cosmiques dans la production des maladies, chez l'homme et chez les animaux.* Montpellier: Boehm et fils.

Dickens, Charles. 1859. "House-Top Telegraphs." *All the Year Round,* November 26. Reproduced in George B. Prescott, *History, Theory, and Practice of the Electric Telegraph* (Boston: Ticknor and Fields), 1860, pp. 355-62.

Dubrov, Aleksandr P. 1978. *The Geomagnetic Field and Life.* New York: Plenum.

Durham, John. 1959. *Telegraphs in Victorian London.* Cambridge: Golden Head Press.

Engel, Hermann. 1913. *Die Beurteilung von Unfallfolgen nach Reichsversicherungsordnung: Ein Lehrbuch für Ärzte.* Berlin: Urban & Schwarzenberg.

Eulenburg, A. 1905. "Über Nerven- und Geisteskrankheiten nach elektrischen Unfällen." *Berliner Klinishe Wochenschrift* 42: 30-33, 68-70.

Fisher, T. W. 1872. "Neurasthenia." *Boston Medical and Surgical Journal* 9(5): 65-72.

Flaskerud, Jacquelyn H. 2007. "Neurasthenia: Here and There, Now and Then." *Issues in Mental Health Nursing* 28(6): 657-59.

Flint, Austin. 1866. *A Treatise on the Principles and Practice of Medicine.* Philadelphia: Henry C. Lea.

Fontègne, J. and E. Solari. 1918. "Le travail de la téléphoniste." *Archives de Psychologie* 17(66): 81-136.

Freedley, Edwin T. 1858. *Philadelphia and its Manufactures.* Philadelphia: Edward Young.

Freud, Sigmund. 1895. "Über die Berechtigung von der Neurasthenie einen bestimmten Symptomencomplex als 'Angstneurose' abzutrennen." *Neurologisches Centralblatt* 14: 50-66. Published in English as "On the Grounds for Detaching a Particular Syndrome from Neurasthenia under the Description 'Anxiety Neurosis,'" in *The Standard Edition of the Complete Psychological Works of Sigmund Freud* (London: The Hogarth Press), 1962, vol. 3, pp. 87-139.

Fulton, Thomas Wemyss. 1884. "Telegraphists' Cramp." *The Edinburgh Clinical and Pathological Journal* 1(17): 369-75.

Gellé, Marie-Ernest. 1889. "Effets nuisibles de l'audition par le téléphone." *Annales des maladies de l'oreille, du larynx, du nez et du pharynx* 1889: 380-81.

Goering, Laura. 2003. "'Russian Nervousness': Neurasthenia and National Identity in Nineteenth-Century Russia." *Medical History* 47: 23-46.

Gosling, Francis George. 1987. *Before Freud: Neurasthenia and the American Medical Community 1870-1910.* Urbana: University of Illinois Press.

Graham, Douglas. 1888. "Local Massage for Local Neurasthenia." *Journal of the American Medical Association* 10(1): 11-15.

Gully, James Manby. 1837. *An Exposition of the Symptoms, Essential Nature, and Treatment of Neuropathy, or Nervousness.* London: John Churchill.

Harlow, Alvin F. 1936. *Old Wires and New Waves: The History of the Telegraph, Telephone, and Wireless.* New York: D. Appleton-Century.

Heijermans, Louis. 1908. *Handleiding tot de kennis der beroepziekten.* Rotterdam: Brusse.

He-Quin, Yan. 1989. "The Necessity of Retaining the Diagnostic Concept of Neurasthenia." *Culture, Medicine, and Psychiatry* 13(2): 139-45.

Highton, Edward. 1851. *The Electric Telegraph: Its History and Progress.* London: John Weale.

Hoffmann, Georg, Siegfried Vogl, Hans Baumer, Oliver Kempski, and Gerhard Ruhenstroth-Bauer. 1991. "Significant Correlations between Certain Spectra of Atmospherics and Different Biological and Pathological Parameters." *International Journal of Biometeorology* 34(4): 247-50.

Hubbard, Geoffrey. 1965. "Cooke and Wheatstone and the Invention of the Electric Telegraph." London: Routledge & Kegan Paul.

Jenness, Herbert T. 1909. *Bucket Brigade to Flying Squadron: Fire Fighting Past and Present.* Boston: George H. Ellis.

Jewell, James S. 1879. "Nervous Exhaustion or Neurasthenia in its Bodily and Mental Relations." *Journal of Nervous and Mental Disease* 6: 45-55, 449-60.

———. 1880. "The Varieties and Causes of Neurasthenia." *Journal of Nervous and Mental Disease* 7: 1-16.

Jones, Alexander. 1852. *Historical Sketch of the Electric Telegraph.* New York: George P. Putnam.

Journal of the American Medical Association. 1885. "Functional Troubles Dependent on Neuasthenia." 5(14): 381-82.

Julliard, Charles. 1910. "Les accidents par l'électricité." *Revue Suisse des Accidents du Travail.* Summarized in *Revue de Médecine Légale* 17(1): 343-45.

Killen, Andreas. 2003. "From Shock to Schreck: Psychiatrists, Telephone Operators and Traumatic Neurosis in Germany, 1900-26." *Journal of Contemporary History* 38(2): 201-20.

Kleinman, Arthur. 1988. "Weakness and Exhaustion in the United States and China." In: Kleinman, *The Illness Narrative* (New York: Basic Books), pp. 100-20.

König, Herbert L. 1971. "Biological Effects of Extremely Low Frequency Electrical Phenomena in the Atmosphere." *Journal of Interdisciplinary Research* 2(3): 317-23.

———. 1974a. "ELF and VLF Signal Properties: Physical Characteristics." In: Michael A. Persinger, ed., *ELF and VLF Electromagnetic Field Effects* (New York: Plenum), pp. 9-34.

———. 1974b. "Behavioral Changes in Human Subjects Associated with ELF Electric Fields." In: Michael A. Persinger, ed., *ELF and VLF Electromagnetic Field Effects* (New York: Plenum), pp. 81-99.

Kowalewsky, P. J. 1890. "Zur Lehre vom Neurasthenia." *Zentralblatt für Nervenheilkunde und Psychiatrie* 13: 241-44, 294-304.

The Lancet. 1862. "The Influence of Railway Travelling on Public Health. Report of the Commission." 1: 15-19, 48-53, 79-84, 107-10, 130-32, 155-58, 231-35, 258, 261.

Le Guillant, Louis, R. Roelens, J. Begoin, P. Béquart, J. Hansen, and M. Lebreton. 1956. "La névrose des téléphonistes." *Presse médicale* 64(13): 274-77.

Levillain, Fernand. 1891. *La Neurasthénie, Maladie de Beard.* Paris: A. Maloine.

Lin, Tsung-yi, Guest Editor. 1989a. "Neurasthenia in Asian Cultures." *Culture, Medicine and Psychiatry* 13(2), June issue.

———. 1989b. "Neurasthenia Revisited: Its Place in Modern Psychiatry." *Culture, Medicine, and Psychiatry* 13(2): 105-29.

Ludwig, H. Wolfgang. 1968. "A Hypothesis Concerning the Absorption Mechanism of Atmospherics in the Nervous System." *International Journal of Biometeorology* 12(2): 93-98.

Lutz, Tom. 1991. *American Nervousness, 1903: An Anecdotal History.* Ithaca, NY: Cornell University Press.

Ming-Yuan, Zhang. 1989. "The Diagnosis and Phenomenology of Neurasthenia: A Shanghai Study." *Culture, Medicine, and Psychiatry* 13(2): 147-61.

Morse, Samuel Finley Breese. 1870. "Telegraphic Batteries and Conductors." *Van Nostrand's Eclectic Engineering Magazine* 2: 602-13.

Müller, Franz Carl. 1893. *Handbuch der Neurasthenie.* Leipzig: F. C. W. Vogel.

Nair, Indira, M. Granger Morgan, and H. Keith Florig. 1989. *Biological Effects of Power Frequency Electric and Magnetic Fields.* Washington, DC: Office of Technology Assessment.

Nature. 1875. "The Progress of the Telegraph." Vol. 11, pp. 390-92, 450-52, 470-72, 510-12; Vol. 12, pp. 30-32, 69-72, 110-13, 149-51, 254-56.

Onimus, Ernest. 1875. "Crampe des Employés au Télégraphe." *Comptes Rendus des Séances et Mémoires de la Société de Biologie,* pp. 120-21.

———. 1878. *Le Mal Télégraphique ou Crampe Télégraphique.* Paris: de Cusset.

———. 1880. "Le Mal Télégraphique ou Crampe Télégraphique." *Comptes Rendus des Séances et Mémoires de la Société de Biologie,* pp. 92-96.

Pacaud, Suzanne. 1949. "Recherches sur le travail des téléphonistes: Étude psychologique d'un métier." *Le travail humain* 1-2: 46-65.

Persinger, Michael A., ed. 1974. *ELF and VLF Electromagnetic Field Effects.* New York: Plenum.

Persinger, Michael A., H. Wolfgang Ludwig, and Klaus-Peter Ossenkopp. 1973. "Psychophysiological Effects of Extremely Low Frequency Electromagnetic Fields: A Review." *Perceptual and Motor Skills* 36: 1131-59.

Politzer, Adam. 1901. *Lehrbuch der Ohrenheilkunde,* 4th ed. Stuttgart: Enke. Pp. 649-50 on telephone operators' illnesses.

Pomme, Pierre. 1763. *Traité des Affections Vaporeuses des Deux Sexes, ou Maladies Nerveuses,* Lyon: Benoit Duplain.

Preece, William Henry. 1876. "Railway Travelling and Electricity." *Popular Science Review* 15: 138-48.

Prescott, George B. 1860. *History, Theory, and Practice of the Electric Telegraph.* Boston: Ticknor and Fields.

———. 1881. *Electricity and the Electric Telegraph,* 4th ed. New York: D. Appleton.

Reid, James D. 1886. *The Telegraph in America.* New York: John Polhemus.

Robinson, Edmund. 1882. "Cases of Telegraphists' Cramp." *British Medical Journal* 2: 880.

Sandras, Claude Marie Stanislas. 1851. *Traité Pratique des Maladies Nerveuses.* Paris: Germer-Baillière.

Savage, Thomas, ed. 1889. *Manual of Industrial and Commercial Intercourse between the United States and Spanish America.* San Francisco: Bancroft. Pages 113-23 on the extent of telegraphs in Central and South America.

Scherf, J. Thomas. 1881. *History of Baltimore City and County.* Philadelphia: Louis H. Everts.

Schilling, Karl. 1915. "Die nervösen Störungen nach Telephonunfällen." *Zeitschrift für die gesamte Neurologie und Psychiatrie* 29(1): 216-51.

Schlegel, Kristian and Martin Füllekrug. 2002. "Weltweite Ortung von Blitzen: 50 Jahre Schumann-Resonanzen." *Physik in unserer Zeit* 33(6): 256-61.

Sheppard, Asher R. and Merril Eisenbud. 1977. *Biological Effects of Electric and Magnetic Fields of Extremely Low Frequency.* New York: NYU Press.

Shixie, Liu. 1989. "Neurasthenia in China: Modern and Traditional Criteria for its Diagnosis." *Culture, Medicine, and Psychiatry* 13(2): 163-86.

Shorter, Edward. 1992. *From Paralysis to Fatigue: A History of Psychosomatic Illness in the Modern Era.* New York: The Free Press.

Sterne, Albert E. 1896. "Toxicity in Hysteria, Epilepsy and Neurasthenia – Relations and Treatment." *Journal of the American Medical Association*, 26(4): 172-74.

Strahan, J. 1885. "Puzzling Conditions of the Heart and Other Organs Dependent on Neurasthenia." *British Medical Journal* 2: 435-37.

Suzuki, Tomonori. 1989. "The Concept of Neurasthenia and Its Treatment in Japan." *Culture, Medicine, and Psychiatry* 13(2): 187-202.

Thébault, M. V. 1910. "La névrose des téléphonistes." *Presse médicale* 18: 630-31.

Thompson, H. Theodore and J. Sinclair. 1912. "Telegraphists' Cramp." *Lancet* 1: 888-90, 941-44.

Tommasi, Jacopo. 1904. "Le lesioni professionali e traumatiche nell'orecchio. Otopathie nei telefonisti." *Atti del settimo congresso della società italiana di Laringologia, d'Otologia e di Rinologia*, Rome, October 29-31, 1903, pp. 97-100. Napoli: E. Pietrocola.

Tourette, Georges Gilles de la. 1889. "Deuxième leçon: Les états neurasthéniques et leur traitement." In: Gilles de la Tourette, *Leçons de clinique thérapeutique sur les maladies du système nerveux* (Paris: E. Plon, Nourrit), pp. 58-127.

Trotter, Thomas. 1807. *A View of the Nervous Temperament.* London: Longman, Hurst, Rees, and Orme.

Trowbridge, John. 1880. "The Earth as a Conductor of Electricity." *American Journal of Science*, 120: 138-41.

Turnbull, Laurence. 1853. *The Electro-Magnetic Telegraph.* Philadelphia: A. Hart.

Wallbaum, G. W. 1905. "Ueber funktionelle nervöse Störungen bei Telephonistinnen nach elektrischen Unfällen." *Deutsche medizinische Wochenschrift* 31(18): 709-11.

Webber, Samuel Gilbert. 1888. "A Study of Arterial Tension in Neurasthenia." *Boston Medical and Surgical Journal* 118(18): 441-45.

Whytt, Robert. 1768. *Observations on the Nature, Causes, and Cure of those Disorders which are commonly called Nervous, Hypochondriac or Hysteric.* In: *The Works of Robert Whytt, M.D.* (Edinburgh: Balfour, Auld, and Smellie), pp. 487-713.

Winter, Thomas. 2004. "Neurasthenia." In: Michael S. Kimmel and Amy Aronson, eds., *Men and Masculinities: A Social, Cultural, and Historical Encyclopedia* (Santa Barbara: ABC-CLIO), pp. 567-69.

World Psychiatric Association. 2002. *Neurasthenia – A Technical Report from the World Psychiatric Association Group of Experts*, Beijing, April 1999, printed in Melbourne, Australia in June 2002.

Yassi, Annalee, John L. Weeks, Kathleen Samson, and Monte B. Raber. 1989. "Epidemic of 'Shocks' in Telephone Operators: Lessons for the Medical Community." *Canadian Medical Association Journal* 140: 816-20.

Young, Derson. 1989. "Neurasthenia and Related Problems." *Culture, Medicine, and Psychiatry* 13(2): 131-38.

Chapter 6

Beccaria, Giambatista. 1775. *Della Elettricità Terrestre Atmosferica a Cielo Sereno.* Torino.

Bertholon, Pierre Nicholas. 1783. *De l'Électricité des Végétaux.* Paris: P. F. Didot Jeune.

Blackman, Vernon H. 1924. "Field Experiments in Electro-Culture." *Journal of Agricultural Science* 14(2): 240-67.

Blackman, Vernon H., A. T. Legg, and F. G. Gregory. 1923. "The Effect of a Direct Electric Current of Very Low Intensity on the Rate of Growth of the Coleoptile of Barley." *Proceedings of the Royal Society of London B* 95: 214-28.

Bose, Georg Mathias. 1747. *Tentamina electrica tandem aliquando hydraulicae chymiae et vegetabilibus utilia.* Wittenberg: Johann Joachim Ahlfeld.

Bose, Jagadis Chunder. 1897. "On the Determination of the Wavelength of Electric Radiation by a Diffraction Grating." *Proceedings of the Royal Society of London* 60: 167-78.

———. 1899. "On a Self-Recovering Coherer and the Study of the Cohering Action of Different Metals." *Proceedings of the Royal Society of London* 65: 166-73.

———. 1900. "On Electric Touch and the Molecular Changes Produced in Matter by Electric Waves." *Proceedings of the Royal Society of London* 66: 452-74.

———. 1902. "On the Continuity of Effect of Light and Electric Radiation on Matter." *Proceedings of the Royal Society of London* 70: 154-74.

———. 1902. "On Electromotive Wave Accompanying Mechanical Disturbance in Metals in Contact with Electrolyte." *Proceedings of the Royal Society of London* 70: 273-94.

———. 1906. *Plant Response.* London: Longmans, Green.

———. 1907. *Comparative Electro-Physiology.* London: Longmans, Green.

———. 1910. *Response in the Livng and Non-Living.* London: Longmans, Green.

———. 1913. *Researches on Irritability of Plants.* London: Longmans, Green.

———. 1915. "The Influence of Homodromous and Heterodromous Electric Currents on Transmission of Excitation in Plant and Animal." *Proceedings of the Royal Society of London B* 88: 483-507.

———. 1919. *Life Movements in Plants.* Transactions of the Bose Research Institute, Calcutta, vol. 2. Calcutta: Bengal Government Press.

———. 1923. *The Physiology of the Ascent of Sap.* London: Longmans, Green.

———. 1926. *The Nervous Mechanism of Plants.* London: Longmans, Green.

———. 1927a. *Collected Physical Papers.* London: Longmans, Green.

———. 1927b. *Plant Autographs and Their Revelations.* London: Longmans, Green.

Bose, Jagadis Chunder and Guru Prasanna Das. 1925. "Physiological and Anatomical Investigations on *Mimosa pudica*." *Proceedings of the Royal Society of London B* 98: 290-312.

Browning, John. 1746. "Part of a Letter concerning the Effect of Electricity on Vegetables." *Philosophical Transactions* 44: 373-75.

Crépeaux, Constant. 1892. "L'électroculture." *Revue Scientifique* 51: 524-32.

Emerson, Darrel T. 1997. "The Work of Jagadis Chandra Bose: 100 Years of Millimeter-wave Research" *IEEE Transactions on Microwave Theory and Techniques* 45(12): 2267-73.

Gardini, Giuseppe Francesco. 1784. *De influxu electricitatis atmosphæricae in vegetantia.* Torino: Giammichele Briolo.

Geddes, Patrick. 1920. *The Life and Work of Sir Jagadis C. Bose.* London: Longmans, Green.

Goldsworthy, Andrew. 1983. "The Evolution of Plant Action Potentials." *Journal of Theoretical Biology* 103: 645-48.

———. 2006. "Effects of Electrical and Electromagnetic Fields on Plants and Related Topics." In: Alexander Volkov, ed., *Plant Electrophysiology* (Heidelberg: Springer), pp. 247-67.

Gorgolewski, Stanisław. 1996. "The Importance of Restoration of the Atmospheric Electrical Environment in Closed Bioregenerative Life Supporting Systems." *Advances in Space Research* 18(4-5): 283-85.

Gorgolewski, Stanisław and B. Rozej. 2001. "Evidence for Electrotropism in Some Plant Species." *Advances in Space Research* 28(4): 633-38.

Hicks, W. Wesley. 1957. "A Series of Experiments on Trees and Plants in Electrostatic Fields." *Journal of the Franklin Institute* 264(1): 1-5.

Hull, George S. 1898. *Electro-Horticulture.* New York: Knickerbocker.

Ingen-Housz, Jean. 1789. "Effet de l'Électricité sur le Plantes." In: Ingen-Housz, *Nouvelles Expériences et Observations Sur Divers Objets de Physique* (Paris: Théophile Barrois le jeune), vol. 2, pp. 181-226.

Ishikawa, Hideo and Michael L. Evans. 1990. "Electrotropism of Maize Roots." *Plant Physiology* 94: 913-18.

Jallabert, Jean. 1749. *Expériences sur l'Électricité.* Paris: Durand et Pissot.

Krueger, Albert Paul, A. E. Strubbe, Michael G. Yost, and E. J. Reed. 1978. "Electric Fields, Small Air Ions and Biological Effects." *International Journal of Biometeorology* 22(3): 202-12.

Kunkel, A. J. 1881. "Electrische Untersuchungen an pflanzlichen und thierischen Gebilden." *Archiv für die gesamte Physiologie des Menschen und der Tiere* 25(1): 342-79.

Lemström, Selim. 1904. *Electricity in Agriculture and Horticulture.* London: "The Electrician."

Marat, Jean-Paul. 1782. *Recherches Physiques sur l'Électricité.* Paris: Clousier.

Marconi, Giuglielmo. 1902. "Note on a Magnetic Detector of Electric Waves, Which Can Be Employed as a Receiver for Space Telegraphy." *Proceedings of the Royal Society of London* 70: 341-44.

Molisch, Hans. 1929. "Nervous Impulse in *Mimosa pudica.*" *Nature* 123: 562-63.

Murr, Lawrence E. 1966. "The Biophysics of Plant Growth in a Reversed Electrostatic Field: A Comparison with Conventional Electrostatic and Electrokinetic Field Growth Responses." *International Journal of Biometeorology* 10(2): 135-46.

Nakamura, N., A. Fukushima, H. Iwayama, and H. Suzuki. 1991. "Electrotropism of Pollen Tubes of Camellia and Other Plants." *Sexual Plant Reproduction* 4: 138-43.

Nollet, Jean Antoine (Abbé). 1753. *Recherches sur les Causes Particulières des Phénomènes Électriques.* Paris: Guérin.

Nozue, Kazunari and Masamitsu Wada. 1993. "Electrotropism of *Nicotiana* Pollen Tubes." *Plant and Cell Physiology* 34(8): 1291-96.

Paulin, le Frère. 1890. *De l'influence de l'électricité sur la végétation*. Montbrison: E. Brassart.

Pohl, Herbert A. 1977. "Electroculture." *Journal of Biological Physics* 5(1): 3-23.

Pozdnyakov, Anatoly and Larisa Pozdnyakova. 2006. "Electro-tropism in 'Soil-Plant System.'" *18th World Congress of Soil Science*, July 9-15, Philadelphia, poster 116-29.

Rathore, Keerti S. and Andrew Goldsworthy. 1985a. "Electrical Control of Growth in Plant Tissue Cultures." *Nature Biotechnology* 3: 253-54.

———. 1985b. "Electrical Control of Shoot Regeneration in Plant Tissue Cultures." *Nature Biotechnology* 3: 1107-9.

Sibaoka, Takao. 1962. "Physiology of Rapid Movements in Higher Plants." *Annual Review of Plant Physiology* 20: 165-84.

———. 1966. "Action Potentials in Plant Organs." *Symposia of the Society for Experimental Biology* 20: 49-73.

Sidaway, G. Hugh. 1975. "Some Early Experiments in Electro-culture." *Journal of Electrostatics* 1: 389-93.

Smith, Edwin. 1870. "Electricity in Plants." *Journal of the Franklin Institute* 89: 69-71.

Stahlberg, Rainer. 2006. "Historical Introduction to Plant Electrophysiology." In: Alexander G. Volkov, ed., *Plant Electrophysiology* (Heidelberg: Springer), pp. 3-14.

Stenz, Hans-Gerhard and Manfred H. Weisenseel. 1993. "Electrotropism of Maize (*Zea mays* L.) Roots." *Plant Physiology* 101: 1107-11.

Stone, George E. 1911. "Effect of Electricity on Plants." In: L. H. Bailey, ed., *Cyclopedia of American Agriculture*, 3rd ed. (London: Macmillan), vol. 2. pp. 30-35.

Chapter 7

Althaus, Julius. 1891. "On the Pathology of Influenza, with Special Reference to its Neurotic Character." *Lancet* 2: 1091-93, 1156-57.

———. 1893. "On Psychoses after Influenza." *Journal of Mental Science* 39: 163-76.

Andrewes, Christopher H. 1951. "Epidemiology of Influenza in the Light of the 1951 Outbreak." *Proceedings of the Royal Society of Medicine* 44(9): 803-4.

Appleyard, Rollo. 1939. *The History of the Institution of Electrical Engineers (1871-1931)*. London: Institution of Electrical Engineers.

Arbuthnot, John. 1751. *An Essay Concerning the Effects of Air on Human Bodies*. London: J. and R. Tonson.

Bell, J. A., J. E. Craighead, R. G. James, and D. Wong. 1961. "Epidemiologic Observations on Two Outbreaks of Asian Influenza in a Children's Institution." *American Journal of Hygiene* 73: 84-89.

Beveridge, William Ian. 1978. *Influenza: The Last Great Plague*. New York: Prodist.

Birkeland, Jorgen. 1949. *Microbiology and Man*. New York: Appleton-Century-Crofts.

Blumenfeld, Herbert L., Edwin D. Kilbourne, Donald B. Louria, and David E. Rogers. 1959. "Studies on Influenza in the Pandemic of 1957-1958. I. An Epidemiologic, Clinical and Serologic Investigation of an Intrahospital Epidemic, with a Note on Vaccination Efficacy." *Journal of Clinical Investigation* 38: 199-212.

Boone, Stephanie A. and Charles P. Gerba. 2005. "The Occurrence of Influenza A on Household and Day Care Center Fomites." *Journal of Infection* 51: 103-09.

Borchardt, Georg. 1890. "Nervöse Nachkrankheiten der Influenza." Berlin: Gustav Schade.

Bordley, James III and A. McGehee Harvey. 1976. *Two Centuries of American Medicine, 1776-1976*. Philadelphia: W. B. Saunders.

Bossers, Adriaan Jan. 1894. *Die Geschichte der Influenza und ihre nervösen und psychischen Nachkrankheiten*. Leiden: Eduard Ijdo.

Bowie, John. 1891. "Influenza and Ear Disease in Central Africa." *Lancet* 2: 66-68.

Brakenridge, David J. 1890. "The Present Epidemic of So-called Influenza." *Edinburgh Medical Journal*, 35 (part 2): 996-1005.

Brankston, Gabrielle, Leah Gitterman, Zahir Hirji, Camille Lemieux, and Michael Gardam. 2007. "Transmission of Influenza A in Human Beings." *Lancet Infectious Diseases* 7(4): 257-65.

Bright, Arthur A., Jr. 1949. *The Electric-Lamp Industry: Technological Change and Economic Development from 1800 to 1947*. New York: Macmillan.

Bryson, Louise Fiske 1890. "The Present Epidemic of Influenza." *Journal of the American Medical Association* 14: 426-28.

———. 1890. "The Present Epidemic of Influenza." *New York Medical Journal* 51: 120-24.

Buzorini, Ludwig. 1841. *Luftelectricität, Erdmagnetismus und Krankheitsconstitution*. Constanz: Belle-Vue.

Cannell, John Jacob, Michael Zasloff, Cedric F. Garland, Robert Scragg, and Edward Giovannucci. 2008. "On the Epidemiology of Influenza." *Virology Journal* 5: 29.

Cantarano, G. 1890. "Sui rapporti tra l'influenza e le malattie nervose e mentali." *La Psichiatria* 8: 158-68.

Casson, Herbert N. 1910. *The History of the Telephone*. Chigago: A. C. McClurg.

Chizhevskiy, Aleksandr Leonidovich. 1934. "L'action de l'activité périodique solaire sur les épidémies." In: Marius Piéry, *Traité de Climatologie Biologique et Médicale* (Paris: Masson) vol. 2, pp. 1034-41.

———. 1936. "Sur la connexion entre l'activité solaire, l'électricité atmosphérique et les épidémies de la grippe." *Gazette des Hôpitaux* 109(74): 1285-86.

———. 1937. "L'activité corpusculaire, électromagnétique et périodique du soleil et l'électricité atmosphérique, comme régulateurs de la distribution, dans la suite des temps, des maladies épidémiques et de la mortalité générale." *Acta Medica Scandinavica* 91(6): 491-522.

———. 1938. *Les Épidémies et Les Perturbations Électromagnétiques Du Milieu Extérieur*. Paris: Dépôt Général: Le François.

———. 1973. *Zemnoe ekho solnechnykh bur'* ("The Terrestrial Echo of Solar Storms"). Moscow: Mysl' (in Russian).

———. 1995. *Kosmicheskiy pul's zhizni: Zemlia v obiatiyakh Solntsa. Geliotaraksiya* ("Cosmic Pulse of Life: The Earth in the Embrace of the Sun"). Moscow: Mysl' (in Russian). Written in 1931, published in abridged form in 1973 as "The Terrestrial Echo of Solar Storms."

Clemow, Frank Gerard. 1903. *The Geography of Disease*, 3 vols. Cambridge: University Press.

Clouston, Thomas Smith. 1892. *Clinical Lectures on Mental Diseases*. London: J. & A. Churchill. Page 647 on influenza.

———. 1893. "Eightieth Annual Report of the Royal Edinburgh Asylum for the Insane, 1892." *Journal of Nervous and Mental Disease*, new ser., 18(12): 831-32.

Creighton, Charles. 1894. "Influenza and Epidemic Agues." In: Creighton, *A History of Epidemics in Britain* (Cambridge: Cambridge University Press), vol. 2, pp. 300-433.

Crosby, Oscar T. and Louis Bell. 1892. *The Electric Railway in Theory and Practice*. New York: W. J. Johnston.

Dana, Charles Loomis. 1889. "Electrical Injuries." *Medical Record* 36(18): 477-78.

———. 1890. "The Present Epidemic of Influenza." *Journal of the American Medical Association* 14(12): 426-27.

Davenport, Fred M. 1961. "Pathogenesis of Influenza." *Bacteriological Reviews* 25(3): 294-300.

D'Hercourt, Gillebert. 1867. *Plan d'études simultanées de Nosologie et de Météorologie, ayant pour but de rechercher le rôle des agents cosmiques dans le production des maladies, chez l'homme et chez les animaux*. Montpellier: Boehm et fils.

Dimmock, Nigel J. and Sandy B. Primrose. 1994. *Introduction to Modern Virology*, 4th ed. Oxford: Blackwell Science.

Dixey, Frederick Augustus. 1892. *Epidemic Influenza*. Oxford: Clarendon Press.

Dominion Bureau of Statistics. 1958. *Influenza in Canada: Some Statistics on its Characteristics and Trends*. Ottawa: Queen's Printer.

DuBoff, Richard B. 1979. *Electric Power in American Manufacturing, 1889-1958*. New York: Arno Press.

Dunsheath, Percy. 1962. *A History of Electrical Power Engineering*. Cambridge, MA: MIT Press.

Eddy, John A. 1976. "The Maunder Minimum." *Science* 192: 1189-1202.

———. 1983. "The Maunder Minimum: A Reappraisal." *Solar Physics* 89: 195-207.

Edison, Thomas Alva. 1891. "Vital Energy and Electricity." *Scientific American* 65(23): 356.

Edström, Gunnar O. 1935. "Studies in National and Artificial Atmospheric Electric Ions." *Acta Medica Scandinavica. Supplementum* 61: 1-83.

Electrical Review. 1889. "Proceedings of the Ninth Convention of the National Electric Light Association." March 2, pp. 1-2.

———. 1890a. "Manufacturing and Central Station Companies." August 30, p. 1.

———. 1890b. "The Cape May Convention." August 30, pp. 1-2.

Electrical Review and Western Electrician. 1913. "Public Street Lighting in Chicago." 63: 453-59.

Erlenmeyer, Albrecht. 1890. "Jackson'sche Epilepsie nach Influenza." *Berliner klinische Wochenschrift* 27(13): 295-97.

Field, C. S. 1891. "Electric Railroad Construction and Operation." *Scientific American*, 65(12): 176.

Firstenberg, Arthur. 1998. "Is Influenza an Electrical Disease?" *No Place To Hide* 1(4): 2-6.

Fisher-Hinnen, Jacques. 1899. *Continuous-Current Dynamos in Theory and Practice*. London: Biggs.

Fleming, D. M., M. Zambon, and A. I. M. Bartelds. 2000. "Population Estimates of Persons Presenting to General Practitioners with Influenza-like Illness, 1987-96: A Study of the Demography of Influenza-like Illness in Sentinel Practice Networks in England and Wales, and in the Netherlands." *Epidemiology & Infection* 124: 245-63.

Friedlander, Amy. 1996. *Power and Light: Electricity in the U.S. Energy Infrastructure, 1870-1940*. Reston, VA: Corp. for National Research Initiatives.

Gill, Clifford Allchin. 1928. *The Genesis of Epidemics and the Natural History of Disease.* New York: William Wood.

Glezen, W. Paul and Lone Simonsen. 2006. "Commentary: Benefits of Influenza Vaccine in U.S. Elderly – New Studies Raise Questions." *International Journal of Epidemiology* 35: 352-53.

Gordon, Charles Alexander. 1884. *An Epitome of the Reports of the Medical Officers To the Chinese Imperial Maritime Customs Service, from 1871 to 1882*. London: Baillière, Tindall, and Cox.

Halley, Edmund. 1716. "An Account of the late surprizing Appearance of the Lights seen in the Air, on the sixth of March last; With an Attempt to explain the Principal Phaenomena thereof." *Philosophical Transactions* 29: 406-28.

Hamer, William H. 1936. "Atmospheric Ionization and Influenza." *British Medical Journal* 1: 493-94.

Harlow, Alvin F. 1936. *Old Wires and New Waves: The History of the Telegraph, Telephone, and Wireless*. New York: Appleton-Century.

Harries, H. 1892. "The Origin of Influenza Epidemics." *Quarterly Journal of the Royal Meteorological Society* 18(82): 132-42.

Harrington, Arthur H. 1890. "Epidemic Influenza and Insanity." *Boston Medical and Surgical Journal* 123: 126-29.

Hedges, Killingworth. 1892. *Continental Electric Light Central Stations*. London: E. & F. N. Spon.

Heinz, F., B. Tůmová, and H. Scharfenoorth. 1990. "Do Influenza Epidemics Spread to Neighboring Countries?" *Journal of Hygiene, Epidmiology, Microbiology, and Immunology* 34(3): 283-88.

Hellpach, Willy Hugo. 1911, 1923. *Die geopsychischen Erscheinungen: Wetter, Klima und Landschaft in ihrem Einfluss auf das Seelenleben*. Leipzig: Wilhelm Engelmann.

Hering, Carl. 1892. *Recent Progress in Electric Railways*. New York: W. J. Johnston.

Hewetson, W. M. 1936. "Atmospheric Ionization and Influenza." *British Medical Journal* 1: 667.

Higgins, Thomas James. 1945. "Evolution of the Three-phase 60-cycle Alternating System." *American Journal of Physics* 13(1): 32-36.

Hirsch, August. 1883. "Influenza." In: Hirsch, *Handbook of Geographical and Historical Pathology* (London: New Sydenham Society), vol. 1, pp. 7-54.

Hogan, Linda. 1995. *Solar Storms*. New York: Simon & Schuster.

Hope-Simpson, Robert Edgar. 1978. "Sunspots and Flu: A Correlation." *Nature* 275: 86.

———. 1979. "Epidemic Mechanisms of Type A Influenza." *Journal of Hygiene (Cambridge)* 83(1): 11-25.

———. 1981. "The Role of Season in the Epidemiology of Influenza." *Journal of Hygiene (London)* 86(1): 35-47.

———. 1984. "Age and Secular Distributions of Virus-Proven Influenza Patients in Successive Epidemics 1961-1976 in Cirencester: Epidemiological Significance Discussed." *Journal of Hygiene, (Cambridge)* 92: 303-36.

———. 1992. *The Transmission of Epidemic Influenza*. New York: Plenum.

Hoyle, Fred and N. Chandra Wickramasinghe. 1990. "Sunspots and Influenza." *Nature* 43: 3-4.

Hughes, C. H. 1892. "The Epidemic Inflammatory Neurosis, or, Neurotic Influenza." *Journal of the American Medical Association* 18(9): 245-49.

Hughes, Thomas P. 1983. *Networks of Power: Electrification in Western Society, 1880-1930*. Baltimore: Johns Hopkins University Press.

Hutchings, Richard H. 1896. "An Analysis of Forty Cases of Post Influenzal Insanity." *State Hospitals Bulletin* 1(1): 111-19.

Jefferson, Tom. 2006. "Influenza Vaccination: Policy Versus Evidence." *British Medical Journal* 333: 912-15.

Jefferson, Tom, C. D. Pietrantonj, M. G. Debalini, A. Rivetti, and V. Demicheli. 2009. "Relation of Study Quality, Concordance, Take Home Message, Funding, and Impact in Studies of Influenza Vaccines: Systematic Review." *British Medical Journal* 338: 354-58.

Jones, Alexander. 1826. "Observations on the Influenza or Epidemic Catarrh, as it Prevailed in Georgia during the Winter and Spring of 1826." *Philadelphia Journal of the Medical and Physical Sciences*, new ser., 4(7): 1-30.

Jordan, Edwin O. 1927. *Epidemic Influenza: A Survey*. Chicago: American Medical Association.

Jordan, William S., Jr. 1961. "The Mechanism of Spread of Asian Influenza." *American Review of Respiratory Disease* 83(2): 29-40.

Jordan, William S., Jr., Floyd W. Denny, Jr., George F. Badger, Constance Curtis, John H. Dingle, Robert Oseasohn, and David A. Stevens. 1958. "A Study of Illness in a Group of Cleveland Families. XVII. The Occurrence of Asian Influenza." *American Journal of Hygiene* 68: 190-212.

Journal of the American Medical Association. 1890a. "The Influenza Epidemic of 1889." 14(1): 24-25.

———. 1890b. "Influenza and Cholera." 14(7): 243-44.

Journal of the Statistical Society of London. 1848. "Previous Epidemics of Influenza in England." 11: 173-79.

Kilbourne, Edwin D. 1975. *The Influenza Viruses and Influenza*. New York: Academic.

———. 1977. "Influenza Pandemics in Perspective." *JAMA* 237(12): 1225-28.

Kirn, Ludwig. 1891. "Die nervösen und psychischen Störungen der Influenza." *Sammlung Klinischer Vorträge*, new ser., no. 23 (*Innere Medicin*, no. 9), pp. 213-44.

Kraepelin, Emil. 1890b. "Über Psychosen nach Influenza." *Deutsche medicinische Wochenschrift* 16(11): 209-12.

Ladame, Paul-Louis. 1890. "Des psychoses après l'influenza." *Annales médico-psychologiques*, 7th ser., 12: 20-44.

Lancet. 1919. "Medical Influenza Victims in South Africa." 1: 78.

Langmuir, Alexander D. 1964. "The Epidemiological Basis for the Control of Influenza." *American Journal of Public Health* 54(4): 563-71.

Lee, Benjamin. 1891. "An Analysis of the Statistics of Forty-One Thousand Five Hundred Cases of Epidemic Influenza." *Journal of the American Medical Association* 16(11): 366-68.

Leledy, Albert. 1891. *La Grippe et l'Alienation Mentale*. Paris: J.-B. Baillière et Fils.

Local Government Board. 1893. *Further Report and Papers on Epidemic Influenza, 1889-1892*. London.

Mackenzie, Morell. 1891. "Influenza." *Fortnightly Review* 55: 877-86.

Macphail, S. Rutherford. 1896. "Post-Influenzal Insanity." *British Medical Journal* 2: 810-11.

Mann, P. G., M. S. Pereira, J. W. G. Smith, R. J. C. Hart, and W. O. Williams. 1981. "A Five-Year Study of Influenza in Families." *Journal of Hygiene (Cambridge)* 87(2): 191-200.

Marian, Christine and Grigore Mihăescu. 2009. "Diversification of Influenza Viruses." *Bacteriologia, Virusologia, Parazitologia, Epidemiologia* 54: 117-23 (in Romanian).

Mathers, George. 1917. "Etiology of the Epidemic Acute Respiratory Infections Commonly Called Influenza." *Journal of the American Medical Association* 68(9): 678-80.

McGrew, Roderick E. 1985. *Encyclopedia of Medical History*. New York: McGraw-Hill.

Meyer, Edward Bernard. 1916. *Underground Transmission and Distribution for Electric Light and Power*. New York: McGraw-Hill.

Mispelbaum, Franz. 1890. "Ueber Psychosen nach Influenza." *Allgemeine Zeitschrift für Psychiatrie* 47(1): 127-53.

Mitchell, Weir. 1893. Paper read at the National Academy of Sciences, Washington. Cited in Johannes Mygge, "Om Saakaldte Barometermennesker: Bidrag til Blysning af Vejrneurosens Patogenese," *Ugeskrift for Læger* 81(31): 1239-59, at p. 1251.

Morrell, C. Conyers. 1936. "Atmospheric Ionization and Influenza." *British Medical Journal* 1: 554-55.

Munter, D. 1890. "Psychosen nach Influenza." *Allgemeine Zeitschrift für Psychiatrie* 47: 156-65.

Mygge, Johannes. 1919. "Om Saakaldte Barometermennesker: Bidrag til Belysning af Vejrneurosens Patogenese." *Ugeskrift for Læger* 81(31): 1239-59.

———. 1930. "Étude sur l'éclosion épidémique de l'influenza." *Acta Medica Scandinavica. Supplementum* 32: 1-145.

National Institutes of Health. 1973. "Epidemiology of Influenza – Summary of Influenza Workshop IV." *Journal of Infectious Diseases* 128(3): 361-99.

Ozanam, Jean-Antoine-François. 1835. *Histore medicale générale et particulière des maladies épidémiques, contagieuses et épizootiques*, 2 vols. Lyon: J. M. Boursy.

Parsons, Franklin. 1891. *Report on the Influenza Epidemic of 1889-1890*. London: Local Government Board.

Patterson, K. David. 1986. *Pandemic influenza 1700-1900*. Totowa, NJ: Rowman & Littlefield.

Peckham, W. C. 1892. "Electric Light for Magic Lantern." *Scientific American* 66(12): 183.

Perfect, William. 1787. *Select Cases in the Different Species of Insanity*. Rochester: Gilman. Pages 126-31 on insanity from influenza.

Preece, William Henry and Julius Maier. 1889. *The Telephone*. London: Whittaker.

Reckenzaun, A. 1887. "On Electric Street Cars, with Special Reference to Methods of Gearing." *Proceedings of the American Institute of Electrical Engineers* 5(1): 2-32.

Revilliod, L. 1890. "Des formes nerveuses de la grippe." *Revue Médicale de la Suisse Romande* 10(3): 145-53.

Ribes, J. C. and E. Nesme-Ribes. 1993. "The Solar Sunspot Cycle in the Maunder Minimum AD 1645 to AD 1715." *Astronomy and Astrophysics* 276: 549-63.

Richter, C. M. 1921. "Influenza Pandemics Depend on Certain Anticyclonic Weather Conditions for their Development." *Archives of Internal Medicine* 27(3): 361-86.

Ricketson, Shadrach. 1808. *A Brief History of the Influenza*. New York.

Rorie, George A. 1901. "Post-Influenzal Insanity in the Cumberland and Westmoreland Asylum, with Statistics of Sixty-Eight Cases." *Journal of Mental Science* 47: 317-26.

Schmitz, Anton. 1891-92. "Ueber Geistesstörung nach Influenza." *Allgemeine Zeitschrift für Psychiatrie* 47: 238-56; 48: 179-83.

Schnurrer, Friedrich. 1823. *Die Krankheiten des Menschen-Geschlechts*. Tübingen: Christian Friedrich Osiander.

Schönlein, Johann Lucas. 1840. *Allgemeine und specielle Pathologie und Therapie*, 5th ed., 4 vols. St. Gallen: Litteratur-Comptoir. Vol. 2, pp. 100-3 on influenza.

Schrock, William M. 1892. "The Progress of Electrical Science." *Scientific American* 66(7) :100.

Schweich, Heinrich. 1836. *Die Influenza: Ein historischer und ätiologischer Versuch*. Berlin: Theodor Christian Friedrich Enslin.

Science. 1888a. "Electric Street Railways." 12: 246-47.

———. 1888b. "The Westinghouse Company's Extentions." 12: 247.

———. 1888c. "Electric-Lighting." 12: 270.

———. 1888d. "The Edison Electric-Lighting System in Berlin." 12: 270.

———. 1888e. "Trial of an Electric Locomotive at Birmingham, England." 12: 270.

———. 1888f. "An Electric Surface Road in New York." 12: 270-71.

———. 1888g. "Electric Propulsion." 12: 281-82.

———. 1888h. "Electric Power-Distribution." 12: 282-84.

———. 1888i. "The Sprague Electric Road at Boston." 12: 324-25.

———. 1888j. "The Advances in Electricity in 1888." 12: 328-29.

———. 1889. "Westinghouse Alternating-Current Dynamo." 13: 451-52.

———. 1890a. "A Big Road Goes in for Electricity." 15: 153.

———. 1890b. "The Electric Light in Japan." 15: 153.

Scientific American. 1889a. "The Danger of Electric Distribution." 60(2): 16.

———. 1889b. "Edison Electric Light Consolidation." 60(3): 34.

———. 1889c. "The Advances of Electricity in 1888" 60(6): 88.

———. 1889d. "Progress of Electric Illumination." 60(12): 176-77.

———. 1889e. "Progress of Electric Installations in London." 60(13): 196.

———. 1889f. "Electricity in the United States." 61(12): 150.

———. 1889g. "The National Electric Light Association Meeting." 61(14): 184.

———. 1889h. "The Westinghouse Electric Company." 61(20): 311.

———. 1890a. "Progress of Electric Lighting in London." 62(3): 40-41.

———. 1890b. "The Westinghouse Alternating Current System of Electrical Distribution." 62(8): 117, 120-21.

———. 1890c. "The National Electric Lighting Association." 62(8): p. 118.

———. 1890d. "The Growth of the Alternating System." 62(17): 57.

———. 1890e. "Electricity in the Home." 62(20): 311.

———. 1890f. "Electrical Notes." 63(7): 97.

———. 1890g. "Long Distance Electrical Power." 63(8): 120.

———. 1890h. "Local Interests Improved by Electricity." 63(12): 182.

———. 1890i. "History of Electric Lighting." 63(14): 215.

———. 1891a. "Meeting of the National Electric Light Association." 64(9): 128.

———. 1891b. "The Electric Transmission of Power." 64(14): 209.

———. 1891c. "Electricity in Foreign Countries." 64(15): 229.

———. 1891d. "Electricity for Domestic Purposes." 64(20): 310.

———. 1891e. "The Edison Electric Illuminating Co.'s Central Station in Brooklyn, N.Y." 64(24): 373.

———. 1891f. "Long Distance Electrical Power." 65(19): 293.

———. 1892a. "Electric Lights for Rome, Italy." 66(2): 25.

———. 1892b. "What is Electricity?" 66(6): 89.

———. 1892c. "The Electrical Transmission of Power between Lauffen on the Neckar and Frankfort on the Main." 66(7): 102.

Shope, Richard E. 1958. "Influenza: History, Epidemiology, and Speculation" *Public Health Reports* 73(2): 165-78.

Solbrig, Dr. 1890. "Neurosen und Psychosen nach Influenza." *Neurologisches Centralblatt* 9(11): 322-25.

Soper, George A. 1919. "Influenza in Horses and in Man." *New York Medical Journal* 109(17): 720-24.

Stuart-Harris, Sir Charles H., Geoffrey C. Schild, and John S. Oxford. 1985. *Influenza: The Viruses and the Disease*, 2nd ed. Edward Arnold: London.

Tapping, Ken F., R. G. Mathias, and D. L. Surkan. 2001. "Influenza Pandemics and Solar Activity." *Canadian Journal of Infectious Diseases* 12(1): 61-62.

Taubenberger, J. K. and D. M. Morens. 2009. "Pandemic Influenza – Including a Risk Assessment of H5N1." *Revue scientifique et tecnnique* 28(1): 187-202.

Thompson, Theophilus. 1852. *Annals of Influenza or Epidemic Catarrhal Fever in Great Britain From 1510 to 1837*. London: Sydenham Society.

Trevert, Edward. 1892. *Electric Railway Engineering*. Lynn, MA: Bubier.

———. 1895. *How to Build Dynamo-Electric Machinery*. Lynn, MA: Bubier.

Tuke, Daniel Hack. 1892. "Mental Disorder Following Influenza." In: Tuke, *A Dictionary of Psychological Medicine* (London: J. & A. Churchill), vol. 2, pp. 688-91.

United States Department of Commerce and Labor, Bureau of the Census. 1905. *Central Electric Light and Power Stations 1902*. Washington, DC: Government Printing Office.

van Tam, Jonathan and Chloe Sellwood. 2010. *Introduction to Pandemic Influenza*. Wallingford, UK: CAB International.

Vaughan, Warren T. 1921. *Influenza: An Epidemiologic Study*. Baltimore: American Journal of Hygiene.

von Niemeyer, Felix. 1874. *A Text-book of Practical Medicine*. New York: D. Appleton. Pages 61-62 on influenza.

Watson, Thomas. 1857. *Lectures on the Principles and Practice of Physic*, 4th ed. London: John W. Parker. Vol. 2, pp. 41-52 on influenza.

Webster, J. H. Douglas. 1940. "The Periodicity of Sun-spots, Influenza and Cancer." *British Medical Journal* 2: 339.

Webster, Noah. 1799. *A Brief History of Epidemic and Pestilential Diseases*, 2 vols. New York: Burt Franklin.

Whipple, Fred H. 1889. *The Electric Railway*. Detroit: Orange Empire Railway Museum.

Widelock, Daniel, Sarah Klein, Olga Simonovic, and Lenore R. Peizer. 1959. "A Laboratory Analysis of the 1957-1958 Influenza Outbreak in New York City." *American Journal of Public Health* 49(7): 847-56.

Yeung, John W. K. 2006. "A Hypothesis: Sunspot Cycles May Detect Pandemic Influenza A in 1700-2000 A.D." *Medical Hypotheses* 67: 1016-22.

Zinsser, Hans. 1922. "The Etiology and Epidemiology of Influenza." *Medicine* 1(2): 213-309.

Chapter 8

Alexanderson, Ernst F. W. 1919. "Transatlantic Radio Communication." *Proceedings of the American Institute of Electrical Engineers* 38(6): 1077-93.

All Hands. 1961. "Flying the Atlantic Barrier." April, pp. 2-5.

Anderson, John. 1930. "'Isle of Wight Disease' in Bees." *Bee World* 11(4): 37-42.

Annual Report of the Surgeon General, U.S. Navy. 1919. Washington, DC: Government Printing Office. "Report on Influenza," pp. 358-449.

Archer, Gleason L. 1938. *History of Radio*. New York: American Historical Society.

Armstrong, D. B. 1919. "Influenza: Is it a Hazard to be Healthy? Certain Tentative Considerations." *Boston Medical and Surgical Journal* 180(3): 65-67.

Ayres, Samuel, Jr. 1919. "Post-Influenzal Alopecia." *Boston Medical and Surgical Journal* 180(17): 464-68.

Baker, William John. 1971. *A History of the Marconi Company*. New York: St. Martins.

Bailey, Leslie 1964. "The 'Isle of Wight Disease': The Origin and Significnce of the Myth." *Bee World* 45(1): 32-37, 18.

Beauchamp, Ken. 2001. *History of Telegraphy*. Hertfordshire, UK: Institution of Electrical Engineers.

Beaussart, P. "Orchi-Epididymitis with Meningitis and Influenza." 1918. *Journal of the American Medical Association* 70(26): 2057.

Berman, Harry. 1918. "Epidemic Influenza in Private Practice." *Journal of the American Medical Association* 71(23): 1934-35.

Beveridge, William Ian. 1978. *Influenza: The Last Great Plague*. New York: Prodist.

Bircher, E. "Influenza Epidemic." 1918. *Journal of the American Medical Association* 71(23): 1946.

Blaine, Robert Gordon. 1903. *Aetheric or Wireless Telegraphy*. London: Biggs and Sons.

Bouchard, Joseph F. 1999. "Guarding the Cold War Ramparts." *Naval War College Review* 52(3): 111-35.

Bradfield, W. W. 1910. "Wireless Telegraphy for Marine Inter-Communication." *The Electrician – Marine Issue*. June 10, pp. 135 ff.

Brittain, James E. 1902. *Alexanderson: Pioneer in American Electrical Engineering*. Baltimore: Johns Hopkins University Press.

Bucher, Elmer Eustice. 1917. *Practical Wireless Telegraphy*. New York: Wireless Press.

Carr, Elmer G. 1918. "An Unusual Disease of Honey Bees." *Journal of Economic Entomology* 11(4): 347-51.

Carter, Charles Frederick. 1914. "Getting the Wireless on Board Train." *Technical World Magazine* 20(6): 914-18.

Chauvois, Louis. 1937. *D'Arsonval: Soixante-cinq ans à travers la Science*. Paris: J. Oliven.

Conner, Lewis A. 1919. "The Symptomatology and Complications of Influenza." *Journal of the American Medical Association* 73(5): 321-25.

Coutant, A. Francis. 1918. "An Epidemic of Influenza at Manila, P.I." *Journal of the American Medical Association* 71(19): 1566-67.

Cowie, David Murray and Paul Webley Beaven. 1919. "On the Clinical Evidence of Involvement of the Suprarenal Glands in Influenza and Influenzal Pneumonia." *Archives of Internal Medicine* 24(1): 78-88.

Craft. E. B. and E. H. Colpitts. 1919. "Radio Telephony." *Proceedings of the American Institute of Electrical Engineers* 38(1): 337-75.

Crawley, Charles G. 1996. *How Did the Evolution of Communications Affect Command and Control of Airpower: 1900-1945?* Maxwell Air Force Base, AL.

Crosby, Alfred W., Jr. 1976. *Epidemic and Peace, 1918*. Westport, CT: Greenwood.

d'Arsonval, Jacques Arsène. 1892a. "Recherches d'électrothérapie. La voltaïsation sinusoïdale." *Archives de physiologie normale et pathologique* 24: 69-80.

———. 1892b. "Sur les effets physiologiques comparés des divers procédés d'électrisation." *Bulletin de l'Académie de Médecine* 56: 424-33.

———. 1893a. "Action physiologique des courants alternatifs a grande fréquence." *Archives de physiologie normale et pathologique* 25: 401-8.

———. 1983b. "Effets physiologiques de la voltaïsation sinusoïdale." *Archives de physiologie normale et pathologique* 25: 387-91.

———. 1893c. "Expériences faites au laboratoire de médecine du Collège de France." *Archives de physiologie normale et pathologique* 25: 789-90.

———. 1893d. "Influence de la fréquence sur les effets physiologiques des courants alternatifs." *Comptes rendus hebdomadaires des séances de l'Académie des Sciences* 116: 630-33.

———. 1896a. "À propos de l'atténuation des toxines par la haute fréquence." *Comptes rendus hebdomadaires des séances et mémoires de la Société de Biologie* 48: 764-66.

———. 1896b. "Effets thérapeutiques des courants à haute fréquence." *Comptes rendus hebdomadaires des séances de l'Académie des Sciences* 123: 23-29.

d'Arsonval, Jacques Arsène and Albert Charrin. 1893a. "Influence de l'électricitè sur la cellule microbienne." *Archives de physiologie normale et pathologique*, 5th ser., 5: 664-69.

———. 1893b. "Électricité et Microbes." *Comptes rendus hebdomadaires des séances et mémoires de la Société de Biologie* 45: 467-69, 764-65.

———. 1896a. "Action des diverses modalités électriques sur les toxines bactériennes." *Comptes rendus hebdomadaires des séances et mémoires de la Société de Biologie* 48: 96-99.

———. 1896b. "Action de l'électricitè sur les toxines bactériennes." *Comptes rendus hebdomadaires des séances et mémoires de la Société de Biologie* 48: 121-23.

———. 1896b. "Action de l'électricitè sur les toxines et les virus." *Comptes rendus hebdomadaires des séances et mémoires de la Société de Biologie* 48: 153-54.

del Pont, Antonino Marcó. 1918. "Historia de las Epidemias de Influenza." *La Semana Médica* 25(27): 1-10.

Eccles, William Henry. 1933a. *Wireless Telegraphy and Telephony*, 2nd ed. London: Benn Brothers.

———. 1933b. *Wireless*. London: Thornton Butterworth.

Ehrenberg, L. 1919. "Transmission of Influenza." *Journal of the American Medical Association* 72(25): 1880.

Elwell, Cyril Frank. 1910. "The Poulsen System of Wireless Telephony and Telegraphy." *Journal of Electricity, Power and Gas* 24(14): 293-97.

———. 1920. "The Poulsen System of Radiotelegraphy. History of Development of Arc Methods." *The Electrician* 84: 596-600.

Erlendsson, V. 1919. "Influenza in Iceland." *Journal of the American Medical Association* 72(25): 1880.

Erskine, Arthur Wright and B. L. Knight. 1918. "A Preliminary Report of a Study of the Coagulability of Influenzal Blood." *Journal of the American Medical Association* 71(22): 1847.

Erskine-Murray, J. 1920. "The Transmission of Electromagnetic Waves About the Earth." *Radio Review* 1: 237-39.

Fantus, Bernard. 1918. "Clinical Observations on Influenza." *Journal of the American Medical Association* 71(21): 1736-39.

Firstenberg, Arthur. 1997. *Microwaving Our Planet*. New York: Cellular Phone Taskforce.

———. 2001. "Radio Waves, the Blood-Brain Barier, and Cerebral Hemorrhage." *No Place To Hide* 3(2): 23-24.

Friedlander, Alfred, Carey P. McCord, Frank J. Sladen, and George W. Wheeler. 1918. "The Epidemic of Influenza at Camp Sherman, Ohio." *Journal of the American Medical Association* 71(20): 1652-56.

Frost, W. H. 1919. "The Epidemiology of Influenza." *Journal of the American Medical Association* 73(5): 313-18.

Goldoni, J. 1990. "Hematological Changes in Peripheral Blood of Workers Occupationally Exposed to Microwave Radiation." *Health Physics* 58(2): 205-7.

Grant, John. 1907. "Experiments and Results in Wireless Telephony." *American Telephone Journal* 15(4): 49-51.

Harris, Wilfred. 1919. "The Nervous System in Influenza." *The Practitioner* 102: 89-100.

Harrison, Forrest Martin. 1919. "Influenza Aboard a Man-of-War: A Clinical Summary." *Medical Record* 95(17): 680-85.

Headrick, Daniel R. 1988. *The Tentacles of Progress: Technology Transfer in the Age of Imperialism, 1850-1940*. New York: Oxford University Press.

———. 1991. *The Invisible Weapon: Telecommunications and International Politics, 1851-1945*. New York: Oxford University Press.

Hewlett, A. W. and W. M. Alberty. 1918. "Influenza at Navy Base Hospital in France." *Journal of the American Medical Association* 71(13): 1056-58.

Hirsch, Edwin F. 1918. "Epidemic of Bronchopneumonia at Camp Grant, Ill." *Journal of the American Medical Association* 71(21): 1735-36.

Hong, Sungook. 2001. *Wireless: From Marconi's Black-Box to the Audion*. Cambridge, MA: MIT Press.

Hopkins, Albert A. and A. Russell Bond, eds. 1905. "Wireless Telegraphy." *Scientific American Reference Book* (New York: Munn & Co.), pp. 199-205.

Howe, George William Osborn. 1920a. "The Upper Atmosphere and Radio Telegraphy." *Radio Review* 1: 381-83.

———. 1920b. "The Efficiency of Aerials." *Radio Review* 1: 540-43.

———. 1920c. "The Power Required for Long Distance Transmission." *Radio Review* 1:598-608.

Howeth, Linwood S. 1963. *History of Communications – Electronics in the United States Navy*. Washington, DC: Bureau of Ships and Office of Naval History.

Huurdeman, Anton A. 2003. *The Worldwide History of Telecommunications*. Hoboken, NJ: Wiley.

Imms, Augustus Daniel. 1907. "Report on a Disease of Bees in the Isle of Wight." *Journal of the Board of Agriculture* 14(3): 129-40.

Jordan, Edwin O. 1918. Discussion in: "The Etiology of Influenza," Proceedings of the American Public Health Association, Forty-Sixth Annual Meeting, Chicago, December 8-11, 1918. *Journal of the American Medical Association* 71(25): 2097.

———. 1922. "Interepidemic Influenza." *American Journal of Hygiene* 2(4): 325-45.

———. 1927. *Epidemic Influenza: A Survey*. Chicago: American Medical Association.

Journal of the American Medical Association. 1918a. "Spanish Influenza." 71(8): 660.

———. 1918b. "The Epidemic of Influenza." 71(13): 1063-64.

———. 1918c. "Epidemic Influenza." 71(14): 1136-37.

———. 1918d. "The Present Epidemic of Influenza." 71(15): 1223.

———. 1918e. "Abstracts on Influenza." 71(19): 1573-80.

———. 1918f. "Influenza in Mexico." 71(20): 1675.

———. 1918g. "Paris Letter. The Influenza Epidemic." 71(20): 1676.

———. 1918h. "The Influenza Epidemic." 71(24): 2009-10.

———. 1918i. "Influenza." 71(25): 2088.

———. 1918j. "Mexico Letter." 71(25): 2089.

———. 1918k. "Febrile Epidemic [in Peru]." 71(25): 2090.

———. 1918l. "The Etiology of Influenza." 71(25): 2097-2100, 2173-75.

———. 1919a. "Unsuccessful Attempts to Transmit Influenza Experimentally." 72(4): 281.

———. 1919b. "Heart Block and Bradycardia Following Influenza." 73(11): 868.

———. 1920a. "The 1920 Influenza." 74(9): 607.

———. 1920b. "Influenza in Alaska." 74(12): 796.

———. 1920c. "Influenza in the Navy Personnel." 74(12): 813.

———. 1920d. "After Effects of Influenza." 75(1): 61.

———. 1920e. "The Influenza Pandemic in India." 75(9): 619-20.

———. 1920f. "Eye Disease Following Influenza Epidemic." 75(10): 709.

Keegan, J. J. 1918. "The Prevailing Epidemic of Influenza." *Journal of the American Medical Association* 71(13): 1051-55.

Keeton, Robert W. and A. Beulah Cushman. 1918. "The Influenza Epidemic in Chicago." *Journal of the American Medical Association* 71(24): 1962-67.

Kilbourne, Edwin D. 1975. *The Influenza Viruses and Influenza.* New York: Academic.

Klessens, J. J. H. M. 1920. "Nervous Manifestations Complicating Influenza." *Journal of the American Medical Association* 74(3): 216.

Kuksinskiy, V. E. 1978. "Coagulation Properties of the Blood and Tissues of the Cardiovascular System Exposed to an Electromagnetic Field." *Kardiologiya* 18(3): 107-11 (in Russian).

Kyuntsel', A. A. and V. I. Karmilov. 1947. "The Effect of an Electromagnetic Field on the Blood Coagulation Rate." *Klinicheskaya Meditsina* 25(3): 78 (in Russian).

La Fay, Howard. 1958. "DEW Line: Sentry of the Far North." *National Geographic* 114(1): 128-46.

Leake, J. P. 1919. "The Transmission of Influenza." *Boston Medical and Surgicial Journal* 181(24): 675-79.

Logwood, C. V. 1916. "High Speed Radio Telegraphy." *The Electrical Experimenter*, June, p. 99.

Loosli, Clayton G., Dorothy Hamre, and O. Warner. 1958. "Epidemic Asian A Influenza in Naval Recruits." *Proceedings of the Society for Experimental Biology and Medicine* 98(3): 589-92.

Lyle, Eugene P., Jr. 1905. "The Advance of 'Wireless.'" *World's Work*, February, pp. 5843-48.

MacNeal, Ward J. 1919. "The Influenza Epidemic of 1918 in the American Expeditionary Forces in France and England." *Archives of Internal Medicine* 23(6): 657-88.

Maestrini, D. 1919. "The Blood in Influenza." *Journal of the American Medical Association* 72(11): 834.

Marconi, Degna. 2001. *My Father, Marconi*, 2nd ed. Toronto: Guernica.

Marconi, Maria Cristina. 1999. *Marconi My Beloved.* Boston: Dante University of America Press.

Marshall, C. J. 1957. "North America's Distant Early Warning Line." *Geographical Magazine* 29(12): 616-28.

Martin, Donald H. 1991. *Communication Satellites 1958-1992.* El Segundo, CA: The Aerospace Corporation.

Menninger, Karl A. 1919a. "Psychoses Associated with Influenza." *Journal of the American Medical Association* 72(4): 235-41.

———. 1919b. "Influenza and Epileptiform Attacks." *Journal of the American Medical Association* 73(25): 1896.

Ministry of Health. 1920. *Report on the Pandemic of Influenza, 1918-19.* Reports on Public Health and Medical Subjects, no. 4. London.

Morenus, Richard. 1957. *DEW Line.* New York: Rand McNally.

Navy Department, Bureau of Equipment. 1906. *List of Wireless-Telegraph Stations of the World.* Washingon: Government Printing Office.

Nicoll, M., Jr. 1918. "Organization of Forces against Influenza." American Public Health Association, Forty-Sixth Annual Meeting, Chicago, Dec. 8-11, 1918, *Journal of the American Medical Association* 71(26): 2173.

Nuzum, John W., Isadore Pilot, F. H. Stangl, and B. E. Bonar. 1918. "Pandemic Influenza and Pneumonia in a Large Civil Hospital." *Journal of the American Medical Association* 71(19): 1562-65.

Oliver, Wade W. 1919. "Influenza – the Sphinx of Diseases." *Scientific American* 120(9): 200, 212-13.

Persson, Bertil R. R., Leif G. Salford, and Arne Brun. 1997. "Blood-brain Barrier Permeability in Rats Exposed to Electromagnetic Fields Used in Wireless Communication." *Wireless Networks* 3: 455-61.

Pettigrew, Eileen. 1983. *The Silent Enemy: Canada and the Deadly Flu of 1918.* Saskatoon: Western Producer Prairie Books.

Pflomm, Erich. 1931. "Experimentelle und klinische Untersuchungen über die Wirkung ultrakurzer elektrischer Wellen auf die Entzündung." *Archiv für klinische Chirurgie* 166: 251-305.

Phillips, Ernest F. 1925. "The Status of Isle of Wight Disease in Various Countries." *Journal of Economic Entomology* 18: 391-95.

Prince, C. E. 1920. "Wireless Telephony on Aeroplanes." *Radio Review* 1: 281-83, 341.

Public Health Reports. 1919. "Some Interesting Though Unsuccessful Attempts to Transmit Influenza Experimentally." 34(2): 33-39.

———. 1919. "Influenza Among American Indians." 34: 1008-9.

Radio Review. 1919. "The Transmission of Electromagnetic Waves Around the Earth." 1: 78-80.

———. 1920. "The Generation of Large Powers at Radio Frequencies." 1: 490-91.

Reid, Ann H., Thomas G. Fanning, Johan V. Hultin, and Jeffery K. Taubenberger. 1999. "Origin and Evolution of the 1918 'Spanish' Influenza Virus Hemagglutinin Gene." *Proceedings of the National Academy of Sciences* 96(4): 1651-56.

Richardson, Alfred W. 1959. "Blood Coagulation Changes Due to Electromagnetic Microwave Irradiations." *Blood* 14: 1237-43.

Robertson, H. E. 1918. "Influenzal Sinus Disease and its Relation to Epidemic Influenza." *Journal of the American Medical Association* 70(21): 1533-35.

Rosenau, Milton J. 1919. "Experiments to Determine Mode of Spread of Influenza." *Journal of the American Medical Association* 73(5): 311-13.

Rusyaev, V. P. and V. E. Kuksinskiy. 1973. "Study of Electromagnetic Field Effect on Coagulative and Fibrinolytic Properties of Blood." *Biofizika* 11(1): 160-63 (in Russian with English abstract).

Saleeby, C. W. 1920. "Mapping the Influenza." *Literary Digest*, May 29, p. 32.

Schaffel, Kenneth. 1991. *The Emerging Shield: The Air Force and the Evolution of Continental Air Defense 1945-1960.* Washington, DC: United States Air Force.

Scheips, Paul J., ed. 1980. *Military Signal Communications*, 2 vols. New York: Arno Press.

Schliephake, Erwin. 1935. *Short Wave Therapy: The Medical Uses of Electrical High Frequencies.* London: Actinic Press.

———. 1960. *Kurzwellentherapie*, 6th ed. Stuttgart: Gustav Fischer.

Scriven, George P. 1914. "Report of the Chief Signal Officer, U.S. Army, 1914." *Annual Reports of the War Department*, pp. 505-56. Reproduced in Scheips 1980, vol. 1.

Sierra, Álvarez. 1921. "Particularidades clínicas de la última epidemia gripal." *El Siglo Médico* 68: 765-66.

Simici, D. 1920. "The Heart in Influenza." *Journal of the American Medical Association* 75(10): 703.

Sofre, G. 1918. "Influenza." *Journal of the American Medical Association* 71(21): 1782.

Soper, George A. 1918. "The Pandemic in the Army Camps." *Journal of the American Medical Association* 71(23): 1899-1909.

Staehelin, R. 1918. "The Influenza Epidemic." *Journal of the American Medical Association* 71(14): 1176.

Stuart-Harris, Charles H. 1965. *Influenza and Other Virus Infections of the Respiratory Tract*, 2nd ed. Baltimore: Williams & Wilkins.

Symmers, Douglas. 1918. "Pathologic Similarity between Pneumonia of Bubonic Plague and of Pandemic Influenza." *Journal of the American Medical Association* 71(18): 1482-85.

Synnott, Martin J. and Elbert Clark. 1918. "The Influenza Epidemic at Camp Dix, N.J." *Journal of the American Medical Association* 71(22): 1816-21.

Taubenberger, Jeffery K., Ann H. Reid, Amy E. Krafft, Karen E. Bijwaard, and Thomas G. Fanning. 1997. "Initial Genetic Characterization of the 1918 'Spanish' Influenza Virus." *Science* 275: 1793-96.

Thompson, George Raynor. 1965. "Radio Comes of Age in World War I." In: Max L. Marshall, ed., *The Story of the U.S. Army Signal Corps* (New York: Watts), pp. 157-66. Reproduced in Scheips 1980, vol. 1.

Turner, Laurence Beddome. 1921. *Wireless Telegraphy and Telephony*. Cambridge: University Press.

————. 1931. *Wireless: A Treatise on the Theory and Practice of High-Frequency Electric Signalling*. Cambridge: Cambridge University Press.

Underwood, Robyn M. and Dennis vanEngelsdorp. 2007. "Colony Collapse Disorder: Have We Seen This Before?" *Bee Culture* 35: 13-18.

United States Signal Corps. 1917. *Radiotelegraphy*. Washington, DC: Government Printing Office.

Vandiver, Ronald Wayne. 1995. *Reflections on the Signal Corps: The Power of Paradigms in Ages of Uncertainty*. Maxwell Air Force Base, AL.

van Hartesveldt, Fred R. 1992. "The 1918-1919 Pandemic of Influenza." Lewiston, NY: Edwin Mellen.

Vaughan, Warren T. 1921. *Influenza: An Epidemiologic Study*. Baltimore: American Journal of Hygiene.

Watkins-Pitchford, Herbert. 1917. "An Enquiry into the Horse Disease Known as Septic or Contagious Pneumonia." *Veterinary Journal* 73: 345-62.

Weightman, Gavin. 2003. *Signor Marconi's Magic Box*. Cambridge, MA: Da Capo Press.

Wiedbrauk, Danny L. 1997. "The 1996-1997 Influenza Season – A View From the Benches." *Pan American Society for Clinical Virology Newsletter* 23(1).

Zinsser, Hans. 1922. "The Etiology and Epidemiology of Influenza." *Medicine* 1(2): 213-309.

Chapter 9

Adams, A. J. S. 1886. "Earth Conduction." *Van Nostrand's Engineering Magazine* 35: 249-52.

Alfvén, Hannes Olof Gösta. 1950. "Discussion of the Origin of the Terrestrial and Solar Magnetic Fields." *Tellus* 2(2): 74-82.

———. 1955. "Electricity in Space." In: *The New Astronomy* (New York: Scientific American Books), pp. 74-79.

———. 1969. *Atom, Man, and Universe: The Long Chain of Complications.* San Francisco: W. H. Freeman.

———. 1981. *Cosmic Plasma.* Dordrecht: D. Reidel.

———. 1984. "Cosmology: Myth or Science?" *Journal of Astrophysics and Astronomy* 5: 79-98.

———. 1986a. "Double Layers and Circuits in Astrophysics." *IEEE Transactions on Plasma Science* PS-14(6): 779-93.

———. 1986b. "Model of the Plasma Universe." *IEEE Transactions on Plasma Science* PS-14(6): 629-38.

———. 1986c. "The Plasma Universe." *Physics Today*, September, pp. 22-27.

———. 1987. "Plasma Universe." *Physica Scripta* T18: 20-28.

———. 1988. "Memoirs of a Dissident Scientist." *American Scientist* 76: 249-51.

———. 1990. "Cosmology in the Plasma Universe: An Introductory Exposition." *IEEE Transactions on Plasma Science* PS-18(1): 5-10.

Alfvén, Hannes and Gustaf Arrhenius. 1976. *Evolution of the Solar System.* Washington, DC: National Aeronautics and Space Administration.

Alfvén, Hannes and Carl-Gunne Fälthammar. 1963. *Cosmical Electrodynamics*, 2nd ed. Oxford: Clarendon Press.

Ando, Yoshiaki and Masashi Hayakawa. 2002. "Theoretical Analysis on the Penetration of Power Line Harmonic Radiation into the Ionosphere." *Radio Science* 37(6): 5-1 to 5-12.

Arnoldy, Roger L. and Paul M. Kintner. 1989. "Rocket Observations of the Precipitation of Electrons by Ground VLF Transmitters." *Journal of Geophysical Research* 94(A6): 6825-32.

Arrhenius, Svante. 1897. "Die Einwirkung kosmischer Einflüsse auf physiologische Verhältnisse." *Skandinavisches Archiv für Physiologie* 8(1): 367-416.

———. 1905. "On the Electric Charge of the Sun." *Terrestrial Magnetism and Atmospheric Electricity* 10(1): 1-8.

Avijgan, Majid and Mahtab Avijgan. 2013. "Can the Primo Vascular System (Bong Han Duct System) Be a Basic Concept for Qi Production?" *International Journal of Integrative Medicine* 1(20): 1-10.

Baik, Ku-Youn, Eun Sung Park, Byung-Cheon Lee, Hak-Soo Shin, Chunho Choi, Seung-Ho Yi, Hyun-Min Johng, Tae Jeong Nam, Kyung-Soon Soh, Yong-Sam Nahm, Yeo Sung Yoon, In-Se Lee, Se-Young Ahn, and Kwang-Sup Soh. 2004. "Histological Aspect of Threadlike Structure Inside Blood Vessel." *Journal of International Society of Life Information Science* 22(2): 473-76.

Baik, Ku-Youn, Baeckkyoung Sung, Byung-Cheon Lee, Hyeon-Min Johng, Vyacheslava Ogay, Tae Jung Nam, Hak-Soo Shin, and Kwang-Sup Soh. 2004. "Bonghan Ducts and Corpuscles with DNA-contained Granules on the Internal Surfaces of Rabbits." *Journal of International Society of Life Information Science* 22(2): 598-601.

Bailey, V. A. and David Forbes Martyn. 1934. "Interaction of Radio Waves." *Nature* 133: 218.

Balser, Martin and Charles A. Wagner. 1960. "Observations of Earth-Ionosphere Cavity Resonances." *Nature* 188: 638-41.

Barr, Richard. 1979. "ELF Radiation from the New Zealand Power System." *Planetary and Space Science* 27: 537-40.

Barr, Richard, D. Llanwyn Jones, and Craig J. Rodger. 2000. "ELF and VLF Radio Waves." *Journal of Atmospheric and Solar-Terrestrial Physics* 62(17-18): 1689-1718.

Bauer, Louis A. 1921. "Measures of the Electric and Magnetic Activity of the Sun and the Earth, and Interrelations." *Terrestrial Magnetism and Atmospheric Electricity* 26(1-2): 33-68.

Beard, George Miller. 1874. "Atmospheric Electricity and Ozone: Their Relation to Health and Disease." *Popular Science Monthly* 4: 456-69.

Becker, Robert Otto. 1963. "The Biological Effects of Magnetic Fields – A Survey." *Medical Electronics and Biological Engineering* 1(3): 293-303.

Becker, Robert O., Maria Reichmanis, Andrew A. Marino, and Joseph A. Spadaro. 1976. "Electrophysiological Correlates of Acupuncture Points and Meridians." *Psychoenergetic Systems* 1: 105-12.

Becquerel, Antoine César. 1851. "On the Causes of the Disengagement of Electricity in Plants, and upon Vegeto-terrestrial Currents." *American Journal of Science and Arts*, 2nd ser., 12: 83-97. Translation from: "Sur les causes qui dégagent de l'électricité dans les végétaux, et sur les courants végétaux-terrestres," *Annales de Chimie et de Physique*, 3rd ser., 31: 40-67.

Bell, Timothy F. 1976. "ULF Wave Generation through Particle Precipitation Induced by VLF Transmitters." *Journal of Geophysical Research* 81(19): 3316-26.

Belyaev, G. G., V. M. Chmyrev, and N. G. Kleimenova. 2003. "Hazardous ULF Electromagnetic Environment of Moscow City." *Physics of Auroral Phenomena*. Proceedings of the 26th Annual Seminar, Apatity, pp. 249-52.

Bering, Edgar A., III, Arthur A. Few, and James R. Benbrook. 1998. "The Global Electric Circuit." *Physics Today*, October, pp. 24-30.

Boerner, Wolfgang M., James B. Cole, William R. Goddard, Michael Z. Tarnawecky, Lotfallah Shafai, and Donald H. Hall. 1983. "Impacts of Solar and Auroral Storms on Power Line Systems." *Space Science Reviews* 35: 195-205.

Bowen, Melissa M., Antony C. Fraser-Smith, and Paul R. McGill. 1992. *Long-Term Averages of Globally-Measured ELF/VLF Radio Noise*. Space, Telecommunication, and RadioScience Laboratory, Stanford University. Technical Report E450-2.

Bradley, Philip B. and Joel Elkes. 1957. "The Effects of Some Drugs on the Electrical Activity of the Brain." *Brain* 80: 77-117.

Brazier, Mary A. B. 1977. *The Electrical Activity of the Nervous System*, 4th ed. Baltimore: Williams & Wilkins.

Brewitt, Barbara. 1996. "Quantitative Analysis of Electrical Skin Conductance in Diagnosis: Historical and Current Views of Bioelectric Medicine." *Journal of Naturopathic Medicine* 6(1): 66-75.

Bullough, Ken. 1983. "Satellite Observations of Power Line Harmonic Radiation." *Space Science Reviews* 35: 175-83.

————. 1995. "Power Line Harmonic Radiation: Sources and Environmental Effects." In: Hans Volland, ed., *Handbook of Atmospheric Electrodynamics*, (CRC Press: Boca Raton, FL), vol. 2, pp. 291-332.

Bullough, Ken, Thomas Reeve Kaiser, and Hal J. Strangeways. 1985. "Unintentional Man-made Modification Effects in the Magnetosphere." *Journal of Atmopheric and Terrestrial Physics* 47(12): 1211-23.

Bullough, Ken , Adrian R. L. Tatnall, and M. Denby. 1976. "Man-made E.L.F./V.L.F. Emissions and the radiation belts." *Nature* 260: 401-3.

Burbank, J. E. 1905. "Earth-Currents: And a Proposed Method for Their Investigation." *Terrestrial Magnetism and Atmospheric Electricity* 10: 23-49.

Cannon, P. S. and Michael J. Rycroft. 1982. "Schumann Resonance Frequency Variations during Sudden Ionospheric Disturbances." *Journal of Atmospheric and Terrestrial Physics* 44(2): 201-6.

Cherry, Neil. 2002. "Schumann Resonances, a Plausible Biophysical Mechanism for the Human Health Effects of Solar/Geomagnetic Activity." *Natural Hazards* 26(3): 279-331.

Chevalier, Gaetan. 2007. *The Earth's Electrical Surface Potential: A Summary of Present Understanding*. Encinitas, CA: California Institute for Human Science.

Cho, Sung-Jin, Byeong-Soo Kim, and Young-Seok Park. 2004. "Thread-like Structures in the Aorta and Coronary Artery of Swine." *Journal of International Society of Life Information Science* 22(2): 609-11.

Cresson, John C. 1836. "History of Experiments on Atmospheric Electricity." *Journal of the Franklin Institute* 22: 166-72.

Davis, John R. 1974. "A Quest for a Controllable ULF Wave Source." *IEEE Transactions on Communications* COM-22(4): 578-86.

de Vernejoul, Pierre, Pierre Albarède, and Jean-Claude Darras. 1985. "Étude des méridiens d'acupuncture par les traceurs radioactifs." *Bulletin de l'Académie Nationale de Médecine* 169(7): 1071-75.

Dolezalek, Hans. 1972. "Discussion of the Fundamental Problem of Atmospheric Electricity." *Pure and Applied Geophysics* 100(1): 8-43.

Dowden, R. L. and B. J. Fraser. 1984. "Waves in Space Plasmas: Highlights of a Conference Held in Hawaii, 7-11 February 1983." *Space Science Reviews* 39: 227-53.

Fälthammar, Carl-Gunne. 1986. "Magnetosphere-Ionosphere Interactions – Near-Earth Manifestations of the Plasma Universe." *IEEE Transactions on Plasma Science* PS-14(6): 616-28.

Faust, Volker. 1978. *Biometeorologie: Der Einfluss von Wetter und Klima auf Gesunde und Kranke*. Stuttgart: Hippokrates.

Fraser-Smith, Antony C. 1979. "A Weekend Increase in Geomagnetic Activity." *Journal of Geophysical Research* 84(A5): 2089-96.

————. 1981. "Effects of Man on Geomagnetic Activity and Pulsations." *Advances in Space Research* 1: 455-66.

Fraser-Smith, Antony C. and Peter R. Bannister. 1998. "Reception of ELF Signals at Antipodal Distances." *Radio Science* 33(1): 83-88.

Fraser-Smith, Antony C. and Melissa M. Bowen. 1992. "The Natural Background Levels of 50/60 Hz Radio Noise." *IEEE Transactions on Electromagnetic Compatibility* 34(3): 330-37.

Fraser-Smith, Antony C., D. M. Bubenick, and Oswald G. Villard, Jr. 1977. *Air/Undersea Communication at Ultra-Low-Frequencies Using Airborne Loop Antennas.* Technical Report 4207-6, Radio Science Laboratory, Stanford Electronics Laboratories, Department of Electrical Engineering, June 1977, SEL-77-013.

Fraser-Smith, Antony C. and D. B. Coates. 1978. "Large-amplitude ULF fields from BART." *Radio Science* 13(4): 661-68.

Fraser-Smith, Antony C., Paul R. McGill, A. Bernardi, Robert A. Helliwell, and M. E. Ladd. 1992. *Global Measurements of Low-Frequency Radio Noise.* Space, Telecommunications and Radioscience Laboratory, Stanford University. Final Technical Report E450-1.

Frölich, O. 1895. "Kompensationsvorrichtung zum Schutze physikalischer Institute gegen die Einwirkung elektrischer Bahnen." *Elektrotechnische Zeitschrift* no. 47, pp. 745-48.

———. 1896. "Demonstration der Kompensationsvorrichtung zum Schutz physikalischer Institute gegen elektrische Bahnen." *Elektrotechnische Zeitschrift,* no. 3, pp. 40-44.

Fujiwara, Satoru and Sun-Bong Yu. 2012. "A Follow-up Study on the Morphological Characteristics in Bong-Han Theory: An Interim Report." In: Kwang-Sup Soh, Kyung A. Kang, and David K. Harrison, eds., *The Primo Vascular System* (New York: Springer), pp. 19-21.

Füllekrug, Martin. 1995. "Schumann Resonances in Magnetic Field Components." *Journal of Atmorpheric and Terrestrial Physics* 57(5): 479-84.

Gerland, E. 1886. "On the Origin of Atmospheric Electricity." *Van Nostrand's Engineering Magazine* 34: 158-60.

Guglielmi, A. and O. Zotov. 2007. "The Human Impact on the Pc1 Wave Activity." *Journal of Atmospheric and Solar-Terrestrial Physics* 69: 1753-58.

Hamer, James R. 1965. *Biological Entrainment of the Human Brain by Low Frequency Radiation.* NSL 65-199, Northrop Space Labs.

Harrison, R. Giles. 2004. "The Global Atmospheric Electrical Circuit and Climate." *Surveys in Geophysics* 25(5-6): 441-84.

———. 2013. "The Carnegie Curve." *Surveys in Geophysics* 34: 209-32.

Hayashi, K., T. Oguti, T. Watanabe, K. Tsuruda, S. Kokubun, and R. E. Horita. 1978. "Power Harmonic Radiation Enhancement during the Sudden Commencement of a Magnetic Storm." *Nature* 275: 627-29.

Helliwell, Robert A. 1965. *Whistlers and Related Ionospheric Phenomena.* Stanford, CA: Stanford University Press.

———. 1977. "Active Very Low Frequency Experiments on the Magnetosphere from Siple Station, Antarctica." *Philosophical Transactions of the Royal Society B* 279: 213-24.

Helliwell, Robert A. and John P. Katsufrakis. 1974. "VLF Wave Injection into the Magnetosphere from Siple Station, Antarctica." *Journal of Geophysical Research* 79(16): 2511-18.

Helliwell, Robert A., John P. Katsufrakis, Timothy F. Bell, and Rajagopalan Raghuram. 1975. "VLF Line Radiation in the Earth's Magnetosphere and Its Association with Power System Radiation." *Journal of Geophysical Research* 80(31): 4249-58.

Hess, Victor F. 1928. *The Electrical Conductivity of the Atmosphere and its Causes.* London: Constable.

Ho, A. M.-H., Antony C. Fraser-Smith, and Oswald G. Villard, Jr. 1979. "Large-Amplitude ULF Magnetic Fields Produced by a Rapid Transit System: Close-Range Measurements." *Radio Science* 14(6): 1011-15.

Hu, X., X. Huang, J. Xu, and B. Wu. 1993. "Distribution of Low Skin Impedance Points Along Meridians over the Medial Side of Forearm." *Zhen Ci Yan Jiu* ("Acupuncture Research") 18(2): 94-97 (in Chinese).

Hu, X., B. Wu, J. Xu, X. Huang, and J. Hau. 1993. "Studies on the Low Skin Impedance Points and the Feature of its Distribution Along the Channels by Microcomputer. II. Distribution of LSIPs Along the Channels." *Zhen Ci Yan Jiu* ("Acupuncture Research") 18(2): 163-67 (in Chinese).

Huang, X., J. Xu, B. Wu, and X. Hu. 1993. "Observation on the Distribution of LSIPs Along Three Yang Meridians as Well as Ren and Du Meridians." *Zhen Ci Yan Jiu* ("Acupuncture Research") 18(2): 98-103 (in Chinese).

Imhof, W. L., H. D. Voss, M. Walt, E. E. Gaines, J. Mobilia, D. W. Datlowe, and J. B. Reagan. 1986. "Slot Region Electron Precipitation by Lightning, VLF Chorus, and Plasmaspheric Hiss." *Journal of Geophysical Research* 91(A8): 8883-94.

Itoh, Shinji, Keisuke Tsujioka, and Hiroo Saito. 1959. "Blood Clotting Time under Metal Cover (Biological P-Test)." *International Journal of Bioclimatology and Biometeorology* 3(1): 269-70.

Jenssen, Matz. 1950. "On Radiation From Overhead Transmission Lines." *Proceedings of the IEE*, part III, 97(47): 166-78.

Jiang, Xiaowen, Byung-Cheon Lee, Chunho Choi, Ku-Youn Baik, Kwang-Sup Soh, Hee-Kyeong Kim, Hak-Soo Shin, Kyung-Soon Soh, and Byeung-Soo Cheun. 2002. "Threadlike Bundle of Tubules Running Inside Blood Vessels: New Anatomical Structure." *arXiv:physics/0211085*.

Johng, Hyeon-Min, Hak-Soo Shin, Jung Sun Yoo, Byung-Cheon Lee, Ku-Youn Baik, and Kwang-Sup Soh. 2004. "Bonghan Ducts on the Surface of Rat Liver." *Journal of International Society of Life Information Science* 22(2): 469-72.

Johng, Hyeon-Min, Jung-Sun Yoo, Tae-Jong Yoon, Hak-Soo Shin, Byung-Cheon Lee, Changhoon Lee, Jin-Kyu Lee, and Kwang-Sup Soh. 2006. "Use of Magnetic Nanoparticles to Visualize Threadlike Structures Inside Lymphatic Vessels of Rats." *Evidence-Based Complementary and Alternative Medicine* 4: 77-82.

Karinen, A., K. Mursula, Th. Ulich, and J. Manninen. 2002. "Does the Magnetosphere Behave Differently on Weekends?" *Annales Geophysicae* 20: 1137-42.

Kikuchi, Hiroshi. 1983a. "Overview of Power-Line Radiation and its Coupling to the Ionosphere and Magnetosphere." *Space Science Reviews* 35: 33-41.

———. 1983b. "Power Line Transmission and Radiation." *Space Science Reviews* 35: 59-80.

Kim, Bong Han. 1963. "On the Kyungrak System." *Journal of the Academy of Medical Sciences of the Democratic People's Republic of Korea*, vol. 1963, no. 5.

———. 1964. *On the Kyungrak System*. Pyongyang, Democratic People's Republic of Korea: Foreign Languages Publishing House.

Kim, Soyean, Kyu Jae Lee, Tae Eul Jung, Dan Jin, Dong Hui Kim, and Hyun-Won Kim. 2004. "Histology of Unique Tubular Structures Believed to Be Meridian Line." *Journal of International Society of Life Information Science* 22(2): 595-97.

Klemm, William R. 1969. *Animal Electroencephalography*. New York: Academic.

Kolesnik, A. G. 1998. "Electromagnetic Background and Its Role in Environmental Protection and Human Ecology." *Russian Physics Journal* 41(8): 839-50.

König, Herbert L. 1971. "Biological Effects of Extremely Low Frequency Electrical Phenomena in the Atmosphere." *Journal of Interdisciplinary Cycle Research* 2(3): 317-23.

———. 1974a. "ELF and VLF Signal Properties: Physical Characteristics." In: Michael A. Persinger, ed., *ELF and VLF Electromagnetic Field Effects* (New York: Plenum), pp. 9-34.

———. 1974b. "Behavioral Changes in Human Subjects Associated with ELF Electric Fields." In: Michael A. Persinger, ed., *ELF and VLF Electromagnetic Field Effects* (New York: Plenum), pp. 81-99.

———. 1975. *Unsichtbare Umwelt: Der Mensch im Spielfeld elektromagnetischer Kräfte.* München: Heinz Moos.

Kornilov, I. A. 2000. "VLF Emissions and Electron Precipitations Stimulated by Emissions of Power Transmission Line Harmonics." *Geomagnetism and Aeronomy* 40(3): 388-92.

Lanzerotti, Louis J. and Giovanni P. Gregori. 1986. "Telluric Currents: The Natural Environment and Interactions with Man-made Systems." In: Geophysics Study Committee, National Research Council, *The Earth's Electrical Environment* (Washington, DC: National Academy Press), pp. 232-57.

Larkina, V. I., O. A. Maltseva and O. A. Molchanov. 1983. "Satellite Observations of Signals from a Soviet Mid-latitude VLF Transmitter in the Magnetic-Conjugate Region." *Journal of Atmospheric and Terrestrial Physics* 45(2/3): 115-19.

Larsen, Adrian P. 2004. *Ryodoraku Acupuncture Measurement and Treatment.* Doctoral thesis, Logan College of Chiropractic, Chesterfield, MO.

Lee, Byung-Cheon, Jung Sun Yoo, Ku Youn Baik, Baeckkyoung Sung, Jawoong Lee, and Kwang-Sup Soh. 2008. "Development of a Fluorescence Stereomicroscope and Observation of Bong-Han Corpuscles Inside Blood Vessels." *Indian Journal of Experimental Biology* 46: 330-35.

Lee, Byung-Cheon, Ki-Hoon Uhm, Kyoung-Hee Bae, Dae-In Kang, and Kwang-Sup Soh. 2009. "Visualization of Potential Acupuncture Points in Rat and Nude Mouse and DiI Tracing Method." *Journal of Pharmacopuncture* 12(3): 25-30.

Lee, Jong-Su. 2004. "Bonghan System and Hypothesis on Oncogenesis." *Journal of International Society of Life Information Science* 22(2): 606-8.

Lee, Sanghun, Yeonhee Ryu, Yungju Yun, Sungwon Lee, Ohsang Kwon, Jaehyo Kim, Inchul Sohn, and Seonghun Ahn. 2010. "Anatomical Discrimination of the Differences between Torn Mesentery Tissues and Internal Organ-surface Primo-vessels." *Journal of Acupunture and Meridian Studies* 3(1): 10-15.

Lerner, Eric J. 1991. *The Big Bang Never Happened.* New York: Times Books.

Lim, Chae Jeong, So Yeong Lee, and Pan Dong Ryu. 2015. "Identification of Primo-Vascular System in Abdominal Subcutaneous Tissue Layer of Rats." *Evidence-Based Complementary and Alternative Medicine*, article ID 751937.

Lin, Hsiao-Tsung. 2008. "Physics Model of Internal Chi System." *Journal of Accord Integrative Medicine* 4(1): 78-83.

Lovering, Joseph. 1854. "Atmospheric Electricity." *American Almanac*, 1854, pp. 70-82.

Lowes, Frank J. 1982. "On Magnetic Observations of Electric Trains." *The Observatory* 102: 44.

Ludwig, Wolfgang and Reinhard Mecke. 1968. "Wirkung künstlicher Atmospherics auf Säuger." *Archiv für Meteorologie, Geophysik und Bioklimatologie*, ser. B, 16: 251-61.

Luette, James Paul, Chung G. Park, and Robert A. Helliwell. 1977. "Longitudinal Variations of Very-Low-Frequency Chorus Activity in the Magnetosphere: Evidence of Excitation By Electrical Power Transmission Lines." *Geophysical Research Letters* 4(7): 275-78.

———. 1979. "The Control of the Magnetosphere by Power Line Radiation." *Journal of Geophysical Research* 84: 2657-60.

Lyman, Charles P. and Regina C. O'Brien. 1977. "A Laboratory Study of the Turkish Hamster *Mesocricetus brandti*." *Breviora* 442: 1-27.

Makarova, L. N. and A. V. Shirochkov. 2000. "Magnetopause Position as an Important Index of the Space Weather." *Physics and Chemistry of the Earth C* 25(5-6): 495-98.

———. 2005. "Atmospheric Electrodynamics Modulated by the Solar Wind." *Advances in Space Research* 35(8): 1480-83.

Markson, Ralph and Michael Muir. 1980. "Solar Wind Control of the Earth's Electric Field." *Science* 208: 979-90.

Mathias, Émile, Jean Bosler, Pierre Loisel, Raphaël Dongier, Charles Maurain, G. Girousse, and René Mesny. 1924. *Traité d'Électricité Atmosphérique et Tellurique*. Paris: Presses Universitaires de France.

Matteucci, Carlo. 1869. *On the Electrical Currents of the Earth*. Washington, DC: Smithsonian Institution.

Matthews, J. P. and Keith H. Yearby. 1981. "Magnetospheric VLF Line Radiation Observed at Halley, Antarctica." *Planetary and Space Science* 29(1): 95-112.

Maurain, Charles. 1905. "Influence perturbatrice des lignes de tramway électriques sur les appareils de mésures électriques et magnétiques: moyens de défense." *Revue Électrique* 4(45): 257-63.

Molchanov, Oleg and Michel Parrot. 1995. "PLHR Emissions Observed on Satellites." *Journal of Atmospheric and Terrestrial Physics* 57(5): 493-505.

Molchanov, Oleg, Michel Parrot, Mikhail M. Mogilevsky, and François Lefeuvre. 1991. "A Theory of PLHR Emissions to Explain the Weekly Variation of ELF Data Observed by a Low-Altitude Satellite." *Annales Geophysicae* 9: 669-80.

Moore-Ede, Martin C., Scott S. Campbell, and Russel J. Reiter, eds. 1992. *Electromagnetic Fields and Circadian Rhythmicity*. Boston: Birkhäuser.

National Research Council, Geophysics Study Comittee. 1986. *The Earth's Electrical Environment*. Washington, DC: National Academy Press.

Němec, František, Ondřej Santolík, Michel Parrot, and Jean-Jacques Berthelier. 2007. "On the Origin of Magnetospheric Line Radiation." *WDS '07 Proceedings of Contributed Papers*, part 2, pp. 64-70.

———. 2007. "Power Line Harmonic Radiation: A Systematic Study Using DEMETER Spacecraft." *Advances in Space Research* 40: 398-403.

Nunn, D., J. Manninen, T. Turunen, V. Trakhtengerts, and N. Erokhin. 1999. "On the Nonlinear Triggering of VLF Emissions by Power Line Harmonic Radiation." *Annales Geophysicae* 17: 79-94.

Ogawa, Toshio, Yoshikazu Tanaka, Teruo Miura, and Michihiro Yasuhara. 1966. "Observations of Natural ELF and VLF Electromagnetic Noises by Using Ball Antennas." *Journal of Geomagnetism and Geoelectricity* 18(4): 443-54.

Ortega, Pascal, Anirban Guha, Earle Williams, and Gabriella Satori. 2014. "Schumann Resonance Observations from the Central Pacific Ocean." Paper presented at XV International Conference on Atmospheric Electricity, 15-20 June 2014, Norman, OK.

Palmer, C. W. 1935. "The 'Luxembourg Effect' in Radio." *Radio-Craft*, February, pp. 467, 499.

Park, Chung. G. and D. C. D. Chang. 1978. "Transmitter Simulation of Power Line Radiation Effects in the Magnetosphere." *Geophysical Research Letters* 5(10): 861-64.

Park, Chung G. and Robert A. Helliwell. 1978. "Magnetospheric Effects of Power Line Radiation." *Science* 200: 727-30.

Park, Chung G., Robert A. Helliwell, and François Lefeuvre. 1983. "Ground Observations of Power Line Radiation Coupled to the Ionosphere and Magnetosphere." *Space Science Reviews* 35: 131-37.

Park, Eun-sung, Hee Young Kim, and Dong-ho Youn. 2013. "The Primo Vascular Structures Alongside Nervous System: Its Discovery and Functional Limitation." *Evidence-Based Complementary and Alternative Medicine*, article ID 538350.

Park, Joong Wha, In Soo Hong, Jin Ha Yoon, and Hyun-Won Kim. 2004. "Migration of Lipiodol Along the Meridian Line." *Journal of International Society of Life Information Science* 22(2): 592-94.

Parrot, Michel, Oleg A. Molchanov, Mikhail M. Mogilevski, and François Lefeuvre. 1991. "Daily Variations of ELF Data Observed by a Low-altitude Satellite." *Geophysical Research Letters* 18(6): 1039-42.

Parrot, Michel, František Němec, Ondřej Santolík, and Jean-Jacques Berthelier. 2005. "ELF Magnetospheric Lines Observed by DEMETER." *Annales Geophysicae* 23: 3301-11.

Parrot, Michel and Youri Zaslavski. 1996. "Physical Mechanisms of Man-Made Influences on the Magnetosphere." *Surveys in Geophysics* 17: 67-100.

Pellegrino, Fernando C. and Roberto E. P. Sica. 2004. "Canine Electroencephalographic Recording Technique: Findings in Normal and Epileptic Dogs." *Clinical Neurophysiology* 115: 477-87.

Peratt, Anthony L. 1989a. "Plasma Cosmology. Part I. Interpretations of the Visible Universe." *The World and I*, August, pp. 294-301.

———. 1989b. "Plasma Cosmology. Part II. The Universe is a Sea of Electrically Charged Particles." *The World and I*, September, pp. 307-17.

———. 1990. "Not with a Bang." *The Sciences*, January/February, pp. 24-32.

———. 1992. *Physics of the Plasma Universe*. New York: Springer.

———. 1995. "Introduction to Plasma Astrophysics and Cosmology." *Astrophysics and Space Science* 227: 3-11.

Persinger, Michael A., ed. 1974. *ELF and VLF Electromagnetic Field Effects*. New York: Plenum.

Persinger, Michael A., H. Wolfgang Ludwig, and Klaus-Peter Ossenkopp. 1973. "Psychophysiological Effects of Extremely Low Frequency Electromagnetic Fields: A Review." *Perceptual and Motor Skills* 36: 1131-59.

Planté, Gaston. 1878. "Electrical Analogies with Natural Phenomena." *Nature* 17: 226-29, 385-87.

Pouillet, Claude Servais Mathias. 1853. *Éléments de Physique expérimentale et de Météorologie*, 6th ed. Paris: L. Hachette.

Preece, William Henry. 1894. "Earth Currents." *Nature* 49: 554.

Randall, Walter and Walter S. Moos. 1993. "The 11-Year Cycle in Human Births." *International Journal of Biometeorology* 37(2): 72-77.

Randall, Walter. 1990. "The Solar Wind and Human Birth Rate: A Possible Relationship Due to Magnetic Disturbances." *International Journal of Biometeorology* 34(1): 42-48.

Reichmanis, Maria, Andrew A. Marino, and Robert O. Becker. 1979. "Laplace Plane Analysis of Impedance on the H Meridian." *American Journal of Chinese Medicine* 7(2): 188-93.

Reiter, Reinhold. 1954. "Umwelteinflüsse auf die Reaktionszeit des gesunden Menschen." *Münchener medizinische Wochenschrift* 96(17, 18): 479-81, 526-29.

———. 1969. "Solar Flares and Their Impact on Potential Gradient and Air-Earth Current Characteristics at High Mountain Stations." *Pure and Applied Geophysics* 72(1): 259-67.

———. 1976. "The Electric Potential of the Ionosphere as Controlled by the Solar Magnetic Sector Structure." *Naturwissenschaften* 63(4): 192-93.

Rheinberger, Margaret B. and Herbert H. Jasper. 1937. "Electrical Activity of the Cerebral Cortex in the Unanesthetized Cat." *American Journal of Physiology* 119: 186-96.

Robinson, G. H. 1966. "Harmonic Phenomena Associated with the Benmore-Haywards H.V.D.C. Transmission Scheme." *New Zealand Engineering*, January 15, pp. 16-28.

Roble, R. G. 1991. "On Modeling Component Processes in the Earth's Global Electric Circuit." *Journal of Atmospheric and Terrestrial Physics* 53(9): 831-47.

Rooney, W. J. 1939. "Earth Currents." In: J. A. Fleming, ed., *Terrestrial Magnetism and Electricity* (New York: McGraw-Hill), pp. 270-307.

Rosenberg, Theodore J., Robert A. Helliwell, and John P. Katsufrakis. 1971. "Electron Precipitation Associated with Discrete Very-Low-Frequency Emissions." *Journal of Geophysical Research* 76(34): 8445-52.

Ruckebusch, Y. 1963. "L'électroencéphalogramme normal du chien." *Revue de Médecine Vétérinaire* 114(1): 119-34.

Rycroft, Michael J. 1965. "Resonances of the Earth-Ionosphere Cavity Observed at Cambridge, England." *Radio Science* 69D(8): 1071-81.

———. 2006. "Electrical Processes Coupling the Atmosphere and Ionosphere: An Overview." *Journal of Atmospheric and Solar-Terrestrial Physics* 68: 445-56.

Sá, Luiz Alexandre Nogueira de. 1990. "A Wave-Particle-Wave Interaction Mechanism as a Cause of VLF Triggered Emissions." *Journal of Geophysical Research* 95(A8): 12,277-86.

Schlegel, Kristian and Martin Füllekrug. 2002. "Weltweite Ortung von Blitzen: 50 Jahre Schumann-Resonanzen." *Physik in unserer Zeit* 33(6): 256-61.

Schulz, Nicolas. 1961. "Lymphocytose relative et l'activité solaire." *Revue médicale de Nancy* 6: 541-44.

Schumann, Winfried O. and Herbert L. König. 1954. "Über die Beobachtung von 'Atmospherics' bei geringsten Frequenzen." *Naturwissenschaften* 41(8): 183-84.

Shin, Hak-Soo, Hyeon-Min Johng, Byung-Cheon Lee, Sung-Il Cho, Kyung-Soon Soh, Ku-Youn Baik, Jung-Sun Yoo, and Kwang-Sup Soh. 2005. "Feulgen Reaction Study of Novel Threadlike Structures (Bonghan Ducts) on the Surface of Mammalian Organs." *Anatomical Record* 284B: 35-40.

Soh, Kwang-Sup, Kyung A. Kang, and David K. Harrison, eds. 2012. *The Primo Vascular System*. New York: Springer.

Soh, Kwang-Sup, Kyung A. Kang, and Yeon Hee Ryu. 2013. "50 Years of Bong-Han Theory and 10 Years of Primo Vascular System." *Evidence-Based Complementary and Alternative Medicine*, article ID 587827.

Starwynn, Darren. 2002. "Electrophysiology and the Acupuncture Systems." *Medical Acupuncture* 13(1): article 7.

Stiles, Gardiner S. and Robert A. Helliwell. 1975. "Frequency-Time Behavior of Artificially Stimulated VLF Emissions." *Journal of Geophysical Research* 80(4): 608-18.

Stoupel Eliyahu, J. Abramson, Stanislava Domarkiene, Michael Shimshoni, and Jaqueline Sulkes. 1997. "Space Proton Flux and the Temporal Distribution of Cardiovascular Deaths." *International Journal of Biometeorolgoy* 40(2): 113-16.

Stoupel, Eliyahu, Helena Frimer, Zvi Appelman, Ziva Ben-Neriah, Hanna Dar, Moshe D. Fejgin, Ruth Gershoni-Baruch, Esther Manor, Gad Barkai, Stavit Shalev, Zully Gelman-Kohan, Orit Reish, Dorit Lev, Bella Davidov, Boleslaw Goldman, and Mordechai Shohat. 2005. "Chromosome Aberration and Environmental Physical Activity: Down Syndrome and Solar and Cosmic Ray Activity, Israel, 1990-2000." *International Journal of Biometeorology* 50(1): 1-5.

Stoupel, Eliyahu, Jadviga Petrauskiene, Ramunė Kalėdienė, Evgeny Abramson, and Jacqueline Sulkes. 1995. "Clinical Cosmobiology: The Lithuanian Study 1990-1992." *International Journal of Biometeorology* 38(4): 204-8.

Stoupel, Eliyahu, Ramunė Kalėdienė, Jadvyga Petrauskienė, Skirmantė Starkuvienė, Evgeny Abramson, Peter Israelevich, and Jacqueline Sulkes. 2007. "Clinical Cosmobiology: Distribution of Deaths During 180 Months and Cosmophysical Activity. The Lithuanian Study, 1990-2004: The Role of Cosmic Rays." *Medicina (Kaunas)* 43(10): 824-31.

Sulman, Felix Gad. 1976. *Health, Weather and Climate*. Basel: Karger.

———. 1980. *The Effect of Air Ionization, Electric Fields, Atmospherics and Other Electric Phenomena on Man and Animal*. Springfield, Ill.: Charles C. Thomas.

———. 1982. *Short- and Long-Term Changes in Climate*, 2 vols. Boca Raton, FL: CRC Press.

Szarka, László. 1988. "Geophysical Aspects of Man-Made Electromagnetic Noise in the Earth – A Review." *Surveys in Geophysics* 9: 287-318.

Tait, Peter Guthrie. 1884. "On Various Suggestions as to the Source of Atmospheric Electricity." *Nature* 29: 517.

Tatnall, Adrian R. L., J. P. Matthews, Ken Bullough, and Thomas Reeve Kaiser. 1983. "Power-Line Harmonic Radiation and the Electron Slot." *Space Science Reviews* 35(2): 139-73.

Tomizawa, Ichiro and Takeo Yoshino. 1984. "Power Line Radiation over Northern Europe Observed on the Balloon B_{15}-1N." *Memoirs of the National Institute of Polar Research*, Special Issue 31: 115-23.

Tomizawa, Ichiro, Hayato Nishida, and Takeo Yoshino. 1995. "A New-Type Source of Power Line Harmonic Radiation Possibly Located on the Kola Peninsula." *Journal of Geomagnetism and Geoelectricity* 47: 213-29.

Tomizawa, Ichiro, Takeo Yoshino, and Hayato Sasaki. 1985. "Geomagnetic Effect on Electromagnetic Field Strength of Power Line Radiation Over Northern Europe Observed on the Balloons B_{15}-1N and B_{15}-2N." *Memoirs of the National Institute of Polar Research*, Special Issue 36: 181-90.

Trakhtengerts, Victor Y. and Michael J. Rycroft. 2000. "Whistler-Electron Interactions in the Magnetosphere: New Results and Novel Approaches." *Journal of Atmospheric and Solar-Terrestrial Physics* 62: 1719-33.

Tromp, Solco W. 1963. *Medical Biometeorology: Weather, Climate and the Living Organism*. Amsterdam: Elsevier.

Trowbridge, John. 1880. "The Earth as a Conductor of Electricity." *American Journal of Science*, 3rd ser., 20: 138-41.

Vampola, Alfred L. 1987. "Electron Precipitation in the Vicinity of a VLF Transmitter." *Journal of Geophysical Research* 92(A5): 4525-32.

Vampola, Alfred L. and C. D. Adams. 1988. "Outer Zone Electron Precipitation Produced by a VLF Transmitter." *Journal of Geophysical Research* 93(A3): 1849-58.

Van Nostrand's Engineering Magazine. 1874. "Terrestrial Electricity." 10: 440-42.

Villante, U., M. Vellante, A. Piancatelli, A. Di Cienzo, T. L. Zhang, W. Magnes, V. Wesztergom, and A. Meloni. 2004. "Some Aspects of Man-made Contamination on ULF Measurements." *Annales Geophysicae* 22: 1335-45.

Vodyanoy, Vitaly, Oleg Pustovyy, Ludmila Globa, and Iryna Sorokulova. 2015. "Primo-Vascular System as Presented by Bong Han Kim." *Evidence-Based Complementary and Alternative Medicine*, article ID 361974.

Volland, Hans, ed. 1982. *Handbook of Atmospherics*, 2 vols. Boca Raton, FL: CRC Press.

———. 1987. "Electromagnetic Coupling between Lower and Upper Atmosphere." *Physica Scripta* T18: 289-97.

———. 1995. *Handbook of Atmospheric Electrodynamics*, 2 vols. Boca Raton, FL: CRC Press.

Watt, A. D. and E. L. Maxwell. 1957. "Characteristics of Atmospheric Noise from 1 to 100 KC." *Proceedings of the IRE* 45: 787-94.

Wei, Jianzi, Huijuan Mao, Yu Zhou, Lina Wang, Sheng Liu, and Xueyong Shen. 2012. "Research on Nonlinear Feature of Electrical Resistance of Acupuncture Points." *Evidence-Based Complementary and Alternative Medicine*, article ID 179657.

Wever, Rütger A. 1973. "Human Circadian Rhythms under the Influence of Weak Electric Fields and the Different Aspects of These Studies." *International Journal of Biometeorology* 17(3): 227-32.

———. 1974. "ELF-Effects on Human Circadian Rhythms." In: Michael A. Persinger, ed., *ELF and VLF Electromagnetic Field Effects* (New York: Plenum), pp. 101-44.

———. 1992. "Circadian Rhythmicity of Man under the Influence of Weak Electromagnetic Fields." In: Martin C. Moore-Ede, Scott S. Campbell, and Russel J.

Reiter, eds., *Electromagnetic Fields and Circadian Rhythmicity* (Boston: Birkhäuser), pp. 121-39.

Williams, Earle R. 2009. "The Global Electrical Circuit: A Review." *Atmospheric Research* 91(2-4): 140-52.

Wu, B., X. Hu, and J. Xu. 1993. "Effect of Increase and Decrease of Measurement Voltage on Skin Impedance." *Zhen Ci Yan Jiu* ("Acupuncture Research") 18(2): 104-7 (in Chinese).

Yearby, Keith H., Andy J. Smith, Thomas Reeve Kaiser, and Ken Bullough. 1983. "Power Line Harmonic Radiation in Newfoundland." *Journal of Atmospheric and Terrestrial Physics* 45(6): 409-19.

Xiang, Zhu Zong, Xu Rui Ming, Xie Jung Guo, and Yu Shu Zhuang. 1984. "Experimental Meridian Line of Stomach and Its Low Impedance Nature." *Acupuncture and Electro-therapeutics Research* 9(3): 157-64.

Zhang, Weibo, Ruimin Xu, and Zongxian Zhu. 1999. "The Influence of Acupuncture on the Impedance Measured by Four Electrodes on Meridians." *Acupuncture and Electro-therapeutics Research* 24(3-4): 181-88.

Chapter 10

Aartsma, Thijs J. and Jan Amesz. 1996. "Reaction Center and Antenna Processes in Photosynthesis at Low Temperature." *Photosynthesis Research* 48: 99-106.

Abdelmelek, H., A. El-May Ben Hamouda, Mohamed Ben Salem, Jean-Marc Pequignot, and Mohsen Sakly. 2003. "Electrical Conduction through Nerve and DNA." *Chinese Journal of Physiology* 46(3): 137-41.

Adey, William Ross. 1993. "Whispering Between Cells: Electromagnetic Fields and Regulatory Mechanisms in Tissue." *Frontier Perspectives* 3(2): 21-25.

Adler, Alan D. 1970. "Solid State Possibilities of Porphyrin Structures." *Journal of Polymer Science: Part C* 29: 73-79.

———. 1973. "Porphyrins as Model Systems for Studying Structural Relationships." *Annals of the New York Academy of Sciences* 206: 7-17.

Adler, Alan D., Veronika Váradi, and Nancy Wilson. 1975. "Porphyrins, Power, and Pollution." *Annals of the New York Academy of Sciences* 244: 685-94.

Alley, Michael C., Eva K. Killam, and Gerald L. Fisher. 1981. "The Influence of d-Penicillamine Treatment upon Seizure Activity and Trace Metal Status in the Senegalese Baboon, *Papio Papio*." *Journal of Pharmacology and Experimental Therapeutics* 217(1): 138-46.

Andant, Christophe, Hervé Puy, Jean Faivre, and Jean-Charles Deybach. 1998. "Acute Hepatic Porphyrias and Primary Liver Cancer." *New England Journal of Medicine* 338(25): 1853-54.

Apeagyei, Eric, Michael S. Bank, and John D. Spengler. 2011. "Distribution of Heavy Metals in Road Dust Along an Urban-Rural Gradient in Massachusetts." *Atmospheric Environment* 45: 2310-23.

Aramaki, Shinji, Ruichi Yoshiyama, Masayoshi Sakai, and Noboru Ono. 2005. "P-19: High Performance Porphyrin Semiconductor for Transistor Applications." *SID 05 Digest*: 296-99.

Arnold, William. 1965. "An Electron-Hole Picture of Photosynthesis." *Journal of Physical Chemistry* 69(3): 788-91.

Arnold, William and Roderick K. Clayton. 1960. "The First Step in Photosynthesis: Evidence for Its Electronic Nature." *Proceedings of the National Academy of Sciences* 46(6): 769-76.

Arnold, William and Helen K. Sherwood. 1957. "Are Chloroplasts Semiconductors?" *Proceedings of the National Academy of Sciences* 43(1): 105-14.

Asbury, Arthur K., Richard L. Sidman, and Merrill K. Wolf. 1966. "Drug-Induced Porphyrin Accumulation in the Nervous System." *Neurology* 16(3): 299. Abstract.

Assaf, S. Y. and Shin-Ho Chung. 1984. "Release of Endogenous Zn^{2+} from Brain Tissue during Activity." *Nature* 308: 734-36.

Athenstaedt, Herbert. 1974. "Pyroelectric and Piezoelectric Properties of Vertebrates." *Annals of the New York Academy of Sciences* 238: 68-94.

Barbeau, André. 1974. "Zinc, Taurine, and Epilepsy." *Archives of Neurology* 30: 52-58.

Bassham, James A. and Melvin Calvin. 1955. *Photosynthesis.* U.S. Atomic Energy Commission, report no. UCRL-2853.

Baum, Larry, Iris Hiu Shuen Chan, Stanley Kwok-Kuen Cheung, William B. Goggins, Vincent Mok, Linda Lam, Vivian Leung, Elsie Hui, Chelsia Ng, Jean Woo, Helen Fung Kum Chiu, Benny Chung-Ying Zee, William Cheng, Ming-Houng Chan, Samuel Szeto, Victor Lui, Joshua Tsoh, Ashley I. Bush, Christopher Wai Kei Lam, and Timothy Kwok. 2010. "Serum Zinc is Decreased in Alzheimer's Disease and Serum Arsenic Correlates Positively with Cognitive Ability." *Biometals* 23: 173-79.

Becker, David Morris and Sidney Kramer. 1977. "The Neurological Manifestations of Porphyria: A Review." *Medicine* 56(5): 411-23.

Becker, David Morris and Frederick Wolfgram. 1978. "Porphyrins in Myelin- and Non-myelin Fractions of Bovine White Matter." *Journal of Neurochemistry* 31: 1109-11.

Becker, Robert Otto. 1960. "The Bioelectric Field Pattern in the Salamander and Its Simulation by an Electronic Analog." *IRE Transactions on Medical Electronics* ME-7(3): 202-7.

———. 1961a. "Search for Evidence of Axial Current Flow in Peripheral Nerves of Salamander." *Science* 134: 101-2.

———. 1961b. "The Bioelectric Factors in Amphibian-Limb Regeneration." *Journal of Bone and Joint Surgery* 43-A(5): 643-56.

Becker, Robert O. and Andrew A. Marino. 1982. *Electromagnetism and Life.* Albany: State University of New York Press.

Becker, Robert O. and Gary Selden. 1985. *The Body Electric: Electromagnetism and the Foundation of Life.* New York: William Morrow.

Berman, J. and T. Bielický. 1956. "Einige äußere Faktoren in der Ätiologie der Porphyria cutanea tarda und des Diabetes mellitus mit besonderer Berücksichtigung der syphilitischen Infektion und ihrer Behandlung." *Dermatologica* 113: 78-87.

Bernal, John Desmond. 1949. "The Physical Basis of Life." *Proceedings of the Physical Society, Section A*, vol. 62, part 9, no. 357A, pp. 537-58.

Blanshard, T. Paul. 1953. "Isolation from Mammalian Brain of Coproporphyrin III and a Uro-Type Porphyrin." *Proceedings of the Society for Experimental Biology and Medicine* 83: 512-13.

Bonkowsky, Herbert L., Donald P. Tschudy, Eugene C. Weinbach, Paul S. Ebert, and Joyce M. Doherty. 1975. "Porphyrin Synthesis and Mitochondrial Respiration in Acute Intermittent Porphyria: Studies Using Cultured Human Fibroblasts." *Journal of Laboratory and Clinical Medicine* 85(1): 93-102.

Bonkowsky, Herbert L. and Wolfgang Schady. 1982. "Neurologic Manifestations of Acute Porphyria." *Seminars in Liver Disease* 2(2): 108-24.

Borgens, Richard B. 1982. "What is the Role of Naturally Produced Electric Current in Vertebrate Regeneration and Healing?" *International Review of Cytology* 76: 245-98.

Borgens, Richard B., Kenneth R. Robinson, Joseph W. Vanable, Jr., and Michael E. McGinnis. 1989. *Electric Fields in Vertebrate Repair: Natural and Applied Voltages in Vertebrate Regeneration and Healing.* New York: Alan R. Liss.

Boyle, Neil J. and Donal P. Murray. 1993. "Unusual Presentation of Porphyria Cutanea Tarda." *Lancet* 2: 186.

Brodie, Martin J., George G. Thompson, Michael R. Moore, Alistair D. Beattie, and Abraham Goldberg. 1977. "Hereditary Coproporphyria." *Quarterly Journal of Medicine* 46: 229-41.

Brown, Glenn H. and Jerome J. Wolken. 1979. *Liquid Crystals and Biological Structures.* New York: Academic.

Burr, Harold Saxton. 1940. "Electrical Correlates of the Menstrual Cycle in Women." *Yale Journal of Biology and Medicine* 12(4): 335-44.

———. 1942. "Electrical Correlates of Growth in Corn Roots." *Yale Journal of Biology and Medicine* 14(6): 581-88.

———. 1943. "An Electrometric Study of Mimosa." *Yale Journal of Biology and Medicine* 15(6): 823-29.

———. 1944a. "Moon-Madness." *Yale Journal of Biology and Medicine* 16(3): 249-56.

———. 1944b. "Potential Gradients in Living Systems and Their Measurements." In: Otto Glasser, ed., *Medical Physics* (Chicago: Yearbook), pp. 1117-21.

———. 1944c. "The Meaning of Bio-electric Potentials." *Yale Journal of Biology and Medicine* 16(4): 353-60.

———. 1945a. "Variables in DC Measurement." *Yale Journal of Biology and Medicine* 17(3): 465-78.

———. 1945b. "Diurnal Potentials in the Maple Tree." *Yale Journal of Biology and Medicine* 17(6): 727-34.

———. 1950. "Electro-cyclic Phenomena: Recording Life Dynamics of Oak Trees." *The Yale Scientific Magazine*, December, pp. 9-10, 32-36, 38, 40.

———. 1956. "Effect of a Severe Storm on Electrical Properties of a Tree and the Earth." *Science* 124: 1204-5.

———. 1972. *Blueprint for Immortality: The Electric Patterns of Life.* Saffron Walden, England: C. W. Daniel.

Burr, Harold Saxton, R. T. Hill, and Edgar Allen. 1935. "Detection of Ovulation in the Intact Rabbit." *Proceedings of the Society for Experimental Biology and Medicine* 33: 109-11.

Burr, Harold Saxton and Carl Iver Hovland. 1937. "Bio-Electric Correlates of Development in Amblystoma." *Yale Journal of Biology and Medicine* 9(6): 541-49.

Burr, Harold Saxton and Cecil Taverner Lane. 1935. "Electrical Characteristics of Living Systems." *Yale Journal of Biology and Medicine* 8(1): 31-35.

Burr, Harold Saxton and Dorothy S. Barton. 1938. "Steady-State Electrical Properties of the Human Organism during Sleep." *Yale Journal of Biology and Medicine* 10(3): 271-74.

Burr, Harold Saxton and Luther K. Musselman. 1936. "Bio-electric Phenomena Associated with Menstruation." *Yale Journal of Biology and Medicine* 9(2): 155-58.

———. 1938. "Bio-Electric Correlates of the Menstrual Cycle in Women." *American Journal of Obstetrics and Gynecology* 35(5): 743-51.

Burr, Harold Saxton, Luther K. Musselman, Dorothy S. Barton, and Naomi B. Kelly. 1937a. "Bio-electric Correlates of Human Ovulation." *Yale Journal of Biology and Medicine* 10(2): 155-60.

———. 1937b. "A Bio-electric Record of Human Ovulation." *Science* 86: 312.

Bush, Ashley I. and Rudolph E. Tanzi. 2008. "Therapeutics for Alzheimer's Disease Based on the Metal Hypothesis." *Neurotherapeutics* 5(3): 421-32.

Bush, Ashley I., Warren H. Pettingell, Gerd Multhaup, Marc d. Paradis, Jean-Paul Vonsattel, James F. Gusella, Konrad Beyreuther, Colin L. Masters, and Rudolph E. Tanzi. 1994. "Rapid Induction of Alzheimer Aβ Amyloid Formation by Zinc." *Science* 265: 1464-67.

Bylesjö, Ingemar. 2008. "Epidemiological, Clinical and Pathogenetic Studies of Acute Intermittent Porphyria." Medical dissertation, Family Medicine, Dept. of Public Health and Clinical Medicine, Umeå University, Sweden.

Calvin, Melvin. 1958. "From Microstructure to Macrostructure and Function in the Photochemical Apparatus." *Brookhaven Symposia in Biology* 11: 160-79.

Cardew, Martin H. and Daniel Douglas Eley. 1959. "The Semiconductivity of Organic Substances. Part 3 – Haemoglobin and Some Amino Acids." *Discussions of the Faraday Society* 27: 115-28.

Chisolm, J. Julian, Jr. 1992. "The Porphyrias." In: Richard E. Behrman, ed., *Nelson Textbook of Pediatrics*, 14th ed. (Philadelphia: W. B. Saunders), pp. 384-90.

Choi, D. W., M. Yokoyama, and J. Koh. 1988. "Zinc Neurotoxicity in Cortical Cell Culture." *Neuroscience* 24(1): 67-79.

Chung, Yong-Gu, Jon A. Schwartz, Raymond E. Sawayo, and Steven L. Jacques. 1997. "Diagnostic Potential of Laser-Induced Autofluorescence Emission in Brain Tissue." *Journal of Korean Medical Science* 12: 135-42.

Clayton, Roderick K. 1962. "Recent Developments in Photosynthesis." *Microbiology and Molecular Biology Reviews* 26 (2 parts 1-2): 151-64.

Cope, Freeman Widener. 1970. "The Solid-State Physics of Electron and Ion Transport in Biology." *Advances in Biological and Medical Physics* 13: 1-42.

———. 1973. "Supramolecular Biology: A Solid State Physical Approach to Ion and Electron Transport." *Annals of the New York Academy of Scinces* 204: 416-33.

———. 1975. "A Review of the Applications of Solid State Physics Concepts to Biological Systems." *Journal of Biological Physics* 3(1): 1-41.

———. 1979. "Semiconduction as the Mechanism of the Cytochrome Oxidase Reaction. Low Activation Energy of Semiconduction Measured for Cytochrome Oxidase Protein. Solid State Theory of Cytochrome Oxidase Predicts Observed Kinetic Peculiarities." *Physiological Chemistry and Physics* 11: 261-62.

Crane, Eva E. 1950. "Bioelectric Potentials, Their Maintenance and Function." *Progress in Biophysics and Biophysical Chemistry* 1: 85-136.

Crile, George Washington. 1926. *A Bipolar Theory of Living Processes*. New York: Macmillan.

———. 1936. *The Phenomena of Life: A Radio-Electric Interpretation*. New York: W. W. Norton.

Cristóvão, Joana S., Renata Santos, and Cláudio M. Gomes. 2016. "Metals and Neuronal Metal Binding Proteins Implicated in Alzheimer's Disease." *Oxidative Medicine and Cellular Longevity*, article ID 9812178.

Crimlisk, Helen L. 1997. "The Little Imitator – Porphyria: A Neuropsychiatric Disorder." *Journal of Neurology, Neurosurgery, and Psychiatry* 62(4): 319-28.

Cuajungco, Math P., Kyle Y. Fagét, Xudong Huang, Rudolph E. Tanzi, and Ashley I. Bush. 2000. "Metal Chelation as a Potential Therapy for Alzheimer's Disease." *Annals of the New York Academy of Sciences* 920: 292-304.

Darus, Fairus Muhamad, Rabiatul Adawiyah Nasir, Siti Mariam Sumari, Zitty Sarah Ismail, and Nur Aliah Omar. 2012. "Heavy Metals Composition of Indoor Dust in Nursery Schools Building." *Procedia – Social and Behavioral Sciences* 38: 169-75.

Dolphin, David, ed. 1978-79. *The Porphyrias*, 7 vols. New York: Academic.

Donald, G. F., G. A. Hunter, W. Roman, and Adelheid E. J. Taylor. 1965. "Cutaneous Porphyria: Favourable Results in Twelve Cases Treated by Chelation." *American Journal of Dermatology* 8(2): 97-115.

Dorfman, W. A. 1934. "Electrical Polarity of the Amphibian Egg and Its Reversal Through Fertilization." *Protoplasma* 21(2): 245-57.

Downey, David C. 1992. "Fatigue Syndromes: New Thoughts and Reinterpretation of Previous Data." *Medical Hypotheses* 39: 185-90.

———. 1994. "Hereditary Coproporphyria." *British Journal of Clinical Practice* 48(2): 97-99.

Durkó, Irene, Jósef Engelhardt, János Szilárd, Krisztina Baraczka, and György Gál. 1984. "The Effect of Haemodialysis on the Excretion of the Mauve Factor in Schizophrenia." *Journal of Orthomolecular Psychiatry* 13(4): 222-32.

Eilenberg, M. D. and B. A. Scobie. 1960. "Prolonged Neuropsychiatric Disability and Cardiomyopathy in Acute Intermittent Porphyria." *British Medical Journal* 1: 858-59.

Elbagermi, M. A., H. G. M. Edwards, and A. I. Alajtal. 2013. "Monitoring of Heavy Metals Content in Soil Collected from City Centre and Industrial Areas of Misurata, Libya." *International Journal of Analytical Chemistry*, article ID 312581.

Eley, Daniel Douglas and D. I. Spivey. 1960. "The Semiconductivity of Organic Substances. Part 6 – A Range of Proteins." *Transactions of the Faraday Society* 56: 1432-42.

———. 1962. "The Semiconductivity of Organic Substances. Part 8. Porphyrins and Dipyrromethenes." *Transactions of the Faraday Society* 58: 405-10.

Ellefson, Ralph D. and R. E. Ford. 1996. "The Porphyrias: Characteristics and Laboratory Tests." 1996. *Regulatory Toxicology and Pharmacology* 24: S119-S125.

Felitsyn, Natalia, Colin McLeod, Albert L. Shroads, Peter W. Stacpoole, and Lucia Notterpek. 2008. "The Heme Precursor Delta-Aminolevulinate Blocks Peripheral Myelin Formation." *Journal of Neurochemistry* 106(5): 2068-79.

Fisch, Michael R. 2004. *Liquid Crystals, Laptops and Life*. Singapore: World Scientific.

Fishbein, Alf, John C. Thornton, Ruth Lilis, José A. Valciukas, Jonine Bernstein, and Irving J. Selikoff. 1980. "Zinc Protoporphyrin, Blood Lead and Clinical Symptoms in Two Occupational Groups with Low-Level Exposure to Lead." *American Journal of Industrial Medicine* 1: 391-99.

Flinn, J. M., D. Hunter, D. H. Linkous, A. Lanzirotti, L. N. Smith, J. Brightwell, and B. F. Jones. 2005. "Enhanced Zinc Consumption Causes Memory Deficits and Increased Brain Levels of Zinc." *Physiology and Behavior* 83: 793-803.

Frederickson, Christopher J., Wolfgang Maret, and Math P. Cuajungco. 2004. "Zinc and Excitotoxic Brain Injury: A New Model." *Neuroscientist* 10(1): 19-25.

Frey, Allan H. 1971. "Biological Function as Influenced by Low Power Modulated RF Energy." *IEEE Transactions on Microwave Theory and Techniques* MTT-19(2): 153-64.

———. 1988. "Evolution and Results of Biological Research with Low-Intensity Nonionizing Radiation." In: Andrew A. Marino, ed., *Modern Bioelectricity* (New York: Marcel Dekker), pp. 785-837.

Fukuda, Eiichi. 1974. "Piezoelectric Properties of Organic Polymers." *Annals of the New York Academy of Sciences* 238: 7-25.

Garrett, C. G. B. 1959. "Organic Semiconductors." In: N. B. Hannay, ed., *Semiconductors* (New York: Reinhold Publishing Corp.), pp. 634-75.

Gibney, G. N., I. H. Jones, and J. H. Meek. 1972. "Schizophrenia in association with erythropoietic protoporphyria – report of a case." *British Journal of Psychiatry* 121: 79-81.

Gilyarovskiy, V. A., I. M. Liventsev, Yu. Ye. Segal', and Z. A. Kirillova. 1958. *Electroson (kliniko-fiziologicheskoye issledovaniye).* Moscow. In English Translation as *Electric Sleep (A Clinical-Physiological Investigation).* JPRS 2278.

Goldberg, Abraham. 1959. "Acute Intermittent Porphyria: A Study of 50 Cases." *Quarterly Journal of Medicine* 28: 183-209.

Goldberg, Abraham and Michael R. Moore, eds. 1980. *The Porphyrias.* Vol. 9, no. 2 of *Clinics in Haematology.*

Granick, S. and H. Gilder. 1945. "The Structure, Function and Inhibitory Action of Porphyrins." *Science* 101: 540.

Hagemann, Ole and Frederik Krebs. 2013. "Syntheses of Asymmetric Porphyrins for Photovoltaics." Polymer Solar Cell Initiative, Danish Polymer Centre, Risø National Laboratory, Roskilde, Denmark. www.risoe.dk/solarcells.

Halpern, R. M. and H. G. Copsey. 1946. "Acute Idiopathic Porphyria; Report of a Case." *Medical Clinics of North America* 30: 385-96.

Hamadani, Jena D., George J. Fuchs, Saskia J. M. Osendarp, F. Khatun, Syed N. Huda, and Sally M. Grantham-McGregor. 2001. "Randomized Controlled Trial of the Effect of Zinc Supplementation on the Mental Development of Bangladeshi Infants." *American Journal of Clinical Nutrition* 74: 381-86.

Hamadani, Jena D, George J. Fuchs, Saskia J. M. Osendarp, Syed N. Huda, and Sally M. Grantham-McGregor. 2002. "Zinc Supplementation During Pregnancy and Effects on Mental Development and Behaviour of Infants: A Follow-up Study." *Lancet* 360: 290-94.

Hancock, Sara M., David I. Finkelstein, and Paul A. Adlard. 2014. "Glia and Zinc in Ageing and Alzheimer's Disease: A Mechanism for Cognitive Decline?" *Frontiers in Aging Neuroscience* 6: 137.

Hardell, Lennart, Nils-Olof Bengtsson, U. Jonsson, S. Eriksson, and Lars-Gunnar Larsson. 1984. "Aetiological Aspects on Primary Liver Cancer with Special Regard to Alcohol, Organic Solvents and Acute Intermittent Porphyria – an Epidemiological Investigation." *British Journal of Cancer* 50: 389-97.

Hargittai, Pál T. and Edward M. Lieberman. 1991. "Axon-Glia Interactions in the Crayfish: Glial Cell Oxygen Consumption is Tightly Coupled to Axon Metabolism." *Glia* 4(4): 417-23.

Hashim, Zawiah, Leslie Woodhouse, and Janet C. King. 1996. "Interindividual Variation in Circulating Zinc Concentrations among Healthy Adult Men and Women." *International Journal of Food Sciences and Nutrition* 47: 393-90.

Hengstman, G. H., K. F. de Laat, B. Jacobs, and B. G. van Engelen. 2009. "Sensorimotor Axonal Polyneuropathy without Hepatic Failure in Erythropoietic Protoporphyria." *Journal of Clinical Neuromuscular Disease* 11(2):72-76.

Herrick, Ariane L., B. Miles Fisher, Michael R. Moore, Sylvia Cathcart, Kenneth E. L. McColl, and Abraham Goldberg. 1990. "Elevation of Blood Lactate and Pyruvate Levels in Acute Intermittent Porphyria – A Reflection of Haem Deficiency?" *Clinica Chimica Acta* 190(3): 157-62.

Ho, Mae-Wan. 1993. *The Rainbow and the Worm: The Physics of Organisms.* Singapore: World Scientific.

———. 1996. "Bioenergetics and Biocommunication." In: R. Cuthbertson, M. Holcombe, and R. Paton, eds., *Computation in Cellular and Molecular Biological Systems* (Singapore: World Scientific), pp. 251-64.

———. 2003. "From 'Molecular Machines' to Coherent Organisms." In: Francesco Musumeci, Larissa S. Brizhik, and Mae-Wan Ho, eds., *Energy and Information Transfer in Biological Systems* (Singapore: World Scientific), pp. 63-81.

———. 2008. *The Rainbow and the Worm: The Physics of Organisms*, 3rd ed. Singapore: World Scientific.

Ho, Mae-Wan, Julian Haffegee, Richard Newton, Yu-Ming Zhou, John S. Bolton, and Stephen Ross. 1996. "Organisms as Polyphasic Liquid Crystals." *Bioelectrochemistry and Bioenergetics* 41: 81-91.

Hoffer, A. and H. Osmond. 1963. "Malvaria: A New Psychiatric Disease." *Acta Psychiatrica Scandinavica* 39: 335-66.

Holtmann, W. and Ch. Xenakis. 1978. "Neurologische und psychiatrische Störungen bei Porphyria cutanea tarda." *Nervenarzt* 49: 282-84.

———. 1979. "Stellungnahme zum Kommentar von C.A. Pierach über die Arbeit von W. Holtman und Ch. Xenakis: 'Neurologische und psychiatrische Störungen bei Porphyria cutanea tarda.'" *Nervenarzt* 50: 542-43.

Hunt, Tam. 2013. "The Rainbow and the Worm: Establishing a New Physics of Life." *Communicative and Integrative Biology* 6(2): e23149.

Huszák, I., Irene Durkó, and K. Karsai. 1972. "Experimental Data to the Pathogenesis of Cryptopyrrole Excretion in Schizophrenia, I." *Acta Physiologica Academiae Scientiarum Hungaricae* 42(1): 79-86.

Ichimura, Shoji. 1960. "The Photoconductivity of Chloroplasts and the Far Red Light Effect." *Biophysical Journal* 1: 99-109.

Irvine, Donald G. and Lennart Wetterberg. 1972. "Kryptopyrrole-like Sybstance in Acute Intermittent Porphyria." *Lancet* 2: 1201.

Jerman, Igor. 1998. "Electromagnetic Origin of Life." *Electro- and Magnetobiology* 17(3): 401-13.

Johnson, Phyllis E., Curtiss D. Hunt, David B. Milne, and Loanne K. Mullen. 1993. "Homeostatic Control of Zinc Metabolism in Men: Zinc Excretion and Balance in Men Fed Diets Low in Zinc." *American Journal of Clinical Nutrition* 57: 557-65.

Katz, E. 1949. "Chlorophyll Fluorescence as an Energy Flowmeter for Photosynthesis." In: James Franck and Walter E. Loomis, eds., *Photosynthesis in Plants* (Ames, IA: Iowa State College Press), pp. 287-92.

Kauppinen, Raili and Pertti Mustajoki. 1988. "Acute Hepatic Porphyria and Hepatocellular Carcinoma." *British Journal of Cancer* 57: 117-20.

Kim, Hooi-Sung, Chun-Ho Kim, Chang-Sik Ha, and Jin-Kook Lee. 2001. "Organic Solar Cell Devices Based on PVK/Porphyrin System." *Synthetic Metals* 117(1-3): 289-91.

King, Janet C., David M. Shames, and Leslie R. Woodhouse. 2000. "Zinc Homeostasis in Humans." *Journal of Nutrition* 130: 1360S-1366S.

Klüver, Heinrich. 1944a. "On Naturally Occurring Porphyrins in the Central Nervous System." *Science* 99: 482-84.

———. 1944b. "Porphyrins, the Nervous System, and Behavior." *Journal of Psychiatry* 17: 209-27.

———. 1967. "Functional Differences between the Occipital and Temporal Lobes." In: Lloyd A. Jeffress, ed., *Cerebral Mechanisms in Behavior – the Hixon Symposium* (New York: Hafner), pp. 147-82.

Kohl, Peter. 2003. "Heterogeneous Cell Coupling in the Heart: An Electrophysiological Role for Fibroblasts." *Circulation Research* 93: 381-83.

Kordač, Václav, Michaela Kozáková, and Pavel Martásek. 1989. "Changes of Myocardial Functions in Acute Hepatic Porphyrias: Role of Heme Arginate Administration." *Annals of Medicine* 21(4): 273-76.

Krijt, Jan, Pavla Stranska, Pavel Maruna, Martin Vokurka, and Jaroslav Sanitrak. 1997. "Herbicide-Induced Experimental Variegate Porphyria in Mice: Tissue Porphyrinogen Accumulation and Response to Porphyrogenic Drugs." *Canadian Journal of Physiology and Pharmacology* 75: 1181-87.

Kuffler, Stephen W. and David D. Potter. 1964. "Glia in the Leech Central Nervous System: Physiological Properties and Neuron-Glia Relationship." *Journal of Neurophysiology* 27: 290-320.

Kulvietis, Vytautas, Eugenijus Zakarevičius, Juozas Lapienis, Gražina Graželienė, Violeta Žalgevičienė, and Ričardas Rotomskis. 2007. "Accumulation of Exogenous Sensitizers in Rat Brain." *Acta Medica Lituanica* 14(3): 219-24.

Labbé, Robert F. 1967. "Metabolic Anomalies in Porphyria: The Result of Impaired Biological Oxidation?" *Lancet* 1: 1361-64.

Lagerwerff, J. V. and A. W. Specht. 1970. "Contamination of Roadside Soil and Vegetation with Cadmium, Nickel, Lead, and Zinc." *Environmental Science and Technology* 4(7): 583-86.

Labbé, Robert F., Hendrik J. Vreman, and David K. Stevenson. 1999. "Zinc Protoporphyrin: A Metabolite with a Mission." *Clinical Chemistry* 45(12): 2060-72.

Laiwah, A. C. Yeung, Abraham Goldberg, and Michael R. Moore. 1983. "Pathogenesis and Treatment of Acute Intermittent Porphyria: Discussion Paper." *Journal of the Royal Society of Medicine* 76: 386-92.

Laiwah, A. C. Yeung, Graeme J. A. Macphee, P. Boyle, Michael R. Moore, and Abraham Goldberg. 1985. "Autonomic Neuropathy in Acute Intermittent Porphyria." *Journal of Neurology, Neurosurgery, and Psychiatry* 48: 1025-30.

Lee, G. Richard. 1993. "Porphyria." In: G. Richard Lee and Maxwell Myer Wintrobe, eds., *Wintrobe's Clinical Hematology*, 9th ed. (Philadelphia: Lea & Febiger), pp. 1272-97..

Lehmann, Otto. 1908. *Flüssige Kristalle und die Theorien des Lebens*. Leipzig: Johann Ambrosius Barth.

Li, Xiaoyan, Shuting Zhang, and Mei Yang. 2014. "Accumulation and Risk Assessment of Heavy Metals in Dust in Main Living Areas of Guiyang City, Southwest China." *Chinese Journal of Geochemistry* 33(3): 272-76.

Libet, Benjamin and Ralph W. Gerard. 1941. "Steady Potential Fields and Neurone Activity." *Journal of Neurophysiology* 4(6): 438-55.

Linet, Martha S., Gloria Gridley, Olof Nyrén, Lene Mellemkjaer, Jørgen H. Olsen, Shannon Keehn, Hans-Olov Admi, and Joseph F. Fraumeni, Jr. 1999. "Primary Liver Cancer, Other Malignancies, and Mortality Risks following Porphyria: A Cohort Study in Denmark and Sweden." *American Journal of Epidemiology* 149(11): 1010-15.

Ling, Gilbert Ning. 1962. *A Physical Theory of the Living State: the Association-Induction Hypothesis*. Waltham, MA: Blaisdell.

———. 1965. "The Physical State of Water in Living Cell and Model Systems." *Annals of the New York Academy of Sciences* 125: 401-17.

———. 1992. *A Revolution in the Physiology of the Living Cell*. Malabar, FL: Krieger.

———. 1994. "The New Cell Physiology." *Physiological Chemistry and Physics and Medical NMR* 26(2): 121-203.

———. 2001. *Life at the Cell and Below-Cell Level: The Hidden History of a Fundamental Revolution in Biology*. Melville, NY: Pacific Press.

Ling, Gilbert Ning, Christopher Miller, and Margaret M. Ochselfeld. 1973. "The Physical State of Solutes and Water in Living Cells According to the Association-Induction Hypothesis." *Annals of the New York Academy of Sciences* 204: 6-50.

Livshits, V. A. and L. A. Blyumenfel'd. 1968. "Semiconductor Properties of Porphyrins." *Journal of Structural Chemistry* 8(3): 383-88.

Lund, Elmer J. 1947. *Bioelectric Fields and Growth*. Austin: University of Texas Press.

Macy, Judy A., John Gilroy, and Jane C. Perrin. 1991. "Hereditary Coproporphyria: An Imitator of Multiple Sclerosis." *Archives of Physical Medicine and Rehabilitation* 72(9): 703-4.

Markovitz, Meyer. 1954. "Acute Intermittent Porphyria: A Report of Five Cases and a Review of the Literature." *Annals of Internal Medicine* 41(6): 1170-88.

Marshall, Clyde and Ralph G. Meader. 1937. "Studies on the Electrical Potentials of Living Organisms: I. Base-lines and Strain Differences in Mice." *Yale Journal of Biology and Medicine* 10(1): 65-78.

————. 1938. "Studies in the Electrical Potentials of Living Organisms: III. Effects of Elevated Body Temperatures in Normal Unanesthetized Mice." *Yale Journal of Biology and Medicine* 11(2): 123-26.

Mason, Verne R., Cyril Courville and Eugene Ziskind. 1933. "The Porphyrins in Human Disease." *Medicine* 12(4): 355-438.

Maxwell, Kate and Giles N. Johnson. 2000. "Chloropyll Fluorescence – a Practical Guide." *Journal of Experimental Botany* 51: 659-68.

McCabe, Donald Lee. 1983. "Kryptopyrroles." *Journal of Orthomolecular Psychiatry* 12(1): 2-18.

McGinnis, Woody R, Tapan Audhya, William J. Walsh, James A. Jackson, John McLaren-Howard, Allen Lewis, Peter H. Lauda, Douglas M. Bibus, Frances Jurnak, Roman Lietha, and Abram Hoffer. 2008a. "Discerning the Mauve Factor, Part 1." *Alternative Therapies* 14(2): 40-50.

————. 2008b. "Discerning the Mauve Factor, Part 2." *Alternative Therapies* 14(3): 50-56.

McLachlan, D. R. Crapper, A. J. Dalton, T. P. A. Kruck, M. Y. Bell, W. L. Smith, W. Kalow, and D. F. Andrews. 1991. "Intramuscular Desferrioxamine in Patients with Alzheimer's Disease." *Lancet* 1: 1304-8.

Meader, Ralph G. and Clyde Marshall. 1938. "Studies on the Electrical Potentials of Living Organisms: II. Effects of Low Temperatures in Normal Unanesthetized Mice." *Yale Journal of Biology and Medicine* 10(4): 365-78.

Mikirova, Nina. 2015. "Clinical Test of Pyrroles: Usefulness and Association with Other Biochemical Markers." *Clinical Medical Reviews and Case Reports* 2: 027.

Milne, David B., Janet R. Mahalko, and Harold H. Sandstead. 1983. "Effect of Dietary Zinc on Whole Body Surface Loss of Zinc: Impact on Estimation of Zinc Retention by Balance Method." *American Journal of Clinical Nutrition* 38: 181-86.

Moore, Michael R. 1990. "The Pathogenesis of Acute Porphyria." *Molecular Aspects of Medicine* 11(1-2): 49-57.

Moore, Michael R., Kenneth E. L. McColl, Claude Rimington, and Abraham Goldberg. 1987. *Disorders of Porphyrin Metabolism*. New York: Plenum.

Morelli, Alessandro, Silvia Ravera, and Isabella Panfoli. 2011. "Hypothesis of an Energetic Function for Myelin." *Cell Biochemistry and Biophysics* 61: 179-87.

Morelli, Alessandro, Silvia Ravera, Daniela Calzia, and Isabella Panfoli. 2012. "Impairment of Heme Synthesis in Myelin as Potential Trigger of Multiple Sclerosis." *Medical Hypotheses* 78: 707-10.

Morton, William E. 1995. "Redefinition of Abnormal Susceptibility to Environmental Chemicals." In: Barry L. Johnson, Charles Xintaras, and John S. Andrews, Jr., eds., *Hazardous Waste Impacts on Human and Ecological Health* (Princeton, NJ: Princeton Scientific), pp. 320-27.

————. 1998. "Chemical-Induced Porphyrinopathy and Its Relation to Multiple Chemical Sensitivity (MCS)." Paper presented at Gordon Research Conference on Chemistry and Biology of Tetrapyrroles, Salve Regina University, Newport, RI, July 13.

————. 2000a. "The Nature of Harderoporphyria?" Paper presented at Gordon Research Conference on the Chemistry and Biology of Tetrapyrroles, Salve Regina University, Newport, RI, July 17.

————. 2000b. "Fecal Porphyrin Measurements are Crucial for Adequate Screening for Porphyrinopathy." *Archives of Dermatology* 136: 554.

————. 2001. "Porphyrinopathy Can Explain Symptoms of Multiple Chemical Sensitivity (MCS)." Paper presented at MCS 2001 Conference, Santa Fe, NM, August 14.

Nazzal, Y., Habes Ghrefat, and Marc A. Rosen. 2014. "Heavy Metal Contamination of Roadside Dusts: A Case Study for Selected Highways of the Greater Toronto Area, Canada Involving Multivariate Geostatistics." *Research Journal of Environmental Sciences* 8(5): 259-73.

Nordenström, Björn E. W. 1983. *Biologically Closed Electric Circuits. Clinical, Experimental and Theoretical Evidence for an Additional Circulatory System*. Stockholm: Nordic Medical.

Northrop, Filmer S. C. and Harold Saxton Burr. 1937. "Experimental Findings Concerning the Electro-dynamic Theory of Life and an Analysis of Their Physical Meaning." *Growth* 1(1): 78-88.

Ovchinnikova, Kate and Gerald H. Pollack. 2009. "Can Water Store Charge?" *Langmuir* 25(1): 542-47.

Painter, Joseph T. and Edwin J. Morrow. 1959. "Porphyria: Its Manifestations and Treatment with Chelating Agents." *Texas State Journal of Medicine* 55(10): 811-18.

Pei, Yinquan, Dayao Zhao, Jianyi Huang, and Longguan Cao. 1983. "Zinc-induced Seizures: A New Experimental Model of Epilepsy." *Epilepsia* 24: 169-76.

Perlroth, Mark G. 1988. "The Porphyrias." In: Edward Rubenstein and Daniel D. Federman, eds., *Scientific American Medicine* (New York: Scientific American), 9V: 1-12.

Peters, Henry A. 1961. "Trace Minerals, Chelating Agents and the Porphyrias." *Federation Proceedings* 20 (3 part 2) (suppl. 10): 227-34.

————. 1993. "Acute Hepatic Porphyria." In: Richard T. Johnson and John W. Griffin, eds., *Current Therapy in Neurologic Disease*, 4th ed. (St. Louis: B. C. Decker), pp. 317-22.

Peters, Henry A., Derek J. Cripps, Ayhan Göcmen, George Bryan, Erdogan Ertürk, and Carl Morris. 1987. "Turkish Epidemic Hexachlorobenzene Porphyria." *Annals of the New York Academy of Sciences* 514: 183-89.

Peters, Henry A., Derek J. Cripps, and Hans H. Reese. 1974. "Porphyria: Theories of Etiology and Treatment." *International Review of Neurobiology* 16: 301-55.

Peters, Henry A., Peter L. Eichman, and Hans H. Reese. 1958. "Therapy of Acute, Chronic and Mixed Hepatic Porphyria Patients with Chelating Agents." *Neurology* 8: 621-32.

Peters, Henry A., Sherwyn Woods, Peter L. Eichman, and Hans H. Reese. 1957. "The Treatment of Acute Porphyria with Chelating Agents: A Report of 21 Cases." *Annals of Internal Medicine* 47(5): 889-99.

Pethig, Ronald. 1979. *Dielectric and Electronic Properties of Biological Materials*. Chichester, UK: John Wiley & Sons.

Petrov, Alexander G. 1999. *The Lyotropic State of Matter: Molecular Physics and Living Matter Physics*. Amsterdam: Gordon & Breach.

Petrova, E. A. and N. P. Kuznetsova. 1972. "The Conditions of the Autonomic Nervous System in Patients with Porphyria Cutanea Tarda." *Vestnik Dermatologii Venerologii* 46: 31-34 (in Russian).

Pfeiffer, Carl Claus. 1975. "Mauve-factor Patients." In: Pfeiffer, *Mental and Elemental Nutrients: A Physician's Guide to Nutrition and Health Care* (New Canaan, CT: Keats), pp. 402-8.

Pierach, Claus A. 1979. "Kommentar zur Arbeit von W. Holtman und Ch. Xenakis: "Neurologische and psychiatrische Störungen bei Porphyria cutanea tarda." *Nervenarzt* 50: 540-1.

Pohl, Herbert A., Peter R. C. Gascoyne, and Albert Szent-Györgyi. 1977. "Electron Spin Resonance Absorption of Tissue Constituents." *Proceedings of the National Academy of Sciences* 74(4): 1558-60.

Pollack, Gerald H. 2001. *Cells, Gels, and the Engines of Life*. Seattle: Ebner & Sons.

———. 2006. "Cells, Gels, and Mechanics." In: Mohammad R. K. Mofrad and Roger D. Kamm, eds., *Cytoskeletal Mechanics* (New York: Cambridge University Press), pp. 129-51.

———. 2010. "Water, Energy and Life: Fresh Views from the Water's Edge." *International Journal of Design & Nature and Ecodynamics* 5(1): 27-29.

———. 2013. *The Fourth Phase of Water: Beyond Solid, Liquid, and Vapor*. Seattle: Ebner & Sons.

Pollack, Gerald H., Xavier Figueroa, and Qing Zhao. 2009. "Molecules, Water, and Radiant Energy: New Clues for the Origin of Life." *International Journal of Molecular Sciences* 10(4): 1419-29.

Popp, Friz Albert, Günther Becker, Herbert L. König, and Walter Peschka, eds. 1979. *Electromagnetic Bio-Information*. München: Urban & Schwarzenberg.

Popp, Fritz Albert, Ulrich Warnke, Herbert L. König, and Walter Peschka, eds. 1989. *Electromagnetic Bio-Information*, 2nd ed. München: Urban & Schwarzenberg.

Popp, Fritz Albert and Lev Beloussov, eds. 2003. *Integrative Biophysics*. Dordrecht: Kluwer.

Que, Emily L., Dylan W. Domaille, and Christopher J. Chang. "Metals in Neurobiology: Probing Their Chemistry and Biology with Molecular Imaging." *Chemical Reviews* 108: 1517-49.

Randolph, Theron G. 1987. *Environmental Medicine – Beginnings and Bibliographies of Clinical Ecology*. Fort Collins, CO: Clinical Ecology Publications.

Ravera, Silvia, Martina Bartolucci, Enrico Adriano, Patrizia Garbati, Sara Ferrando, Paola Ramoino, Daniela Calzia, Alessandro Morelli, Maurizio Balestrino, and Isabella Panfoli. 2015. "Support of Nerve Conduction by Respiring Myelin Sheath: Role of Connexons." *Molecular Neurobiology* [Epub ahead of print].

Ravera, Silvia, Martina Bartolucci, Daniela Calzia, Maria Grazia Aluigi, Paola Ramoino, Alessandro Morelli, and Isabella Panfoli. 2013. "Tricarboxylic Acid Cycle-Sustained Oxidative Phosphorylation in Isolated Myelin Vesicles." *Biochimie* 95: 1991-98.

Ravera, Silvia, Lucilla Nobbio, Davide Visigalli, Martina Bartolucci, Daniela Calzia, Fulvia Fiorese, Gianluigi Mancardi, Angelo Schenone, Alessandro Morelli, and Isabella Panfoli. 2013. "Oxidative Phosphorylation in Sciatic Nerve Myelin and

Its Impairment in a Model of Dysmyelinating Peripheral Neuropathy." *Journal of Neurochemistry* 126: 82-92.

Ravera, Silvia and Isabella Panfoli. 2015. "Role of Myelin Sheat Energy Metabolism in Neurodegenerative Diseases." *Neural Regeneration Research* 10(10): 1570-71.

Ravera, Silvia, Isabella Panfoli, Daniela Calzia, Maria Grazia Aluigi, Paolo Bianchini, Alberto Diaspro, Gianluigi Mancardi, and Alessandro Morelli. 2009. "Evidence for Aerobic ATP Synthesis in Isolated Myelin Vesicles." *International Journal of Biochemistry and Cell Biology* 41: 1581-91.

Ravitz, Leonard J. 1953. "Electrodynamic Field Theory in Psychiatry." *Southern Medical Journal* 46(7): 650-60.

———. 1962. "History, Measurement, and Applicability of Periodic Changes in the Electromagnetic Field in Health and Disease." *Annals of the New York Academy of Sciences* 98: 1144-1201.

Reboul, J., H. B. Friedgood, and H. Davis. 1937. "Electrical Detection of Ovulation." *American Journal of Physiology* 119: 387.

Regland, B., W. Lehmann, I. Abedini, K. Blennow, M. Jonsson, I. Karlsson, M. Sjögren, A. Wallin, M. Xilinas, and C.-G. Gottfries. 2001. "Treatment of Alzheimer's Disease with Clioquinol." *Dementia and Geriatric Cognitive Disorders* 12(6): 408-14.

Religa, D., D. Strozyk, Robert A. Cherny, Irene Volitakis, V. Haroutunian, B. Winblad, J. Naslund, and Ashley I. Bush. 2006. "Elevated Cortical Zinc in Alzheimer Disease." *Neurology* 67: 69-75.

Riccio, P., S. Giovannelli, A. Bobba, E. Romito, A. Fasano, T. Bleve-Zacheo, R. Favilla, E. Quagliarello, and P. Cavatorta. 1995. "Specificity of Zinc Binding to Myelin Basic Protein." *Neurochemical Research* 20(9): 1107-13.

Ridley, Alan. 1969. "The Neuropathy of Acute Intermittent Porphyria." *Quarterly Journal of Medicine* 38: 307-33.

———. 1975. "Porphyric Neuropathy." In: Peter James Dyck, P. K. Thomas, and Edward H. Lambert, eds., *Peripheral Neuropathy* (Philadelphia: W. B. Saunders), pp. 942-55.

Ritchie, Craig W., Ashley I. Bush, Andrew Mackinnon, Steve Macfarlane, Maree Mastwyk, Lachlan MacGregor, Lyn Kiers, Robert Cherny, Qiao-Xin Li, Amanda Tammer, Darryl Carrington, Christine Mavros, Irene Volitakis, Michel Xilinas, David Ames, Stephen Davis, Konrad Beyreuther, Rudolph E. Tanzi, and Colin L. Masters. 2003. "Metal-Protein Attenuation with Iodochlorhydroxyquin (Clioquinol) Targeting Aβ Amyloid Deposition and Toxicity in Alzheimer Disease." *Archives of Neurology* 60: 1685-91.

Rivera, Hiram, J. Kent Pollock, and Herbert A. Pohl. 1985. "The AC Field Patterns About Living Cells." *Cell Biophysics* 7: 43-55.

Rock, John, Jean Reboul, and Harold C. Wiggers. 1937. "The Detection and Measurement of the Electrical Concomitant of Human Ovulation by Use of the Vacuum-Tube Potentiometer." *New England Journal of Medicine* 217(17): 654-58.

Roman, W. 1969. "Zinc in Porphyria." *American Journal of Clinical Nutrition* 22(10): 1290-1303.

Rook, Arthur and Robert H. Champion. 1960. "Porphyria Cutanea Tarda and Diabetes." *British Medical Journal* 1: 860-61.

Rose, Florence C. and Sylvan Meryl Rose. 1965. "The Role of Normal Epidermis in Recovery of Regenerative Ability in Xrayed Limbs of *Triturus.*" *Growth* 29: 361-93.

Rose, Sylvan Meryl. 1970. *Regeneration.* New York: Appleton-Century-Crofts.

———. 1978. "Regeneration in Denervated Limbs of Salamanders After Induction by Applied Direct Currents." *Bioelectrochemistry and Bioenergetics* 5: 88-96.

Rose, Sylvan Meryl and Florence C. Rose. 1974. "Electrical Studies on Normally Regenerating, on X-Rayed, and on Denervated Limb Stumps of *Triturus.*" *Growth* 38: 363-80.

Ross, Stephen, Richard Newton, Yu-Ming Zhou, Julian Haffegee, Mae-Wan Ho, John P. Bolton, and David Knight. 1997. "Quantitative Image Analysis of Birefringent Biological Material." *Journal of Microscopy* 187(1): 62-67.

Runge, Walter and Cecil J. Watson. 1962. "Experimental Production of Skin Lesions in Human Cutaneous Porphyria." *Proceedings of the Society for Experimental Biology and Medicine* 109: 809-11.

Saint, Eric G., D. Curnow, R. Paton, and John B. Stokes. 1954. "Diagnosis of Acute Porphyria." *British Medical Journal* 1: 1182-84.

Sedlak, Włodzimierz. 1970. "Biofizyczne aspekty ekologii" ("Biophysical Aspects of Ecology"). *Wiadomości Ekologiczne* 16(1): 43-53.

———. 1973. "Ochrona środowiska człowieka w zakresie niejonizującego promieniowania." *Wiadomości Ekologiczne* 19(3): 223-37.

———. 1979. *Bioelektronika: 1967-1977.* Warsaw: PAX.

———. 1980. *Bioelektronika – Środowisko – Człowiek* ("Bioelectronics – Environment – Man"). Wrocław: Zakład Narodowy Imienia Ossolińskich.

———. 1984. *Postępy fizyki życia* ("Progress in the Physics of Life") Warsaw: PAX.

Silbergeld, Ellen K. and Bruce A. Fowler, eds. 1987. *Mechanisms of Chemical-Induced Porphyrinopathies.* Vol. 514 of *Annals of the New York Academy of Sciences.*

Soldán, M. Mateo Paz and Istvan Pirko. 2012. "Biogenesis and Significance of Central Nervous System Myelin." *Seminars in Neurology* 32(1): 9-14.

Solomon, Harvey M. and Frank H. J. Figge. 1958. "Occurrence of Porphyrins in Peripheral Nerves." *Proceedings of the Society for Experimental Biology and Medicine* 97: 329-30.

Stein, Jeffrey A. and Donald P. Tschudy. 1970. "Acute Intermittent Porphyria: A Clinical and Biochemical Study of 46 Patients." *Medicine* 49(1): 1-16.

Sterling, Kenneth, Marvin Silver, and Henry T. Ricketts. 1949. "Development of Porphyria in Diabetes Mellitus." *Archives of Internal Medicine* 84: 965-75.

Szent-Györgyi, Albert. 1941. "Towards a New Biochemistry." *Science* 93: 609-11.

———. 1957. *Bioenergetics.* New York: Academic.

———. 1960. *Introduction to a Submolecular Biology.* New York: Academic.

———. 1968. *Bioelectronics: A Study in Cellular Regulations, Defense, and Cancer.* New York: Academic.

———. 1969. "Molecules, Electrons and Biology." *Transactions of the New York Academy of Sciences,* 2nd ser., 31(4): 334-40.

———. 1971. "Biology and Pathology of Water." *Perspectives in Biology and Medicine* 14(2): 239-49.

———. 1972. *The Living State: With Observations on Cancer.* New York: Academic.

———. 1976. *Electronic Biology and Cancer.* New York: Marcel Dekker.

————. 1977. "The Living State and Cancer." *Proceedings of the National Academy of Sciences* 74(7): 2844-47.

————. 1978. *The Living State and Cancer.* New York: Marcel Dekker.

————. 1980a. "The Living State and Cancer." *International Journal of Quantum Chemistry* 18(S7): 217-22.

————. 1980b. "The Living State and Cancer." *Physiological Chemistry and Physics* 12: 99-110.

Tamrakar, Chirika Shova and Pawan Raj Shakya. 2011. "Assessment of Heavy Metals in Street Dust in Kathmandu Metropolitan City and Their Possible Impacts on the Environment." *Pakistani Journal of Analytical and Environmental Chemistry* 12(1-2): 32-41.

Taylor, Caroline M., Jeffrey R. Bacon, Peter J. Aggett, and Ian Bremner. 1991. "Homeostatic Regulation of Zinc Absorption and Endogenous Losses in Zinc-deprived Men." *American Journal of Clinical Nutrition* 53(3): 755-63.

Tefferi, Ayalew, Laurence A. Solberg, Jr., and Ralph D. Ellefson. 1994. "Porphyrias: Clinical Evaluation and Interpretation of Laboratory Tests." *Mayo Clinic Proceedings* 69: 289-90.

Tefferi, Ayalew, Joseph P. Colgan, and Laurence A. Solberg, Jr.. 1994. "Acute Porphyrias: Diagnosis and Management." *Mayo Clinic Proceedings* 69: 991-95.

Terzuolo, Carlo A. and Theodore H. Bullock. 1956. "Measurement of Imposed Voltage Gradient Adequate to Modulate Neuronal Firing." *Proceedings of the National Academy of Sciences* 42(9): 687-94.

Todd, Tweedy John. 1823. "On the Process of Regeneration of the Members of the Aquatic Salamander." *Quarterly Journal of Science, Literature and the Arts* 16: 84-96.

Trampusch, H. A. L. 1964. "Nerves as Morphogenetic Mediators in Regeneration." *Progress in Brain Research* 13: 214-27.

Vacher, Monique, Claude Nicot, Mollie Pflumm, Jeremy Luchins, Sherman Beychok, and Marcel Waks. 1984. "A Heme Binding Site on Myelin Basic Protein: Characterization, Location, and Significance." *Archives of Biochemistry and Biophysics* 231(1): 86-94.

Vass, Imre. 2003. "The History of Photosynthetic Thermoluminescence." *Photosynthesis Research* 76: 303-18.

Vernon, Leo P. and Gilbert R. Seely, eds. 1966. *The Chlorophylls.* New York: Academic.

Vgontzas, Alexandros N., Joyce D. Kales, James O. Ballard, Antonio Vela-Bueno, and Tjiauw-Ling Tan. 1993. "Porphyria and Panic Disorder with Agoraphobia." *Psychosomatics* 34(5): 440-43.

Virchow, Rudolf Ludwig Carl. 1854. "Ueber das ausgebreitete Vorkommen einer dem Nervenmark analogen Substanz in den thierischen Geweben." *Archiv für pathologische Anatomie und Physiologie und für klinische Medicin* 6: 562-72.

Voyatzoglou, Vassilis, Theodore Mountokalakis, Vassiliki Tsata-Voyatzoglou, Anton Koutselinis, and Gregory Skalkeas. 1982. "Serum Zinc Levels and Urinary Zinc Excretion in Patients with Bronchogenic Carcinoma." *American Journal of Surgery* 144(3): 355-58.

Waldenström, Jan. 1937. "Studien über Porphyrie." *Acta Medica Scandinavica. Supplementum,* vol. 82.

———. 1957. "The Porphyrias as Inborn Errors of Metabolism." *American Journal of Medicine* 22(5): 758-72.

Walker, Franklin D. and Walther J. Hild. 1969. "Neuroglia Electrically Coupled to Neurons." *Science* 165: 602-3.

Watson, Cecil James and Evrel A. Larson. 1947. "The Urinary Coproporphyrins in Health and Disease." *Physiological Reviews* 27(3): 478-510.

Waxman, Alan D., Don S. Schalch, William D. Odell, and Donald P. Tschudy. 1967. "Abnormalities of Carbohydrate Metabolism in Acute Intermittent Porphyria." *Journal of Clinical Investigation* 46 (part 1): 1129. Abstract.

Wei, Ling Y. 1966. "A New Theory of Nerve Conduction." *IEEE Spectrum* 3(9): 123-27.

Whetsell, William O., Jr., Shigeru Sassa, and Attallah Kappas. 1984. "Porphyrin-Heme Biosynthesis in Organotypic Cultures of Mouse Dorsal Root Ganglia: Effects of Heme and Lead on Porphyrin Synthesis and Peripheral Myelin." *Journal of Clinical Investigation* 74: 600-7.

With, Torben K. 1980. "A Short History of Porphyrins and the Porphyrias." *International Journal of Biochemistry* 11: 189-200.

Wnuk, Marian. 1987. *Rola układów porfirynowych w ewolucji życia* ("The Role of Porphyrin Systems in the Evolution of Life"). Warsaw: Akademia Teologii Katolickiej (in Polish with English summary).

———. 1996. *Istota procesów życiowych w świetle koncepcji elektromagnetycznej natury życia: Bioelektromagnetyczny model katalizy enzymatycznej wobec problematyki biosystemogenezy* ("The Essence of Life Processes in Light of the Concept of the Electromagnetic Nature of Life: Bioelectromagnetic Model of Enzyme Catalysis in View of the Problems of the Origin of Biosystems"). Lublin: John Paul II Catholic University of Lublin.

———. 2001. "The Electromagnetic Nature of Life – The Contribution of W. Sedlak to the Understanding of the Essence of Life." *Frontier Perspectives* 10(1): 32-35.

Wong, J. W. C. 1996. "Heavy Metal Contents in Vegetables and Market Garden Soils in Hong Kong." *Environmental Technology* 17(4): 407-14.

Wong, J. W. C. and N. K. Mak. 1997. "Heavy Metal Pollution in Children Playgrounds in Hong Kong and Its Health Implications." *Environmental Technology* 18(1): 109-15.

Xu, Jiancheng, Qi Zhou, Gilbert Liu, Yi Tan, and Lu Cai. 2013. "Analysis of Serum and Urinal Copper and Zinc in Chinese Northeast Population with the Prediabetes or Diabetes with and without Complications." *Oxidative Medicine and Cellular Longevity*, article ID 635214.

Yntema, Chester L. 1959. "Regeneration in Sparsely Innervated and Aneurogenic Forelimbs of Amblystoma Larvae." *Journal of Experimental Zoology* 140(1): 101-24.

Yokoyama, M., J. Koh, and D. W. Choi. 1986. "Brief Exposure to Zinc is Toxic to Cortical Neurons." *Neuroscience Letters* 71: 351-55.

York, J. Lyndal. 1972. *The Porphyrias*. Springfield, IL: Charles C. Thomas.

Zhou, Xiaoli. 2009. "Synthesis and Characterization of Novel Discotic Liquid Crystal Porphyrins for Organic Photovoltaics." Ph.D. dissertation, Kent State University, Kent, OH.

Zon, Józef Roman. 1976. "Wpływ naturalnego środowiska elektromagnetycznego na człowieka" ("The Effect of the Natural Electromagnetic Environment on Man"). *Roczniki Filozoficzne* 23(3): 89-100.

———. 1979. "Physical Plasma in Biological Solids: A Possible Mechanism for Resonant Interactions between Low Intensity Microwaves and Biological Systems." *Physiological Chemistry and Physics* 11: 501-6.

———. 1980. "The Living Cell as a Plasma Physical System." *Physiological Chemistry and Physics* 12: 357-64.

———. 1983. "Electronic Conductivity in Biological Membranes". *Roczniki Filozoficzne* 31(3): 165-183.

———. 1986a. "Bioelectronics: A Background Area for Biomicroelectronics in the Sciences of Bioelectricity." *Roczniki Filozoficzne* 34(3): 183-201.

———. 1986b. *Plazma elektronowa w błonach biologicznych* ("Electronic Plasma in Biological Membranes"). Lublin: Catholic University of Lublin.

———. 1994. "Bioelektromagnetyka i etyka: Niektóre kwestie moralne związane ze skażeniem elektromagnetycznym środowiska" ("Bioelectromagnetic and Ethics: Some Moral Questions Related to the Electromagnetic Pollution of the Environment"). *Ethos* 7(1-2): 135-50.

———. 2000. "Bioplazma i plazma fizyczna w układach żywych: Studium przyrodnicze i filozoficzne." ("Bioplasma and Physical Plasma in Living Systems: A Study in Science and Philosophy"). Lublin: Catholic University of Lublin.

Zon, Józef Roman and H. Ti Tien. "Electronic Properties of Natural and Modeled Bilayer Membranes." In: Andrew A. Marino, ed., *Modern Bioelectricity* (New York: Marcel Dekker), pp. 181-241.

Zs.-Nagy, Imre. 1995. "Semiconduction of Proteins as an Attribute of the Living State: The Ideas of Albert Szent-Györgyi Revisited in Light of the Recent Knowledge Regarding Oxygen Free Radicals." *Experimental Gerontology* 30(3/4): 327-35.

———. 2001. "On the True Role of Oxygen Free Radicals in the Living State, Aging, and Degenerative Disorders." *Annals of the New York Academy of Sciences* 928: 187-99.

Sulfonal

Bresslauer, Hermann. 1891. "Ueber die schädlichen und toxischen Wirkungen des Sulfonal." *Wiener medizinischer Blätter* 14: 3-4, 19-20.

Erbslöh, W. 1903. "Zur Pathologie und pathologischen Anatomie der toxischen Polyneuritis nach Sulfonalgebrauch." *Zeitschrift für Nervenheilkunde* 23: 197-204.

Fehr, Johann Heinrich Maria Christian. 1891. "Et Par Tilfælde af Sulfonalforgiftning." *Hospitals-Tidende*, 3rd ser., 9: 1121-38.

Geill, Christian. 1891. "Sulfonal og Sulfonalforgiftning." *Hospitals-Tidende*, 3rd ser., 9: 797-812, 821-35.

Hammond, Græme M. 1891. "Sulfonal in Affections of the Nervous System." *Journal of Nervous and Mental Disease*, new ser., 16: 440-42.

Hay, C. M. 1889. "A Clinical Study of Paraldehyde and Sulphonal." *American Journal of the Medical Sciences*, new ser., 98: 34-43.

Ireland, W. W. 1889. "Marandon de Montyel and Others on the Dangers of Sulfonal." *London Medical Recorder* 2: 499-500.

Leech, D. J. 1888. "Sulfonal." *Medical Chronicle* 9: 146-50.

Marandon de Montyel, E. 1889. "Recherches cliniques sur le sulfonal chez les aliénés."
    La France Médicale 36: 1566-70, 1577-82, 1589-93, 1602-8, 1613-17.
Matthes, M. 1888. "Beitrag zur hypnotischen Wirkung des Sulfonals." *Centralblatt für
    Klinische Medicin* 9(40): 723-27.
Morel, Jules. 1893. "Accidents produits par le sulfonal." *Bulletin de la Société de Médecine
    Mentale de Belgique* 68: 120-23.
Revue des Sciences Médicales. 1889. "Thérapeutique." 34: 502-3.
Rexford, C. M. 1889. "Some Experiences with Sulfonal." *The Medical Record* 35(13):
    348.

Chapter 11

Abbate, Mara, Giovanni Tinè, and Luigi Zanforlin. 1996. "Evaluation of Pulsed
    Microwave Influence on Isolated Hearts." *IEEE Transactions on Microwave Theory
    and Techniques* MTT-44(10): 1935-41.
Adams, Ronald L. and R. A. Williams. 1976. *Biological Effects of Electromagnetic Radi-
    ation (Radiowaves and Microwaves) – Eurasian Communist Countries (U).* Defense
    Intelligence Agency, DST-1810S-074-76.
Afrikanova, Lena Andreevna and Yury Grigorievich Grigoriev. 1996. "Vliyanie elek-
    tromagnitnogo izlucheniya razlichnykh rezhimov na serdechnuyu deyatel'nost' (v
    ekcperimente)" ("Effects of various regimes of electromagnetic radiation on car-
    diac activity (by experiment)"). *Radiatsionnaya biologiya. Radioekologiya* 36(5): 691-99.
Ammari, Mohamed, Anthony Lecomte, Mohsen Sakly, Hafedh Abdelmelek, and René
    de Sèze. 2008. "Exposure to GSM 900 MHz Electromagnetic Fields Affects Cere-
    bral Cytochrome C Oxidase Activity." *Toxicology* 250(1): 70-74.
Appleby, Paul N., Margaret Thorogood, Jim I. Mann, and Timothy J. A. Key. 1999.
    "The Oxford Vegetarian Study: an Overview." *American Journal of Clinical Nutrition*
    70(3): 525S-531S.
Arora, Sameer, George A. Stouffer, Anna M. Kucharska-Newton, Arman Qamar,
    Muthiah Vaduganathan, Ambarish Pandey, Deborah Porterfield, Ron Blankstein,
    Wayne D. Rosamond, Deepak L. Bhatt, and Melissa C. Caughey. 2019. "Twenty
    Year Trends and Sex Differences in Young Adults Hospitalized With Acute Myo-
    cardial Infarction: The ARIC Community Surveillance Study." *Circulation* 139:
    1047-56.
Aschenheim, Erich. 1915. "Über Störungen der Herztätigkeit." *Münchener mediz-
    inische Wochenschrift* 62(20): 692-93.
Aubertin, Charles. 1916. "La récupération des faux cardiaques." *Presse médicale* 24:
    92-93.
Bachurin, V. I. 1979. "Influence of Small Doses of Electromagnetic Waves on Some
    Human Organs and Systems." *Vrachebnoye Delo* 1979(7): 95-97. JPRS 75515 (1980),
    pp. 36-39.
Bajwa, Waheed K., Gregory M. Asnis, William C. Sanderson, Ahman Irfan, and Her-
    man M. van Praag. 1992. "High Cholesterol Levels in Patients with Panic Disor-
    der." *American Journal of Psychiatry* 149(3): 376-78.
Barański, Stanisław and Przemysław Czerski. 1976. "Health Status of Personnel Occu-
    pationally Exposed to Microwaves, Symptoms of Microwave Overexposure." In:

Barański and Czerski, *Biological Effects of Microwaves* (Stroudsburg, PA: Dowden, Hutchinson & Ross), pp. 153-69.

Barlow, David H. 2002. *Anxiety and its Disorders*, 2nd ed. New York: Guilford.

Barron, Charles I., Andrew A. Love, and Albert A. Baraff. 1955. "Physical Evaluation of Personnel Exposed to Microwave Emanations." *Journal of Aviation Medicine* 26(6): 442-52.

Bates, David W., Dedra Buchwald, Joshua Lee, Phalla Kith, Teresa Doolittle, Cynthia Rutherford, W. Hallowell Churchill, Peter H. Schur, Mark Wener, Donald Wybenga, James Winkelman, and Anthony L. Komaroff. 1995. "Clinical Laboratory Test Findings in Patients with Chronic Fatigue Syndrome." *Archives of Internal Medicine* 155(1): 97-103.

Beall, Robert T. 1940. "Rural Electrification." In: Gove Hambidge, ed., *Farmers in a Changing World* (Washington, DC: U.S. Department of Agriculture), pp. 790-809.

Beattie, A. D., Michael R. Moore, Abraham Goldberg, and R. L. Ward. 1973. "Acute Intermittent Porphyria: Response of Tachycardia and Hypertension to Propranolol." *British Medical Journal* 3: 257-60.

Behan, W. M. H., I. A. R. More, and P. O. Behan. 1991. "Mitochondrial Abnormalitieis in the Postviral Fatigue Syndrome." *Acta Neuropathologica* 83: 61-65.

Beitman, Bernard D., Imad Basha, Greg Flaker, Lori DeRosear, Vaskar Mukerji, and Joseph Lamberti. 1987. "Non-Fearful Panic Disorder: Panic Attacks without Fear." *Behaviour Research and Therapy* 25(6): 487-92.

Blank, Martin and Reba Goodman. 2009. "Electromagnetic Fields Stress Living Cells." *Pathophysiology* 16(2-3): 71-78.

Blom, Dirk. 2011. "Secondary Dyslipidaemia." *South African Family Practice* 53(4): 317-23.

Blom, Gaston E. 1951. "A Review of Electrocardiographic Changes in Emotional States." *Journal of Nervous and Mental Disease* 113(4): 283-300.

Bonkowsky, Herbert L., Donald P. Tschudy, Eugene C. Weinbach, Paul S. Ebert, and Joyce M. Doherty. 1975. "Porphyrin Synthesis and Mitochondrial Respiration in Acute Intermittent Porphyria: Studies Using Cultured Human Fibroblasts." *Journal of Laboratory and Clinical Medicine* 85(1): 93-102.

Bortkiewicz, A., M. Zmyslony, E. Gadzicka, and W. Szymczak. 1996. "Evaluation of Selected Parameters of Circulatory System Function in Various Occupational Groups Exposed to High Frequency Electromagnetic Fields. II. Electrocardiographic Changes." *Medycyna Pracy* 47(3): 241-52 (in Polish).

Bowen, Rudy Cecil, Ambikaipakan Senthilselvan, and Anthony Barale. 2000. "Physical Illness as an Outcome of Chronic Anxiety Disorders." *Canadian Journal of Psychiatry* 45(5): 459-64.

Bowlby, Anthony A., Howard H. Tooth, Cuthbert Wallace, John E. Calverley, and Surgeon-Major Kilkelly. 1901. *A Civilian War Hospital: Being an Account of the Work of the Portland Hospital, and of Experience of Wounds and Sickness in South Africa, 1900.* London: John Murray. Pages 128-29 on neurasthenia.

Brasch, Dr. 1915. "Herzneurosen mit Hauthyperästhesie." *Münchener medizinische Wochenschrift* 62(20): 693-95.

Braun, Ludwig. 1915. "Ueber die Konstatierung bie Herzkranken." *Wiener klinische Wochenschrift* 28(46): 1249-51.

Brodeur, Paul. 1977. *The Zapping of America*. New York: W. W. Norton.

Brown, Louis. 1999. *A Radar History of World War II*. Bristol, UK: Institute of Physics.

Burr, Michael L. and Peter M. Sweetnam. 1982. "Vegetarianism, Dietary Fiber, and Mortality." *American Journal of Clinical Nutrition* 36(5): 873-77.

Canadian Medical Association Journal. 1916. "Soldier's Heart and the Hampstead Hospital." 6(7): 613-18.

Caruthers, B. M., M. I. van de Sande, K. L. De Meirleir, N. G. Klimas, G. Broderick, T. Mitchell, D. Staines, A. C. P. Powles, N. Speight, R. Vallings, L. Bateman, B. Baumgarten-Austrheim, D. S. Bell, N. Carlo-Stella, J. Chia, A. Darragh, D. Jo, D. Lewis, A. R. Light, S. Marshal-Gradisbik, I. Mena, J. A. Mikovits, K. Miwa, M. Murovska, M. L. Pall, and S. Stevens. 2011. "Myalgic Encephalomyelitis: International Consensus Criteria." *Journal of Internal Medicine* 270(4): 327-38.

Chadha, S. L., N. Gopinath, and S. Shekhawat. 1997. "Urban-Rural Differences in the Prevalence of Coronary Heart Disease and Its Risk Factors in Delhi." *Bulletin of the World Health Organization* 75(1): 31-38.

Chapman, William P., Mandel E. Cohen, and Stanley Cobb. 1946. "Measurements Related to Pain in Neurocirculatory Asthenia, Anxiety Neurosis, or Effort Syndrome: Levels of Heat Stimulus Perceived as Painful and Producing Wince and Withdrawal Reactions." *Journal of Clinical Investigation* 25: 890-96.

Chernysheva, O. N. and F. A. Kolodub. 1976. "Effect of a Variable Magnetic Field of Industrial Frequency (50 Hz) on Metabolic Processes in the Organs of Rats." *Gigiyena truda i professional'nyye zabolevaniya* 1975(11): 20-23. In: *Effects of Non-Ionizing Electromagnetic Radiation*, JPRS L/5615, February 10, 1976, pp. 33-37.

Chin, Kazuo, Kouichi Shimizu, Takaya Nakamura, Noboru Narai, Hiroaki Masuzaki, Yoshihiro Ogawa, Michiaki Mishima, Takashi Nakamura, Kazuwa Nakao, and Motoharu Ohi. 1999. "Changes in Intra-Abdominal Visceral Fat and Serum Leptin Levels in Patients with Obstructve Sleep Apnea Syndrome Following Nasal Continuous Positive Airway Pressure Therapy." *Circulation* 100: 706-12.

Cleary, Stephen F., ed. 1970. *Biological Effects and Health Implications of Microwave Radiation. Symposium Proceedings*, Richmond, Virginia, September 17-19, 1969. Rockville, MD: U.S. Department of Health, Education and Welfare. Publication BRH/DBE 70-2.

Cobb, Stanley, Mandel E. Cohen, and Daniel W. Badal. 1946. "Capillaries of the Nail Fold in Patients with Neurocirculatory Asthenia (Effort Syndrome, Anxiety Neurosis)." *Archives of Neurology and Psychiatry* 56: 643-50.

Cohen, Anne Hamlen, ed. 2003. "In Memoriam – Mandel E. Cohen, M.D. (March 8, 1907 – November 19, 2000)." *Annals of Clinical Psychiatry* 15(3/4): 149-59.

Cohen, Mandel Ettelson. 1949. "Neurocirculatory Asthenia (Anxiety Neurosis, Neurasthenia, Effort Syndrome, Cardiac Neurosis." *Medical Clinics of North America* 33(9): 1343-64.

Cohen, Mandel E., Daniel W. Badal, Alice Kilpatrick, Eleanor W. Reed, and Paul D. White. 1951. "The High Familial Prevalence of Neurocirculatory Asthenia (Anxiety Neurosis, Effort Syndrome)." *American Journal of Human Genetics* 3: 126-58.

Cohen, Mandel E., Frank Consolazio, and Robert E. Johnson. 1947. "Blood Lactate Response during Modern Exercise in Neurocirculatory Asthenia, Anxiety Neurosis, or Effort Syndrome." *Journal of Clinical Investigation* 26: 339-42.

Cohen, Mandel E., Robert E. Johnson, William P. Chapman, Daniel W. Badal, Stanley Cobb, and Paul D. White. 1946. *A Study of Neurocirculatory Asthenia, Anxiety Neurosis, Effort Syndrome*. Final Report. Contract OEM-cmr 157. Committee on Medical Research of the Office of Scientific Research and Development.

Cohen, Mandel E., Robert E. Johnson, Stanley Cobb, William P. Chapman, and Paul D. White. 1948. "Studies of Work and Discomfort in Patients with Neurocirculatory Asthenia." *Journal of Clinical Investigation* 27: 934. Abstract.

Cohen, Mandel E., Robert E. Johnson, Frank Consolazio, and Paul D. White. 1946. "Low Oxygen Consumption and Low Ventilatory Efficiency during Exhausting Work in Patients with Neurocirculatory Asthenia, Effort Syndrome, Anxiety Neurosis." *Journal of Clinical Investigation* 25: 920. Abstract.

Cohen, Mandel E. and Paul D. White. 1947. "Studies of Breathing, Pulmonary Ventilaton and Subjective Awareness of Shortness of Breath (Dyspnea) in Neurocirculatory Asthenia, Effort Syndrome, Anxiety Neurosis." *Journal of Clinical Investigation* 26: 520-29.

———. 1951. "Life Situations, Emotions, and Neurocirculatory Asthenia (Anxiety Neurosis, Neurasthenia, Effort Syndrome)." *Psychosomatic Medicine* 13(6): 335-57.

———. 1972. "Neurocirculatory Asthenia: 1972 Concept." *Military Medicine* 137: 142-44.

Cohen, Mandel E., Paul D. White, and Robert E. Johnson. 1948. "Neurocirculatory Asthenia, Anxiety Neurosis or the Effort Syndrome." *Archives of Internal Medicine* 81(3): 260-81.

Cohn, Alfred E. 1919. "The Cardiac Phase of the War Neuroses." *American Journal of the Medical Sciences* 158(4): 453-70.

Conner, Lewis A. 1919. "Cardiac Diagnosis in the Light of Experiences with Army Physical Examinations." *American Journal of the Medical Sciences* 158(6): 773-82.

Corcoran, A. P. 1917. "Wireless in the Trenches." *Popular Science Monthly* 90: 795-99.

Coryell, William, Russell Noyes, and John Clancy. 1982. "Excess Mortality in Panic Disorder." *Archives of General Psychiatry* 39: 701-3.

Coryell, William, Russell Noyes, and J. Daniel House. 1986. "Mortality Among Outpatients with Anxiety Disorders." *American Journal of Psychiatry* 143(4): 508-10.

Coryell, William. 1988. "Panic Disorder and Mortality." *Psychiatric Clinics of North America* 11(2): 433-40.

Cotton, Thomas F., D. L. Rapport, and Thomas Lewis. 1917. "After Effects of Exercise on Pulse Rate and Systolic Blood Pressure in Cases of 'Irritable Heart.'" *Heart* 6: 269-84.

Coughlin, Steven R., Lynn Mawdsley, Julie A. Mugarza, Peter M. A. Calverley, and John P. H. Wilding. 2004. "Obstructuve Sleep Apnoea is Independently Associated with an Increased Prevalence of Metabolic Syndrome." *European Heart Journal* 25: 735-41.

Cowdry, Edmund V. 1933. *Arteriosclerosis: A Survey of the Problem*. New York: Macmillan.

Craig, Henry R. and Paul D. White. 1934. "Etiology and Symptoms of Neurocirculatory Asthenia." *Archives of Internal Medicine* 53(5): 633-48.

Crimlisk, Helen L. 1997. "The Little Imitator – Porphyria: A Neuropsychiatric Disorder." *Journal of Neurology, Neurosurgery, and Psychiatry* 62: 319-28.

Csaba, B. M. 2006. "Anxiety as an Independent Cardiovascular Risk." *Neuropsycophar-macologia Hungarica* 8(1): 5-11 (in Hungarian).

Çuhadaroğlu, Çağlar, Ayfer Utkusavaş, Levent Öztürk, Serpil Salman, and Turhan Ece. 2009. "Effects of Nasal CPAP Treatment on Insulin Resistance, Lipid Profile, and Plasma Leptin in Sleep Apnea." *Lung* 187: 75-81.

Cutler, David M. and Elizabeth Richardson. 1997. "Measuring the Health of the U.S. Population." *Brookings Papers on Economic Activity* 28: 217-82.

Czerski, Przemysław, Kazimierz Ostrowski, Morris L. Shore, Charlotte Silverman, Michael J. Suess, and Berndt Waldeskog, eds. 1974. *Biologic Effects and Health Hazards of Microwave Radiation: Proceedings of an International Symposium, Warsaw, 15-18 October 1973*. Warsaw: Polish Medical Publishers.

Da Costa, Jacob Mendes. 1871. "On Irritable Heart: a Clinical Study of a Form of Functional Cardiac Disorder and its Consequences." *American Journal of the Medical Sciences*, new ser., 61: 17-52.

Daily, L. Eugene. 1943. "A. Clinical Study of the Results of Exposure of Laboratory Personnel to Radio and High Frequency Radar." *U.S. Naval Medical Bulletin* 41(4): 1052-56.

Dawber, Thomas R., Felix E. Moore, and George V. Mann. 1957. "Coronary Heart Disease in the Framingham Study." *American Journal of Public Health* 47 (4 part 2): 4-24.

Devoto, L. 1915. "Il cuore stanco nei militari poco alienati." *Il Lavoro* 8: 138-47.

Dodge, Christopher H. 1970. "Clinical and Hygienic Aspects of Exposure to Electromagnetic Fields (A Review of Soviet and Eatern European Literature)." In: Stephen F. Cleary, ed., *Biological Effects and Health Implications of Microwave Radiation. Symposium Proceedings* (Rockville, MD: U.S. Department of Health, Education and Welfare), Publication BRH/DBE 70-2, pp. 140-49.

Dorkova, Zuzana, Darina Petrasova, Angela Molcanyiova, Marcela Popovnakova, and Ruzena Tkacova. 2008. "Effects of Continuous Positive Airway Pressure on Cardiovascular Risk Profile in Patients with Severe Obstructuve Sleep Apnea and Metabolic Syndrome." *Chest* 134(4): 686-92.

Doyle, Joseph T., A. Sandra Heslin, Herman E. Hilleboe, Paul F. Formel, and Robert F. Korns. 1957. "A Prospective Study of Degenerative Cardiovascular Disease in Albany: Report of Three Years' Experience – 1. Ischemic Heart Disease." *American Journal of Public Health* 47(4 part 2): 25-32.

Drager, Luciano F., Jonathan Jun, and Vsevolod Y. Polotsky. 2010. "Obstructive Sleep Apnea and Dyslipidemia: Implications for Atherosclerosis." *Current Opinion in Endocrinology , Diabetes and Obesity* 17(2): 161-65.

Drogichina, E. A. 1960. "The Clinic of Chronic UHF Influence on the Human Organism." In: A. A. Letavet and Z. V. Gordon, eds., *The Biological Action of Ultra-high Frequencies* (Moscow: Academy of Medical Sciences), JPRS 12471, pp. 22-24.

Drury, Alan N. 1920. "The Percentage of Carbon Dioxide in the Alveolar Air, and the Tolerance to Accumulating Carbon Dioxide, in Cases of So-Called 'Irritable Heart' of Soldiers." *Heart* 7: 165-73.

Dry, Thomas J. 1938. "The Irritable Heart and Its Accompaniments." *Journal of the Arkansas Medical Society* 34: 259-64.

Dumanskiy, Yury D. and V. F. Rudichenko. 1976. "Dependence of the Functional Activity of Liver Mitochondria on Microwave Radiation." *Gigiyena i Sanitariya* 1976(4): 16-19. JPRS 72606 (1979), pp. 27-32.

Dumanskiy, Yury D. and Mikhail G. Shandala. 1974. "The Biologic Action and Hygienic Significance of Electromagnetic Fields of Superhigh and Ultrahigh Frequencies in Densely Populated Areas." In: P. Czerski et al., eds., *Biologic Effects and Health Hazards of Microwave Radiation: Proceedings of an International Symposium, Warsaw, 15-18 October 1973* (Warsaw: Polish Medical Publishers), pp. 289-93.

Dumanskiy, Yury D. and Lyudmila A. Tomashevskaya. 1978. "Investigation of the Activity of Some Enzymatic Systems in Response to a Superhigh Frequency Electromagnetic Field." *Gigiyena i Sanitariya* 1978(8): 23-27. JPRS 72606 (1979), pp. 1-7.

———. 1982. "Hygienic Evaluation of 8-mm Wave Electromagnetic Fields." *Gigiyena i Sanitariya* 1982(6): 18-20. JPRS 81865, pp. 6-9.

Eaker, Elaine D., Joan Pinsky, and William P. Castelli. 1992. "Myocardial Infarction and Coronary Death among Women: Psychosocial Predictors from a 20-Year Follow-up of Women in the Framingham Study." *American Journal of Epidemiology* 135(8): 854-64.

Eaker, Elaine D., Lisa M. Sullivan, Margaret Kelly-Hayes, Ralph B. D'Agostino, and Emilia J. Benjamin. 2005. "Tension and Anxiety and the Prediction of the 10-Year Incidence of Coronary Heart Disease, Atrial Fibrillation, and Total Mortality: The Framingham Offspring Study." *Psychosomatic Medicine* 67: 692-96.

Edison Electric Institute. 1940. *The Electric Light and Power Industry in the United States. Year 1939.* Statistical Bulletin no. 7.

Edison Electric Institute. 1941. *The Electric Light and Power Industry in the United States. Year 1940.* Statistical Bulletin no. 8.

Ehret, Hermann. 1915. "Zur Kenntnis der Herzschädigungen bei Kriegsteilnehmern." *Münchener medizinische Wochenschrift* 62: 689-92.

Eilenberg, M. D. and B. A. Scobie. 1960. "Prolonged Neuropsychiatric Disability and Cardiomyopathy in Acute Intermittent Porphyria." *British Medical Journal* 1: 858-59.

Fang, Jing, George A. Mensah, Janet B. Croft, and Nora L. Keenan. 2008. "Heart Failure-Related Hospitalization in the U.S., 1979 to 2004." *Journal of the American College of Cardiology* 52(6): 428-34.

Fattal, Omar, Jessica Link, Kathleen Quinn, Bruce H. Cohen, and Kathleen Franco. 2007. "Psychiatric Comorbidity in 36 Adults with Mitochondrial Cytopathies." *CNS Spectrums* 12(6): 429-38.

Fava, G. A., C. Magelli, G. Savron, S. Conti, G. Bartolucci, S. Grandi, F. Semprini, F. M. Saviotti, P. Belluardo, and B. Magnani. 1994. "Neurocirculatory Asthenia: A Reassessment Using Modern Psychosomatic Criteria." *Acta Psychiatrica Scandinavica* 89(5): 314-19.

Feinleib, Manning, William B. Kannel, Cesare G. Tedeschi, Thomas K. Landau, and Robert J. Garrison. 1979. "The Relation of Antemortem Characteristics to Cardiovascular Findings at Necropsy: The Framingham Study." *Atherosclerosis* 34: 145-57.

Fernández-Miranda C., M. De La Calle, S. Larumbe, T. Gómez-Izquierdo, A. Porres, J. Gómez-Gerique, and R. Enríquez de Salamanca. 2000. "Lipoprotein

Abnormalities in Patients with Asymptomatic Acute Porphyria." *Clinica Chimica Acta* 294(1-2): 37-43.

Fisher, Irving. 1899. "Mortality Statistics of the United States Census." In: *The Federal Census. Criticial Essays by Members of the American Economic Association*, Publications of the American Economic Association, new ser., no. 2, March 1899, pp. 121-69.

Flint, Austin. 1866. *A Treatise on the Principles and Practice of Medicine*. Philadelphia: Henry C. Lea.

Fones, Edgar and Simon Wessely. 1999. "Case of Chronic Fatigue Syndrome after Crimean War and Indian Mutiny." *British Medical Journal* 319: 1645-47.

Fox, Herbert. 1921. "Comparative Pathology of the Heart as Seen in the Captive Animals at the Philadelphia Zoölogical Garden." *Transactions of the College of Physicians of Philadelphia*, 3rd ser., no. 43, pp. 130-45.

———. 1923. *Disease in Captive Wild Mammals and Birds*. Philadelphia: J. B. Lippincott.

Fraser, Allan and Allan H. Frey. 1968. "Electromagnetic Emission at Micron Wavelengths from Active Nerves." *Biophysical Journal* 8: 731-34.

Fraser, Gary E. 1999. "Associations between Diet and Cancer, Ischemic Heart Disease, and All-Cause Mortality in Non-Hispanic White California Seventh-day Adventists." *American Journal of Clinical Nutrition* 70(3): 532S-538S.

———. 2009. "Vegetarian Diets: What Do We Know of Their Effects on Common Chronic Diseases?" *American Journal of Clinical Nutrition* 89(5): 1607S-1612S.

Frasure-Smith, Nancy and François Lespérance. 2008. "Depression and Anxiety as Predictors of 2-Year Cardiac Events in Patients with Stable Coronary Artery Disease." *Archives of General Psychiatry* 65(1): 62-71.

Freedman, David S., Tim Byers, Drue H. Barrett, Nancy E. Stroup, Elaine Eaker, and Heather Monroe-Blum. 1995. "Plasma Lipid Levels and Psychologic Characteristics in Men." *American Journal of Epidemiology* 141(6): 507-17.

Frentzel-Beyme, R., J. Claude, and U. Eilber. 1988. "Mortality Among German Vegetarians: First Results after Five Years of Follow-up." *Nutrition and Cancer* 11(2): 117-26.

Freud, Sigmund. 1895. "Ueber die Berechtigung von der Neurasthenie einen bestimmten Symptomencomplex als 'Angstneurose' abzutrennen." *Neurologisches Centralblatt* 14: 50-66. Published in English as "On the Grounds for Detaching a Particular Syndrome from Neurasthenia under the Description 'Anxiety Neurosis,'" in James Strachey, ed., *The Standard Edition of the Complete Psychological Works of Sigmund Freud* (London: Hogarth), 1962, vol. 3, pp. 87-139.

Frey, Allan H. 1961. "Auditory System Response to Radio Frequency Energy." *Aerospace Medicine* 32: 1140-42.

———. 1962. "Human Auditory System Response to Modulated Electromagnetic Energy." *Journal of Applied Physiology* 17(4): 689-92.

———. 1963. "Some Effects on Human Subjects of Ultra-High-Frequency Radiation." *American Journal of Medical Electronics* 2: 28-31.

———. 1965. "Behavioral Biophysics." *Psychological Bulletin* 63: 322-37.

———. 1967. "Brain Stem Evoked Responses Associated with Low-Intensity Pulsed UHF Energy." *Journal of Applied Physiology* 23(6): 984-88.

———. 1968. "Some Effects on Human Subjects of Ultrahigh Frequency Radiation." *American Journal of Medical Electronics*, January-March, pp. 28-31.

————. 1970. "Effects of Microwave and Radio Frequency Energy on the Central Nervous System." In: Stephen F. Cleary, ed., *Biological Effects and Health Implications of Microwave Radiation. Symposium Proceedings* (Rockville, MD: U.S. Department of Health, Education and Welfare), Publication BRH/DBE 70-2, pp. 134-139.

————. 1971. "Biological Function as Influenced by Low Power Modulated RF Energy." *IEEE Transactions on Microwave Theory and Techniques* MTT-19(2): 153-64.

————. 1985. "Data Analysis Reveals Significant Microwave-Induced Eye Damage in Humans." *Journal of Microwave Power* 20(1): 53-55.

————. 1988. "Evolution and Results of Biological Research with Low-Intensity Nonionizing Radiation." In: Andrew A. Marino, ed., *Modern Bioelectricity* (New York: Marcel Dekker), pp. 785-837.

Frey, Allan H. and Edwin S. Eichert. 1986. "Modification of Heart Function with Low Intensity Electromagnetic Energy." *Electromagnetic Biology and Medicine* 5(2): 201-10.

Frey, Allan H. and S. R. Feld. 1975. "Avoidance by Rats of Illumination with Low Power Nonionizing Electromagnetic Energy." *Journal of Comparative and Physiological Psychology* 89(2): 183-88.

Frey, Allan H., Sondra Feld, and Barbara Frey. 1975. "Neural Function and Behavior: Defining the Relationship." *Annals of the New York Academy of Sciences* 247: 433-39.

Frey, Allan H. and Rodman Messenger, Jr. 1973. "Human Perception of Illumination with Pulsed Ultrahigh-Frequency Electromagnetic Energy." *Science* 181: 356-58.

Frey, Allan H. and Elwood Seifert. 1968. "Pulse Modulated UHF Energy Illumination of the Heart Associated with Change in Heart Rate." *Life Sciences* 7 (part 2): 505-12.

Frey, Allan H. and Jack Spector. 1976. "Exposure to RF Electromagnetic Energy Decreases Aggressive Behavior." In: U.S. National Committee of the International Union of Radio Science, Program and Abstracts, URSI 1979 Spring Meeting, June 18-22 (Washington, DC: USNC-URSI), p. 456.

Frey, Allan H. and Lee S. Wesler. 1979. "Modification of Tail Pinch Consummatory Behavior in Microwave Energy Exposure." *Aggressive Behavior* 12(4): 285-91.

Friedman, Meyer. 1947. *Functional Cardiovascular Disease*. Baltimore: Williams and Wilkins.

Galli, G. 1916. "Il cuore dei soldati." *Il Policlinico, Sezione Pratica* 23: 489-91.

Gardner, Ann, Anna Johansson, Rolf Wibom, Inger Nennesmo, Ulrika von Döbeln, Lars Hagenfeldt, and Tore Hällström. 2003. "Alterations of Mitochondrial Function and Correlations with Personality Traits in Selected Major Depressive Disorder Patients." *Journal of Affective Disorders* 76: 55-68.

Gardner, Ann and Richard G. Boles. 2008. "Symptoms of Somatization as a Rapid Screening Tool for Mitochondrial Dysfunction in Depression." *BioPsychoSocial Medicine* 2: 7.

————. 2011. "Beyond the Serotonin Hypothesis: Mitochondria, Inflammation and Neurodegeneration in Major Depression and Affective Spectrum Disorders." *Progress in Neuro-Psychopharmocology & Biological Psychiatry* 35: 730-43.

Garssen, Bert, Mariete Buikhuisen, Doctorandus, and Richard van Dyck. 1996. "Hyperventilation and Panic Attacks." *American Journal of Psychiatry* 153(4): 513-18.

Gembitskiy, Ye. V. 1970. "Changes in the Functions of the Internal Organs of Personnel Operating Microwave Generators." In: I. R. Petrov. ed., *Influence of Microwave*

Radiation on the Organism of Man and Animals (Leningrad: "Meditsina"), in English translation, 1972 (Washington, DC: NASA), report no. TTF-708, pp. 106-25.

Ghali, Jalal K., Richard Cooper, and Earl Ford. 1990. "Trends in Hospitalization Rates for Heart Failure in the United States, 1973-1986." *Archives of Internal Medicine* 150: 769-73.

Glaser, Zorach R. 1971-1976. *Bibliography of Reported Biological Phenomena ("Effects") and Clinical Manifestations Attributed to Microwave and Radio-Frequency Radiation.* Bethesda, MD: Naval Medical Research Institute. NTIS reports nos. AD 734391, AD 750271, AD 770621, AD 784007, AD A015622, AD A025354, and AD A029430.

————. 1977. *Bibliography of Reported Biological Phenomena ("Effects") and Clinical Manifestations Attributed to Microwave and Radio-Frequency Radiation: Ninth Supplement to Bibliography of Microwave and RF Biologic Effects.* Cincinnato, OH: National Institute for Occupational Safety and Health. NTIS report no. PB83176537.

Goldberg, Abraham. 1959. "Acute Intermittent Porphyria: a Study of 50 Cases." *Quarterly Journal of Medicine* 28: 183-209.

Goldberg, Abraham, D. Doyle, A. C. Yeung Laiwah, Michael R. Moore, and Kenneth E. L. McColl. 1985. "Relevance of Cytochrome-c-Oxidase Deficiency to Pathogenesis of Acute Porphyria." *Quarterly Journal of Medicine* 57: 799. Abstract.

Gordon, Zinaida V. 1966. *Voprosy gigieny truda i biologicheskogo deistviya elektromagnitnykh polei sverkhvysokikh chastot.* Leningrad: "Meditsina." In English translation as *Biological Effect of Microwaves in Occupational Hygiene* (Jerusalem: Israel Program for Scientific Translations), 1970.

Gordon, Zinaida V., ed. 1973. *O biologicheskom deystvii elektromagnitnykh poley radiochastot,* 4th ed. Moscow. In English translation as *Biological Effects of Radiofrequency Electromagnetic Fields,* JPRS 63321 (1974).

Gorman, Jack M., M. R. Fyer, R. R. Goetz., J. Askanazi, M. R. Liebowitz, A. J. Fyer, J. Kinney, and D. F. Klein. 1988. "Ventilatory Physiology of Patients with Panic Disorder." *Archives of General Psychiatry* 45: 31-39.

Gozal, David, Oscar Sans Capdevila, and Leila Kheirandish-Gozal. 2008. "Metabolic Alterations and Systemic Inflammation in Obstructve Sleep Apnea among Nonobese and Obese Prepubertal Children." *American Journal of Respiratory and Critical Care Medicine* 177: 1142-49.

Grace, Sherry L., Susan E. Abbey, Jane Irvine, Zachary M. Shnek, and Donna E. Stewart. 2004. "Prospective Examination of Anxiety Persistence and Its Relationship to Cardiac Symptoms and Recurrent Cardiac Events." *Psychotherapy and Psychosomatics* 73: 344-52.

Grant, Ronald T. 1925. "Observations on the After-Histories of Men Suffering from the Effort Syndrome." *Heart* 12: 121-42.

Graybiel, Ashton and Paul D. White. 1935. "Inversion of the T Wave in Lead I or II of the Electrocardiogram in Young Individuals with Neurocirculatory Asthenia, with Thyrotoxicosis, in Relation to Certain Infections, and Following Paroxysmal Ventricular Tachycardia." *American Heart Journal* 10: 345-54.

Haldane, John Scott. 1922. *Respiration.* New Haven: Yale University Press.

Haldane, John Scott and John Gillies Priestley. 1935. *Respiration.* New Haven: Yale University Press.

Hamman, Louis and Charles W. Wainwright. 1936. "The Diagnosis of Obscure Fever. I. The Diagnosis of Unexplained, Long-continued, Low-grade Fever." *Bulletin of the Johns Hopkins Hospital* 58: 109-33.

Harrison, Tinsley Randolph, F. C. Turley, Edgar Jones, and J. Alfred Calhoun. 1931. "Congestive Heart Failure X: The Measurement of Ventilation as a Test of Cardiac Function." *Archives of Internal Medicine* 48(3): 377-98.

Hartshorne, Henry. 1864. "On Heart Disease in the Army." *American Journal of the Medical Sciences* 48(7): 89-91.

Hatano, Shuichi and Toshihisa Matsuzaki. 1977. "Atherosclerosis in Relation to Personal Attributes of a Japanese Populatiion in Homes for the Aged." In: Schettler G, Y. Gogo, Y. Hata, and G. Klose, eds, *Atherosclerosis IV: Proceedings of the Fourth International Symposium.* (New York: Springer), pp. 116-20.

Hay, John. 1923. "Disorders of the Cardio-Vascular System." In: W. G. MacPherson, W. P. Herringham, T. R. Elliott, and A. Balfour, eds., *History of the Great War* (London: His Majesty's Stationery Office), vol. 1, pp. 504-38.

Hayward, Chris, C. Barr Taylor, Walton T. Roth, Roy King, and W. Stewart Agras. 1989. "Plasma Lipid Levels in Patients with Panic Disorder or Agoraphobia." *American Journal of Psychiatry* 146(7): 917-19.

Healer, Janet. 1970. "Review of Studies of People Occupationally Exposed to Radio-Frequency Radiation." In: Stephen F. Cleary, ed., *Biological Effects and Health Implications of Microwave Radiation. Symposium Proceedings* (Rockville, MD: U.S. Department of Health, Education and Welfare), Publication BRH/DBE 70-2, pp. 90-97.

Herrick, Ariane L., B. Miles Fisher, Michael R. Moore, Sylvia Cathcart, Kenneth E. L. McColl, and Abraham Goldberg. 1990. "Elevation of Blood Lactate and Pyruvate Levels in Acute Intermittent Porphyria – A Reflection of Haem Deficiency?" *Clinica Chimica Acta* 190(3): 157-62.

Hibbert, George and David Pilsbury. 1989. "Hyperventilation: Is It a Cause of Panic Attacks?" *British Journal of Psychiatry* 155(6): 805-9.

Hick, Ford Kimmel. 1936. "Criteria of Oxygen Want with Especial Reference to Neurocirculatory Asthenia." Ph.D. thesis, University of Illinois, Chicago.

Hick, Ford Kimmel, A. W. Christian, and P. W. Smith. 1937. "Criteria of Oxygen Want, with Especial Reference to Neurocirculatory Asthenia." *American Journal of the Medical Sciences* 194: 800-4.

Hill, Ian G. W. and H. A. Dewar. 1945. "Effort Syndrome." *Lancet* 2: 161-64.

Holmes, Gary P., Jonathan E. Kaplan, Nelson M. Gantz, Anthony L. Komaroff, Lawrence B. Schonberger, Stephen E. Straus, James F. Jones, Richard E. Dubois, Charlotte Cunningham-Rundles, Savita Pahwa, Giovanna Tosato, Leonard S. Zegans, David T. Purtilo, Nathaniel Brown, Robert T. Schooley, and Irena Brus. 1988. "Chronic Fatigue Syndrome: A Working Case Definition." *Annals of Internal Medicine* 108: 387-89.

Holmgren, A., B. Jonsson, M. Levander, H. Linderholm, T. Sjöstrand, and G. Ström. 1959. "Ecg Changes in Vasoregulatory Asthenia and the Effect of Physical Training." *Acta Medica Scandinavica* 165(4): 259-71.

Holt, Phoebe E. and Gavin Andrews. 1989. "Hyperventilation and Anxiety in Panic Disorder, Social Phobia, GAD and Normal Controls." *Behaviour Research and Therapy* 27(4): 453-60.

Howell, Joel D. 1985. "'Soldier's Heart': The Redefinition of Heart Disease and Specialty Formation in Early Twentieth-Century Great Britain." *Medical History. Supplement* 5: 34-52.

Hroudová, Jana and Zdeněk Fišar. 2011. "Connectivity between Mitochondrial Functions and Psychiatric Disorders." *Psychiatry and Clinical Neurosciences* 65: 130-41.

Huffman, Jeff C., Mark H. Pollack, and Theodore A. Stern. 2002. "Panic Disorder and Chest Pain: Mechanisms, Morbidity, and Management." *Primary Care Companion, Journal of Clinical Psychiatry* 4(2): 54-62.

Hume, W. E. 1918. "A Study of the Cardiac Disabilities of Soldiers in France (V.D.H. and D.A.H.)." *Lancet* 1: 529-34.

International Labour Office. 1921. *Compensation for War Disabilities in Great Britain and the United States.* Studies and Reports, ser. E, no. 4, December 30. Geneva.

Izmerov, N. F., ed. 2005. *Rossiyskaya entsiklopediya po meditsine truda* ("Russian Encyclopedia of Occupational Medicine"). Moscow: "Meditsina."

———. 2011a. *Professional'naya patologiya: natsional'noe rykovodstvo* ("Occupational Pathology: National Manual"). 2011. Moscow: GEOTAR-Media.

———. 2011b. *Professional'nye bolezni* ("Occupational Diseases"). Moscow: Academia.

Izmerov, N. F. and E. I. Denisov, eds. 2001. *Professional'niy risk* ("Occupational Risk"). Moscow: Sotsizdat.

Izmerov, N. F. and V. F. Kirillova, eds. 2008. *Gigiyena truda* ("Occupational Hygiene"). Moscow: GEOTAR-Media.

Jammes, Y., J. G. Steinberg, O. Mambrini, F. Brégeon, and S. Delliaux. 2005. "Chronic Fatigue Syndrome: Assessment of Increased Oxidative Stress and Altered Muscle Excitability in Response to Incremental Exercise." *Journal of Internal Medicine* 257: 299-310.

Jason, Leonard A., Karina Corradi, Sara Gress, Sarah Williams, and Susan Torres-Harding. 2006. "Causes of Death Among Patients with Chronic Fatigue Syndrome." *Health Care for Women International* 27: 615-26.

Jerabek, Jiri. 1979. "Biological Effects of Magnetic Fields." *Pracovni Lekarstvi* 31(3): 98-106. JPRS 76497 (1980), pp. 1-26.

Johnson, George. 1868. "A Lecture on Dropsy: Its Pathology, Prognosis, and Principles of Treatment." *British Medical Journal* 1: 213-15.

Johnston, William J. 1880. *Telegraphic Tales and Telegraphic History.* New York: W. J. Johnston.

Jones, Maxwell. 1948. "Physiological and Psychological Responses to Stress in Neurotic Patients." *Journal of Mental Science* 94: 392-427.

Jones, Maxwell and Veronica Mellersh. 1946. "A Comparison of the Exercise Response in Anxiety States and Normal Controls." *Psychosomatic Medicine* 8: 180-87.

Jones, Maxwell and Ronald Scarisbrick. 1943. "Effect of Exercise on Soldiers with Effort Intolerance." *Lancet* 2: 331-32.

———. 1946. "The Effect of Exercise on Soldiers with Neurocirculatory Asthenia." *Psychosomatic Medicine* 8: 188-92.

Justeson, Don R. 1979. "Behavioral and Psychological Effects of Microwave Radiation." *Bulletin of the New York Academy of Medicine* 55(11): 1058-78.

Kannel, William B., 1974. "The Role of Cholesterol in Coronary Atherogenesis." *Medical Clinics of North America* 58(2): 363-79.

Kannel, William B., Thomas R. Dawber, and Mandel E. Cohen. 1958. "The Electrocardiogram in Neurocirculatory Asthenia (Anxiety Neurosis or Neurasthenia): A Study of 203 Neurocirculatory Asthenia Patients and 757 Healthy Controls in the Framingham Study." *Annals of Internal Medicine* 49(6): 1351-60.

Kaplan, Peter W. and Darrell V. Lewis. 1986. "Juvenile Acute Intermittent Porphyria with Hypercholesterolemia and Epilepsy: A Case Report and Review of the Literature." *Journal of Child Neurology* 1(1): 38-45.

Katerndahl, David. 2004. "Panic & Plaques: Panic Disorder and Coronary Artery Disease in Patients with Chest Pain." *Journal of the American Board of Family Practice* 17(2): 114-26.

Kawachi, Ichiro, David Sparrow, Pantel S. Vokonas, and Scott T. Weiss. 1994. "Symptoms of Anxiety and Risk of Coronary Heart Disease: The Normative Aging Study." *Circulation* 90(5): 2225-29.

Key, Timothy J., Gary E. Fraser, Margaret Thorogood, Paul N. Appleby, Valerie Beral, Gillian Reeves, Michael L. Burr, Jenny Chang-Claude, Rainer Frentzel-Beyme, Jan W. Kusma, Jim Mann, and Klim McPherson. 1999. "Mortality in vegetarians and Nonvegetarians: Detailed Findings from a Collaborative Analysis of 5 Prospective Studies." *American Journal of Clinical Nutrition* 70: 516S-524S.

Keys, Ancel. 1953. "Atherosclerosis: A Problem in Newer Public Health." *Journal of the Mount Sinai Hospital* 20(2): 118-39.

Kholodov, Yury A. 1966. *The Effect of Electromagnetic and Magnetic Fields on the Central Nervous System.* Translation of *Vliyaniye elektromagnitnykh i magnitnykh poley na tsentral'nuyu nervnuyu sistemu* (Moscow: Nauka). NASA report no. TT-F-465.

Klimková-Deutschová, Eliska. 1974. "Neurologic Findings in Persons Exposed to Microwaves." In: P. Czerski et al., eds., *Biologic Effects and Health Hazards of Microwave Radiation: Proceedings of an International Symposium, Warsaw, 15-18 October 1973* (Warsaw: Polish Medical Publishers), pp. 268-72.

Knickerbocker, G. G., translator. 1975. *Study in the USSR of Medical Effects of Electric Fields on Electric Power Systems.* New York: IEEE Power Engineering Society. Special Publication no. 10.

Kochanek, Kenneth D., Sherry L. Murphy, Jiaquan Xu, and Elizabeth Arias. 2019. "Deaths: Final data for 2017." *National Vital Statistics Reports*, vol. 68, no. 9. Hyattsville, MD: National Center for Health Statistics.

Koller, F. 1962. "The Value of Anticoagulants in the Prophylaxis and Therapy of Ischaemic Heart Disease." *Bulletin of the World Health Organization* 27(6): 659-66.

Kolodub, F. A. and O. N. Chernysheva. 1980. "Special Features of Carbohydrate-Energy and Nitrogen Metabolism in the Rat Brain under the Influence of Magnetic Fields of Commercial Frequency." *Ukrainskiy Biokhimicheskiy Zhurnal* 1980(3): 299-303. JPRS 77393 (1981), pp. 42-44.

Korach, S. 1916. "Über Blutdruckmessungen bei Herzstörungen der Kriegsteilnehmer." *Berliner klinische Wochenschrift* 53(34): 944-45.

Kordač, Václav, Michaela Kozáková, and Pavel Martásek. 1989. "Changes of Myocardial Functions in Acute Hepatic Porphyrias: Role of Heme Arginate Administration." *Annals of Medicine* 21(4): 273-76.

Krutikov, V. N., Yu. I. Bregadze, and A. B. Kruglov, eds. 2003. *Kontrol' fizicheskikh faktorov okruzhayushchey sredy, opasnykh dlya cheloveka* ("Control of Environmental Physical Factors that are Hazardous to People"). "Ekometriya" encyclopedia series. Moscow: IPK Standards Press.

Krutikov, V. N., N. V. Rubtsova, Y. I. Bregadze, and A. B. Kruglov, eds. 2004. *Vozdeystviye na organizm cheloveka opasnykh i vrednykh proizvodstvennykh faktorov. Mediko-biologicheskiye i metrologicheskiye aspekty* ("The Effect of Dangerous and Injurious Occupational Factors on the Human Body. Medical, Biological and Metrological Aspects"). "Ekometriya" encyclopedia series, 2 vols. Moscow: IPK Standards Press.

Kudryashov, Yu. B., Yu. F. Perov, and A. B. Rubin. 2008. *Radiatsionnaya biofizika: radiochastotnye i mikrovolnovye elektromagnitnye izlucheniya* ("Radiation Biophysics: Radiofrequency and Microwave Electromagnetic Radiation"). Moscow: Fizmatlit.

Kumar, Neelima, Sonika Sangwan, and Pooja Badotra. 2011. "Exposure to Cell Phone Radiations Produces Biochemical Changes in Worker Honey Bees." *Toxicology International* 18(1): 70-72.

Lary, Darrel and Nora Goldschlager. 1974. "Electrocardiographic Changes during Hyperventilation Resembling Myocardial Ischemia in Patients with Normal Coronary Arteriograms." *American Heart Journal* 87(3): 383-90.

Lazarev, V. I., V. F. Vinogradov, and V. V. Trotsiuk. 1989. "Blood Lipid Levels in Patients with Neurocirculatory Asthenia of the Cardiac Type." *Kardiologiya* 29(7): 74-77 (in Russian).

Lees, Robert S., Chull S. Song, Richard D. Levere, and Attallah Kappas. 1970. "Hyper-beta-Lipoproteinemia in Acute Intermittent Porphyria – Preliminary Report." *New England Journal of Medicine* 282: 432-33.

Lefebvre, B., J.-L. Pépin, J.-P. Baguet, R. Tamisier, M. Roustit, K. Riedweg, G. Bessard, P. Lévy, and F. Stanke-Labesque. 2008. "Leukotriene B_4: Early Mediator of Atherosclerosis in Obstructve Sleep Apnoea?" *European Respiratory Journal* 32: 113-20.

Leibowitz, Joshua Otto. 1970. *The History of Coronary Heart Disease*. Berkeley: University of California Press.

Leonhardt, K. F. 1981. "Kardiovaskuläre Störungen bei der akuten intermittierenden Porphyrie (AIP)." *Wiener klinische Wochenschrift* 93(18): 580-84.

Lerner, A. Martin, Claudine Lawrie and Howard S. Dworkin. 1993. "Repetitively Negative Changing T Waves at 24-h Electrocardiographic Monitors in Patients with the Chronic Fatigue Syndrome." *Chest* 104(5): 1417-21.

Letavet, A. A. and Zinaida V. Gordon, eds. 1960. *O biologicheskom vozdeystvii sverkhvysokikh chastot*. Moscow: Academy of Medical Sciences. In English translation, 1962, as *The Biological Action of Ultrahigh Frequencies*, JPRS 12471.

Levander-Lindgren, Maj. 1962. "Studies in NeurocirculatoryAsthenia (Da Costa's Syndrome). I. Variations with Regard to Symptoms and Some Pathophysiological Signs." *Acta Medica Scandinavica* 172(6): 665-76.

————. 1963. "Studies in Neurocirculatory Asthenia. III. On the Etiology and Pathogenesis of Signs in the Work Test and Orthostatic Test." *Acta Medica Scandinavica* 173(5): 631-37.

Levitina, N. A. 1966. "Nonthermal Action of Microwaves on the Cardiac Rhythm of a Frog." *Bulletin of Experimental Biology and Medicine* 62(6): 1386-87.

Levy, Robert L., Howard G. Bruenn, and Dorothy Kurtz. 1934. "Facts on Disease of Coronary Arteries. Based on a Survey of Clinical and Pathologic Records of Seven Hundred and Sixty-Two Cases." *American Journal of the Medical Sciences* 187(3): 376-90.

Lewis, Thomas. 1918a. "Report on Neuro-Circulatory Asthenia and Its Management." *Military Surgeon* 42: 409-26, 711-19.

————. 1918b. *The Soldier's Heart and the Effort Syndrome.* London: Shaw and Sons.

————. 1940. *The Soldier's Heart and the Effort Syndrome,* 2nd ed. London: Shaw and Sons.

Lewis, Thomas, Thomas F. Cotton, J. Barcroft, T. R. Milroy, D. Dufton, and T. R. Parsons. 1916. "Breathlessness in Soldiers Suffering from Irritable Heart." *British Medical Journal* 2: 517-19.

Li, Jianguo, Laura N. Thorne, Naresh M. Punjabi, Cheuk-Kwan K. Sun, Alan R. Schwartz, Philip L. Smith, Rafael L. Marino, Annabelle Rodriguez, Walter C. Hubbard, Christopher P. O'Donnell, and Vsevolod Y. Polotsky. 2005. "Intermittent Hypoxia Induces Hyperlipidemia in Lean Mice." *Circulation Research* 97(7): 698-706.

Li, Jianguo, Vladimir Savransky, Ashika Nanayakkara, Phillip L. Smith, Christopher P. O'Donnell, and Vsevolod Y. Polotsky. 2007. "Hyperlipidemia and Lipid Peroxidation are Dependent on the Severity of Chronic Intermittent Hypoxia." *Journal of Applied Physiology* 102(2): 557-63.

Lian, Camille. 1916. "Les palpitations par hypertension artérielle aux armées." *Presse médicale,* 24(29): 228-29.

Lin, James C. 1978. *Microwave Auditory Effects and Applications.* Springfield, IL: Charles C. Thomas.

Logue, Robert Bruce, James Fletcher Hanson, and William A. Knight. 1944. "Electrocardiographic Studies in Neurocirculatory Asthenia." *American Heart Journal* 28(5): 574-77.

Lopez, Alan D., Colin D. Mathers, Majid Ezzati, Dean T. Jamiston, and Christopher J. L. Murray. 2006. *Global Burden of Disease and Risk Factors.* Oxford University Press.

MacFarlane, Andrew. 1918. "Neurocirculatory Myasthenia." *Journal of the American Medical Association* 71(9): 730-33.

MacKenzie, James. 1916a. "The Soldier's Heart." *British Medical Journal* 1: 117-19.

————. 1916b. "Discussion on the Soldier's Heart." *Proceedings of the Royal Society of Medicine,* Therapeutical and Pharmacological Section, 9: 27-60.

Makolkin, V. I., E. A. Sokova, and S. A. Abbakumov. 1984. "The Oxygen Supply in Patients with Neurocirculatory Asthenia during Exercise." *Kardiologiya* 24(11): 71-76 (in Russian).

Mäntysaari, Matti J., Kari J. Antila, and Tuomas E. Peltonen. 1988. "Blood Pressure Reactivity in Patients with Neurocirculatory Asthenia." *American Journal of Hypertension* 1(2): 132-39.

Marazziti, D., S. Baroni, M. Picchetti, P. Landi, S. Silvestri, E. Vatteroni and M. Catena Dell'Osso. 2011. "Mitochondrial Alterations and Neuropsychiatric Disorders." *Current Medicinal Chemisry* 18: 4715-21.

Marha, Karel. 1970. "Maximum Admissible Values of HF and UHF Electromagnetic Radiation at Work Places in Czechoslovakia." In: Stephen F. Cleary, ed., *Biological Effects and Health Implications of Microwave Radiation. Symposium Proceedings* (Rockville, MD: U.S. Department of Health, Education and Welfare), Publication BRH/DBE 70-2, pp. 188-96.

Marha, Karel, Jan Musil, and Hana Tuhá. 1971. *Electromagnetic Fields and the Life Environment.* Berkeley: San Francisco Press.

Maron, Barry J., Joseph J. Doerer, Tammy S. Haas, David M. Tierney, and Frederick O. Mueller. 2009. "Sudden Deaths in Young Competitive Athletes: Analysis of 1866 Deaths in the United States, 1980-2006." *Circulation* 119: 1085-92.

Martens, Elisabeth J., Peter de Jonge, Beeya Na, Beth E. Cohen, Heather Lett, and Mary A. Whooley. 2010. "Scared to Death? Generalized Anxiety Disorder and Cardiovascular Events in Patients with Stable Coronary Heart Disease: The Heart and Soul Study." *Archives of General Psychiatry* 67(7): 750-58.

Martin, Linda G., Vicki A. Freedman, Robert F. Schoeni, and Patricia M. Andreski. 2009. "Health and Functioning Among Baby Boomers Approaching 60." *Journal of Gerontology: Social Sciences* 64B(3): 369-77.

Master, Arthur M. 1943. "Effort Syndrome or Neurocirculatory Asthenia in the Navy." *United States Naval Medical Bulletin* 41(3): 666-69.

Mathers, Colin, Ties Boerma, and Doris Ma Fat. 2008. *The Global Burden of Disease, 2004 Update.* Geneva: World Health Organization.

McArdle, Nigel, David Hillman, Lawrie Beilin, and Gerald Watts. 2007. "Metabolic Risk Factors for Vascular Disease in Obstructve Sleep Apnea." *American Journal of Respiratory and Criticial Care Medicine* 175: 190-95.

McCullough, Peter A., Edward F. Philbin, John A. Spertus, Scott Kaatz, Keisha R. Sandberg, W. Douglas Weaver. 2002. "Confirmation of a Heart Failure Epidemic: Findings from the Resource Utilization Among Congestive Heart Failure (REACH) Study." *Journal of the American College of Cardiology* 39(1): 60-69.

McCully, Kevin K., Benjamin H. Natelson, Stefano Iotti, Sueann Sisto, and John S. Leigh. 1996. "Reduced Oxidative Muscle Metabolism in Chronic Fatigue Syndrome." *Muscle & Nerve* 19: 621-25.

McFarland, Ross Armstrong. 1932. "The Psychological Effects of Oxygen Deprivation (Anoxemia) on Human Behavior." *Archives of Psychology*, no. 145.

———. 1941. "The Internal Environment and Behavior." *American Journal of Psychiatry* 97: 858-77.

McGovern, Paul G., David R. Jacobs, Jr., Eyal Shahar, Donna K. Arnett, Aaron R. Folsom, Henry Blackburn, and Russell V. Luepker. 2001. "Trends in Acute Coronary Heart Disease Mortality, Morbidity, and Medical Care from 1985 through 1997: The Minnesota Heart Survey." *Circulation* 104: 19-24.

McLaughlin, John T. 1962. "Health Hazards from Microwave Radiation." *Western Medicine* 3(4): 126-30.

McLeod, K. 1898. "Tropical Heart." *Journal of Tropical Medicine* 1: 3-4.

McMurray, John J. and Simon Stewart. 2000. "Epidemiology, Aetiology, and Prognosis of Heart Failure." *Heart* 83: 596-602.

McRee, Donald I. "Review of Soviet/Eastern European Research on Health Aspects of Microwave Radiation." 1979. *Bulletin of the New York Academy of Medicine* 55(11): 1133-51.

———. 1980. "Soviet and Eastern European Research on Biological Effects of Microwave Radiation." *Proceedings of the IEEE* 68(1): 84-91.

McRee, Donald I., Michael J. Galvin, and Clifford L. Mitchell. 1988. "Microwave Effects on the Cardiovascular System: A Model for Studying the Responsivity of the Autonomic Nervous System to Microwaves." In: Mary Ellen O'Connor and Richard H. Lovely, eds., *Electromagnetic Fields and Neurobehavioral Function* (New York: Alan R. Liss), pp. 153-77.

Meade, Thomas W. 2001. "Cardiovascular Disease—Linking Pathology and Epidemiology." *International Journal of Epidemiology* 30: 1179-83.

Menawat, Anand S., R. B. Panwar, D. K. Kochar, and C. K. Joshi. 1979. "Propranolol in Acute Intermittent Porphyria." *Postgraduate Medical Journal* 55: 546-47.

Merkel, Friedrich. 1915. "Ueber Herzstörungen im Kriege." *Münchener medizinische Wochenschrift* 62(20): 695-96.

Michaels, Leon. 1966. "Ætiology of Coronary Artery Disease: An Historical Approach." *British Heart Journal* 28: 258-64.

Mild, Kjell Hansson, Monica Sandström, and Eugene Lyskov, eds. 2001. *Clinical and Physiological Investigations of People Highly Exposed to Electromagnetic Fields*. Umeå, Sweden: National Institute for Working life. Arbetslivsrapport 3.

Milham, Samuel. 1979. "Cancer in Aluminum Reduction Plant Workers." *Journal of Occupational and Environmental Medicine* 7: 475-80.

———. 1982. "Mortality from Leukemia in Workers Exposed to Electrical and Magnetic Fields." *New England Journal of Medicine* 307(4): 249.

———. 1985a. "Mortality in Workers Exposed to Electromagnetic Fields." *Environmental Health Perspectives* 62: 297-300.

———. 1985b. "Silent Keys: Leukaemia Mortality in Amateur Radio Operators." *Lancet* 1: 812.

———. 1988a. "Increased Mortality in Amateur Radio Operators Due to Lymphatic and Hematopoietic Malignancies." *American Journal of Epidemiology* 127(1): 50-54.

———. 1988b. "Mortality by License Class in Amateur Radio Operators." *American Journal of Epidemiology* 128(5): 1175-76.

———. 1996. "Increased Cancer Incidence in Office Workers Exposed to Strong Magnetic Fields." *American Journal of Industrial Medicine* 30(6): 702-4.

———. 2010a. "Historical Evidence that Electrification Caused the 20th Century Epidemic of 'Diseases of Civilization.'" *Medical Hypotheses* 74: 337-45.

———. 2010b. *Dirty Electricity: Electrification and the Diseases of Civilization*. New York: iUniverse.

Milham, Samuel and Eric M. Ossiander. 2001. "Historical Evidence that Residential Electrification Caused the Emergence of the Childhood Leukemia Peak." *Medical Hypotheses* 56(3): 290-95.

Miwa, Kunihisa and Masatoshi Fujita. 2009. "Cardiac Function Fluctuates during Exacerbation and Remission in Young Adults with Chronic Fatigue Syndrome and 'Small Heart.'" *Journal of Cardiology* 54(1): 29-35.

Moir, Raymond A. and K. Shirley Smith. 1946. "Cardiovascular Diseases in the British Army Overseas." *British Heart Journal* 8(2): 110-14.

Moore, Julie L., indexer. 1984. *Cumulated Index to the Bibliography of Reported Biological Phenomena ("Effects") and Clinical Manifestations Attributed to Microwave and Radio-Frequency Radiation*, compiled by Zorach R. Glaser. Riverside, CA: Julie Moore & Associates.

Moore, Michael R. 1990. "The Pathogenesis of Acute Porphyria." *Molecular Aspects of Medicine* 11(1-2): 49-57.

Morris, Jeremiah Noah. 1951. "Recent History of Coronary Disease." *Lancet* 1: 1-7, 69-73.

———. 1961/2. "Epidemiological Aspects of Ischaemic Heart Disease." *Yale Journal of Biology and Medicine* 34: 359-69.

Munroe, H. E. 1919. "Observations on Flying Sickness, with Special Reference to its Diagnosis." *Canadian Medical Association Journal* 9(10): 883-95.

Murphy, Sherry L., Jiaquan Xu, and Kenneth D. Kochanek. 2012. "Deaths: Preliminary Data for 2010." *National Vital Statistics Reports*, vol. 60, no. 4. Hyattsville, MD: National Center for Health Statistics.

Murray, Christopher J. L. and Alan D. Lopez, eds. 1996. *The Global Burden of Disease*. Cambridge, MA: Harvard University Press.

Myhill, Sarah, Norman E. Booth, and John McLaren-Howard. 2009. "Chronic Fatigue Syndrome and Mitochondrial Dysfunction." *International Journal of Clinical and Experimental Medicine* 2: 1-16.

Nadeem, Rashid, Mukesh Singh, Mahwish Nida, Sarah Kwon, Hassan Sajid, Julie Witkowski, Elizabeth Pahomov, Kruti Shah, William Park, and Dan Champeau. 2014. "Effect of CPAP Treatment for Obstructve Sleep Apnea Hypopnea Syndrome on Lipid Profile: A Meta-Regression Analysis." *Journal of Clinical Sleep Medicine* 10(12): 1295-1302.

Naghavi, Mohsen, Haidong Wang, Rafael Lozano, Adrian Davis, Xiaofeng Liang, Maigeng Zhou, Stein Emil Vollset, et al. 2015. "Global, Regional, and National Age-Sex Specific All-Cause and Cause-Specific Mortality for 240 Causes of Death, 1990–2013: A Systematic Analysis for the Global Burden of Disease Study 2013." *Lancet* 385: 117-71.

National Center for Health Statistics, National Vital Statistics System. 1999. "Worktable I. Deaths from Each Cause, by 5-Year Age Groups, Race, and Sex." Atlanta: Centers for Disease Control and Prevention.

National Center for Health Statistics, National Vital Statistics System. 2006. "Worktable I. Deaths from Each Cause, by 5-Year Age Groups, Race, and Sex." Atlanta: Centers for Disease Control and Prevention.

National Electric Light Association. 1932. *The Electric Light and Power Industry 1931*. Statistical Bulletin no. 8.

National Electric Light Association. 1931. *The Electric Light and Power Industry 1930*. Statistical Bulletin no. 7.

Navas-Nacher, Elena L., Laura Colangelo, Craig Beam, and Philip Greenland. 2001. "Risk Factors for Coronary Heart Disease in Men 18 to 39 Years of Age." *Annals of Internal Medicine* 134(6): 433-39.

Neaton, James D. and Deborah Wentworth. 1992. "Serum Cholesterol, Blood Pressure, Cigarette Smoking, and Death from Coronary Heart Disease: Overall Findings and Differences by Age for 316,099 White Men." *Archives of Internal Medicine* 152: 56-64.

Neuhof, Selian. 1919. "The Irritable Heart in General Practice: A Comparison between It and the Irritable Heart of Soldiers." *Archives of Internal Medicine* 24(1): 51-64.

Newman, Anne B., F. Javier Nieto, Ursula Guidry, Bonnie K. Lind, Susan Redline, Eyal Shahar, Thomas G. Pickering, and Stuart F. Quan. 2001. "Relation of Sleep-disordered Breathing to Cardiovascular Disease Risk Factors: The Sleep Heart Health Study." *American Journal of Epidemiology* 154(1): 50-59.

Nikitina, Valentina N. 2001. "Hygienic, Clinical and Epidemiological Analysis of Disturbances Induced by Radio Frequency EMF Exposure in Human Body." In: Kjell Hansson Mild, Monica Sandström, and Eugene Lyskov, eds., *Clinical and Physiological Investigations of People Highly Exposed to Electromagnetic Fields* (Umeå, Sweden: National Institute for Working life), Arbetslivsrapport 3, pp. 32-38.

Njølstad, Inger, Egil Arnesen, and Per G. Lund-Larsen. 1996. "Smoking, Serum Lipids, Blood Pressure, and Sex Differences in Myocardial Infarction: A 12-Year Follow-up of the Finnmark Study." *Circulation* 93: 450-6.

Novitskiy, Yu. I., Zinaida V. Gordon, Aleksandr S. Presman, and Yury A. Kholodov. 1970. "Radio Frequencies and Microwaves. Magnetic and Electrical Fields." Vol. 2, part 1, chap. 1 of *Osnovy kosmicheskoy biologii i meditsiny* ("Foundations of Space Biology and Medicine"). Moscow: Academy of Sciences USSR. English translation by Scientific Translation Service (Washington, DC: NASA), 1971, report no. TT-F-14,021.

Nutzinger, D. O. 1992. "Hertz und Angst: Herzbezogene Ängste und kardiovaskuläres Morbiditätsrisiko bei Patienten mit einer Angststörung." *Der Nervenarzt* 63(3): 187-91.

Okumiya, Noriya, Kenzo Tanaka, Kazuo Ueda, and Teruo Omae. 1985. "Coronary Atherosclerosis and Antecedent Risk Factors: Pathologic and Epidemiologic Study in Hisayama, Japan." *American Journal of Cardiology* 56: 62-66.

Olafiranye, O., G. Jean-Louis, F. Zizi, J. Nunes, and M. T. Vincent. 2011. "Anxiety and Cardiovascular Risk: Review of Epidemiological and Clinical Evidence." *Mind Brain* 2(1): 32-37.

Orlova, A. A. 1960. "The Clinic of Changes of the Internal Organs under the Influence of UHF." In: A. A. Letavet and Z. V. Gordon, eds. *The Biological Action of Ultrahigh Frequencies* (Moscow: Academy of Medical Sciences), JPRS 12471, pp. 30-35.

Parikh, Nisha I., Philimon Gona, Martin G. Larson, Caroline S. Fox, Emelia J. Benjamin, Joanne M. Murabito, Christopher J. O'Donnell, Ramachandran S. Vasan, and Daniel Levy. 2009. "Long-term Trends in Myocardial Infarction Incidence and Case-Fatality in the National Heart, Lung, and Blood Institute's Framingham Heart Study." *Circulation* 119(9): 1203-10.

Park, Mi Ran, Jeong Kee Seo, Jae Sung Ko, Ju Young Chang, and Hye Ran Yang. 2011. "Acute Intermittent Porphyria Presented with Recurrent Abdominal Pain and Hypertension." *Korean Journal of Pediatric Gastroenterology and Nutrition* 14: 81-85.

Parkinson, John. 1941. "Effort Syndrome in Soldiers." *British Medical Journal* 1: 545-49.

Paterniti, Sabrina, Mahmoud Zureik, Pierre Ducimetière, Pierre-Jean Touboul, Jean-Marc Fève, and Annick Alpérovitch. 2001. "Sustained Anxiety and 4-Year Progression of Carotid Atherosclerosis." *Arteriosclerosis, Thrombosis, and Vascular Biology* 21(1): 136-41.

Paul, Oglesby. 1987. "Da Costa's Syndrome or Neurocirculatory Asthenia." *British Heart Journal* 58: 306-15.

Peckerman, Arnold, John J. Lamanca, Kristina A. Dahl, Rahul Chemitiganti, Bushra Qureishi, and Benjamin H. Natelson. 2003. "Abnormal Impedance Cardiography Predicts Symptom Severity in Chronic Fatigue Syndrome." *American Journal of the Medical Sciences* 326(2): 55-60.

Pervushin, V. Yu. 1957. "Changes Occurring in the Cardiac Nervous Apparatus Due to the Action of Ultra-High-Frequency Field." *Bulletin of Experimental Biology and Medicine* 43(6): 734-40.

Peter, Helmut, Philipp Goebel, Susanne Müller, and Iver Hand. 1999. "Clinically Relevant Cholesterol Elevation in Anxiety Disorder: A Comparison with Normal Controls." *International Journal of Behavioral Medicine* 6(1): 30-39.

Petrov, Ioakim Romanovich, ed. 1970a. *Vliyaniye SVCh-izlucheniya na organism cheloveka i zhivotnykh*. Leningrad: "Meditsina." In English translation as *Influence of Microwave Radiation on the Organism of Man and Animals* (Washington, DC: NASA), report no. TTF-708, 1972.

Phillips, Anna C., G. David Batty, Catharine R. Gale, Ian J. Deary, David Osborn, Kate MacIntyre, and Douglas Carroll. 2009. "Generalized Anxiety Disorder, Major Depressive Disorder, and Their Comorbidity as Predictors of All-Cause and Cardiovascular Mortality: The Vietnam Experience Study." *Psychosomatic Medicine* 71: 395-403.

Phillips, Roland L, Frank R. Lemon, W. Lawrence Beeson, and Jan W. Kuzma. 1978. "Coronary Heart Disease Mortality among Seventh-day Adventists with Differing Dietary Habits: A Preliminary Report." *American Journal of Clinical Nutrition* 31 (10 suppl.): S191-S198.

Pitts, Ferris N., Jr. and James N. McClure, Jr. 1967. "Lactate Metabolism in Anxiety Neurosis." *New England Journal of Medicine* 277(25): 1329-36.

Plum, William Rattle. 1882. *The Military Telegraph during the Civil War in the United States*, 2 vols. Chicago: Jansen, McClurg.

Popular Science Monthly. 1918. "How the Zeppelin Raiders Are Guided by Radio Signals." 92: 632-34.

Presman, Aleksandr Samuilovich. 1970. *Electromagnetic Fields and Life*. New York: Plenum. Translation of *Elektromagnitnye polya i zhivaya priroda* (Moscow: Nauka), 1968.

Presman, Aleksandr Samuilovich and N. A. Levitina. 1962a. "Nonthermal Action of Microwaves on Cardiac Rhythm. Communication I. A Study of the Action of Continuous Microwaves." *Bulletin of Experimental Biology and Medicine* 53(1): 36-39.

————. 1962b. "Nonthermal Action of Microwaves on the Rhythm of Cardiac Contractions in Animals. Report II. Investigation of the Action of Impulse Microwaves." *Bulletin of Experimental Biology and Medicine* 53(2): 154-57.

Ratcliffe, Herbert L. 1963a. "Editorial: Environmental Factors and Coronary Disease." *Circulation* 27: 481-83.

————. 1963b. "Phylogenetic Considerations in the Etiology of Myocardial Infarction." In: Thomas N. James and John W. Keyes, eds., *The Etiology of Myocardial Infarction* (Boston: Little, Brown), pp. 61-89.

————. 1965. "Age and Environment as Factors in the Nature and Frequency of Cardiovascular Lesions in Mammals and Birds in the Philadelphia Zoological Garden." *Comparative Cardiology* 127: 715-35.

Ratcliffe, Herbert L. and M. T. I. Cronin. 1958. "Changing Frequency of Arteriosclerosis in Mammals and Birds at the Philadelphia Zoological Garden: Review of Autopsy Records." *Circulation* 18: 41-52.

Ratcliffe, Herbert L., T. G. Yerasimides and G. A. Elliott. 1960. "Changes in the Character and Location of Arterial Lesions in Mammals and Birds in the Philadelphia Zoological Garden." *Circulation* 21: 730-38.

Ravnskov, Uffe. 2000. *The Cholesterol Myths*. Washington, DC: New Trends.

Reed Dwayne M., Jack P. Strong, Joseph Resch, and Takuji Hayashi. 1989. "Serum Lipids and Lipoproteins as Predictors of Atherosclerosis: An Autopsy Study." *Arteriosclerosis, Thrombosis, and Vascular Biology* 9: 560-64.

Reeves, William C., James F. Jones, Elizabeth Maloney, Christine Heim, David C. Hoaglin, Roumiana S. Boneva, Marjorie Morrissey, and Rebecca Devlin. 2007. "Prevalence of Chronic Fatigue Syndrome in Metropolitan, Urban, and Rural Georgia." *Population Health Metrics* 5: 5.

Reyes, Michele, Rosane Nisenbaum, David C. Hoaglin, Elizabeth R. Unger, Carol Emmons, Bonnie Randall, John A. Stewart, Susan Abbey, James F. Jones, Nelson Gantz, Sarah Minden, and William C. Reeves. 2003. "Prevalence and Incidence of Chronic Fatigue Syndrome in Wichita, Kansas." *Archives of Internal Medicine* 163: 1530-36.

Rhoads, George G, William C. Blackwelder, Grant N. Stemmermann, Takuji Hayashi, and Abraham Kagan. 1978. "Coronary Risk Factors and Autopsy Findings in Japanese-American Men." *Laboratory Investigation* 38(3): 304-11.

Ridley, Alan. 1969. "The Neuropathy of Acute Intermittent Porphyria." *Quarterly Journal of Medicine* 38: 307-33.

————. 1975. "Porphyric Neuropathy." In: Peter James Dyck, P. K. Thomas, and Edward H. Lambert, eds., *Peripheral Neuropathy* (Philadelphia: W. B. Saunders), pp. 942-55.

Rigg, Kathleen J., R. Finlayson, C. Symons, K. R. Hill, and R. N. T-W-Fiennes. 1960. "Degenerative Arterial Disease of Animals in Captivity with Special Reference to the Comparative Pathology of Atherosclerosis." *Proceedings of the Zoological Society of London* 135(2): 157-64.

Robey, William H. and Ernst P. Boas. 1918. "Neurocirculatory Asthenia." *Journal of the American Medical Association* 71(7): 525-29.

Robinson, G. V., J. C. T. Pepperell, H. C. Segal, R. J. O. Davies, and J. R. Stradling. 2004. "Circulating Cardiovascular Risk Factors in Obstructive Sleep Apnoea: Data from Randomised Controlled Trials." *Thorax* 59: 777-82.

Rodríguez-Artalejo, F., P. Guallar-Castillón, J. R. Banegas Banegas, and J. del Rey Calero. 1997. "Trends in Hospitalization and Mortality for Heart Failure in Spain, 1980-1993." *European Heart Hournal* 18: 1771-79.

Roger, Véronique L., Susan A. Weston, Margaret M. Redfield, Jens P. Hellermann-Homan, Jill Killian, Barbara P. Yawn, and Steven J. Jacobsen. 2004. "Trends in Heart Failure Incidence and Survival in a Community-Based Population." *JAMA* 292(3): 344-50.

Rothenbacher, Dietrich, Harry Hahmann, Bernd Wüsten, Wolfgang Koenig, and Hermann Brenner. 2007. "Symptoms of Anxiety and Depression in Patients with Stable Coronary Heart Disease: Prognostic Value and Consideration of Pathogenetic Links." *European Journal of Cardiovascular Prevention and Rehabilitation* 14: 547-54.

Rothschild, Marcus A. 1930. "Neurocirculatory Asthenia." *Bulletin of the New York Academy of Medicine* 6(4): 223-42.

Rozanski, Alan, James A. Blumenthal, and Jay Kaplan. 1999. "Impact of Psychological Factors on the Pathogenesis of Cardiovascular Disease and Implications for Therapy." *Circulation* 99: 2192-2217.

Rural Electrification Administration, U.S. Dept. of Agriculture. January 1940. *Rural Electrification in Utah*. Washington, DC.

———. 1941. *Report of the Administrator of the Rural Electrification Administration*. Washington, DC.

Ryle, John A. and W. T. Russell. 1949. "The Natural History of Coronary Disease." *British Heart Journal* 11(4): 370-89.

Sadchikova, Maria N. 1960. "State of the Nervous System under the Influence of UHF." In: A. A. Letavet and Z. V. Gordon, eds., *The Biological Action of Ultrahigh Frequencies* (Moscow: Academy of Medical Sciences), JPRS 12471, pp. 25-29.

———. 1974. "Clinical Manifestations of Reactions to Microwave Irradiation in Various Occupational Groups." In: P. Czerski et al., eds., *Biologic Effects and Health Hazards of Microwave Radiation: Proceedings of an International Symposium, Warsaw, 15-18 October 1973* (Warsaw: Polish Medical Publishers), pp. 261-67.

Sadchikova, Maria N. and K. V. Glotova. 1973. "The Clinic, Pathogenesis, Treatment, and Outcome of Radiowave Sickness." In: Z. V. Gordon, ed., *Biological Effects of Radiofrequency Electromagnetic Fields*, JPRS 63321 (1974), pp. 54-62.

Sadchikova, Maria N., S. F. Kharlamova, N. N. Shatskaya, and N. V. Kuznetsova. 1980. "Significance of Blood Lipid and Electrolyte Disturbances in the Development of Some Reactions to Microwaves." *Gigiyena truda i professional'nyye zabolevaniya* 1980(2): 38-39. JPRS 77393 (1981), pp. 37-39.

Saint, Eric G., D. Curnow, and R. Paton. 1954. "Diagnosis of Acute Porphyria." *British Medical Journal* 1: 1182-84.

Sanders, Aaron P., William T. Joines, and John W. Allis. 1984. "The Differential Effects of 200, 591, and 2,450 MHz Radiation on Rat Brain Energy Metabolism." *Bioelectromagnetics* 5: 419-33.

Savransky, Vladimir, Ashika Nanayakkara, Jianguo Li, Shannon Bevans, Philip L. Smith, Annabelle Rodriguez, and Vsevolod Y. Polotsky. 2007. "Chronic Intermittent Hypoxia Induces Atherosclerosis." *American Journal of Respiratory and Critical Care Medicine* 175: 1290-97.

Scherrer, Jeffrey F., Timothy Chrusciel, Angelique Zeringue, Lauren D. Garfield, Paul J. Hauptman, Patrick J. Lustman, Kenneth E. Freedland, Robert M. Carney, Kathleen K. Bucholz, Richard Owen, and William R. True. 2010. "Anxiety Disorders Increase Risk for Incident Myocardial Infarction in Depressed and Nondepressed Veterans Administration Patients." *American Heart Journal* 159(5): 772-79.

Schott, Theodor. 1915. "Beobachtungen über Herzaffektionen bei Kriegsteilnehmern." *Münchener medizinische Wochenschrift* 62(20): 677-79.

Scriven, George P. 1915. "Notes on the Organization of Telegraph Troops in Foreign Armies. Great Britain." In: Scriven, *The Service of Information, United States Army*, (Washington, DC: Government Printing Office), pp. 127-32. Reproduced in Paul J. Scheips, ed., *Military Signal Communications* (New York: Arno Press), 1980, vol. 2.

Seldenrijk, Adrie, Nicole Vogelzangs, Hein P. J. van Hout, Harm W. J. van Marwijk, Michaela Diamant, and Brenda W. J. H. Penninx. 2010. "Depression and Anxiety Disorders and Risk of Subclinical Atherosclerosis: Findings from the Netherlands Study of Depression and Anxiety (NESDA)." *Journal of Psychosomatic Research* 69: 203-10.

Sharrett, A. R., C. M. Ballantyne, S. A. Coady, G. Heiss, P. D. Sorlie, D. Catellier, and W. Patsch. 2001. "Coronary Heart Disease Prediction From Lipoprotein Cholesterol Levels, Triglycerides, Lipoprotein(a), Apolipoproteins A-I and B, and HDL Density Subfractions: The Atherosclerosis Risk in Communities (ARIC) Study." *Circulation* 104: 1108-13.

Shibeshi, Woldecherkos A., Yinong Young-Xu, and Charles M. Blatt. 2007. "Anxiety Worsens Prognosis in Patients with Coronary Artery Disease." *Journal of the American College of Cardiology* 49(20): 2021-27.

Shiue, J. W., F. Y. Lee, K. J. Hsiao, Y. T. Tsai, S. D. Lee, and S. J. Wu. 1989. "Abnormal Thyroid Function and Hypercholesterolemia in a Case of Acute Intermitten Porphyria." *Taiwan Yi Xue Hui Za Zhi* (Journal of the Formosa Medical Association) 88(7): 729-31.

Shorter, Edward. 1992. *From Paralysis to Fatigue: A History of Psychosomatic Illness in the Modern Era*. New York: Free Press.

———. 1997. *A History of Psychiatry*. New York: John Wiley & Sons.

Shutenko, O. I., I. P. Kozarin and I. I. Shvayko. 1981. "Effects of Superhigh Frequency Electromagnetic Fields on Animals of Different Ages." *Gigiyena i Sanitariya* 1981(10): 35-38. JPRS 81300 (1982), pp. 85-90.

Siekierzyński, Maksymilian. 1974. "A Study of the Health Status of Microwave Workers." In: P. Czerski et al., eds., *Biologic Effects and Health Hazards of Microwave Radiation: Proceedings of an International Symposium, Warsaw, 15-18 October 1973* (Warsaw: Polish Medical Publishers), pp. 273-80.

Siekierzyński, Maksymilian, Przemysław Czerski, Halina Milczarek, Andrej Gidyński, Czesław Czarnecki, Eugeniusz Dziuk, and Wiesław Jedrzejczak. 1974. "Health Surveillance of Personnel Occupationally Exposed to Microwaves. II. Functional Disturbances." *Aerospace Medicine* 45(10): 1143-45.

Sijbrands, Eric J. G., Rudi G. J. Westendorp, Joep C. Defesche, Paul H. E. M. de Meier, Augustinus H. M. Smelt, and John J. P. Kastelein. 2001. "Mortality Over Two Centuries in Large Pedigree with Familial Hypercholesterolaemia: Family Tree Mortality Study." *British Medical Journal* 322: 1019-23.

Silverman, Charlotte. 1979. "Epidemiologic Approach to the Study of Microwave Effects." *Bulletin of the New York Academy of Medicine* 55(11): 1166-81.

Smart, Charles. 1888. "Cardiac Diseases." In: Smart, *The Medical and Surgical History of the War of the Rebellion*, part III, vol. I, *Medical History* (Washington, DC: Government Printing Office), pp. 860-69.

Snowdon, David A. 1988. "Animal Product Consumption and Mortality Because of All Causes Combined, Coronary Heart Disease, Stroke, Diabetes, and Cancer in Seventh-day Adventists." *American Journal of Clinical Nutrition* 48: 739-48.

Soares-Filho, Gastão L. F., Oscar Arias-Carrión, Gaetano Santulli, Adriana C. Silva, Sergio Machado, Alexandre M. Valença, and Antonio E. Nardi. 2014. "Chest Pain, Panic Disorder and Coronary Artery Disease: A Systematic Review." *CNS & Neurological Disorders – Drug Targets* 13(6): 992-1001.

Solberg, Lars A., Jack P. Strong, Ingar Holme, Anders Helgeland, Ingvar Hjermann, Paul Leren, and Svein Børre Mogensen. 1985. "Stenoses in the Coronary Arteries: Relation to Atherosclerotic Lesions, Coronary Heart Disease, and Risk Factors. The Oslo study." *Laboratory Investigation* 53: 648-55.

Sonimo, N., G. A. Fava, M. Boscaro, and F. Fallo. 1998. "Life Events and Neurocirculatory Asthenia. A Controlled Study." *Journal of Internal Medicine* 244: 523-28.

Spinhoven, Philip, E. J. Onstein, P. J. Sterk, and D. Le Haen-Versteijnen. 1992. "The Hyperventilation Provocation Test in Panic Disorder." *Behaviour Research and Therapy* 30(5): 453-61.

Stamler, Jeremiah, Deborah Wentworth, and James D. Neaton. 1986. "Is Relationship between Serum Cholesterol and Risk of Premature Death from Coronary Heart Disease Continuous and Graded?" *JAMA* 256(20): 2823-28.

Stamler, Jeremiah, Martha L. Daviglus, Daniel B. Garside, Alan R. Dyer, Philip Greenland, and James D. Neaton. 2000. "Relationship of Baseline Serum Cholesterol Levels in 3 Large Cohorts of Younger Men to Long-term Coronary, Cardiovascular, and All-Cause Mortality and to Longevity." *JAMA* 284: 311-18.

Statistical Report of the Health of the Navy for the Year 1915. 1922. London: His Majesty's Stationery Office.

Stein, Jeffrey A. and Donald P. Tschudy. 1970. "Acute Intermittent Porphyria: A Clinical and Biochemical Study of 46 Patients." *Medicine* 49(1): 1-16.

Steiropoulous, Paschalis, Venetia Tsara, Evangelia Nena, Christina Fitili, Margarita Kataropoulou, Marios Froudarakis, Pandora Christaki, and Demosthenes Bouros. 2007. "Effect of Continuous Positive Airway Pressure Treatment on Serum Cardiovascular Risk Factors in Patients with Obstructuve Sleep Apnea-Hypopnea Syndrome." *Chest* 132(3): 843-51.

Stephenson, G. V. and Kenneth Cameron. 1943. "Anxiety States in the Navy: A Clinical Survey and Impression." *British Medical Journal* 2: 603-7.

Stewart, S., K. MacIntyre, M. M. C. MacLeod, A. E. M. Bailey, S. Capewell, and J. J. V. McMurray. 2001. "Trends in Hospitalization for Heart Failure in Scotland, 1990-1996." *European Heart Journal* 22: 209-17.

Subbota, A. G. 1970. "Changes in Functions of Various Systems of the Organism." In: I. R. Petrov. ed., *Influence of Microwave Radiation on the Organism of Man and Animals* (Leningrad: "Meditsina"), in English translation, 1972 (Washington, DC: NASA), report no. TTF-708, pp. 66-87.

Suvorov, G. A. and N. F. Izmerov. 2003. *Fizicheskiye faktory proizvodstvennoy i prirodnoy credy* ("Physical Factors of Occupational and Natural Environment"). Moscow: "Meditsina."

Taddeini, Luigi, Karen L. Nordstrom, and C. J. Watson. 1974. "Hypercholesterolemia in Experimental and Human Hepatic Porphyria." *Metabolism* 13: 691-701.

Tamburello, C. C., L. Zanforlin, G. Tiné, and A. E. Tamburello. 1991. "Analysis of Microwave Effects on Isolated Hearts." *IEEE MTT-S Digest* (IEEE Microwave Theory and Techniques Symposium, Boston), pp. 805-8.

Thorogood, Margaret, Jim Mann, Paul Appleby, and Klim McPherson. 1994. "Risk of Death from Cancer and Ischaemic Heart Disease in Meat and Non-Meat Eaters." *British Medical Journal* 308: 1667-71.

Thunell, Stig. 2000. "Porphyrins, Porphyrin Metabolism and Porphyrias. I. Update." *Scandinavian Journal of Clinical and Laboratory Investigation* 60: 509-40.

Tomashevskaya, Lyudmila A. and E. A. Solenyi. 1986. "Biologicheskoye deystviye i gigiyenicheskoye znacheniye elektromagnitnogo polya, sozdavayemogo beregovimi radiolokatsionnimi sredstvami" ("Biological Action and Hygienic Significance of the Electromagnetic Field Created by Coastal Radar Facilities"). *Gigiyena i Sanitariya* 1986(7): 34-36.

Tomashevskaya, Lyudmila A. and Yury D. Dumanskiy. 1988. "Gigiyenicheskaya otsenka biologicheskogo deystviya impul'snykh elektromagnitnykh poley 850-2750 MGts" ("Hygienic Evaluation of the Biological Effect of Pulsed Electromagnetic Fields or 850-2750 MHz"). *Gigiyena i Sanitariya* 1988(9): 21-24.

———. 1989. "Influence of Low-Intensity 8-mm Wave EMF on Some Exchange Processes." In: *Fundamental and Applied Aspects of Use of Millimeter Electromagnetic Radiation in Medicine: Proceedings of the First All-Union Symposium with International Participation* (Kiev: VNK "Otlik"), pp. 135-37.

Tourniaire, A., M. Tartulier, J. Blum, and F. Deyrieux. 1961. "Confrontation des données fonctionnelles respiratoires et hémodynamiques cardiaques dans les névroses tachycardiaques et chez les sportifs." *Presse médicale* 69(16): 721-23.

Treupel, G. 1915. "Kriegsärztliche Herzfragen." *Medizinische Klinik (Berlin)* 62(11): 356-59.

Tuomilehto Jaakko and Kari Kuulasmaa. 1989. "WHO MONICA Project: Assessing CHD Mortality and Morbidity." *International Journal of Epidemiology* 18: S38-S45.

Tyagin, Nikolay Vasil'evich. 1971. *Klinicheskiye aspekty oblucheniy SVCh-diapazona* ("Clinical Aspects of Irradiation in the SHF-range"). Leningrad: "Meditsina."

Tzivoni, Dan, Zvi Stern, Andre Keren, and Shlomo Stern. 1980. "Electrocardiographic Characteristics of Neurocirculatory Asthenia during Everyday Activities." *British Heart Journal* 44: 426-32.

van Rensburg, S. J., F. C. Potocnik, T. Kiss, F. Hugo, P. van Zijl, E. Mansvelt, and M. E. Carstens. 2001. "Serum Concentrations of Some Metals and Steroids in Patients with Chronic Fatigue Syndrome with Reference to Neurological and Cognitive Abnormalities." *Brain Research Bulletin* 55(2): 319-25.

Vastesaeger, Marcel M. and R. Delcourt. 1962. "The Natural History of Atherosclerosis." *Circulation* 26: 841-55.

Verschuren, W. M. Monique, David R. Jacobs, Bennie P. M. Bloemberg, Daan Kromhout, Alessandro Menotti, Christ Aravanis, Henry Blackburn, Ratko Buzina, Anastasios S. Dontas, Flaminio Fidanza, Martti J. Karvonen, Srećko Nedeljković, Aulikki Nissinen, and Hironori Toshima. 1995. "Serum Total Cholesterol and Long-Term Coronary Heart Disease Mortality in Different Cultures: Twenty-five-Year Follow-up of the Seven Countries Study." *JAMA* 274(2): 131-36.

Vogelzangs, Nicole, Adrie Seldenrijk, Aartjan T. F. Beekman, Hein P. J. van Hout, Peter de Jonge, and Brenda W. J. H. Penninx. 2010. "Cardiovascular Disease in Persons with Depressive and Anxiety Disorders." *Journal of Affective Disorders* 215: 241-48.

von Dziembowski, C. 1915. "Die Vagotonie, eine Kriegskrankheit." *Therapie der Gegenwart* 56: 405-13.

von Romberg, Ernst. 1915. "Beobachtungen über Herz- und Gefässkrankheiten während der Kriegszeit." *Münchener medizinische Wochenschrift* 62(20): 671-72.

Vural, M. and E. Başar. 2007. "Anxiety Disoder as a Potential for Sudden Death." *Anadolu Kardiyoloji Dergisi* 7(2): 179-83 (in Turkish).

Watson, Raymond C., Jr. 2009. *Radar Origins Worldwide*. Victoria, BC: Trafford.

Weissman, Myrna M., Jeffrey S. Markowitz, Robert Ouellette, Steven Greenwald, and Jeffrey P. Kahn. 1990. "Panic Disorder and Cardiovascular/Cerebrovascular Problems: Results from a Community Survey." *American Journal of Psychiatry* 147: 1504-8.

Wendkos, Martin H. 1944. "The Influence of Autonomic Imbalance on the Human Electrocardiogram." *American Heart Journal* 28(5): 549-67.

Wheeler, Edwin O., Paul D. White, Eleanor W. Reed, and Mandel E. Cohen. 1950. "Neurocirculatory Asthenia (Anxiety Neurosis, Effort Syndrome, Neurasthenia): A Twenty Year Follow-up Study of One Hundred and Seventy-three Patients." *Journal of the American Medical Association* 142(12): 878-89.

White, Paul Dudley. 1920. "The Diagnosis of Heart Disease in Young People." *Journal of the American Medical Association* 74(9): 580-82.

———. 1938. *Heart Disease*, 2nd ed. New York: Macmillan.

———. 1957. "The Cardiologist Enlists the Epidemiologist." *American Journal of Public Health*, vol. 47, no. 4, part 2, pp. 1-3.

———. 1971. *My Life and Medicine: An Autobiographical Memoir*. Boston: Gambit.

Whitelaw, Andrew G. L. 1974. "Acute Intermittent Porphyria, Hypercholesterolaemia, and Renal Impairment." *Archives of Disability in Childhood* 49: 406-7.

Wilson, Peter W. F., Ralph B. D'Agostino, Daniel Levy, Albert M. Belanger, Halit Silbershatz, and William B. Kannel. 1998. "Prediction of Coronary Heart Disease Using Risk Factor Categories." *Circulation* 97: 1837-47.

Wilson, Robert McNair. 1916. "The Irritable Heart of Soldiers." *British Medical Journal* 1: 119-20.

Wong, Roger, Gary Lopaschuk, Gang Zhu, Dorothy Walker, Dianne Catellier, David Burton, Koon Teo, Ruth Collins-Nakai, and Terrence Montague. 19 92. "Skeletal Muscle Metabolism in the Chronic Fatigue Syndrome." *Chest* 102(6): 1716-22.

Wooley, Charles F. 1976. "Where are the Diseases of Yesteryear? DaCosta's Syndrome, Soldier's Heart, the Effort Syndrome, Neurocirculatory Asthenia – And the Mitral Valve Prolapse Syndrome." *Circulation* 53(5): 749-51.

———. 1985. "From Irritable Heart to Mitral Valve Prolapse: British Army Medical Reports, 1860 to 1870." *American Journal of Cardiology* 55(8): 1107-9.

———. 1988. "Lewis A. Conner, MD (1867-1950), and Lessons Learned from Examining Four Million Young Men in World War I." *American Journal of Cardiology* 61: 900-3.

Worts, George F. 1915. "Directing the War by Wireless." *Popular Mechanics*, May, pp. 647-50.

York, J. Lyndal. 1972. *The Porphyrias*. Springfield, IL: Charles C. Thomas.

Zalyubovskaya, N. P. and R. I. Kiselev. 1978. "Biological Oxidation in Cells Exposed to Microwaves in the Millimeter Range." *Tsitologiya i Genetika* 12(3): 232-36 (in Russian).

Zalyubovskaya, N. P., R. I. Kiselev, and L. N. Turchaninova. 1977. "Effects of Electromagnetic Waves of the Millimetric Range on the Energy Metabolism of Liver Mitochondria." *Biologicheskiye Nauki* 1977(6): 133-34. JPRS 70107, pp. 51-52.

Zhang, X., A. Patel, H. Horibe, Z. Wu, F. Barzi, A. Rodgers, S. MacMahon, and M. Woodward. 2003. "Cholesterol, Coronary Heart Disease, and Stroke in the Asia Pacific Region." *International Journal of Epidemiology* 32(4): 563-72.

Zheng, Zhi-Jie, Janet B. Croft, Wayne H. Giles, and George A. Mensah. 2005. "Out-of-Hospital Cardiac Deaths in Adolescents and Young Adults in the United States, 1989 to 1998." *American Journal of Preventive Medicine* 29 (5S1): 36-41.

Chapter 12

Allen, Frederick M. 1914. "Studies Concerning Diabetes." *Journal of the American Medical Association* 63(11): 939-43.

———. 1915. "Metabolic Studies in Diabetes." *New York State Journal of Medicine* 15(9): 330-33.

———. 1916. "Investigative and Scientific Phases of the Diabetic Question." *Journal of the American Medical Association* 66(20): 1525-32.

———. 1922. "Observations on the Progressiveness of Diabetes." *Medical Clinics of North America* 6(3): 465-74.

Antoun, Ghadi, Fiona McMurray, A. Brianne Thrush, David A. Patten, Alyssa C. Peixoto, Ruth S. Slack, Ruth McPherson, Robert Dent, and Mary-Ellen Harper. 2015. "Impaired Mitochondrial Oxidative Phosphorylation and Supercomplex Assembly in Rectus Abdominis Muscle of Diabetic Obese Individuals." *Diabetologia* 58(12): 2861-66.

Bartoníček, V. and Eliska Klimková-Deutschová. 1964. "Effect of Centimeter Waves on Human Biochemistry." *Casopís Lékařů Ceských* 103(1): 26-30 (in Czech). English

Translation in G. L. Khazan, ed., *Biological Effects of Microwaves*, ATD Report P-65-68, September 17, 1965 (Washington, DC: Dept. of Commerce), pp. 13-14.

Belokrinitskiy, Vasily S. 1982. "Hygienic Evaluation of Biological Effects of Nonionizing Microwaves." *Gigiyena i Sanitariya* 1982(6): 32-34. JPRS 81865, pp. 1-5.

Belokrinitskiy, Vasily S. and A. N. Grin'. 1983. "Nature of Morphofunctional Renal Changes in Response to SHF Field-Hypoxia Combination." *Vrachebnoye Delo* 1983(1): 112-15. JPRS 84221, pp. 27-31.

Bielski, J. and M. Sikorski. 1996. "Disturbances of Glucose Tolerance in Workers Exposed to Electromagnetic Radiation." *Medycyna Pracy* 47(3): 227-31 (in Polish).

Brown, John. 1790. *The Elements of Medicine*. Philadelphia: T. Dobson.

Bruce, Clinton R., Mitchell J. Anderson, Andrew L. Carey, David G. Newman, Arend Bonen, Adamandia D. Kriketos, Gregory J. Cooney, and John A. Hawley. 2003. "Muscle Oxidative Capacity Is A Better Predictor of Insulin Sensitivity than Lipid Status." *Journal of Clinical Endocrinology and Metabolism* 88(11): 5444-51.

Brun, J. F., C. Fedou, and J. Mercier. 2000. "Postprandial Reactive Hypoglycemia." *Diabetes & Metabolism (Paris)* 26: 337-51.

Casson, Herbert N. 1910. *The History of the Telephone*. Chicago: A. C. McClurg.

Centers for Disease Control and Prevention. 2011. "Long-Term Trends in Diagnosed Diabetes." Atlanta.

———. 2014a. "Long-term Trends in Diabetes." Atlanta.

———. 2014b. "National Diabetes Statistics Report." Atlanta.

———. 2017. "National Diabetes Statistics Report." Atlanta.

Czerski, Przemysław, Kazimierz Ostrowski, Morris L. Shore, Charlotte Silverman, Michael J. Suess, and Berndt Waldeskog, eds. 1974. *Biologic Effects and Health Hazards of Microwave Radiation: Proceedings of an International Symposium, Warsaw, 15-18 October 1973*. Warsaw: Polish Medical Publishers.

DeLany, James P., John J. Dubé, Robert A. Standley, Giovanna Distefano, Bret H. Goodpaster, Maja Stefanovic-Racic, Paul M. Coen, and Frederico G. S. Toledo. 2014. "Racial Differences in Peripheral Insulin Sensitivity and Mitochondrial Capacity in the Absence of Obesity." *Journal of Clinical Endocrinology and Metabolism* 99(11): 4307-14.

Diabetes Care. 2002. "Report of the Expert Committee on the Diagnosis and Classification of Diabetes Mellitus." 25 (supp. 1): S5-S20.

Dodge, Christopher H. 1970. "Clinical and Hygienic Aspects of Exposure to Electromagnetic Fields." In: Stephen F. Cleary, ed., *Biological Effects and Health Implications of Microwave Radiation. Symposium Proceedings* (Rockville, MD: U.S. Department of Health, Education and Welfare), Publication BRH/DBE 70-2, pp. 140-49.

Dufty, William. 1975. *Sugar Blues*. Radnor, PA: Chilton.

Dumanskiy Yury D., N. G. Nikitina, Lyudmila A. Tomashevskaya, F. R. Kholyavko, K. S. Zhypakhin, and V. A. Yurmanov. 1982. "Meteorological Radar as Source of SHF Electromagnetic Field Energy and Problems of Environmental Hygiene." *Gigiyena i Sanitariya* 1982(2): 7-11. JPRS 81300, pp. 58-63.

Dumanskiy, Yury D. and V. F. Rudichenko. 1976. "Dependence of the Functional Activity of Liver Mitochondria on Microwave Radiation." *Gigiyena i Sanitariya* 1976(4): 16-19. JPRS 72606 (1979), pp. 27-32.

Dumanskiy, Yury D. and M. G. Shandala. 1974. "The Biologic Action and Hygienic Significance of Electromagnetic Fields of Superhigh and Ultrahigh Frequencies in Densely Populated Areas." In: P. Czerski et al., eds., *Biologic Effects and Health Hazards of Microwave Radiation: Proceedings of an International Symposium, Warsaw, 15-18 October 1973* (Warsaw: Polish Medical Publishers), pp. 289-93.

Dumanskiy Yury D. and Lyudmila A. Tomashevskaya. 1978. "Investigation of the Activity of Some Enzymatic Systems in Response to a Super-high Frequency Electromagnetic Field." *Gigiyena i Sanitariya* 1978(8): 23-27. JPRS 72606 (1979), pp. 1-7.

———. 1982. "Hygienic Evaluation of 8-mm Wave Electromagnetic Fields." *Gigiyena i Sanitariya* 1982(6): 18-20. JPRS 81865, pp. 6-9.

Felber, Jean-Pierre and Alfredo Vannotti. 1964. "Effects of Fat Infusion on Glucose Tolerance and Insulin Plasma Levels." *Medicina Experimentalis* 10: 153-56.

Flegal, Katherine M., Margaret D. Carroll, Robert J. Kuczmarski, and Clifford L. Johnson. 1998. "Overweight and Obesity in the United States: Prevalence and Trends, 1960-1994." *International Journal of Obesity* 22: 39-47.

Flegal, Katherine M., Margaret D. Carroll, Cynthia L. Ogden, and Clifford L. Johnson. 2002. "Prevalence and Trends in Obesity Among US Adults, 1999-2000." *JAMA* 288(14): 1723-27.

Flegal, Katherine M., Margaret D. Carroll, Cynthia L. Ogden, and Lester R. Curtin. 2010. "Prevalence and Trends in Obesity Among US Adults, 1999-2008." *JAMA* 303(3): 235-41.

Fothergill, J. Milner. 1884. "The Diagnosis of Diabetes." *North Carolina Medical Journal* 13: 146-47 (reprinted from *Philadelphia Medical Times*).

Gabovich, P. D., O. I. Shutenko, I. P. Kozyarin, and I. I. Shvayko. 1979. "Effects from Combined Exposure to Infrasound and Superhigh Frequency Electromagnetic Fields in Experiment." *Gigiyena i Sanitariya* 1979(10): 12-14. JPRS 75515 (1980), pp. 30-35.

Gel'fon, I. A. and Maria N. Sadchikova. 1960. "Protein Fractions and Histamine of the Blood under the Influence of UHF and HF." In: A. A. Letavet and Z. V. Gordon, eds., *The Biological Action of Ultrahigh Frequencies* (Moscow: Academy of Medical Sciences), JPRS 12471, pp. 42-46.

Gembitskiy, Ye. V. 1970. "Changes in the Functions of the Internal Organs of Personnel Operating Microwave Generators." In: I. R. Petrov. ed., *Influence of Microwave Radiation on the Organism of Man and Animals* (Leningrad: "Meditsina"), in English translation, 1972 (Washington, DC: NASA), report no. TTF-708, pp. 106-25.

Gerbitz, Klaus-Dieter, Klaus Gempel, and Dieter Brdiczka. 1996. "Mitochondria and Diabetes: Genetic, Biochemical, and Clinical Implications of the Cellular Energy Circuit." *Diabetes* 45(2): 113-26.

Gohdes, Dorothy. 1995. "Diabetes in North American Indians and Alaska Natives." In: M. I. Harris et al., eds., *Diabetes in America*, 2nd ed. (Bethesda, MD: National Institute of Diabetes and Digestive and Kidney Diseases), NIH publication no. 95-1468, pp. 683-702.

Gordon, Zinaida V., ed. 1973. *O biologicheskom deystvii elektromagnitnykh poley radiochastot*, 4th ed. Moscow. In English translation as *Biological Effects of Radiofrequency Electromagnetic Fields*, JPRS 63321 (1974).

Gray, Charlotte. 2006. *Reluctant Genius: The Passionate Life and Inventive Mind of Alexander Graham Bell.* Toronto: HarperCollins.

Hales, Craig M., Cheryl D. Fryar, Margaret D. Carroll, David S. Freedman, and Cynthia L. Ogden. 2018. "Trends in Obesity and Severe Obesity Prevalence in US Youth and Adults by Sex and Age, 2007-2008 to 2015-2016." *JAMA* 319(16): 1723-25.

Harris, Maureen I., Catherine C. Cowie, Michael P. Stern, Edward J. Boyko, Gayle E. Reiber, and Peter H. Bennet, eds. 1995. *Diabetes in America*, 2nd ed. Bethesda, MD: National Institute of Diabetes and Digestive and Kidney Diseases. NIH publication no. 95-1468.

Harris, Seale. 1924. "Hyperinsulinism and Dysinsulinism." *Journal of the American Medical Association* 83(10): 729-33.

Hirsch, August. 1883, 1885, 1886. *Handbook of Geographical and Historical Pathology*, 3 vols. London: New Sydenham Society.

Howe, Hubert S. 1931. "Edison Lost Will to Live, Doctor Says." *Pittsburgh Post-Gazette*, October 19, p. 2.

Hurley, Dan. 2011. *Diabetes Rising: How a Rare Disease Became a Modern Pandemic, and What To Do About It.* New York: Kaplan.

Israel, Paul. 1998. *Edison: A Life of Invention.* New York: Wiley.

Jerabek, Jiri. 1979. "Biological Effects of Magnetic Fields." *Pracovni Lekarstvi* 31(3): 98-106. JPRS 76497 (1980), pp. 1-25.

Jones, Francis Arthur. 1907. *Thomas Alva Edison: Sixty Years of an Inventor's Life.* New York: Thomas Y. Crowell.

Joslin, Elliott Proctor. 1917. *The Treatment of Diabetes Mellitus*, 2nd ed. Philadelphia: Lea & Febiger.

———. 1924. "The Treatment of Diabetes Mellitus." *Canadian Medical Association Journal* 14(9): 808-11.

———. 1927. "The Outlook for the Diabetic." *California and Western Medicine* 26(2): 177-82, 26(3): 328-31.

———. 1943. "The Diabetic." *Canadian Medical Association Journal* 48: 488-97.

———. 1950. "A Half-Century's Experience in Diabetes Mellitus." *British Medical Journal* 1: 1095-98.

Joslin Diabetes Clinic, paid advertisement. "Edison Lived His Last 50 Years with Diabetes," *Pittsburgh Press*, April 14, 1990; October 14, 1990; *Pittsburgh Post-Gazette*, April 18, 1990; April 25, 1990; May 23, 1990; June 22, 1990; September 19, 1990; October 17, 1990.

Josephson, Matthew. 1959. *Edison: A Biography.* McGraw-Hill, NY.

Kelley, David E., Bret Goodpaster, Rena R. Wing and Jean-Aimé Simoneau. 1999. "Skeletal Muscle Fatty Acid Metabolism in Association with Insulin Resistance, Obesity, and Weight Loss." *American Journal of Physiology – Endocrinology and Metabolism* 277: E1130-41.

Kelley, David E. and Lawrence J. Mandarino. 2000. "Fuel Selection in Human Skeletal Muscle in Insulin Resistance: A Reexamination." *Diabetes* 49: 677-83.

Kelley, David E., Jing He, Elizabeth V. Menshikova, and Vladimir B. Ritov. 2002. "Dysfunction of Mitochondria in Human Skeletal Muscle in Type 2 Diabetes." *Diabetes* 51: 2944-50.

Kelley, David E. and Jean-Aimé Simoneau. 1994. "Impaired Free Fatty Acid Utilization by Skeletal Muscle in Non-Insulin-dependent Diabetes Mellitus." *Journal of Clinical Investigation* 94: 2349-56.

Kim, Juhee, Karen E. Peterson, Kelley S. Scanlon, Garrett M. Fitzmaurice, Aviva Must, Emily Oken, Sheryl L. Rifas-Shiman, Janet W. Rich-Edwards, and Matthew W. Gillman. 2006. "Trends in Overweight from 1980 through 2001 among Preschool-Aged Children Enrolled in a Health Maintenance Organization." *Obesity* 14(7): 1-6.

Kleinfield, N. R. 2006. "Diabetes and Its Awful Toll Quietly Emerge as a Crisis." *New York Times*, January 9, 2006.

Klimentidis, Yann C., T. Mark Beasley, Hui-Yi Lin, Giulianna Murati, Gregory E. Glass, Marcus Guyton, Wendy Newton, Matthew Jorgensen, Steven B. Heymsfield, Joseph Kemnitz, Lynn Fairbanks, and David B. Allison. 2011. "Canaries in the Coal Mine: a Cross-Species Analysis of the Plurality of Obesity Epidemics." *Proceedings of the Royal Society B* 278: 1626-32.

Klimková-Deutschová, Eliska. 1974. "Neurologic Findings in Persons Exposed to Microwaves." In: P. Czerski et al., eds., *Biologic Effects and Health Hazards of Microwave Radiation: Proceedings of an International Symposium, Warsaw, 15-18 October 1973* (Warsaw: Polish Medical Publishers), pp. 269-72.

Kochanek, Kenneth D., Sherry L. Murphy, Jiaquan Xu, and Elizabeth Arias. 2019. "Deaths: Final data for 2017." *National Vital Statistics Reports*, vol. 68, no. 9. Hyattsville, MD: National Center for Health Statistics.

Kolodub, F. A. and O. N. Chernysheva. 1980. "Special Features of Carbohydrate-energy and Nitrogen Metabolism in the Rat Brain under the Influence of Magnetic Fields of Commercial Frequency." *Ukrainskiy Biokhemicheskiy Zhurnal* 1980(3): 299-303. JPRS 77393 (1981), pp. 42-44.

Koo, Won W. and Richard D. Taylor. 2011. "2011 Outlook of the U.S. and World Sugar Markets, 2010-2020." *Agribusiness & Applied Economics*, no. 679.

Kuczmarski, Robert J., Katherine M. Flegal, Stephen M. Campbell, and Clifford L. Johnson. 1994. "Increasing Prevalence of Overweight Among US Adults: The National Health and Nutrition Examination Surveys, 1960 to 1991." *JAMA* 272(3): 205-11.

Kwon, Myoung Soo, Victor Vorobyev, Sami Kännälä, Matti Laine, Juha O. Rinne, Tommi Toivonen, Jarkko Johansson, Mika Teräs, Harri Lindholm, Tommi Alanko, and Heikki Hämäläinen. 2011. "GSM Mobile Phone Radiation Suppresses Brain Glucose Metabolism." *Journal of Cerebral Blood Flow and Metabolism*, 31(12): 2293-2301.

Levy, Renata Bertazzi, Rafael Moreira Claro, Daniel Henrique Bandoni, Lenise Mondini, and Carlos Augusto Monteiro. 2012. "Availability of Added Sugars in Brazil: Distribution, Food Sources and Time Trends." *Revista Brasileira de Epidemiologia* 15(1): 3-12.

Li, De-Kun, Jeannette R. Ferber, Roxana Odouli, and Charles P. Quesenberry, Jr. 2012. "A Prospective Study of *In-utero* Exposure to Magnetic Fields and the Risk of Childhood Obesity." *Scientific Reports* 2: 540.

Lorenzo, Carlos and Steven M. Haffner. 2010. "Performance Characteristics of the New Definition of Diabetes: The Insulin Resistance Atherosclerosis Study." *Diabetes Care* 33(2): 335-37.

Mann, Devin M., April P. Carson, Daichi Shimbo, Vivian Fonseca, Caroline S. Fox, and Paul Muntner. 2010. "Impact of A1C Screening Criterion on the Diagnosis of Pre-Diabetes Among U.S. Adults." *Diabetes Care* 33(10): 2190-95.

Mazur, Allan. 2011. "Why Were 'Starvation Diets' Promoted for Diabetes in the Pre-Insulin Period?" *Nutrition Journal* 10: 23.

Morino, Katsutaro, Kitt Falk Petersen, and Gerald I. Shulman. 2006. "Molecular Mechanisms of Insulin Resistance in Humans and Their Potential Links with Mitochondrial Dysfunction." *Diabetes* 55 (suppl. 2): S9-S15.

Morris, Jeremiah Noah. 1995. "Obesity in Britain: Lifestyle Data Do Not Support Sloth Hypothesis." *British Medical Journal* 311: 1568-69.

Navakatikian, Mikhail A. and Lyudmila A. Tomashevskaya. 1994. "Phasic Behavioral and Endocrine Effects of Microwaves of Nonthermal Intensity." In: David O. Carpenter and Sinerik Ayrapetyan, eds., *Biological Effects of Electric and Magnetic Fields* (New York: Academic), vol. 1, pp. 333-42.

Nikitina, Valentina N. 2001. "Hygienic, Clinical and Epidemiological Analysis of Disturbances Induced by Radio Frequency EMF Exposure in Human Body." In: Kjell Hansson Mild, Monica Sandström, and Eugene Lyskov, eds., *Clinical and Physiological Investigations of People Highly Exposed to Electromagnetic Fields* (Umeå, Sweden: National Institute for Working life), Arbetslivsrapport 3, pp. 32-38.

Ogden, Cynthia L., Margaret D. Carroll, Brian K. Kit, and Katherine M. Flegal. 2012. "Prevalence of Obesity in the United States, 2009-2010." NCHS Data Brief no. 82, January 2012. Atlanta: National Center for Health Statistics, Centers for Disease Control and Prevention.

Ogden, Cynthia L., Katherine M. Flegal, Margaret D. Carroll, and Clifford L. Johnson. 2002. "Prevalence and Trends in Overweight Among US Children and Adolescents, 1999-2000." JAMA 288(14): 1728-32.

Patti, Mary-Elizabeth and Silvia Corvera. 2010. "The Role of Mitochondria in the Pathogenesis of Type 2 Diabetes." *Endocrine Reviews* 31(3): 364-95.

Petrov, Ioakim Romanovich, ed. 1970a. *Vliyaniye SVCh-izlucheniya na organism cheloveka i zhivotnykh*. Leningrad: "Meditsina." In English translation as *Influence of Microwave Radiation on the Organism of Man and Animals* (Washington, DC: NASA), report no. TTF-708, 1972.

———. 1970b. "Problems of the Etiology and Pathogenesis of the Pathological Processes Caused by Microwave Radiation." In: Petrov, ed., *Influence of Microwave Radiation on the Organism of Man and Animals*, pp. 147-165.

Prentice, Andrew M. and Susan A. Jebb. 1995. "Obesity in Britain: Gluttony or Sloth?" *British Medical Journal* 311: 437-39.

Presman, Aleksandr Samuilovich. 1970. *Electromagnetic Fields and Life*. New York: Plenum.

Randle, Philip J. 1998. "Regulatory Interactions between Lipids and Carbohydrates: The Glucose Fatty Acid Cycle After 35 Years." *Diabetes/Metabolism Reviews* 14: 263-83.

Randle, Philip J., P. B. Garland, C. N. Hales, and E. A. Newsholme. 1963. "The Glucose Fatty-Acid Cycle." *Lancet* 1: 785-89.

Reynolds, C. and D. B. Orchard. 1977. "The Oral Glucose Tolerance Test Revisited and Revised." *CMA Journal* 116: 1223-24.

Richardson, Benjamin Ward. 1876. *Diseases of Modern Life*. New York: D. Appleton.

Ritov, Vladimir B., Elizabeth V. Menshikova, Koichiro Azuma, Richard Wood, Frederico G. S. Toledo, Bret H. Goodpaster, Neil B. Ruderman, and David E. Kelley. 2010. "Deficiency of Electron Transport Chain in Human Skeletal Muscle Mitochondria in Type 2 Diabetes Mellitus and Obesity." *American Journal of Physiology – Endocrinology and Metabolism* 298: E49-58.

Rollo, John. 1798. *Cases of the Diabetes Mellitus*, 2nd ed. London: C. Dilly.

Sadchikova, Maria N. 1974. "Clinical Manifestations of Reactions to Microwave Irradiation in Various Occupational Groups." In: P. Czerski et al., eds., *Biologic Effects and Health Hazards of Microwave Radiation: Proceedings of an International Symposium, Warsaw, 15-18 October 1973* (Warsaw: Polish Medical Publishers), pp. 261-67.

Sadchikova, Maria N. and K. V. Glotova. 1973. "The Clinic, Pathogenesis, Treatment, and Outcome of Radiowave Sickness." In: Z. V. Gordon, ed., *Biological Effects of Radiofrequency Electromagnetic Fields*, JPRS 63321 (1974), pp. 54-62.

Schalch, Don S. and David M. Kipnis. 1965. "Abnormalities in Carbohydrate Tolerance Associated with Elevated Plasma Nonesterified Fatty Acids." *Journal of Clinical Investigation* 44(12): 2010-20.

Scriven, George P. 1915. "Notes on the Organization of Telegraph Troops in Foreign Armies. Great Britain." In: Scriven, *The Service of Information: United States Army*, (Washington, DC: Government Printing Office), pp. 127-32.

Shutenko, O. I., I. P. Kozyarin, and I. I. Shvayko. 1981. "Effects of Superhigh Frequency Electromagnetic Fields on Animals of Different Ages." *Gigiyena i Sanitariya* 1981(10): 35-38. JPRS 81300 (1982), pp. 85-90.

Simoneau, Jean-Aimé, Sheri R. Colberg, F. Leland Thaete, and David E. Kelley. 1995. "Skeletal Muscle Glycolytic and Oxidative Enzyme Capacities are Determinants of Insulin Sensitivity and Muscle Composition in Obese Women." *FASEB Journal* 9: 273-78.

Simoneau, Jean-Aimé and David E. Kelley. 1997. "Altered Glycolytic and Oxidative Capacities of Skeletal Muscle Contribute to Insulin Resistance in NIDDM." *Journal of Applied Physiology* 83: 166-71.

Stalvey, Michael S. and Desmond A. Schatz. 2008. "Childhood Diabetes Explosion." In: D. LeRoith and A. I. Vinik, eds., *Contemporary Endocrinology: Controversies in Treating Diabetes: Clinical and Research Aspects* (Totowa, NJ: Humana), pp. 179-98.

Starr, Douglas. 1998. *Blood: An Epic History of Medicine and Commerce*. New York: Knopf.

Sydenham, Thomas. 1848. *The Works of Thomas Sydenham, M.D.*, London: Sydenham Society.

Syngayevskaya, V. A. 1970. "Metabolic Changes." In: I. R. Petrov, ed., *Influence of Microwave Radiation on the Organism of Man and Animals* (Leningrad: "Meditsina"), in English translation, 1972 (Washington, DC: NASA), report no. TTF-708, pp. 48-60.

Thatcher, Craig D., R. Scott Pleasant, Raymond J. Geor, François Elvinger, Kimberly A. Negrin, J. Franklin, Louisa Gay, and Stephen R. Werre. 2009. "Prevalence of Obesity in Mature Horses: An Equine Body Condition Study." *Journal of Animal Physiology and Animal Nutrition* 92: 222.

The Sun. 1891. "Edison His Own Doctor." May 10, p. 26.

Therapeutic Gazette. 1884. "Sugar in the Urine – What Does it Signify?" 8: 180.

Toledo, Frederico G. S., Elizabeth V. Menshikova, Koichiro Azuma, Zofia Radiková, Carol A. Kelley, Vladimir B. Ritov, and David E. Kelley. 2008. "Mitochondrial Capacity in Skeletal Muscle is Not Stimulated by Weight Loss Despite Increases in Insulin Action and Decreases in Intramyocellular Lipid Content." *Diabetes* 57: 987-94.

Tomashevskaya, Lyudmila A. and E. A. Solenyi. 1986. "Biologicheskoye deystviye i gigiyenicheskoye znacheniye elektromagnitnogo polya, sozdavayemogo beregovimi radiolokatsionnimi sredstvami" ("Biological Action and Hygienic Significance of the Electromagnetic Field Created by Coastal Radar Facilities"). *Gigiyena i Sanitariya* 1986(7): 34-36.

Tomashevskaya, Lyudmila A. and Yuri D. Dumanskiy. 1988. "Gigiyenicheskaya otsenka biologicheskogo deystviya impul'snykh elektromagnitnykh poley" ("Hygienic Evaluation of the Biological Effect of Pulsed Electromagnetic Fields"). *Gigiyena i Sanitariya* 1988(9): 21-24.

Welsh, Jean A., Andrea Sharma, Jerome L. Abramson, Viola Vaccarino, Cathleen Gillespie, and Miriam B. Vos. 2010. "Caloric Sweetener Consumption and Dyslipidemia among US Adults." *JAMA* 303(15): 1490-97.

Whytt, Robert. 1768. *The Works of Robert Whytt, M.D.* Edinburgh: J. Balfour. Reprinted by The Classics of Neurology and Neurosurgery Library, Birmingham, AL, 1984.

Woodyatt, R. T. 1921. "Object and Method of Diet Adjustment in Diabetes." *Archives of Internal Medicine* 28(2): 125-41.

World Health Organization. 2010. *Definition and Diagnosis of Diabetes Mellitus and Intermediate Hyperglycemia: Report of a WHO/IDF Consultation*. Geneva 2010.

―――. 2014. *Global Status Report on Noncommunicable Diseases*. Geneva.

Bhutan

Bhutan Broadcasting Service. 2007. "Diabetes: Emerging Non-communicable Disease in Bhutan." November 13.

Chhetri, Pushkar. 2010. "ADB Grants $21.6 m for Rural Electrification." *Bhutan Observer*, November 10.

Choden, Tshering. 2010. "Be Wary of Lifestyle Disease." *Bhutan Times*, March 21.

Giri, Bhakta Raj, Krishna Prasad Sharma, Rup Narayan Chapagai, and Dorji Palzom. 2013. "Diabetes and Hypertension in Urban Bhutanese Men and Women." *Indian Journal of Community Medicine* 38(3): 138-43.

Pelden, Sonam. 2009. "Diabetes – The Slow Killer." *Kuensel Online* (Bhutan's daily news website), November 18.

United States Agency for International Development. September 2002. *Regional Hydro-power Resources: Status of Development and Barriers: Bhutan*. Prepared by Nexant/South Asia Regional Initiative for Energy.

Wangchuk, Jigme. 2011. "Bhutan Could Be Eating Itself Sick." *Bhutan Observer*, November 19.

Wangdi, Tashi. 2015. "Type 1 Diabetes Mellitus in Bhutan." *Indian Journal of Endocrinology and Metabolism* 19 (suppl. 1): S14-S15.

Chapter 13

Acebo, Paloma, Daniel Giner, Piedad Calvo, Amaya Blanco-Rivero, Álvaro D. Ortega, Pedro L. Fernández, Giovanna Roncador, Edgar Fernández-Malavé, Margarita Chamorro, and José M. Cuezva. 2009. "Cancer Abolishes the Tissue Type-Specific Differences in the Phenotype of Energetic Metabolism." *Translational Oncology* 2(3): 138-45.

Adams, Samuel Hopkins. 1913. "What Can We Do About Cancer?" *Ladies Home Journal*, May, pp. 21-22.

American Lung Association. 2010. *Trends in Lung Cancer Morbidity and Mortality*. Washington, DC.

———. 2011. *Trends in Tobacco Use*. Washington, DC.

Apte, Shireesh P. and Rangaprasad Sarangarajan, eds. 2009a. *Cellular Respiration and Carcinogenesis*. New York: Humana.

———. 2009b. "Metabolic Modulation of Carcinogenesis. In: Apte and Sarangarajan, eds., *Cellular Respiration and Carcinogenesis* (New York: Humana), pp. 103-18.

Barlow, Lotti, Kerstin Westergren, Lars Holmberg, and Mats Talbäck. 2009. "The Completeness of the Swedish Cancer Register – A Sample Survey for Year 1998." *Acta Oncologica* 48: 27-33.

Brière, Jean-Jacques, Paul Bénit, and Pierre Rustin. 2009. "The Electron Transport Chain and Carcinogenesis." In: Shireesh P. Apte and Rangaprasad Sarangarajan, eds., *Cellular Respiration and Carcinogenesis* (New York: Humana), pp. 19-32.

Burk, Dean. 1942. "On the Specificity of Glycolysis in Malignant Liver Tumors as Compared with Homologus Adult or Growing Liver Tissues." In: *A Symposium on Respiratory Enzymes* (Madison: University of Wisconsin Press), pp. 235-45.

Burk, Dean, Mark Woods and Jehu Hunter. 1967. "On the Significance of Glucolysis for Cancer Growth, with Special Reference to Morris Rat Hepatomas." *Journal of the National Cancer Institute* 38(6): 839-63.

Coley, William B. 1910. "The Increase of Cancer." *Southern Medical Journal* 3(5): 287-92.

Cori, Carl F. and Gerty T. Cori. 1925. "The Carbohydrate Metabolism of Tumors. I. The Free Sugar, Lactic Acid, and Glycogen Content of Malignant Tumors." *Journal of Biological Chemistry* 64: 11-22.

———. 1925. "The Carbohydrate Metabolism of Tumors. II. Changes in the Sugar, Lactic Acid, and CO_2-Combining Power of Blood Passing Through a Tumor." *Journal of Biological Chemistry* 65: 397-405.

Cuezva, José M. 2010. "The Bioenergetic Signature of Cancer." *BMC Proceedings* 4 (suppl. 2): 07.

Cuezva, José M., Maryla Krajewska, Mighel López de Heredia, Stanislaw Krajewski, Gema Santamaría, Hoguen Kim, Juan M. Zapata, Hiroyuki Marusawa, Margarita

Chamorro, and John C. Reed. 2002. "The Bioenergetic Signature of Cancer: A Marker of Tumor Progression." *Cancer Research* 62: 6674-81.

Cutler, David M. 2008. "Are We Finally Winning the War on Cancer?" *Journal of Economic Perspectives* 22(4): 3-26.

Czarnecka, Anna and Ewa Bartnik. 2009. "Mitochondrial DNA Mutations in Tumors." In: Shireesh P. Apte and Rangaprasad Sarangarajan, eds., *Cellular Respiration and Carcinogenesis* (New York: Humana), pp. 119-30.

Dang, Chi V. and Gregg L. Semenza. 1999. "Oncogenic Alterations of Metabolism." *Trends in Biochemical Sciences* 24: 68-72.

Fantin, Valeria R., Julie St.-Pierre, and Philip Leder. 2006. "Attenuation of LDH-A Expression Uncovers a Link between Glycolysis, Mitochondrial Physiology, and Tumor Maintenance." *Cancer Cell* 9: 425-34.

Felty, Quentin and Deodutta Roy. 2005. "Estrogen, Mitochondria, and Growth of Cancer and Non-Cancer Cells." *Journal of Carcinogenesis* 4: 1.

Ferreira, Túlio César and Élida Geralda Campos. 2009. "Regulation of Glucose and Energy Metabolism in Cancer Cells by Hypoxia Inducible Factor 1." In: Shireesh P. Apte and Rangaprasad Sarangarajan, eds., *Cellular Respiration and Carcinogenesis* (New York: Humana), pp. 73-90.

Furlow, Bryant. 2007. "VA Withholds Data From Cancer Registries Used to Track Veteran Cancer Rates." *Lancet Oncology* 8(9): 762-63.

Gatenby, Robert A. and Robert J. Gillies. 2004. "Why do Cancers have High Aerobic Glycolysis?" *Nature Reviews. Cancer* 4: 891-99.

Gillies, Robert J., Ian Robey, and Robert A. Gatenby. 2008. "Causes and Consequences of Increased Glucose Metabolism of Cancers." *Journal of Nuclear Medicine* 49(6) (suppl.): 24S-42S.

Giovannucci, Edward, David M. Harlan, Michael C. Archer, Richard M. Bergenstal, Susan M. Gapstur, Laurel A. Habel, Michael Pollak, Judith G. Regensteiner, and Douglas Yee. 2010. "Diabetes and Cancer: A Consensus Report." *Diabetes Care* 33(7): 1674-84.

Goldblatt, Harry and Gladys Cameron. 1953. "Induced Malignancy in Cells from Rat Myocardium Subjected to Intermittent Anaerobiosis during Long Propagation *In vitro*." *Journal of Experimental Medicine* 97: 525-52.

Goldblatt, Harry and Libby Friedman. 1974. "Prevention of Malignant Change in Mammalian Cells during Prolonged Culture *In vitro*." *Proceedings of the National Academy of Sciences* 71(5): 1780-82.

Goldblatt, Harry, Libby Friedman, and Ronald L. Cechner. 1973. "On the Malignant Transformation of Cells during Prolonged Culture Under Hypoxic Conditions *In vitro*." *Biochemical Medicine* 7: 241-52.

Goldhaber, Paul. 1959. "The Influence of Pore Size on Carcinogenicity of Subcutaneously Implanted Millipore Filters." *Proceedings of the American Association for Cancer Research* 3(1): 228. Abstract.

Gonzalez-Cuyar, Luis F., Fabio Tavora, Iusta Caminha, George Perry, Mark A. Smith, and Rudy J. Castellani. 2009. "Cellular Respiration and Tumor Suppressor Genes." In: Shireesh P. Apte and Rangaprasad Sarangarajan, eds., *Cellular Respiration and Carcinogenesis* (New York: Humana), pp. 131-44.

Gordon, Tavia, Margaret Crittendon, and William Haenszel. 1961. "Cancer Mortality Trends in the United States, 1930-1955." In: *End Results and Mortality Trends in Cancer*, National Cancer Institute Monograph no. 6 (Washington, DC: U.S. Dept. of Health, Education, and Welfare), pp. 131-298.

Gover, Mary. 1939. *Cancer Mortality in the United States. I. Trend of Recorded Cancer Mortality in the Death Registration States of 1900 from 1900 to 1935.* Public Health Bulletin no. 248, U.S. Public Health Service. Washington, DC: Government Printing Office.

Guan, Xiaofan and Olle Johansson. 2005. "The Sun-Shined Health." *European Biology and Bioelectromagnetics* 1: 420-23.

Gullino, Pietro M., Shirley H. Clark, and Flora H. Grantham. 1964. "The Interstitial Fluid of Solid Tumors." *Cancer Research* 24: 780-97.

Hallberg, Örjan. 2009. *Facts and Fiction about Skin Melamona.* Farsta, Sweden: Hallberg Independent Research.

Hallberg, Örjan and Olle Johansson. 2002a. "Cancer Trends during the 20th Century." *Journal of the Australasian College of Nutrition and Environmental Medicine* 21(1): 3-8.

———. 2002b. "Melanoma Incidence and Frequency Modulation (FM) Broadcasting." *Archives of Environmental Health* 57(1): 32-40.

———. 2004a. "Malignant Melanoma of the Skin – Not a Sunshine Story!" *Medical Science Monitor* 10(7): CR336-40.

———. 2004b. "1997 – A Curious Year in Sweden." *European Journal of Cancer Prevention* 13: 535-38.

———. 2005. "FM Broadcasting Exposure Time and Malignant Melanoma Incidence." *Electromagnetic Biology and Medicine* 24: 1-8.

———. 2009. "Apparent Decreases in Swedish Public Health Indicators After 1997 – Are They Due to Improved Diagnostics or to Environmental Factors?" *Pathophysiology* 16(1): 43-46.

———. 2010. "Sleep on the Right Side – Get Cancer on the Left?" *Pathophysiology* 17(3): 157-60.

Hardell, Lennart. 2007. "Long-term Use of Cellular and Cordless Phones and the Risk of Brain Tumours." Örebro University, power point presentation, August 31.

Hardell, Lennart and Michael Carlberg. 2009. "Mobile Phones, Cordless Phones and the Risk for Brain Tumours." *International Journal of Oncology* 35: 5-17.

Hardell, Lennart, Michael Carlberg, and Kjell Hansson Mild. 2010. "Mobile Phone Use and the Risk for Malignant Brain Tumors: A Case-Control Study on Deceased Cases and Controls." *Neuroepidemiology* 35: 109-14.

———. 2011a. "Pooled Analysis of Case-control Studies on Malignant Brain Tumours and the Use of Mobile and Cordless Phones Including Living and Deceased Subjects." *International Journal of Oncology* 38: 1465-74.

———. 2011b. "Re-analysis of Risk for Glioma in Relation to Mobile Telephone Use: Comparison with the Results of the Interphone International Case-control Study." *International Journal of Epidemiology* 40(4): 1126-28.

Hardell, Lennart, Michael Carlberg, Fredrik Söderqvist, and Kjell Hansson Mild. 2010. "Re: Time Trends in Brain Tumor Incidence Rates in Denmark, Finland, Norway, and Sweden, 1974-2003." *Journal of the National Cancer Institute* 102(10): 740-41.

Harris, Adrian L. 2002. "Hypoxia – a Key Regulatory Factor in Tumour Growth." *Nature Reviews. Cancer* 2: 38-47.

Harris, David, Nora Kropp, and Paul Pulliam. 2008. "A Comparison of National Cancer Registries in India and the United States of America." 3MC Conference Proceedings, Berlin.

Highton, Edward. 1852. *The Electric Telegraph: Its History and Progress*. London: John Weale.

Hirsch, August. 1886. "Cancer." In: Hirsch, *Handbook of Geographical and Historical Pathology* (London: New Sydenham Society), vol. 3, pp. 502-9.

Hoffman, Frederick Ludwig. 1915. *The Mortality From Cancer Throughout the World*. Newark: Prudential.

Howlader, Nadia, Lynn A. Ries, David G. Stinchcomb, and Brenda K. Edwards. 2009. "The Impact of Underreported Veterans Affairs Data on National Cancer Statistics: Analysis Using Population-Based SEER Registries." *Journal of the National Cancer Institute* 101(7): 533-36.

International Agency for Research on Cancer. *World Cancer Report 2008*. Lyon, France.

Isodoro, Antonio, Enrique Casado, Andrés Redondo, Paloma Acebo, Enrique Espinosa, Andrés M. Alonso, Paloma Cejas, David Hardisson, Juan A. Fresno Vara, Cristóbal Belda-Iniesta, Manuel González-Barón, and José M. Cuezva. 2005. "Breast Carcinomas Fulfill the Warburg Hypothesis and Provide Metabolic Markers of Cancer Prognosis." *Carcinogenesis* 26(12): 2095-2104.

Isidoro, Antonio, Marta Martínez, Pedro L. Fernández, Álvaro D. Ortega, Gema Santamaría, Margarita Chamorro, John C. Reed, and José M. Cuezva. 2004. "Alteration of the Bioenergetic Phenotype of Mitochondria is a Hallmark of Breast, Gastric, Lung and Oesophageal Cancer." *Biochemical Journal* 378: 17-20.

Johansen, Christoffer, John D. Boice, Jr., Joseph K. Mclaughlin, and Jørgen H. Olsen. 2001. "Cellular Telephones and Cancer – a Nationwide Cohort Study in Denmark." *Journal of the National Cancer Institute* 93(3): 203-7.

Johansson, Olle. 2005. "The Effects of Radiation in the Cause of Cancer." *Integrative Cancer and Oncology News* 4(4): 32-37.

Khurana, Vini G., Charles Teo, Michael Kundi, Lennart Hardell, and Michael Carlberg. 2009. "Cell Phones and Brain Tumors: A Review Including the Long-Term Epidemiological Data." *Surgical Neurology* 72(3): 205-14.

Kidd, John G., Richard J. Winzler, and Dean Burk. 1944. "Comparative Glycolytic and Respiratory Metabolism of Homologous Normal, Benign, and Malignant Rabbit Tissues." *Cancer Research* 4: 547-53.

Kim, Jung-whan and Chi V. Dang. 2006. "Cancer's Molecular Sweet Tooth and the Warburg Effect." *Cancer Research* 66(18): 8927-30.

Kochanek, Kenneth D., Sherry L. Murphy, Jiaquan Xu, and Elizabeth Arias. 2019. "Deaths: Final data for 2017." *National Vital Statistics Reports*, vol. 68, no. 9. Hyattsville, MD: National Center for Health Statistics.

Kondoh, Hiroshi. 2009. "The Role of Glycolysis in Cellular Immortalization." In: Shireesh P. Apte and Rangaprasad Sarangarajan, eds., *Cellular Respiration and Carcinogenesis*, (New York: Humana), pp. 91-102.

Kondoh, Hiroshi, Matilde E. Lleonart, Jesus Gil, David Beach, and Gordon Peters. 2005. "Glycolysis and Cellular Immortalization." *Drug Discovery Today: Disease Mechanisms* 2(2): 263-67.

Kondoh, Hiroshi, Matilde E. Lleonart, Jesus Gil, Jing Wang, Paolo Degan, Gordon Peters, Dolores Martinez, Amancio Carnero, and David Beach. 2005. "Glycolytic Enzymes Can Modulate Cellular Life Span." *Cancer Research* 65(1): 177-85.

Krebs, Hans. 1981. *Otto Warburg: Cell Physiologist, Biochemist, and Eccentric.* Oxford: Clarendon Press.

Kroemer, G. 2006. "Mitochondria in Cancer." *Oncogene* 25: 4630-32.

Lombard, Louise S. and Ernest J. Witte. 1959. "Frequency and Types of Tumors in Mammals and Birds of the Philadelphia Zoological Gardens." *Cancer Research* 19(2): 127-41.

López-Ríos, Fernando, María Sánchez-Aragó, Elena García-García, Álvaro D. Ortega, José R. Berrendero, Francisco Pozo-Rodríguez, Ángel López-Encuentra, Claudio Ballestín, and José M. Cuezva. 2007. "Loss of the Mitochondrial Bioenergetic Capacity Underlies the Glucose Avidity of Carcinomas." *Cancer Research* 67(19): 9013-17.

Malmgren, Richard A. and Clyde C. Flanigan. 1955. "Localization of the Vegetative Form of *Clostridium tetani* in Mouse Tumors Following Intravenous Spore Administration." *Cancer Research* 15: 473-78.

Maynard, George Darell. 1910. "A Statistical Study in Cancer Death-Rates." *Biometrika* 7: 276-304.

McFate, Thomas, Ahmed Mohyeldin, Huasheng Lu, Jay Thakar, Jeremy Henriques, Nader D. Halim, Hong Wu, Michael J. Schell, Tsz Mon Tsang, Orla Teahan, Shaoyu Zhou, Joseph A. Califano, Nam Ho Jeoung, Robert A. Harris, and Ajay Verma. 2008. "Pyruvate Dehydrogenase Complex Activity Controls Metabolic and Malignant Phenotype in Cancer Cells." *Journal of Biological Chemistry* 283(33): 22700-8.

Milham, Samuel and Eric M. Ossiander. 2001. "Historical Evidence that Residential Electrification Caused the Emergence of the Childhood Leukemia Peak." *Medical Hypotheses* 56(3): 290-95.

Moffat, Shannon. 1988. "Stanford's Power Line Research Pioneers." *Sandstone and Tile* 12(2-3): 3-7.

Moreno-Sánchez, Rafael, Sara Rodríguez-Enríquez, Álvaro Marín-Hernández and Emma Saavedra. 2007. "Energy Metabolism in Tumor Cells." *FEBS Journal* 274: 1393-1418.

National Cancer Institute. 2009. "New Early Detection Studies of Lung Cancer in Non-Smokers Launched Today." Press release, May 4.

Pascua, Marcelino, Director, Division of Health Statistics, World Health Organization. 1952. "Evolution of Mortality in Europe during the Twentieth Century." *Epidemiological and Vital Statistics Report* 5: 1-144.

Pedersen, Peter L. 1978. "Tumor Mitochondria and the Bioenergetics of Cancer Cells." *Progress in Experimental Tumor Research* 22: 190-274.

Racker, Efraim and Mark Spector. 1956. "Warburg Effect Revisited: Merger of Biochemistry and Molecular Biology." *Science* 213: 303-7.

Richardson, Benjamin Ward. 1876. *Diseases of Modern Life.* New York: D. Appleton.

Ristow, Michael. 2006. "Oxidative Metabolism in Cancer Growth." *Current Opinion in Clinical Nutrition and Metabolic Care* 9: 339-45.

Ristow, Michael and José M. Cuezva. 2009. "Oxidative Phosphorylation and Cancer: The Ongoing Warburg Hypothesis." In: Shireesh P. Apte and Rangaprasad Sarangarajan, eds., *Cellular Respiration and Carcinogenesis* (New York: Humana), pp. 1-18.

Sánchez-Aragó, María, Margarita Chamorro and José M. Cuezva. 2010. "Selection of Cancer Cells with Repressed Mitochondria Triggers Colon Cancer Progression." *Carcinogenesis* 31(4): 567-76.

Scatena, Roberto, Patrizia Bottoni, and Bruno Giardina. 2009. "Cellular Respiration and Dedifferentiation." In: Shireesh P. Apte and Rangaprasad Sarangarajan, eds., *Cellular Respiration and Carcinogenesis* (New York: Humana), pp. 45-54.

Scheers, Isabelle, Vincent Bachy, Xavier Stéphenne, and Étienne Marc Sokal. 2005. "Risk of Hepatocellular Carcinoma in Liver Mitochondrial Respiratory Chain Disorders." *Journal of Pediatrics* 146(3): 414-17.

Schüz, Joachim, Rune Jacobsen, Jørgen H. Olsen, John D. Boice, Jr., Joseph K. McLaughlin, and Christoffer Johansen. 2006. "Cellular Telephone Use and Cancer Risk: Update of a Nationwide Danish Cohort." *Journal of the National Cancer Institute* 98(23): 1707-13.

Semenza, Gregg L. "Foreword." 2009. In: Shireesh P. Apte and Rangaprasad Sarangarajan, eds., *Cellular Respiration and Carcinogenesis* (New York: Humana), pp. v-vi.

Semenza, Gregg L., Dmitri Artemov, Atul Bedi, Zaver Bhujwalla, Kelly Chiles, David Feldser, Erik Laughner, Rajani Ravi, Jonathan Simons, Panthea Taghavi, and Hua Zhong. 2001. "The Metabolism of Tumours: 70 Years Later." In: *The Tumour Microenvironment: Causes and Consequences of Hypoxia and Acidity*. Novartis Foundation Symposium 240 (Chichester, UK: Wiley), pp. 251-64.

Simonnet, Hélène, Nathalie Alazard, Kathy Pfeiffer, Catherine Gallou, Christophe Béroud, Jocelyne Demont, Raymonde Bouvier, Hermann Schägger, and Catherine Godinot. 2002. "Low Mitochondrial Respiratory Chain Content Correlates with Tumor Aggressiveness in Renal Cell Carcinoma." *Carcinogenesis* 23(5): 759-68.

Smith, Lloyd H., Jr. 1985. "Na⁺-H⁺ Exchange, Oncogenes and Growth Regulation in Normal and Tumor Cells." *Western Journal of Medicine* 143(3): 365-70.

Soderqvist, Fredrik, Michael Carlberg, Kjell Hansson Mild, and Lennart Hardell. 2011. "Childhood Brain Tumour Risk and Its Association with Wireless Phones: A Commentary." *Environmental Health* 10: 106.

Srivastava, Sarika and Carlos T. Moraes. 2009. "Cellular Adaptations to Oxidative Phosphorylation Defects in Cancer." In: Shireesh P. Apte and Rangaprasad Sarangarajan, eds., *Cellular Respiration and Carcinogenesis* (New York: Humana), pp. 55-72.

Stein, Yael, Or Levy-Nativ, and Elihu D. Richter. 2011. "A Sentinel Case Series of Cancer Patients with Occupational Exposures to Electromagnetic Non-ionizing Radiation and Other Agents." *European Journal of Oncology* 16(1): 21-54.

Teo, Charlie. 2012. "What If Your Mobile Phone Is Giving You Brain Cancer?" *The Punch*, May 7.

Teppo, Lyly, Eero Pukkala, and Maria Lehtonen. 1994. "Data Quality and Quality Control of a Population-Based Cancer Registry." *Acta Oncologica* 33(4): 365-69.

van Waveren, Corina, Yubo Sun, Herman S. Cheung, and Carlos T. Moraes. 2006. "Oxidative Phosphorylation Dysfunction Modulates Expression of Extracellular Matrix-Remodeling Genes and Invasion." *Carcinogenesis* 27(3): 409-18.

Vaupel, P., O. Thews, D. K. Kelleher, and M. Hoeckel. 1998. "Current Status of Knowledge and Critical Issues in Tumor Oxygenation." In: Antal G. Hudetz and Duane F. Bruley, eds., *Oxygen Transport to Tissue XX* (New York: Plenum), pp. 591-602.

Vigneri, Paolo, Francesco Frasca, Laura Sciacca, Guiseppe Pandini, and Riccardo Vigneri. 2009. "Diabetes and Cancer." *Endocrine-Related Cancer* 16: 1103-23.

Warburg, Otto Heinrich. 1908. "Notes on the Oxidation Processes in the Sea-Urchin's Egg." In: Warburg, *The Metabolism of Tumours* (London: Constable), 1930, pp. 13-25. Originally published as "Beobachtungen über die Oxydationsprozesse im Seeigelei," *Hoppe-Seyler's Zeitschrift für physiologische Chemie* 57(1-2): 1-16.

———. 1925. "The Metabolism of Carcinoma Cells." *Journal of Cancer Research* 9: 148-63.

———. 1928. "The Chemical Constitution of Respiration Ferment." *Science* 68: 437-43.

———. 1930. *The Metabolism of Tumours.* London: Constable.

———. 1956. "On the Origin of Cancer Cells." *Science* 123: 309-14.

———. 1966a. "Oxygen, the Creater of Differentiation." In: Nathan O. Kaplan and Eugene P. Kennedy, eds., *Current Aspects of Biochemical Energetics* (New York: Academic), pp. 103-9.

———. 1966b. *The Prime Cause and Prevention of Cancer.* Lecture at the meeting of the Nobel Laureates, Lindau, Lake Constance, Germany, June 30. English edition by Dean Burk (Würzburg: Konrad Triltsch), 1969.

Warburg, Otto, Karlfried Gawehn, August-Wilhelm Geissler, Detlev Kayser, and Siegfried Lorenz. 1965. "Experimente zur Anaerobiose der Krebszellen." *Klinische Wochenschrift* 43(6): 289-93.

Warburg, Otto, August-Wilhelm Geissler, and Siegfried Lorenz. 1965. "Messung der Sauerstoffdrucke beim Umschlag des embryonalen Stoffwechsels in Krebs-Stoffwechsel." *Zeitschrift für Naturforschung* 7(20b): 1070-3.

———. 1966. "Irreversible Erzeugung von Krebsstoffwechsel im embryonalen Mäusezellen." *Zeitschrift für Naturforschung* 7(21b): 707-8.

Warburg, Otto, Karl Posener and Erwin Negelein. 1924. "Über den Stoffwechsel der Tumoren." *Biochemische Zeitschrift* 152: 309-44. Reprinted in English translation as "The Metabolism of the Carcinoma Cell" in Warburg, *The Metabolism of Tumours* (London: Constable), 1930, pp. 129-69.

Warburg, Otto, Franz Wind, and Erwin Negelein. 1926. "The Metabolism of Tumors in the Body." *Journal of General Physiology* 8: 519-30.

Weinhouse, Sidney. 1956. "On Respiratory Impairment in Cancer Cells." *Science* 124: 267-68. Response by Otto Warburg, pp. 269-70. Response by Dean Burk, pp. 270-71.

Werner, Erica. 2009. "How Cancer Cells Escape Death." In: Shireesh P. Apte and Rangaprasad Sarangarajan, eds., *Cellular Respiration and Carcinogenesis* (New York: Humana), pp. 161-178.

Williams, W. Roger. 1908. *The Natural History of Cancer, with Special Reference to Its Causation and Prevention.* New York: William Wood.

Women's Health Policy and Advocacy Program. 2010. *Out of the Shadows: Women and Lung Cancer*. Boston: Brigham and Women's Hospital.

Wu, Min, Andy Neilson, Amy L. Swift, Rebecca Moran, James Tamagnine, Diane Parslow, Suzanne Armistead, Kristie Lemire, Jim Orrell, Jay Teich, Steve Chomicz, and David A. Ferrick. 2007. "Multiparameter Metabolic Analysis Reveals a Close Link between Attenuated Mitochondrial Bioenergetic Function and Enhanced Glycolysis Dependency in Human Tumor Cells." *American Journal of Physiology – Cell Physiology* 292: C125-36.

Fellingsbro

Ekblom, Adolf E. 1902. "Något statistik från död- och begrafningsböckerna i Fellingsbro 1801-1900 jämte förslag till Sveriges läkare angående samarbete för utredande af kräftsjukdomarnas frekvens." *Hygiea*, 2nd ser., 2(1): 11-21.

Guinchard, J. 1914. "Telegraph Service." In: Guinchard, *Sweden: Historical and Statistical Handbook*, 2nd ed., English issue. Stockholm: Government Printing Office, pp. 643-44.

Radio Towers and Cancer

Anderson, Bruce S. and Alden K. Henderson. 1986. *Cancer Incidence in Census Tracts with Broadcasting Towers in Honolulu, Hawaii*. Environmental Epidemiology Program, State of Hawaii Department of Health.

Cherry, Neil. 2000. *Childhood Cancer Incidence in the Vicinity of the Sutro Tower, San Francisco*. Environmental Management and Design Division, Lincoln University, Canterbury, New Zealand.

Dode, Adilza C., Mônica M. D. Leão, Francisco de A. F. Tejo, Antônio C. R. Gomes, Daiana C. Dode, Michael C. Dode, Cristina W. Moreira, Vânia A. Condessa, Cláudia Albinatti, and Waleska T. Caiaffa. 2011. "Mortality by Neoplasia and Cellular Telephone Base Stations in the Belo Horizonte Municipality, Minas Gerais State, Brazil." *Science of the Total Environment* 409(19): 3649-65.

Dolk, Helen, Gavin Shaddick, Peter Walls, Chris Grundy, Bharat Thakrar, Immo Kleinschmidt, and Paul Elliott. 1997. "Cancer Incidence near Radio and Television Transmitters in Great Britain. I. Sutton Coldfield Transmitter." *American Journal of Epidemiology* 145(1): 1-9.

Dolk, Helen, Paul Elliott, Gavin Shaddick, Peter Walls, and Bharat Thakrar. 1997. "Cancer Incidence near Radio and Television Transmitters in Great Britain. II. All High Power Transmitters." *American Journal of Epidemiology* 145(1): 10-17.

Eger, Horst, Klaus Uwe Hagen, Birgitt Lucas, Peter Vogel, and Helmut Voit. 2004. "Einfluss der räumlichen Nähe von Mobilfunksendeanlagen auf die Krebsinzidenz." *Umwelt-Medizin- Gesellschaft* 17(4): 326-32.

Hocking, Bruce, Ian R. Gordon, Heather L. Grain, and Gifford E. Hatfield. 1996. "Cancer Incidence and Mortality and Proximity to TV Towers." *Medical Journal of Australia* 165(11-12): 601-5.

Morton, William and David Phillips. 1983. *Radioemission Density and Cancer Epidemiology in the Portland Metropolitan Area*. Research Triangle Park, NC: United States Environmental Protection Agency.

Morton, William and David Phillips. 2000. "Cancer Promotion by Radiowave Emissions." *Epidemiology* 11(4): S57. Abstract.

Wolf, Ronni and Danny Wolf. 2004. "Increased Incidence of Cancer near a Cell-Phone Transmitter Station." *International Journal of Cancer Prevention* 1(2): 123-38.

Vatican Radio

Agence France Presse. 2001. "Italian Minister Threatens Hunger Strike over Vatican Radio." April 30.

————. 2003. "La Cour de Cassation Renvoie Radio Vatican Devant un Tribunal." April 9.

Allen, John L., Jr. 2001. "Vatican Radio Officials Charged." *National Catholic Reporter*, March 23.

Bartoli, Ilaria Ciancaleoni. 2006. "I comitati contro l'elettrosmog: la Santa Sede sapeva dei rischi." *E Polis Roma*, November 24, p. 25.

BBC News. April 11, 2003. "Vatican Radio Back in the Dock."

————. May 9, 2005. "Vatican Radio Officials Convicted."

Cinciripini, Giorgio. February 27, 2010. "Vatican Radio Caused Cancers, Must Compensate Victims." esmog.free.italia@gmail.com.

Corriere della Sera. 2002. "In una perizia nesso 'tra onde e casi di leucemia,'" May 10.

Deutsche Press-Agentur. 2003. "Italian Court Okays Trial into Vatican Radio Cancer Claims." April 10.

Gentile, Cecilia. 2002. "Leucemie a Cesano: 'Colpa delle Antenne.'" *La Repubblica*, May 10.

La Corte Suprema di Cassazione (Supreme Court of Cassation). 2011. Sentence no. 376/2011, February 24, Rome.

La Repubblica. 2001. "Radio Vaticana ancora fuorilegge." May 1.

Lavinia, Gianvito. 2011. "Elettrosmog, in procura altri 23 casi di leucemia." *Corriere della Sera*, June 8.

Lombardi, Federico. 2001. "Vatican Radio and the Electromagnetic Pollution." *Vatican Radio*, press release, May 4.

Micheli, Andrea. 2010. *Perizia mediante indagine epidemiologica incidente probatorio*. Procedimento Penale 33642/03, Tribunale Penale di Roma, June 25.

Michelozzi, Paola, Alessandra Capon, Ursula Kirchmayer, Francesco Forastiere, Annibale Biggeri, Alessandra Barca, and Carlo A. Perucci. 2002. "Adult and Childhood Leukemia near a High-power Radio Station in Rome, Italy." *American Journal of Epidmiology* 155(12): 1096-1103.

Michelozzi, Paola, Ursula Kirchmayer, Alessandra Capon, Francesco Forestiere, Annibale Biggeri, Alessandra Barca, C. Ancona, D. Fusco, A. Sperati, P. Papini, A. Pierangelini, R. Rondelli, and Carlo A. Perucci. 2001. "Mortalità per leucemia e incidenza di leucemia infantile in prossimità della stazione di Radio Vaticana di Roma." *Epidemiologia & Prevenzione* 25(6): 249-55.

Pierucci, Adelaide. 2006. "Elettrosmog a Radio Vaticana: perizia sulle morti di leucemia." *E Polis Roma*, November 24.

Stanley, Alessandra. 2001. "In Radio Feud, a Higher Kind of Superpower Irks Italy." *New York Times*, April 13.

Times of India. 2011. "Vatican Seeks to Stave off Trial of Top Radio Officials." February 14.

Chapter 14

Austad, S. N. 1989. "Life Extension by Dietary Restriction in the Bowl and Doily Spider, *Frontinella pyramitela*." *Experimental Gerontology* 24(1): 83-92.

Bacon, Francis. 1605. *The Advancement of Learning.* Translated and edited by Joseph Devey (New York: P. F. Collier and Son), 1901.

———. 1623. *The History of Life and Death.* In: James Spedding, Robert Leslie Ellis, and Douglas Denon Heath, eds., *The Works of Francis Bacon* (Boston: Taggard and Thompson), 1864, volume X, pp. 7-176.

Beard, George Miller. 1880. *A Practical Treatise on Nervous Exhaustion (Neurasthenia).* New York: William Wood.

———. 1881a. *American Nervousness: Its Causes and Consequences.* New York: G. P. Putnam's Sons.

Bodkin, Noni L., Theresa M. Alexander, Heidi K. Ortmeyer, Elizabeth Johnson, and Barbara C. Hansen. 2003. "Mortality and Morbidity in Laboratory-maintained Rhesus Monkeys and Effects of Long-term Dietary Restriction." *Journal of Gerontology: Biological Sciences* 58A(3): 212-19.

Caratero, A., M. Courtade, L. Bonnet, H. Planel, and C. Caratero. 1998. "Effect of a Continuous Gamma Irradiation at a Very Low Dose on the Life Span of Mice." *Gerontology* 44: 272-76.

Carlson, Loren Daniel and Betty H. Jackson. 1959. "The Combined Effects of Ionizing Radiation and High Temperature on the Longevity of the Sprague-Dawley Rat." *Radiation Research* 11: 509-19.

Carlson, Loren Daniel, William J. Scheyer, and B. H. Jackson. 1957. "The Combined Effects of Ionizing Radiation and Low Temperature on the Metabolism, Longevity, and Soft Tissues of the White Rat." *Radiation Research* 7: 190-97.

Chittenden, Russell Henry. 1907. *Physiological Economy in Nutrition.* New York: Frederick A. Stokes.

Chou, Chung-Kwang, Arthur William Guy, Lawrence L. Kunz, Robert B. Johnson, John J. Crowley, and Jerome H. Krupp. 1992. "Long-term, Low-level Microwave Irradiation of Rats." *Bioelectromagnetics* 13(6): 469-96.

Colman, Ricki J., Rozalyn M. Anderson, Sterling C. Johnson, Erik K. Kastman, Kristopher J. Kosmatka, T. Mark Beasley, David B. Allison, Christina Cruzen, Heather A. Simmons, Joseph W. Kemnitz, and Richard Weindruch. 2009. "Caloric Restriction Delays Disease Onset and Mortality in Rhesus Monkeys." *Science* 325: 201-4.

Colman, Ricki J., Mark Beasley, Joseph W. Kemnitz, Sterling C. Johnson, Richard Weindruch, and Rozalyn M. Anderson. 2014. "Caloric Restriction Reduces Age-related and All-cause Mortality in Rhesus Monkeys." *Nature Communications* 5: 557.

Condran, Gretchen A. 1987. "Declining Mortality in the United States in the Late Nineteenth and Early Twentieth Centuries." *Annales de démographie historique*, vol. 1987, pp. 119-41.

Cutler, Richard G. 1981. "Life-Span Extension." In: James L. McGaugh and Sara B. Kiesler, eds., *Aging: Biology and Behavior* (New York: Academic), pp. 31-76.

Ducoff, Howard S. 1972. "Causes of Death in Irradiated Adult Insects." *Biological Reviews* 47: 211-40.

———. 1975. "Form of the Increased Longevity of *Tribolium* after X-irradiation." *Experimental Gerontology* 10: 189-93.

Dunham, H. Howard. 1938. "Abundant Feeding Followed by Restricted Feeding and Longevity in Daphnia." *Physiological Zoölogy* 11(4): 399-407.

Elder, Joseph A. 1994. "Thermal, Cumulative, and Lifespan Effects and Cancer in Mammals Exposed to Radiofrequency Radiation." In: David O. Carpenter and Sinerik Ayrapetyan, eds., *Biological Effects of Electric and Magnetic Fields* (San Diego: Academic), vol. 2, pp. 279-95.

Finot, Jean. 1906. *La Philosophie de la Longévité*, 11th ed. Paris: Félix Alcan.

Fischer-Piette, Édouard. 1939. "Sur la croissance et la longevité de *Patella vulgata* L. en fonction du milieu." *Journal de Conchyliologie* 83: 303-10.

Griffin, Donald Redfield. 1958. *Listening in the Dark: The Acoustic Orientation of Bats and Men*. New Haven, CT: Yale University Press.

Hansson, Artur, Eskil Brännäng, and Olof Claesson. 1953. "Studies on Monozygous Cattle Twins. XIII. Body Development in Relation to Heredity and Intensity of Rearing." *Acta Agriculturæ Scandinavica* 3(1): 61-95.

Hochachka, Peter W. and Michael Guppy. 1987. *Metabolic Arrest and the Control of Biological Time*. Cambridge, MA: Harvard University Press.

Johnson, Thomas E., David H. Mitchell, Susan Kline, Rebecca Kemal, and John Foy. 1984. "Arresting Development Arrests Aging in the Nematode *Caenorhabditis elegans*." *Mechanisms of Ageing and Development* 28: 23-40.

Kagawa, Yasuo. 1978. "Impact of Westernization on the Nutrition of Japanese: Changes in Physique, Cancer, Longevity and Centenarians." *Preventive Medicine* 7: 205-17.

Kannisto, Väinö. 1994. *Development of Oldest-Old Mortality, 1950-1990: Evidence from 28 Developed Countries*. Monographs on Population Aging, 1. Odense, Denmark: Odense University Press.

Kannisto, Väinö, Jens Lauritsen, A. Roger Thatcher, and James W. Vaupel. 1994. "Reductions in Mortality at Advanced Ages: Several Decades of Evidence from 27 Countries." *Population and Development Review* 20(4): 793-810.

Kemnitz, Joseph W. 2011. "Calorie Restriction and Aging in Nonhuman Primates." *ILAR Journal* 52(1): 66-77.

Kirk, William P. 1984. "Life Span and Carcinogenesis." In: Joseph A. Elder and Daniel F. Cahill, eds., *Biological Effects of Radiofrequency Radiation* (Research Triangle Park, NC: U.S. Environmental Protection Agency), report no. EPA-600/8-83-026F, pp. 5-106 to 5-111.

Lane, Mark A., Donald K. Ingram, and George S. Roth. 1999. "Calorie Restriction in Nonhuman Primates: Effects on Diabetes and Cardiovascular Disease Risk." *Toxicological Sciences* 52 (suppl.): 41-48.

Liu, Robert K. and Roy L. Walford. 1972. "The Effect of Lowered Body Temperature on Lifespan and Immune and Non-Immune Processes." *Gerontologia* 18: 363-88.

Loeb, Jacques and John Howard Northrop. 1917. "What Determines the Duration of Life in Metazoa?" *Proceedings of the National Academy of Sciences* 3(5): 382-86.

———. 1917. "On the Influence of Food and Temperature upon the Duration of Life." *Journal of Biological Chemistry* 32: 103-21.

Lorenz, Egon, Joanne Weikel Hollcroft, Eliza Miller, Charles C. Congdon, and Robert Schweisthal. 1955. "Long-term Effects of Acute and Chronic Irradiation in Mice. I. Survival and Tumor Incidence Following Chronic Irradiation of 0.11 r per Day." *Journal of the National Cancer Institute* 15(4): 1049-58.

Lorenz, Egon, Leon O. Jacobson, Walter E. Heston, Michael Shimkin, Allen B. Eschenbrenner, Margaret K. Deringer, Jane Doniger, and Robert Schweisthal. 1954. "Effects of Long-Continued Total Body Gamma Irradiation of Mice, Guinea Pigs, and Rabbits. III. Effects on Life Span, Weight, Blood Picture, and Carcinogenesis and the Role of the Intensity of Radiation." In: Raymond E. Zirkle, ed., *Biological Effects of External X and Gamma Radiation* (New York: McGraw-Hill), part I, pp. 24-148.

Lyman, Charles P., Regina C. O'Brien, G. Cliett Greene, and Elaine D. Papafrangos. 1981. "Hibernation and Longevity in the Turkish Hamster *Mesocricetus brandti*." *Science* 212: 668-70.

Lynn, William S. and James C. Wallwork. 1992. "Does Food Restriction Retard Aging by Reducing Metabolic Rate?" *Journal of Nutrition* 122: 1917-18.

Mattison, Julie A., Mark A. Lane, George S. Roth, and Donald K. Ingram. 2003. "Calorie Restriction in Rhesus Monkeys." *Experimental Gerontology* 38: 35-46.

McCarter, Roger, E. J. Masoro, and Byung P. Yu. 1985. "Does Food Restriction Retard Aging by Reducing the Metabolic Rate?" *American Journal of Physiology – Endocrinology and Metabolism* 248: E488-90.

McKay, Clive M. and Mary F. Crowell. 1934. "Prolonging the Life Span." *Scientific Monthly* 39: 405-14.

McCay, Clive M., Mary F. Crowell, and Leonard A. Maynard. 1935. "The Effect of Retarded Growth upon the Length of Life Span and upon the Ultimate Body Size." *Journal of Nutrition* 10: 63-79.

McKay, Clive M., Leonard A. Maynard, Gladys Sperling, and LeRoy L. Barnes. 1939. "Retarded Growth, Life Span, Ultimate Body Size and Age Changes in the Albino Rat After Feeding Diets Restricted in Calories." *Journal of Nutrition* 18(1): 1-13.

McDonald, Roger B. and Jon J. Ramsey. 2010. "Honoring Clive McCay and 75 Years of Calorie Restriction Research." *Journal of Nutrition* 140(7): 1205-10.

Millward, Robert and Frances N. Bell. 1998. "Economic Factors in the Decline of Mortality in Late Nineteenth Century Britain." *European Review of Economic History* 2: 263-88.

Mitchel, Ronald E. J. 2006. "Low Doses of Radiation are Protective *In vitro* and *In vivo*: Evolutionary Origins." *Dose-Response* 4(2): 75-90.

Okada, M., A. Okabe, Y. Uchihori, H. Kitamura, E. Sekine, S. Ebisawa, M. Suzuki, and R. Okayasu. 2007. "Single Extreme Low-dose/Low Dose Rate Irradiation Causes Alteration in Lifespan and Genome Instability in Primary Human Cells." *British Journal of Cancer* 96: 1707-10.

Ordy, J. Mark, Thaddeus Samorajki, Wolfgang Zeman, and Howard J. Curtis. 1967. "Interaction Effects of Environmental Stress and Deuteron Irradiation of the

Brain on Mortality and Longevity of C57BL/10 Mice." *Proceedings of the Society for Experimental Biology and Medicine* 126(1): 184-90.

Osborne, Thomas B., Lafayette B. Mendel, and Edna L. Ferry. 1917. "The Effect of Retardation of Growth upon the Breeding Period and Duration of Life of Rats." *Science* 45: 294-95.

Pearl, Raymond. 1928. *The Rate of Living*. New York: Alfred A. Knopf.

Perez, Felipe P., Ximing Zhou, Jorge Morisaki, and Donald Jurivich. 2008. "Electromagnetic Field Therapy Delays Cellular Senescence and Death by Enhancement of the Heat Shock Response." *Experimental Gerontology* 43: 307-16.

Pinney, Don O., D. F. Stephens, and L. S. Pope. 1972. "Lifetime Effects of Winter Supplemental Feed Level and Age at First Parturition on Range Beef Cows." *Journal of Animal Science* 34(6): 1067-74.

Ramsey, Jon J., Mary-Ellen Harper, and Richard Weindruch. 2000. "Restriction of Energy Intake, Energy Expenditure, and Aging." *Free Radical Biology and Medicine* 29(10): 946-68.

Rattan, Suresh I. S. 2004. "Aging Intervention, Prevention, and Therapy Through Hormesis." *Journal of Gerontology: Biological Sciences* 59A(7): 705-9.

Reimers, N. 1979. "A History of a Stunted Brook Trout Population in an Alpine Lake: A Life Span of 24 Years." *California Fish and Game* 65: 196-215.

Ross, Morris H. 1961. "Length of Life and Nutrition in the Rat." *Journal of Nutrition* 75(2): 197-210.

———. 1972. "Length of Life and Caloric Intake." *American Journal of Clinical Nutrition* 25(8): 834-38.

Ross, Morris H. and Gerrit Bras. 1965. "Tumor Incidence Patterns and Nutrition in the Rat." *Journal of Nutrition* 87: 245-60.

———. 1971. "Lasting Influence of Early Caloric Restriction on Prevalence of Neoplasms in the Rat." *Journal of the National Cancer Institute* 47(5): 1095-1113.

———. 1973. "Influence of Protein Under- and Overnutrition on Spontaneous Tumor Prevalence in the Rat." *Journal of Nutrition* 103: 944-63.

Rubner, Max. 1908. *Das Problem der Lebensdauer*. München: R. Oldenbourg.

Rudzinska, Maria A. 1952. "Overfeeding and Life Span in *Tokophyra infusionum*." *Journal of Gerontology* 7: 544-48.

Sacher, George A. 1963. "Effects of X-rays on the Survival of *Drosophila* Imagoes." *Physiological Zoölogy* 36(4): 295-311.

———. 1977. "Life Table Modification and Life Prolongation." In: Caleb E. Finch and Leonard Hayflick, eds., *Handbook of the Biology of Aging* (New York: Van Nostrand Reinhold), pp. 582-638.

Simmons, Heather A. and Julie A. Mattison. 2011. "The Incidence of Spontaneous Neoplasia in Two Populations of Captive Rhesus Macaques (*Macaca mulatta*)." *Antioxidants & Redox Signaling* 14(2): 221-27.

Sohal, Rajindar S. 1986. "The Rate of Living Theory: A Contemporary Interpretation." In: K.-G. Collatz and R. S. Sohal, eds., *Insect Aging* (Berlin: Springer), pp. 23-44.

Sohal, Rajindar S. and Robert G. Allen. 1985. "Relationship between Metabolic Rate, Free Radicals, Differentiation and Aging: a Unified Theory." In: Avril D.

Woodhead, Anthony D. Blackett, and Alexander Hollaender, eds., *Molecular Biology of Aging* (New York: Plenum), pp. 75-104.

Spalding, Jonathan F., Robert W. Freyman, and Laurence M. Holland. 1971. "Effects of 800-MHz Electromagnetic Radiation on Body Weight, Activity, Hematopoiesis and Life Span in Mice." *Health Physics* 20: 421-24.

Süsskind, Charles. 1959. *Cellular and Longevity Effects of Microwave Radiation*. Berkeley, CA: University of California, Berkeley. Annual Scientific Report (1958-59) on Contract AF41(657)-114. Institute of Engineering Research, ser. 60, no. 241, June 30. Rome Air Development Center report no. RADC-TR-59-131.

Süsskind, Charles. 1961. *Longevity Study of the Effects of 3-cm Microwave Radiation on Mice*. Berkeley, CA: University of California, Berkeley. Annual Scientific Report (1960-61) on Contract AF41(657)-114. Institute of Engineering Research, ser. 60, no. 382, June 30. Rome Air Development Center report no. RADC-TR-61-205.

Suzuki, Masao, Zhi Yang, Kazushiro Nakano, Fumio Yatagai, Keiji Suzuki, Seiji Kodama, and Masami Watanabe. 1998. "Extension of *In vitro* Life-span of γ-irradiated Human Embryo Cells Accompanied by Chromosome Instability." *Journal of Radiation Research* 39: 203-13.

Tryon, Clarence Archer and Dana P. Snyder. 1971. "The Effect of Exposure to 200 and 400 R of Ionizing Radiation on the Survivorship Curves of the Eastern Chipmunk (*Tamias Striatus*) under Natural Conditions." In: D. J. Nelson, ed., *Radionuclides in Ecosystems: Proceedings of the Third National Symposium on Radioecology, May 10-12, 1971, Oak Ridge, Tennessee*, Oak Ridge National Laboratory, report no. CONF-71501-P2, vol. 2, pp. 1037-41.

Vickery, Hubert Bradford. 1944. *Biographical Memoir of Russell Henry Chittenden 1856-1943*. Washington, DC: National Academy of Sciences.

Wachter, Kenneth W. and Caleb E. Finch, eds. 1997. *Between Zeus and the Salmon: The Biodemography of Longevity*. Washington, DC: National Academy Press.

Walford, Roy L. 1983. *Maximum Life Span*, New York: Norton.

———. 1982. "Studies in Immunogerontology." *Journal of the American Geriatrics Society* 30(10): 617-25.

Weindruch, Richard and Roy L. Walford. 1988. "The Retardation of Aging and Disease by Dietary Restriction." Springfield, IL: Charles C. Thomas.

Wilkinson, Gerald S. and Jason M. South. 2002. "Life History, Ecology and Longevity in Bats." *Aging Cell* 1: 124-31.

Wilmoth, John R. 2000. "Demography of Longevity: Past, Present, and Future Trends." *Experimental Gerontology* 35: 1111-29.

Wilmoth, John R., L. J. Deegan, H. Lundström, and S. Horiuchi. 2000. "Increase of Maximum Life-Span in Sweden, 1861-1999." *Science* 289: 2366-68.

Wilmoth, John R. and Hans Lundström. 1996. "Extreme Longevity in Five Countries." *European Journal of Population* 12: 63-93.

Young, Vernon R. 1979. "Diet as a Modulator of Aging and Longevity." *Federation Proceedings* 38(6): 1994-2000.

Yu, Byung Pal, ed. 1994. *Modulation of Aging Processes by Dietary Restriction*. Boca Raton, FL: CRC Press.

Chapter 15

Cell Phones and Cell Towers

Mild, Kjell Hansson and Jonna Wilén. 2009. "Occupational Exposure in Wireless Communication." In: James C. Lin, ed., *Advances in Electromagnetic Fields in Living Systems*, vol. 5, *Health Effects of Cell Phone Radiation* (New York: Springer), pp. 199-219.

Tuor, Markus, Sven Ebert, Jürgen Schuderer, and Niels Kuster. 2005. "Assessment of ELF Exposure from GSM Handsets and Development of an Optimized RF/ELF Exposure Setup for Studies of Human Volunteers." BAG Reg. No. 2.23.02.-18/02.001778. Zürich: Foundation for Research on Information Technologies in Society.

Electronic Consumer Devices

Stetzer, David A. April 2, 2000. Testimony before the Michigan Public Service Commission.

Zyren, Jim. May 2010. "HomePlug Green PHY Overview." Atheros Technical Paper.

Electromodel of the Ear

Allen, Jont B. 1980. "Cochlear Micromechanics – A Physical Model of Transduction." *Journal of the Acoustical Society of America* 68(6): 1660-70.

Art, Jonathan J. and Robert Fettiplace. 1987. "Variation of Membrane Properties in Hair Cells Isolated from the Turtle Cochlea." *Journal of Physiology* 385: 207-42.

Ashmore, Jonathan F. 1987. "A Fast Motile Response in Guinea-Pig Outer Hair Cells: The Cellular Basis of the Cochlear Amplifier." *Journal of Physiology* 388: 323-47.

———. 2008. "Cochlear Outer Hair Cell Motililty." *Physiological Reviews* 88: 173-210.

Bell, Andrew. 2000. *The Underwater Piano: Revival of the Resonance Theory of Hearing*. Canberra: Australian National University.

———. 2004. "Resonance Theories of Hearing – A History and a Fresh Approach." *Acoustics Australia* 32(3): 95-100.

———. 2005. "The Underwater Piano: A Resonance Theory of Cochlear Mechanics." Doctoral thesis, The Australian National University, Canberra.

———. 2006. "Sensors, Motors, and Tuning in the Cochlea: Interacting Cells Could Form a Surface Acoustic Wave Resonator." *Bioinspiration and Biomimetics* 1: 96-101.

———. 2007. "Detection with Deflection? A Hypothesis for Direct Sensing of Sound Pressure by Hair Cells." *Journal of Biosciences* 32(2): 385-404.

———. 2010. "The Cochlea as a Graded Bank of Independent, Simultaneously Excited Resonators: Calculated Properties of an Apparent 'Travelling Wave.'" *Proceedings of the 20th International Congress on Acoustics, ICA 2010, 23-27 August 2010, Sydney, Australia*, pp. 1-9.

———. 2011. "How Do Middle Ear Muscles Protect the Cochlea? Reconsideration of the Intralabyrinthine Pressure Theory." *Journal of Hearing Science* 1(2): 9-23.

———. 2012. "A Resonance Approach to Cochlear Mechanics." *PLoS ONE* 7(11): e47918.

Bell, DeLamar T., Jr. and Robert C. M. Li. 1976. "Surface-Acoustic-Wave Resonators." *Proceedings of the IEEE* 64(5): 711-21.

Braun, Martin. 1994. "Tuned Hair Cells for Hearing, But Tuned Basilar Membrane for Overload Protection: Evidence from Dolphins, Bats, and Desert Rodents." *Hearing Research* 78: 98-114.

Breneman, Kathryn D., William Brownell, and Richard D. Rabbit. 2009. "Hair Cell Bundles: Flexoelectric Motors of the Inner Ear." *PLoS ONE* 4(4): e5201.

Breneman, Kathryn D. and Richard D. Rabbit. 2009. "Piezo- and Flexoelectric Membrane Materials Underlie Fast Biological Motors in the Ear." *Materials Research Society Symposia Proceedings* 1186E: 1186-JJ06-04.

Brownell, William E. 2006. "The Piezoeletric Outer Hair Cell." In: Ruth Anne Eatock, Richard R. Fay, and Arthur N. Popper, eds., *Vertebrate Hair Cells* (New York: Springer), pp. 313-47.

Brownell, William E., Charles R. Bader, Daniel Bertrand, and Yves de Ribaupierre. 1985. "Evoked Mechanical Responses of Isolated Cochlear Outer Hair Cells." *Science* 227: 194-96.

Canlon, Barbara, Lou Brundin, and Åke Flock. 1988. "Acoustic Stimulation Causes Tonotopic Alterations in the Length of Isolated Outer Hair Cells from Guinea Pig Hearing Organ." *Proceedings of the National Academy of Sciences* 85(18): 7033-35.

Crawford, Andrew C. and Robert Fettiplace. 1981. "An Electrical Tuning Mechanism in Turtle Cochlear Hair Cells." *Journal of Physiology* 312: 377-412.

de Vries, Hessel. 1948a. "Brownian Movement and Hearing." *Physica* 14(1): 48-60.

———. 1948b. "Die Reizschwelle der Sinnesorgane als physikalisches Problem." *Experientia* 4(6): 205-13.

Degens, Egon T., Werner G. Deuser, and Richard L. Haedrich. 1969. "Molecular Structure and Composition of Fish Otoliths." *International Journal on Life in Oceans and Coastal Waters* 2(2): 105-13.

Dimbylow, Peter J. 1988. "The Calculation of Induced Currents and Absorbed Power in a Realistic, Heterogeneous Model of the Lower Leg for Applied Electric Fields from 60 Hz to 30 MHz." *Physics in Medicine and Biology* 33(12): 1453-68.

Dong, Xiao-xia, Mark Ospeck, and Kuni H. Iwasa. 2002. "Piezoelectrical Reciprocal Relationship of the Membrane Motor in the Cochlear Outer Hair Cell." *Biophysical Journal* 82(3): 1254-59.

Fettiplace, Robert and Paul A. Fuchs. 1999. "Mechanisms of Hair Cell Tuning." *Annual Review of Physiology* 61: 809-34.

Ghaffari, Roozbeh, Alexander J. Aranyosi, and Dennis M. Freeman. 2007. "Longitudinally Propagating Traveling Waves of the Mammalian Tectorial Membrane." *Proceedings of the National Academy of Sciences* 104(42): 16510-15.

Gummer, Anthony W., Werner Hemmert, and Hans-Peter Zenner. 1996. "Resonant Tectorial Membrane Motion in the Inner Ear: Its Crucial Role in Frequency Tuning." *Proceedings of the National Academy of Sciences* 93(16): 8727-32.

Gummer, Anthony W. and Serena Preyer. 1997. "Cochlear Amplification and its Pathology: Emphasis on the Role of the Tectorial Membrane." *Ear, Nose, & Throat Journal* 76(3): 151-58.

Hackney, Carole M. and David N. Furness. 1995. "Mechanotransduction in Ververtebrate Hair Cells: Structure and Function of the Stereociliary Bundle." *American Journal of Cell Physiology* 268: C1-C13.

Hallpike, Charles Skinner and Alexander Francis Rawdon-Smith. 1934a. "The 'Wever and Bray Phenomenon': A Study of the Electrical Response in the Cochlea with Especial Reference to its Origin." *Journal of Physiology* 81: 395-408.

———. 1934b. "The Origin of the Wever and Bray Phenomenon." *Journal of Physiology* 83: 243-54.

Hassan, Waled and Peter B. Nagy. 1997. "On the Low-Frequency Oscillation of a Fluid Layer between Two Elastic Plates." *Journal of the Acoustical Society of America* 102(6): 3343-48.

Helmholtz, Hermann Ludwig Ferdinand. 1877. *Die Lehre von den Tonempfindungen als physiologische Grundlage für die Theorie der Musik*. Braunschweig: Friedrich Vieweg und Sohn. Translation by Alexander J. Ellis, *On the Sensations of Tone as a Physiological Basis for the Theory of Music*, 4th ed. (London: Longmans, Green), 1912.

Hoar, William Stewart and David J. Randall, eds. 1971. *Fish Physiology*. Vol. 5: *Sensory Systems and Electric Organs*. New York: Academic.

Holley, Matthew C. and Jonathan F. Ashmore. 1988. "On the Mechanism of a High-Frequency Force Generator in Outer Hair Cells Isolated from the Guinea Pig Cochlea." *Proceedings of the Royal Society of London B* 232: 413-29.

Honrubia, Vicente, David Strelioff, and Stephen Sitko. 1976. "Electroanatomy of the Cochlea: Its Role in Cochlear Potential Measurements." In: Robert J. Ruben, Claus Elberling, and Gerhard Salomon, eds. (Baltimore, MD: University Park Press), pp. 23-39.

Hudspeth, A. James and R. S. Lewis. 1988. "A Model for Electrical Resonance and Frequency Tuning in Saccular Hair Cells of the Bull-Frog, *Rana catesbeiana*." *Journal of Physiology* 400: 275-97.

Iwasa, Kuni H. 2001. "A Two-State Piezoelectric Model for Outer Hair Cell Motility." *Biophysical Journal* 81(5): 2495-2506.

Jákli, Antal. and Nandor Éber. 1993. "Piezoelectric Effects in Liquid Crystals." In: Agnes Buka, ed., *Modern Topics in Liquid Crystals* (Singapore: World Scientific) pp. 235-56.

Jielof, Renske, A. Spoor and Hessel de Vries. 1952. "The Microphonic Activity of the Lateral Line." *Journal of Physiology* 116: 137-57.

Keen, J. A. 1940. "A Note on the Length of the Basilar Membrane in Man and in Various Mammals." *Journal of Anatomy* 75: 524-27.

Konishi, Teruzo, Donald C. Teas, and Joel S. Wernick. 1970. "Effects of Electrical Current Applied to Cochlear Partition on Discharges in Individual Auditory-Nerve Fibers. I. Prolonged Direct-Current Polarization." *Journal of the Acoustical Society of America* 47 (6): 1519-26.

Kostelijk, Pieter Jan. 1950. *Theories of Hearing*. Leiden: Universitaire Pers Leiden.

Lissmann, Hans W. 1958. "On the Function and Evolution of Electric Organs in Fish." *Journal of Experimental Biology* 35: 156-91.

Mamishev, Alexander V., Kishore Sundara-Rajan, Fumin Yang, Yanqing Du, and Markus Zahn. 2004. "Interdigital Sensors and Transducers." *Proceedings of the IEEE* 92(5): 808-45.

Moller, Peter. 1995. *Electric Fishes: History and Behavior*. London: Chapman & Hall.

Mountain, David C. 1986. "Electromechanical Properties of Hair Cells." In: R. A. Altschuler, D. W. Hoffman, and R. P. Bobbin, eds., *Neurobiology of Hearing: The Cochlea* (New York: Raven Press), pp. 77-90.

Mountain, David C. and Allyn E. Hubbard. 1994. "A Piezoelectric Model of Outer Hair Cell Function." *Journal of the Acoustical Society of America* 95(1): 350-54.

Naftalin, Lionel. 1963. "The Transmission of Acoustic Energy from Air to the Receptor Organ in the Cochlea." *Life Sciences* 2(2): 101-6.

———. 1964. "Reply to Criticisms by Mr. A. Tumarkin and Mr. J. D. Gray." *Journal of Laryngology and Otology* 78: 969-71.

———. 1965. "Some New Proposals Regarding Acoustic Transmission and Transduction." *Cold Spring Harbor Symposia on Quantitative Biology* 30: 169-80.

———. 1967. "The Cochlear Geometry as a Frequency Analyser." *Journal of Laryngology and Otology* 81(6): 619-31.

———. 1968. "Acoustic Transmission and Transduction in the Peripheral Hearing Apparatus." *Progress in Biophysics and Molecular Biology* 18: 3-27.

———. 1969. "A Liquid Ion-exchange Resin Microphone." *Life Sciences* 8 (part 2): 223-28.

———. 1970. "Biochemistry and Biophysics of the Tectorial Membrane." In: Michael M. Paparella, ed., *Biochemical Mechanisms in Hearing and Deafness* (Springfield, IL: Charles C. Thomas), pp. 205-10, discussion on pp. 290-93.

———. 1976. "The Peripheral Hearing Mechanism: A Biochemical and Biological Approach." *Annals of Otology, Rhinology and Laryngology* 85: 38-42.

———. 1980. "Frequency Analysis in the Cochlea and the Traveling Wave of von Békésy." *Physiological Chemistry and Physics* 12: 521-26.

———. 1981. "Energy Transduction in the Cochlea." *Hearing Research* 5: 307-15.

Naftalin, Lionel, M. Spence Harrison and A. Stephens. 1964. "The Character of the Tectorial Membrane." *Journal of Laryngology and Otology* 78: 1061-78.

Naftalin, Lionel and G. P. Jones. 1969. "Propagation of Acoustic Waves in Gels with Special Reference to the Theory of Hearing." *Life Sciences* 8 (part 1): 765-68.

Naftalin, Lionel and Michael Mattey. 1995. "The Transmission of Acoustic Energy from Air to the Receptor and Transducer in the Cochlea." Paper presented at conference on "Non-linear Coherent Structures in Physics and Biology," Heriot-Watt University, Edinburgh, July 1995.

Naftalin, Lionel, Michael Mattey, and Eve M. Lutz. 2009. "The Transmission of Acoustic Energy from Air to the Receptor and Transducer Structures within the Cochlea with Special Reference to the Tectorial Membrane." Manuscript submitted to *Hearing Research*.

Naftalin, Lionel and A. Stephens. 1966. "A Protein Electret Microphone." *Life Sciences* 5(3): 223-26.

Neely, S. T. 1989. "A Model for Bidirectional Transduction in Outer Hair Cells." In: J. P. Wilson and D. T. Kemp, eds., *Cochlear Mechanisms* (New York: Plenum), pp. 75-82.

Nowotny, Manuela and Anthony W. Gummer. 2006. "Nanomechanics of the Subtectorial Space Caused by Electromechanics of Cochlear Outer Hair Cells." *Proceedings of the National Academy of Sciences* 103(7): 2120-25.

Offutt, George C. 1968. "Auditory Response in the Goldfish." *Journal of Auditory Research* 8: 391-400.

———. 1970. "A Proposed Mechanism for the Perception of Acoustic Stimuli near Threshold." *Journal of Auditory Research* 10: 226-28.

———. 1974. "Structures for the Detection of Acoustic Stimuli in the Atlantic Codfish, *Gadus morhua*." *Journal of the Acoustical Society of America* 56(2): 665-71.

———. 1984. *The Electromodel of the Auditory System.* Shepherdstown, WV: GoLo Press.

———. 1986. "Wever and Lawrence Revisited: Effects of Nulling Basilar Membrane Movement on Concomitant Whole-Nerve Action Potential." *Journal of Auditory Research* 26: 43-54.

———. 1999. "New Electromodel Hearing Aid." *Resonance: Newsletter of the Bioelectromagnetics SIG* 34: 17-18.

———. 2000. "What is the Basis of Human Hearing?" *Frontier Perspectives* 9(2): 33-36.

———. 2002. "Energy Flow and Basilar Membrane Vibrations (Sound in the Cochlea's Fluids)." Presented at 25th Midwinter Research Meeting of the Association for Research in Otolaryngology, January.

O'Leary, Dennis P. 1970. "An Electrokinetic Model of Transduction in the Semicircular Canal." *Biophysical Journal* 10: 859-75.

Özen, Şükrü 2008. "Low-frequency Transient Electric and Magnetic Fields Coupling to Child Body." *Radiation Protection Dosimetry* 128(1): 62-67.

Parks, Susan E., Darlene R. Ketten, Jennifer T. O'Malley, and Julie Arruda. 2007. "Anatomical Predictions of Hearing in the North Atlantic Right Whale." *Anatomical Record* 290: 734-44.

Pohlman, Augustus G. 1922. "Structural Factors Contributing to Acoustic Insulation of the End Organ." *The Anatomical Record* 23:32. Abstract.

———. 1930. "Correlations Between the Acuity for Hearing Air and Bone Transmitted Sounds in Rinne Negative and Rinne Positive Cases." *Annals of Otology, Rhinology and Laryngology* 39(4): 927-60.

———. 1933. "A Reconsideration of the Mechanics of the Auditory Apparatus." *Journal of Laryngology and Otology* 48: 156-95.

———. 1936. "The Present Status of the Mechanics of Sound Conduction in Its Relation to the Possible Correction of Conduction Deafness." *Journal of the Acoustical Society of America* 8(2): 112-17.

———. 1938. "Objections to the Accepted Interpretation of Cochlear Mechanics." *Acta Oto-Laryngologica* 26: 162-69.

———. 1942. "Further Objections to the Accepted Interpretations of Cochlear Mechanics." *Archives of Otolaryngology* 35: 613-22.

Rabbit, Richard D., Harold E. Ayliffe, Douglas Christensen, Kranti Pamarthy, Carl Durney, Sarah Clifford, and William E. Brownell. 2005. "Evidence of Piezoelectric Resonance in Isolated Outer Hair Cells." *Biophysical Journal* 88: 2257-65.

Raphael, Robert M., Aleksander S. Popel, and William E. Brownell. 2000. "A Membrane Bending Model of Outer Hair Cell Electromotility." *Biophysical Journal* 78: 2844-62.

Richter, Claus-Peter, Gulam Emadi, Geoffrey Getnick, Alicia Quesnel, and Peter Dallos. 2007. "Tectorial Membrane Stiffness Gradients." *Biophysical Journal* 93: 2265-76.

Ross, Muriel D. 1974. "The Tectorial Membrane of the Rat." *American Journal of Anatomy* 139: 449-82.

Russell, Ian J., Alan R. Cody, and Guy P. Richardson. 1986. "The Responses of Inner and Outer Hair Cells in the Basal Turn of the Guinea-Pig Cochlea and in the Mouse Cochlea Grown *In vitro*." *Hearing Research* 22: 199-216.

Russell, Ian J. and Peter M. Sellick. 1978. "Intracellular Studies of Hair Cells in the Mammalian Cochlea." *Journal of Physiology* 284: 261-90.

Santos-Sacchi, Joseph and James P. Dilger. 1988. "Whole Cell Currents and Mechanical Responses of Isolated Outer Hair Cells." *Hearing Research* 35: 143-50.

Spector, William S., ed. 1956. *Handbook of Biological Data*. Philadelphia: W. B. Saunders. Page 323 on cochlear dimensions across species.

Strelioff, David, Åke Flock, and Karl E. Minser. 1985. "Role of Inner and Outer Hair Cells in Mechanical Frequency Selectivity of the Cochlea." *Hearing Research* 18: 169-75.

Tasaki, Ichiji and César Fernández. 1952. "Modification of Cochlear Microphonics and Action Potentials by KCl Solution and by Direct Currents." *Journal of Neurophysiology* 15: 497-512.

Teas, Donald C., Teruzo Konishi, and Joel S. Wernick. 1970. "Effects of Electrical Current Applied to Cochlear Partition on Discharges in Individual Auditory-Nerve Fibers. II. Interaction of Electrical Polarization and Acoustic Stimulation." *Journal of the Acoustical Society of America* 47(6): 1527-37.

Ulfendahl, Mats and Åke Flock. 1998. "Outer Hair Cells Provide Active Tuning in the Organ of Corti." *Physiology* 13: 107-11.

Weitzel, Erik K., Ron Tasker, and William E. Brownell. 2003. "Outer Hair Cell Piezoelectricity: Frequency Response Enhancement and Resonance Behavior." *Journal of the Acoustical Society of America* 114(3): 1462-66.

Wever, Ernest Glen. 1966. "Electrical Potentials of the Cochlea." *Physiological Reviews* 46(1): 102-27.

Wever, Ernest Glen and Charles William Bray. 1930. "Action Currents in the Auditory Nerve in Response to Acoustical Stimulation." *Proceedings of the National Academy of Sciences* 16(5): 344-50.

Zotterman, Yngve. 1943. "The Microphonic Effect of Teleost Labyrinths and its Biological Significance." *Journal of Physiology* 102: 313-18.

Zwislocki, Josef J. 1980. "Theory of Cochlear Mechanics." *Hearing Research* 2: 171-82.

Zwislocki, Josef J. and Lisa K. Cefaratti. 1989. "Tectorial Membrane II: Stiffness Measurements *In vivo*." *Hearing Research* 42: 211-28.

Zwislocki, Josef J. and My Nguyen. 1999. "Place Code for Pitch: A Necessary Revision." *Acta Oto-Laryngologica* 119(2): 140-45.

Zwislocki, Josef J., Norma B. Slepecky, Lisa K. Cefaratti, and Robert L. Smith. 1992. "Ionic Coupling Among Cells in the Organ of Corti." *Hearing Research* 57: 175-94.

Electrophonic Effect

Adrian, Donald J. 1977. "Auditory and Visual Sensations Stimulated by Low-frequency Electric Currents." *Radio Science* 12(6S): 243-50.

Althaus, Julius. 1873. *A Treatise on Medical Electricity*, 3rd ed. Philadelphia: Lindsay and Blakiston.

Augustin, Friedrich Ludwig. 1801. *Vom Galvanismus und dessen medicinischer Anwendung*. Berlin.

————. 1803. *Versuch einer vollständigen systematischen Geschichte der galvanischen Electricität und ihrer medicinischen Anwendung*. Berlin: Felisch.

Bartholow, Roberts. 1881. *Medical Electricity*. Philadelphia: Henry C. Lea's Son.

Bredon, Alan Dale. 1963. *Investigation of Diplexing Transducers for Voice Communications*. Electromagnetic Warfare and Communication Laboratory, Aeronautical Systems Division, Air Force Systems Command, Wright-Patterson Air Force Base, Ohio. Accession no. AD 400487, Technical Documentary Report no. ASD-TDR-63-157.

Brenner, Rudolf. 1868. *Untersuchungen und Beobachtungen über die Wirkung Elektrischer Ströme auf das Gehörorgan in gesunden und kranken Zustande*. Leipzig: Giesecke & Devrient.

Craik, Kenneth J. W., Alexander Francis Rawdon-Smith, and Rowan S. Sturdy. 1937. "Note on the Effect of A.C. on the Human Ear." *Proceedings of the Physiological Society*, May 8, pp. 2P-5P.

Eichhorn, Gustav. 1930. "The Electrostatic 'Radiophon.'" *Radio-Craft*, January, p. 330.

Einhorn, Richard N. 1967. "Army Tests Hearing Aids that Bypass the Ears." *Electronic Design* 15(26): 30-32.

Flanagan, Gillis Patrick. 1962. "Nervous System Excitation Device." U.S. Patent 3,393,279, filed March 13, 1962, issued July 16, 1968.

Flies, Carl Eduard. 1801. "Versuch des Herrn Dr. Flies." In: Carl Johann Christian Grapengiesser, *Versuche den Galvanismus zur Heilung Einiger Krankheiten anzuwenden* (Berlin: Mylius), pp. 241-52.

Flottorp, Gordon. 1953. "Effect of Different Types of Electrodes in Electrophonic Hearing." *Journal of the Acoustical Society of America* 25(2): 236-45.

Gersuni, Grigoryi V. and A. A. Volokhov. 1936. "On the Electrical Excitability of the Auditory Organ on the Effect of Alternating Currents on the Normal Auditory Apparatus." *Journal of Experimental Psychology* 19: 370-82.

Grapengiesser, Carl Johann Christian. 1801. *Versuche den Galvanismus zur Heilung Einiger Krankheiten anzuwenden*. Berlin: Mylius.

Hallpike, Charles Skinner and Hamilton Hartridge. 1937. "On the Response of the Human Ear to Audio-Frequency Electrical Stimulation." *Proceedings of the Royal Society of London B* 123: 177-93.

Harvey, William T. and James P. Hamilton. 1964. "Hearing Sensations in Amplitude Modulated Radio Frequency Fields." Master's thesis, Air Force Institute of Technology, Wright-Patterson Air Force Base, Ohio. Accession no. AD 608889.

Healer, Janet. 1967. "Auditory Response to Audio-Frequency Currents." In: Healer, ed., *Summary Report on a Review of Biological Mechanisms for Application to Instrument Design*, (Washington, DC: National Aeronautics and Space Administration), vol. 5, pp. 5-8 to 5-13. Accession no. N67-40136, Document no. ARA 346-F-2, part 1.

Hellwag, Christoph Friedrich and Maximilian Jacobi. 1802. *Erfahrungen über die Heilkräfte des Galvanismus, und Betrachtungen über desselben chemische und physiologische Wirkungen.* Hamburg: Friedrich Perthes.

Hoshiko, Michael S. 1970. "Electrostimulation of Hearing." In: Norman L. Wulfsohn and Anthony Sances, Jr., eds., *The Nervous System and Electic Currents* (New York: Plenum), pp. 85-88.

Johnson, Patrick Woodruff. 1971. "A Search for the Electrophonic Phenomena in the Microwatt Power Domain." Master's thesis, Naval Postgraduate School, Monterey, CA. Accession no. AD 744911.

Jones, H. Lewis. 1913. *Medical Electricity*, 6th ed. Philadelphia: P. Blakiston's Son.

Jones, R. Clark, Stanley Stephens Stevens, and Moses H. Lurie. 1940. "Three Mechanisms of Hearing by Electrical Stimulation." *Journal of the Acoustical Society of America* 12: 281-90.

Le Roy, Jean Baptiste. 1755. "Ou l'on rend compte de quelques tentatives que l'on a faites pour guérir plusieurs maladies par l'Électricité." *Mémoires de l'Académie Royale des Sciences*, pp. 60-98.

Martens, Franz Heinrich. 1803. *Vollständige Anweisung zur therapeutischen Anwendung des Galvanismus; nebst einer Geschichte dieses Heilmittels.* Weiszenfels: Böse.

Merzdorff, Johann Friedrich Alexander. 1801. Treatment of tinnitus with the galvanic current. In: Carl Johann Christian Grapengiesser, *Versuche den Galvanismus zur Heilung Einiger Krankheiten anzuwenden* (Berlin: Mylius), pp. 131-33.

Morgan, Charles E. 1868. *Electro-Physiology and Therapeutics.* New York: William Wood.

Moxon, Edwin Charles. 1971. "Neural and Mechanical Responses to Electrical Stimulation of the Cat's Inner Ear." Ph.D. dissertation, Massachusetts Institute of Technology.

Puharich, Henry K. and Joseph L. Lawrence. 1964. *Electro-Stimulation Techniques of Hearing.* QRC Branch, Rome Air Development Center, Research and Technology Division, Air Force Systems Command, Griffiss Air Force Base, NY. Accession no. AD 459956, Technical Documentary Report no. RADC-TDR-64-18.

Ritter, Johann Wilhelm. 1802. *Beyträge zur nähern Kentniss des Galvanismus und der Resultate seiner Untersuchung*, vol. 2, part 2. Jena: Friedrich Fromann.

Salmansohn, M. 1969. *Non-Acoustic Audio Coupling to the Head (NAACH).* Warminster, PA: Aero-Electronic Technology Department, Naval Air Development Center Johnsville. Accession no. AD 862280, Report no. NADC-AE-6922.

Salomon, Gerhard and Arnold Starr. 1963. "Sound Sensations Arising from Direct Current Stimulation of the Cochlea in Man." *Danish Medical Bulletin* 10(6-7): 215-16.

Skinner, Garland Frederick. 1968. "The Trans-Derma-Phone – A Research Device for the Investigation of Radio-Frequency Sound Stimulation." Master's thesis, Naval Postgraduate School, Monterey, CA.

Sommer, H. C. and Henning E. von Gierke. 1964. "Hearing Sensations in Electric Fields." *Aerospace Medicine* 35: 834-39.

Sprenger, Johann Justus Anton. 1802. "Anwendungsart der Galvani-Voltaischen Metall-Electricität zur Abhelfung der Taubheit und Harthörigkeit." *Annalen der Physik* 11(7): 354-66.

Stevens, Stanley Smith. 1937. "On Hearing by Electrical Stimulation." *Journal of the Acoustical Society of America* 8: 191-95.

Stevens, Stanley Smith and Hallowell Davis. 1938. *Hearing: Its Psychology and Physiology*. New York: American Institute of Physics.

Stevens, Stanley Smith and R. Clark Jones. 1939. "The Mechanism of Hearing by Electrical Stimulation." *Journal of the Acoustical Society of America* 10(4): 261-69.

Stevens, Stanley Smith and Fred Warshofsky. 1965. *Sound and Hearing*. New York: Time-Life Books.

Struve, Christian August. 1802. *System der medicinischen Elektrizitäts-Lehre mit Rücksicht auf den Galvanismus*. Breslau: Wilhelm Gottlieb Korn.

Tousey, Sinclair. 1921. *Medical Electricity, Röntgen Rays and Radium*, 3rd ed. Philadelphia: W.B. Saunders. Page 469 on auditory effects.

Volta, Alexander. 1800. "On the Electricity excited by the mere Contact of conducting Substances of different Kinds." *Philosophical Magazine* 7 (September): 289-311.

Wolke, Christian Heinrich. 1802. *Nachricht von den zu Jever durch die Galvani-Voltaische Gehör-Gebe-Kunst beglükten Taubstummen und von Sprengers Methode sie durch die Voltaische Elekricität auszuüben*. Oldenburg: Schulz.

Energy Efficient Light Bulbs

National Lighting Product Information Program. June 1999. "Screwbase Compact Fluorescent Lamp Products." *Specifier Reports* 7(1).

National Lighting Product Information Program. May 2000. "Electronic Ballasts." *Specifier Reports* 8(1).

Low Frequency Sounds

Begley, Sharon. 1993. "Do You Hear What I Hear? A Hum in Taos is Driving Dozens of People Crazy." *Newsweek*, May 3, pp. 54-55.

Brodeur, Paul. 1977. *The Zapping of America*. New York: W. W. Norton.

Cooke, Patrick. 1994. "The Hum." *Health*, July/August, pp. 71-75.

Curry, Bill P. and Gretchen V. Fleming. 2003. *RF Radiation Measurements in Selected Locations in Kokomo, Indiana*. Prepared for Acentech, Inc., Cambridge, MA, August 29.

Deming, David. 2004. "The Hum: An Anomalous Sound Heard Around the World." *Journal of Scientific Exploration* 18(4): 571-95.

Federation of American Scientists. 1995. *Submarine Communications Master Plan*. Washington, DC.

Firstenberg, Arthur. 1999. "The Source of the Taos Hum." *No Place To Hide* 2(2): 3-5.

Fox, Barry. 1989. "Low-frequency 'Hum' May Permeate the Environment." *New Scientist*, December 9, p. 27.

Garufi, Frank. 1989. *Loran C Field Strength Contours: Contiguous United States*. Washington, DC: Federal Aviation Administration. Report no. DOT/FAA/CT-TN89/16.

Hubbell, Schatzie. 1995. Hum survey results. Fort Worth, TX, October 6.

Jansky & Bailey, Atlantic Research Corporation. 1962. *The Loran-C System of Navigation*. Washington, DC.

Mullins, Joe H. and James P. Kelly. 1995. *The Elusive Hum in Taos, New Mexico. Acoustical Society Newsletter* 5(3): 1 ff.

Mullins, Joe H., James P. Kelly, and Sherry Robinson. 1993. "Hum Investigation: Source Still Unknown, Questions Raised." Albuquerque: University of New Mexico, August 23.

Samaddar, S. N. 1979. "Theory of Loran-C Ground Wave Propagation – A Review." *Journal of the Institute of Navigation* 26(3): 173-87.

Sheppard, L. and C. Sheppard. 1993. *The Phenomenon of Low Frequency Hums*. Norfolk, England: Norfolk Tinnitus Society.

United States Coast Guard. 1974. *Loran-C User Handbooks* Washington, DC. Publication no. CG-462.

———. 1992. *Loran-C User Handbook*. Washington, DC. Commandant Publication P16562.6.

Microwave Hearing

Chou, Chung-Kwang and Arthur William Guy. 1977. "Characteristics of Microwave-induced Cochlear Microphonics." *Radio Science* 6(S): 221-27.

Elder, Joseph A. and Chung-Kwang Chou. 2003. "Auditory Response to Pulsed Radiofrequency Energy." *Bioelectromagnetics*, suppl. 6: S162-73.

Frey, Allan H. 1961. "Auditory System Response to Radio Frequency Energy." *Aerospace Medicine* 32: 1140-42.

———. 1963. "Some Effects on Human Subjects of Ultra-High-Frequency Radiation." *American Journal of Medical Electronics*, January-March 1963, pp. 28-31.

———. 1967. "Brain Stem Evoked Responses Associated with Low-intensity Pulsed UHF Energy." *Journal of Applied Physiology* 23(6): 984-88.

———. 1970. "Effects of Microwave and Radio Frequency Energy on the Central Nervous System." In: Stephen F. Cleary, ed., *Biological Effects and Health Implications of Microwave Radiation. Symposium Proceedings* (Rockville, MD: U.S. Department of Health, Education and Welfare), Publication BRH/DBE 70-2, pp. 134-39.

———. 1971. "Biological Function as Influenced by Low-power Modulated RF Energy." *IEEE Transactions on Microwave Theory and Techniques* MTT-19(2): 153-64.

———. 1988. "Evolution and Results of Biological Research with Low-intensity Nonionizing Radiation." In: Andrew A. Marino, ed., *Modern Bioelectricity* (New York: Marcel Dekker, pp. 785-837.

Frey, Allan H. and Edwin S. Eichert III. 1972. "The Nature of Electrosensing in the Fish." *Biophysical Journal* 12: 1326-58.

———. 1985. "Psychophysical Analysis of Microwave Sound Perception." *Journal of Bioelectricity* 4(1): 1-14.

Frey, Allan H. and Rodman Messenger, Jr. 1973. "Human Perception of Illumination with Pulsed Ultrahigh-Frequency Electromagnetic Energy." *Science* 181: 356-58.

Justesen, Don R. 1975. "Microwaves and Behavior." *American Psychologist* 30(3): 391-401.

Khizhnyak, E. P., V. V. Tyazhelov, and V. V. Shorokhov. 1979. "Some Peculiarities and Possible Mechanisms of Auditory Sensation Evoked by Pulsed Electromagnetic Irradiation." *Activitas Nervosa Superior* 21(4): 247-51.

Lebovitz, Robert M. and Ronald L. Seaman. 1977. "Single Auditory Unit Responses to Weak, Pulsed Microwave Radiation." *Brain Research* 126: 370-5.

Lin, James C. 1978. *Microwave Auditory Effects and Applications*. Springfield, IL: Charles C. Thomas.

———. 2001. "Hearing Microwaves: The Microwave Auditory Phenomenon." *IEEE Antennas and Propagation Magazine* 43(6): 166-68.

Seaman, Ronald L. 2002. "Transmission of Microwave-induced Intracranial Sound to the Inner Ear is Most Likely Through Cranial Aqueducts." Brooks Air Force Base, TX: Walter Reed Army Institute of Research.

Seaman, Ronald L. and Robert M. Lebovitz. 1989. "Thresholds of Cat Cochlear Nucleus Neurons to Microwave Pulses." *Bioelectromagnetics* 10: 147-60.

Sharp, Joseph C., H. Mark Grove, and Om P. Gandhi. 1974. "Generation of Acoustic Signals by Pulsed Microwave Energy." *IEEE Transactions on Microwave Theory and Techniques* MTT-22(5): 583-84.

Stocklin, Philip L. and Brian F. Stocklin. 1979. "Possible Microwave Mechanisms of the Mammalian Nervous System." *T.-I.-T. Journal of Life Sciences* 9: 29-51.

Taylor, Eugene M. and Bonnie T. Ashleman. 1974. "Analysis of Central Nervous System Involvement in the Microwave Auditory Effect." *Brain Research* 74: 201-8.

Tyazhelov, V. V., R. E. Tigranian, E. O. Khizhniak, and I. G. Akoev. 1979. "Some Peculiarities of Auditory Sensations Evoked by Pulsed Microwave Fields." *Radio Science* 14(6S): 259-63.

Watanabe, Yoshiaki and Toshiyuki Tanaka. 2000. "FDTD Analysis of Microwave Hearing Effect." *IEEE Transactions on Microwave Theory and Techniques* MTT-48(11): 2126-32.

Wilson, Blake S. and William T. Joines. 1985. "Mechanisms and Physiologic Significance of Microwave Action on the Auditory System." *Journal of Bioelectricity* 4(2): 495-525.

Wilson, Blake S., John M. Zook, William T. Joines, and John H. Casseday. 1980. "Alterations in Activity at Auditory Nuclei of the Rat Induced by Exposure to Microwave Radiation: Autoradiographic Evidence Using [^{14}C]2-deoxy-d-Glucose." *Brain Reserch* 187: 291-306.

Power Line Radiation

Kikuchi, Hiroshi. 1972. "Investigations of Electromagnetic Noise and Interference Due to Power Lines in Japan and Some Results from the Aspect of Electromagnetic Theory." *Proceedings of the 1972 Symposium on Electromagnetic Hazards, Pollution and Environmental Quality*, Purdue University, Lafayette, Indiana, May 8-9, pp. 147-62.

———. 1983a. "Overview of Power-Line Radiation and its Coupling to the Ionosphere and Magnetosphere." *Space Science Reviews* 35: 33-41.

———. 1983b. "Power Line Transmission and Radiation." *Space Science Reviews* 35: 59-80.

———, ed. 1983c. *Power Line Radiation and Its Coupling to the Ionosphere and Magnetosphere*. Amsterdam: Reidel.

Vignati, Maurizio and Livio Giuliani. 1997. "Radiofrequency Exposure near High-Voltage Lines." *Environmental Health Perspectives* 105 (suppl. 6): 1569-73.

Saccular Hearing

Akin, Faith Wurm and Owen D. Murnane. 2004. "Vestibular Evoked Myogenic Potentials (VEMP)." *Clinical Topics in Otoneurology*, a publication of GN Otometrics, Copenhagen. April issue.

Bocca, Ettore and G. Perani. 1960. "Further Contributions to the Knowledge of Vestibular Hearing." *Acta Oto-Laryngologica* 51: 260-67.

Cazals, Yves, Jean-Marie Aran, and Jean-Paul Erre. 1982. "Frequency Sensitivity and Selectivity of Acoustically Evoked Potentials After Complete Cochlear Hair Cell Destruction." *Brain Research* 231: 197-203.

———. 1983. "Intensity Difference Thresholds Assessed with Eighth Nerve and Auditory Cortex Potentials: Compared Values from Cochlear and Saccular Responses." *Hearing Research* 10: 263-68.

Cazals, Yves, Jean-Marie Aran, Jean-Paul Erre, Anne Guilhaume, and Catherine Aurousseau. 1983. "Vestibular Acoustic Reception in the Guinea Pig: A Saccular Function?" *Acta Oto-Laryngologica* 95(3-4): 211-17.

Clarke, Andrew H., Uwe Schönfeld, and Kai Helling. 2003. "Unilateral Examination of Utricle and Saccule Function." *Journal of Vestibular Research* 13: 215-25.

Colebatch, James G. 2006. "Assessing Saccular (Otolith) Function in Man." *Journal of the Acoustical Society of America*, 119 (5 part 2): 3432. Abstract.

———. 2014. "Overview of VEMPs (Vestibular-Evoked Myogenic Potentials)." *30th International Congress of Clinical Neurophysiology*, Berlin, p. 53. Abstract.

Colebatch, James G., G. Michael Halmagyi, and Nevell F. Skuse. 1994. "Myogenic Potentials Generated by a Click-Evoked Vestibulocollic Reflex." *Journal of Neurology, Neurosurgery, and Psychiatry* 57(2): 190-97.

Didier, Anne and Yves Cazals. 1989. "Acoustic Responses Recorded from the Saccular Bundle on the Eighth Nerve of the Guinea Pig." *Hearing Research* 37: 123-28.

Emami, Seyede Faranak. 2013. "Is All Human Hearing Cochlear?" *Scientific World Journal*, article ID 147160.

———. 2014a. "Hypersensitivity of Vestibular System to Sound and Pseudoconductive Hearing Loss in Deaf Patients." *ISRN Otolaryngology*, article ID 817123.

———. 2014b. "Vestibular Activation by Sound in Human." *Scholars Journal of Applied Medical Sciences* 2(6H): 3445-51.

Emami, Seyede Faranak and Nasrin Gohari. 2014. "The Vestibular-Auditory Interaction for Auditory Brainstem Response to Low Frequencies." *ISRN Otolaryngology*, article ID 103598.

Emami, Seyede Faranak, Akram Pourbakht, Kianoush Sheykholeslami, Mohammad Kamali, Fatholah Behnoud, and Ahmad Daneshi. 2012. "Vestibular Hearing and Speech Processing." *ISRN Otolaryngology*, article ID 850629.

Guinan, John J., Jr. 2006. "Acoustically Responsive Fibers in the Mammalian Vestibular Nerve." *Journal of the Acoustical Society of America* 119 (5 part 2): 3432. Abstract.

Igarashi, Makoto and Yuho Kato. 1975. "Effect of Different Vestibular Lesions upon Body Equilibrium Function in Squirrel Monkeys." *Acta Oto-Laryngolica. Supplementum* 330: 91-99.

Lenhardt, Martin L. 1999. "Stapedial-Saccular Strut and Method." U.S. Patent 6,368,267, filed October 14, 1999, issued April 9, 2002.

———. 2006. "Saccular Hearing: Turtle Model for a Human Prosthesis." *Journal of the Acoustical Society of America* 119 (5 part 2): 3433-34. Abstract.

McCue, Michael P. and John J. Guinan, Jr. 1994. "Acoustically Responsive Fibers in the Vestibular Nerve of the Cat." *Journal of Neuroscience* 14(10): 6058-70.

———. 1997. "Sound-Evoked Activity in Primary Afferent Neurons of a Mammalian Vestibular System." *American Journal of Otology* 18(3): 355-60.

Meyer, Max F. 1931. "Hearing Without Cochlea?" *Science* 73: 236-37.

Reuter, Tom and Sirpa Nummela. 1998. "Elephant Hearing." *Journal of the Acoustical Society of America* 104 (2 part 1): 1122-23.

Ribarić, Ksenija, Tine S. Prevec, and Vladimir Kozina. 1984. "Frequency-Following Response Evoked by Acoustic Stimuli in Normal and Profoundly Deaf Subjects." *Audiology* 23(4): 388-400.

Robertson, D. D. and Dennis J. Ireland. 1995. "Vestibular Evoked Myogenic Potentials." *Journal of Otolaryngology* 24(1): 3-8.

Rosengren, Sally M., Miriam S. Welgampola, and James G. Colebatch. 2010. "Vestibular Evoked Myogenic Potentials: Past, Present and Future." *Clinical Neurophysiology* 121(5): 636-51.

Ross, Muriel D. 1983. "Gravity and the Cells of Gravity Receptors in Mammals." *Advances in Space Research* 3(9): 179-90.

Sohmer, Haim, Sharon Freeman, and Ronen Perez. 2004. "Semicircular Canal Fenestration – Improvement of Bone- but not Air-conducted Auditory Thresholds." *Hearing Research* 187: 105-10.

Tait, John. 1932. "Is All Hearing Cochlear?" *Annals of Otology, Rhinology and Laryngology* 41: 681-704.

Todd, Neil P. McAngus. 2001. "Evidence for a Behavioral Significance of Saccular Acoustic Sensitivity in Humans." *Journal of the Acoustical Society of America* 110(1): 380-90.

———. 2006. "Is All Hearing Cochlear? – Revisited (Again)." *Journal of the Acoustical Society of America* 119 (5 part 2): 3431-32. Abstract.

Trivelli, Maurizio, Massimiliano Potena, Valeria Frari, Tomassangelo Petitti, Valentina Deidda, and Fabrizio Salvinelli. 2013. "Compensatory Role of Saccule in Deaf Children and Adults: Novel Hypotheses." *Medical Hypotheses* 80(1): 43-46.

Wit, Hero P., J. D. Bleeker, and H. H. Mulder. 1984. "Responses of Pigeon Vestibular Nerve Fibers to Sound and Vibration with Audiofrequencies." *Journal of the Acoustical Society of America* 75(1): 202-8.

Wu, Chen-Chi and Yi-Ho Young. 2002. "Vestibular Evoked Myogenic Potentials Are Intact After Sudden Deafness." *Ear and Hearing* 23(3): 235-38.

Young, Eric D., César Fernández and Jay M. Goldberg. 1977. "Responses of Squirrel Monkey Vestibular Neurons to Audio-Frequency Sound and Head Vibration." *Acta Oto-Laryngologica* 84(5-6): 352-60.

Tinnitus

Del Bo, Luca, Stella Forti, Umberto Ambrosetti, Serena Costanzo, Davide Mauro, Gregorio Ugazio, Berthold Langguth, and Antonio Mancuso. 2008. "Tinnitus Aurium in Persons with Normal Hearing: 55 Years Later." *Otolaryngology – Head and Neck Surgery* 139: 391-94.

Heller, Morris F. and Moe Bergman. 1953. "Tinnitus Aurium in Normally Hearing Persons." *Annals of Otology* 62: 73-83.

Holgers, Kajsa-Mia. 2003. "Tinnitus in 7-year-old Children." *European Journal of Pediatrics* 162: 276-78.

Holgers, Kajsa-Mia and Jolanta Juul. 2006. "The Suffering of Tinnitus in Childhood and Adolescence." *International Journal of Audiology* 45: 267-72.

Holgers, Kajsa-Mia and Bo Pettersson. 2005. "Noise Exposure and Subjective Hearing Symptoms among School Children in Sweden." *Noise and Health* 7(27): 27-37.

Hutter, Hans-Peter, Hanns Moshammer, Peter Wallner, Monika Cartellieri, Doris-Maria Denk-Linnert, Michaela Katzinger, Klaus Ehrenberger, and Michael Kundi. 2010. "Tinnitus and Mobile Phone Use." *Occupational and Environmental Medicine* 67: 804-8.

Juul, Jolanta, Marie-Louise Barrenäs, and Kajsa-Mia Holgers. 2012. "Tinnitus and Hearing in 7-year-old Children." *Archives of Disease in Childhood* 97: 28-30.

Kochkin, Sergei, Richard Tyler, and Jennifer Born. 2011. "MarkeTrak VIII: The Prevalence of Tinnitus in the United States and the Self-reported Efficacy of Various Treatments." *Hearing Review*, November, pp. 10ff.

Møller, Aage R., Berthold Langguth, Dirk DeRidder, and Tobias Kleinjung, eds. 2011. *Textbook of Tinnitus*. New York: Springer.

National Center for Health Statistics. 1982-1996. "Current Estimates From the National Health Interview Survey." Table 57, "Number of Selected Reported Chronic Conditions per 1,000 Persons, by Age: United States." *Vital and Health Statistics*, ser. 10, nos. 150, 154, 160, 164, 166, 173, 176, 181, 184, 189, 190, 193, 199, 200.

Nondahl, David M., Karen J. Cruickshanks, Guan-Hua Huang, Barbara E. K. Klein, Ronald Klein, Ted S. Tweed, and Weihai Zhan. 2012. "Generational Differences in the Reporting of Tinnitus." *Ear and Hearing* 33(5): 640-44.

Shargorodsky, Josef, Gary C. Curhan, and Wildon R. Farwell. 2010. "Prevalence and Characteristics of Tinnitus among US Adults." *American Journal of Medicine* 123(8): 711-18.

Wieske, Clarence W. 1963. "Human Sensitivity to Electric Fields." *Biomedical Sciences Instrumentation* 1: 467-75.

Ultrasonic Hearing

Ball, Geoffrey R. and Bob H. Katz. 1998. "Ultrasonic Hearing System." U.S. Patent 6,217,508 B1, filed August 14, 1998, issued April 17, 2001.

Bance, Manohar, Osama Majdalawieh, Andrew Stewart, Michael Kiefte, and Rene van Wijhe. 2006. "Comparison of Air and Bone Conduction Fine Frequency Hearing Responses." Dalhousie University, Nova Scotia: Ear and Auditory Research Laboratory.

Bellucci, Richard J. and Daniel E. Schneider. 1962. "Some Observations on Ultrasonic Perception in Man." *Annals of Otology, Rhinology and Laryngology* 71: 719-26.

Combridge, J. H. and J. O. Ackroyd. 1945. *The Design of German Telephone Subscribers' Apparatus*. British Intelligence Objectives Sub-Committee. BIOS Final Report no. 606.

Corso, John F. 1963. "Bone-conduction Thesholds for Sonic and Ultrasonic Frequencies." *Journal of the Acoustical Society of America* 35(11): 1738-43.

Corso, John F. and Murray Levine. 1965a. "Pitch-Discrimination at High Frequencies by Air- and Bone-conduction." *American Journal of Psychology* 78(4): 557-66.

———. 1965b. "Sonic and Ultrasonic Equal-Loudness Contours." *Journal of Experimental Psychology* 70(4): 412-16.

Deatherage, Bruce H., Lloyd A. Jeffress, and Hugh C. Blodgett. 1954. "A Note on the Audibility of Intense Ultrasonic Sound." *Journal of the Acoustical Society of America* 26(4): 582.

Dieroff, H. G. and H. Ertel. 1975. "Some Thoughts on the Perception of Ultrasonics by Man." *Archives of Oto-Rhino-Laryngology* 209: 277-90.

Flach, M. and G. Hofmann. 1980. "Ultraschallhören des Menschen: Objektivierung mittels Hirnstammpotential." *Laryngo-Rhino-Otologie.* 59(12): 840-43.

Fujisaka, Yoh-ichi, Seiji Nakagawa, and Mitsuo Tonoike. 2005. "A Numerical Study on the Perception Mechanism for Detecting Pitch in Bone-conducted Ultrasound." Paper presented at the Twelfth International Congress on Sound and Vibration, July 11-15, Lisbon, Portugal.

Gavrilov, L. R., G. V. Gershuni, V. I. Pudov, A. S. Rozenblyum, and E. M. Tsirul'nikov. 1980. "Human Hearing in Connection with the Action of Ultrasound in the Megahertz Range on the Aural Labyrinth." *Soviet Physics – Acoustics.* 26(4): 290-92.

Haeff, Andrew V. and Cameron Knox. 1963. "Perception of Ultrasound." *Science* 139: 590-92.

Hotehama, Takuya and Seiji Nakagawa. 2010. "Modulation Detection for Amplitude-modulated Bone-conducted Sounds with Sinusoidal Carriers in the High- and Ultrasonic-frequency Range." *Journal of the Acoustical Society of America* 128(5): 3011-18.

Imaizumi, Satoshi, Hiroshi Hosoi, Takefumi Sakaguchi, Yoshiaki Watanabe, Norihiro Sadato, Satoshi Nakamura, Atsuo Waki, and Yoshiharu Yonekura. 2001. "Ultrasound Activates the Auditory Cortex of Profoundly Deaf Subjects." *NeuroReport* 12(3): 583-86.

International Organization for Standardization. 2003. *Normal Equal-loudness-level Contours.* ISO 226:2003 – Acoustics, 2nd ed. Geneva.

Kietz, Hans. 1951. "Hörschwellenmessung im Ultraschallgebiet." *Acta Oto-Laryngologica* 39(2-3): 183-87.

Lenhardt, Martin L. 1999. "Upper Audio Range Hearing Apparatus and Method." U.S. Patent 6,731,769, filed October 14, 1999, issued May 4, 2004.

———. 2003. "Ultrasonic Hearing in Humans: Applications for Tinnitus Treatment." *International Tinnitus Journal* 9(2): 69-75.

———. 2006. "A Second Pair of Ears." *Echoes* 16(4): 5-6.

———. 2008. "Ring Transducers for Sonic, Ultrasonic Hearing." U.S. Patent 8,107,647, filed January 3, 2008, issued January 31, 2012.

Lenhardt, Martin, Alex M. Clarke, and William Regelson. 1989. "Supersonic Bone Conduction Hearing Aid and Method." U.S. Patent 4,982,434, filed May 30, 1989, issued January 1, 1991.

Lenhardt, Martin L., Ruth Skellett, Peter Wang, and Alex M. Clarke. 1991. "Human Ultrasonic Speech Perception." *Science* 253: 83-85.

Magee, Timothy R. and Alun H. Davies. 1993. "Auditory Phenomena during Transcranial Doppler Insonation of the Basilar Artery." *Journal of Ultrasound in Medicine* 12: 747-50.

Maggs, James E. 1976. "Coherent Generation of VLF Hiss." *Journal of Geophysical Research* 81(10): 1707-24.

Moller, Henrik and Christian Sejer Pedersen. 2004. "Hearing at Low and Infrasonic Frequncies." *Noise and Health* 6(23): 37-58.

Nishimura, Tadashi, Seiji Nakagawa, Takefumi Sakaguchi, and Hiroshi Hosoi. 2003. "Ultrasonic Masker Clarifies Ultrasonic Perception in Man." *Hearing Research* 175: 171-77.

Nishimura, Tadashi, Tadao Okayasu, Osamu Saito, Ryota Shimokura, Akinori Yamashita, Toshiaki Yamanaka, Hiroshi Hosoi, and Tadashi Kitahara. 2014. "An Examination of the Effects of Broadband Air-conduction Masker on the Speech Intelligibility of Speech-modulated Bone-conduction Ultrasound." *Hearing Research* 317: 41-49.

Nishimura, Tadashi, Tadao Okayasu, Yuka Uratani, Fumi Fukuda, Osamu Saito, and Hiroshi Hosoi. 2011. "Peripheral Perception Mechanism of Ultrasonic Hearing." *Hearing Research* 277: 176-83.

Nishimura, Tadashi, Takefumi Sakaguchi, Seiji Nakagawa, Hiroshi Hosoi, Yoshiaki Watanabe, Mitsuo Tonoike, and Satoshi Imaizumi. 2000. "Dynamic Range for Bone Conduction Ultrasound." In: *Biomag 2000: Proceedings of 12th International Conference on Biomagnetism*, August 13-17, 2000, Helsinki University of Technology, Espoo, Finland, pp. 125-28.

Ohyama, Kenji, Jun Kusakari, and Kazutomo Kawamoto. 1987. "Sound Perception in the Ultrasonic Region." *Acta Oto-Laryngolica. Supplementum.* 435: 73-77.

Oohashi, Tsutomu, Emi Nishina, Manabu Honda, Yoshiharu Yonekura, Yoshitaka Fuwamoto, Norie Kawai, Tadao Maekawa, Satoshi Nakamura, Hidenao Fukuyama, and Hiroshi Shibasaki. 2000. "Inaudible High-Frequency Sounds Affect Brain Activity: Hypersonic Effect." *Journal of Neurophysiology* 83(6): 3548-58.

Ozen, Sukru. 2008. "Low-Frequency Transient Electric and Magnetic Fields Coupling to Child Body." *Radiation Protection Dosimetry* 128(1): 62-67.

Petrie, William. 1963. *Keoeeit: The Story of the Aurora Borealis.* Oxford: Pergamon Press.

Prasch, G. and H. Siegl-Graz. 1969. "Gehörseindrücke durch Einwirkung von tonfrequenten Wechselströmen und amplituden-modulierten Hochfrequenzströmen." *Archiv für klinische und experimentelle Ohren-, Nasen- und Kehlkopfheilkunde* 194(2): 516-21.

Pumphrey, R. J. 1950. "Upper Limit of Frequency for Human Hearing." *Nature* 166: 571.

Qin, Michael K., Derek Schwaller, Matthew Babina, and Edward Cudahy. 2011. "Human Underwater and Bone Conduction Hearing in the Sonic and Ultrasonic Range." *Journal of the Acoustical Society of America* 129 (4 part 2): 2485. Abstract.

Singh, D. K. and R. P. Singh. 2002. "Hiss Emissions during Quiet and Disturbed Periods." *Pramana – Journal of Physics* 59(4): 563-73.

Stanley, Raymond M. and Bruce N. Walker. 2005. "Relative Threshold Curves for Implementation of Auditory Displays on Bone-conduction Headsets in Multiple

Listening Environments." Presented at the 11th International Conference on Auditory Display, Limerick, Ireland, July 6-9.

Wegel, Raymond L., Robert R. Riesz, and Ralph B. Blackman. 1932. "Low Frequency Thresholds of Hearing and of Feeling in the Ear and Ear Mechanisms." *Journal of the Acoustical Society of America* 4(1A): 6.

World Health Organization. 1993. *Environmental Health Criteria 137. Electromagnetic Fields (300 Hz to 300 GHz).* Geneva.

Chapter 16

Balmori, Alfonso. 2014. "Electrosmog and Species Conservation." *Science of the Total Environment* 496: 314-16.

———. 2015. "Anthropogenic Radiofrequency Electromagnetic Fields as an Emerging Threat to Wildlife Orientation." *Science of the Total Environment* 518-519: 58-60.

Amazon Rainforest

da Costa, Thomaz Guedes. 2002. "Brazil's SIVAM: As It Monitors the Amazon, Will It Fulfill Its Human Security Promise?" *ECSP Report* 7: 47-58.

Jensen, David. 2002. "SIVAM: Communication, Navigation and Surveillance for the Amazon." *Avionics,* June 1.

Phillips, Oliver L, Luiz E. O. C. Aragão, Simon L. Lewis, Joshua B. Fisher, Jon Lloyd, Gabriela López-González, Yadvinder Malhi, Abel Monteagudo, Julie Peacock, Carlos A. Quesada, Geertje van der Heijden, Samuel Almeida, Iêda Amaral, Luzmila Arroyo, Gerardo Aymard, Tim R. Baker, Olaf Bánki, Lilian Blanc, Damien Bonal, Paulo Brando, Jerome Chave, Átila Cristina Alves de Oliveira, Nallaret Dávila Cardozo, Claudia I. Czimczik, Ted R. Feldpausch, Maria Aparecida Freitas, Emanuel Gloor, Niro Higuchi, Eliana Jiménez, Gareth Lloyd, Patrick Meir, Casimiro Mendoza, Alexandra Morel, David A. Neill, Daniel Nepstad, Sandra Patiño, Maria Cristina Peñuela, Adriana Prieto, Fredy Ramírez, Michael Schwarz, Javier Silva, Marcos Silveira, Anne Sota Thomas, Hans ter Steege, Juliana Stropp, Rodolfo Vásquez, Przemyslaw Zelazowski, Ésteban Álvarez Dávila, Sandy Andelman, Ana Andrade, Kuo-Jung Chao, Terry Erwin, Anthony Di Fiore, Eurídice Honorio C., Helen Keeling, Tim J. Killeen, William F. Laurance, Antonio Peña Cruz, Nigel C. A. Pitman, Percy Núñez Vargas, Hirma Ramírez-Ángulo, Agustín Rudas, Rafael Salamão, Natalino Silva, John Terborgh, and Armando Torres-Lezama. 2009. "Drought Sensitivity of the Amazon Rainforest." 2009. *Science* 323: 1344-47.

Rohter, Larry. 2002. "Brazil Employs Tools of Spying to Guard Itself." *New York Times,* July 27, p. 1.

Wittkoff, E. Peter. 1999. "Amazon Surveillance System (SIVAM): U.S. and Brazilian Cooperation." Master's thesis, Naval Postgraduate School, Monterey, CA.

Amphibians

Balmori, Alfonso. 2006. "The Incidence of Electromagnetic Pollution on the Amphibian Decline: Is This an Important Piece of the Puzzle?" *Toxicological and Environmental Chemistry* 88(2): 287-89.

———. 2010. "Mobile Phone Mast Effects on Common Frog (*Rana temporaria*) Tadpole: The City Turned into a Laboratory." *Electromagnetic Biology and Medicine* 29: 31-35.

Hallowell, Christopher. 1996. "Trouble in the Lily Pads." *Time*, October 28, p. 87.

Hawk, Kathy. 1996. *Case Study in the Heartland*. Butler, PA.

Hoperskaya, O. A., L. A. Belkova, M. E. Bogdanov, and S. G. Denisov. 1999. "The Action of the 'Gamma-7N' Device on Biological Objects Exposed to Radiation from Personal Computers." *Electromagnetic Fields and Human Health. Proceedings of the Second International Conference*. Moscow, September 20-24, pp. 354-55. Abstract.

Revkin, Andrew C. 2006. "Frog Killer is Linked to Global Warming." *New York Times*, January 12.

Souder, William. 1996. "An Amphibian Horror Story." *New York Newsday*, October 15, pp. B19, B21.

———. 1997. "Deformed Frogs Show Rift Among Scientists." *Houston Chronicle*, November 5, p. 4A.

Stern, John. 1990. "Space Aliens Stealing Our Frogs." *Weekly World News*, April 17, p. 21.

Vogt, Amanda. 1998. "Mutant Frogs Spark a Mega Mystery." *Chicago Tribune*, August 4, sec. 7, p. 3.

Watson, Traci. 1998. "Frogs Falling Silent across USA." *USA Today*, August 12, p. 3A.

Birds

Balmori, Alfonso. 2003. "Aves y telefonía móvil." *El Ecologista* 36: 40-42.

———. 2005. "Possible Effects of Electromagnetic Fields from Phone Masts on a Population of White Stork (*Ciconia ciconia*)." *Electromagnetic Biology and Medicine* 24: 109-19.

Bigu del Blanco, Jaime. 1969. *An Introduction to the Effects of Electromagnetic Radiation on Living Matter with Special Reference to Microwaves*. Laboratory Technical Report LTR-CS-7, Control Systems Laboratory, Division of Mechanical Engineering, National Research Council Canada.

———. 1973. *Interaction of Electromagnetic Fields and Living Systems with Special Reference to Birds*. Laboratory Technical Report LTR-CS-113, Control Systems Laboratory, Division of Mechanical Engineering, National Research Council Canada.

Bigu del Blanco, Jaime and César Romero-Sierra. 1973. *Bird Feathers as Dielectric Receptors of Microwave Radiation*. Laboratory Technical Report LTR-CS-89, Control Systems Laboratory, Division of Mechanical Engineering, National Research Council Canada.

———. 1975. "Microwave Pollution of the Environment and the Ecological Problem." In: Tomáš Dvořák, ed., *Electromagnetic Compatibility 1975: 1st Symposium and Technical Exhibition on Electromagnetic Compatibility, Montreux, May 20-22, 1975*, pp. 127-33.

Bigu del Blanco, Jaime, César Romero-Sierra, and J. Alan Tanner. 1973a. *Environmental Pollution by Microwave Radiation – A Potential Threat to Human Health*. Laboratory Technical Report LTR-CS-98, Control Systems Laboratory, Division of Mechanical Engineering, National Research Council Canada.

————. 1973b. "Radiofrequency Fields: A New Ecological Factor." *1973 IEEE International Electromagnetic Compatibility Symposium Record*, June 20-22, New York, pp. 54-59.

Engels, Svenja, Nils-Lasse Schneider, Nele Lefeldt, Christine Maira Hein, Manuela Zapka, Andreas Michalik, Dana Elbers, Achim Kittel, P. J. Hore, and Henrik Mouritsen. 2014. "Anthropogenic Electromagnetic Noise Disrupts Magnetic Compass Orientation in a Migratory Bird." *Nature* 509: 353-56.

Keeton, William T. 1979. "Avian Orientation and Navigation: A Brief Overview." *British Birds* 72(10): 451-70.

Romero-Sierra, César, Carol Husband, and J. Alan Tanner. 1969. *Effects of Microwave Radiation on Parakeets in Flight*. Laboratory Technical Report LTR-CS-18. Control Systems Laboratory, Division of Mechanical Engineering, National Research Council Canada.

Romero-Sierra, César, Arthur O. Quanbury, and J. Alan Tanner. 1970. *Feathers as Microwave and Infra-Red Filters and Detectors – Preliminary Experiments*. Laboratory Technical Report LTR-CS-40, Control Systems Laboratory, Division of Mechanical Engineering, National Research Council Canada.

Romero-Sierra, César, J. Alan Tanner, and F. Villa. 1969. *EMG Changes in the Limb Muscles of Chickens Subjected to Microwave Radiation*. Laboratory Technical Report LTR-CS-16, Control Systems Laboratory, Division of Mechanical Engineering, National Research Council Canada.

Tanner, J. Alan. 1966. "Effect of Microwave Radiation on Birds." *Nature* 210: 636.

————. 1970. "Bird Feathers as Sensory Detectors of Microwave Fields." In: Stephen F. Cleary, ed., *Biological Effects and Health Implications of Microwave Radiation. Symposium Proceedings* (Rockville, MD: U.S. Department of Health, Education and Welfare), Publication BRH/DBE 70-2, pp. 185-87.

Tanner, J. Alan and César Romero-Sierra. 1971. *Non-Ionizing Electromagnetic Radiation and Pollution of the Atmosphere*. Report no. DMENAE19714, Control Systems Laboratory, Division of Mechanical Engineering, National Research Council Canada.

————. 1982. "The Effects of Chronic Exposure to Very Low Intensity Microwave Radiation on Domestic Fowl." *Journal of Bioelectricity* 1(2): 195-205.

Xenos, Thomas D. and Ioannis N. Magras. 2003. "Low Power Density RF-Radiation Effects on Experimental Animal Embryos and Foetuses." In: Peter Stavroulakis, ed., *Biological Effects of Electromagnetic Fields* (Berlin: Springer), pp. 579-602.

Cedars

Earth Link and Advanced Resources Development S. A. R. L. 2010. "Vulnerability and Adaptation of the Forestry Sector." *Climate Risks, Vulnerability and Adaptation Assessment*, pp. 6-1 to 6-44. Prepared for United Nations Development Programme and Ministry of Environment of Lebanon.

Bentouati, Abdallah and Michel Bariteau. 2006. "Réflexions sur le dépérissement du Cèdre de l'Atlas des Aurès (Algérie)." *Forêt Méditerranéenne* 27(4): 317-22.

Hennon, Paul E., David V. D'Amore, Paul G. Schaberg, Dustin T. Wittwer, and Colin S. Shanley. 2012. "Shifting Climate, Altered Niche, and a Dynamic Conservation Strategy for Yellow-Cedar in the North Pacific Coastal Rainforest." *BioScience* 62(2): 147-58.

Hennon, Paul E., David V. D'Amore, Stefan Zeglen, and Mike Grainger. 2005. *Yellow-Cedar Decline in the North Coast Forest District of British Columbia*. Research Note PNW-RN-549. Juneau, AK: USDA Forest Service, Pacific Northwest Research Station.

Hennon, Paul E. and Charles G. Shaw III. 1994. "Did Climatic Warming Trigger the Onset and Development of Yellow-Cedar Decline in Southeast Alaska?" *European Journal of Forest Pathology* 24: 399-418.

Hennon, Paul E., Charles G. Shaw III, and Everett M. Hansen. 1990. "Dating Decline and Mortality of *Chamaecyparis nootkatensis* in Southeast Alaska." *Forest Science* 36(3): 502-15.

Hennon, Paul E., David V. D'Amore, Dustin T. Wittwer, A. Johnson, Paul G. Schaberg, G. Hawley, C. Beier, S. Sink, and G. Juday. 2006. "Climate Warming, Reduced Snow, and Freezing Injury Could Explain the Demise of Yellow-Cedar in Southeast Alaska, USA." *World Resource Review* 18(2): 427-50.

Masri, Rania. 1995. *The Cedars of Lebanon: Significance, Awareness and Management of the Cedrus libani in Lebanon*. Lecture given at Massachusetts Institute of Technology, November 9.

Navy Department, Bureau of Equipment. August 1, 1907. *Wireless Telegraph Stations of the World*. Washington, DC.

Navy Department, Bureau of Equipment. *Wireless Telegraph Stations of the World. Corrected to October 1, 1908*. Washington, DC.

United States Department of Commerce, Bureau of Navigation. July 1, 1913. *Radio Stations of the United States*. Washington, DC.

Verstege, A., J. Esper, B. Neuwirth, M. Alifriqui, and D. Frank. 2004. "On the Potential of Cedar Forests in the Middle Atlas (Morocco) for Climate Reconstructions." In: E. Jansma, A. Bräuning, H. Gärtner, and G. Schleser, eds., *TRACE – Tree Rings in Archaeology, Climatology and Ecology*, vol. 2, Proceedings of the DENDROSYMPOSIUM, May 1-3, Utrecht, The Netherlands (Forschungszentrum Jülich), pp. 78-84.

Colegio García Quintana

Santiago, Ana. 2012. "El caso García Quintana cumple diez años sin nuevos diagnósticos de cáncer." *El Norte de Castilla*, March 23.

Diario de León. 2004. "El sexto caso de cáncer desata la alarma en un colegio de Valladolid." May 8.

Cantalapiedra, Francisco. 2004. "Aflora otro caso de cáncer en el colegio García Quintana de Valladolid." *El País*, May 8.

El Mundo. 2004. "Un mujer diagnosticada en 2002, sexto caso de cáncer en el colegio de Valladolid." May 7.

Forests

Allen, Craig D., Alison K. Macalady, Haroun Chenchouni, Dominique Bachelet, Nate McDowell, Michel Vennetier, Thomas Kitzberger, Andreas Rigling, David D. Breshears, E. H. Hogg, Patrick Gonzalez, Rod Fensham, Zhen Zhang, Jorge Castro, Natalia Demidova, Jong-Hwan Lim, Gillian Allard, Steven W. Running, Akkin Semerci, and Neil Cobb. 2010. "A Global Overview of Drought and Heat-induced Tree Mortality Reveals Emerging Climate Change Risks for Forests." *Forest Ecology and Management* 259: 660-84.

Balmori, Alfonso. 2004. "¿Pueden afectar las microondas pulsadas emitidas por las antenas de telefonía a los arboles y otros vegetales?" *Ecosistemas* 13(3): 79-87.

Ciesla, William M. and Edwin Donaubauer. 1994. *Decline and Dieback of Trees and Forests: A Global Overview*. Rome: Food and Agriculture Organization of the United Nations. FAO Forestry Paper 120.

Glinz, Franz. 1992. "Der Wald stirbt am Electrosmog." *Auto-illustrierte* 2: 1.

Haggerty, Katie. 2010. "Adverse Influence of Radio Frequency Background on Trembling Aspen Seedlings: Preliminary Observations." *International Journal of Forestry Research*, article ID 836278.

Hertel, Hans Ulrich. 1991. "Der Wald Stirbt und Politiker Sehen Zu." *Raum & Zeit* 9(51): 3-12.

Hommel, H. 1985. "Elektromagnetischer SMOG – Schadfaktor und Stress?" *Forstarchiv* 56: 227-33.

LeBlanc, David C., Dudley J. Raynal, and Edwin H. White. 1987. "Acidic Deposition and Tree Growth: I. The Use of Stem Analysis to Study Historical Growth Patterns." *Journal of Environmental Quality* 16(4): 325-40.

Lohmeyer, Michael. 1991. "Von Mikrowellen verseuchte Umgebung; Richtfunk schneidet Schneisen in Wälder." *Die Presse*, July 31.

Lorenz, M., V. Mues, G. Becher, Ch. Müller-Edzards, S. Luyssaert, H. Raitio, A. Fürst, and D. Langouche. 2003. *Forest Condition in Europe*. Geneva and Brussels: United Nations Economic Commission for Europe and the European Commission.

Melhorn, G., B. J. Francis, and A. R. Wellburn. 1988. "Prediction of the Probability of Forest Decline Damage to Norway Spruce Using Three Simple Site-independent Diagnostic Parameters." *New Phytologist* 110: 525-34.

Robbins, Jim. 2010. "What's Killing the Great Forests of the American West?" *Environment 360*, March 15.

Schütt, Peter and Ellis B. Cowling. 1985. "Waldsterben, A General Decline of Forests in Central Europe: Symptoms, Development and Possible Causes." *Plant Disease* 69(7): 548-58.

Skelly, John M. and John L. Innes. 1994. "Waldsterben in the Forests of Central Europe and Eastern North America: Fantasy or Reality?" *Plant Disease* 78(11): 1021-32.

van Mantgem, Phillip J., Nathan L. Stephenson, John C. Byrne, Lori D. Daniels, Jerry F. Franklin, Peter Z. Fulé, Mark E. Harmon, Andrew J. Larson, Jeremy M. Smith, Alan H. Taylor, and Thomas T. Veblen. 2009. "Widespread Increase of Tree Mortality Rates in the Western United States." *Science* 323: 521-24.

Volkrodt, Wolfgang. 1989. "Electromagnetic Pollution of the Environment." In: Robert Krieps, ed., *Environment and Health: A Holistic Approach* (Aldershot, UK: Avebury), pp. 71-76.

———. 1991. "Mikrowellensmog und Waldschäden – Tut Sich Doch Noch Was in Bonn?" *Raum & Zeit* 9(52): 22-25.

———. 1992. Letter to William H. Smith, Yale University, December 26.

Waldmann-Selsam, Cornelia and Horst Eger. 2013. "Baumschäden im Umkreis von Mobilfunksendeanlagen." *Umwelt-Medizin-Gesellschaft* 26(3): 198-208.

Worrall, James J., Leanne Egeland, Thomas Eager, Roy A. Mask, Erik W. Johnson, Philip A. Kemp, and Wayne D. Shepperd. 2008. "Rapid Mortality of *Populus tremuloides* in Southwestern Colorado, USA." *Forest Ecology and Management* 225: 686-96.

HAARP
Browne, Malcolm W. 1995. "Scope System Also Offers a Tool for Submarines and Soldiers." *New York Times*, November 21, p. C10.

Busch, Lisa. 1997. "Ionosphere Research Lab Sparks Fear in Alaska." *Science* 275: 1060-61.

Microwave News. 1994. "U.S. Military Plans Powerful RF 'Heater' for Ionospheric Studies." May/June, pp. 10-11.

Papadopoulos, Dennis, Paul A. Bernhardt, Herbert C. Carlson, Jr., William E. Gordon, Alexander V. Gurevich, Michael C. Kelley, Michael J. Keskinen, Roald Z. Sagdeev, and Gennady M. Milikh. 1995. *HAARP: Research and Applications. A Joint Program of Phillips Laboratory and the Office of Naval Research. Executive Summary*. Washington, DC: Naval Research Laboratory.

Weinberger, Sharon. 2008. "Heating Up the Heavens." *Nature* 452: 930-32.

Williams, Richard. 1988. "Atmospheric Threat." *Physics and Society* 17(2): 16.

Zickuhr, Clare and Gar Smith. 1994. "Project HAARP: The Military's Plan to Alter the Ionosphere." *Earth Island Journal*, Fall 1994, pp. 21-23.

Homing Pigeons
Armas, Genaro C. 1998. "The Homing Pigeons That Didn't." *Seattle Times*, October 9.

Chaudhary, Vivek. 2004. "Phone Masts Blamed for Pigeons' Lost Art." *The Guardian*, January 23.

Elston, Laura. 2004. "Phone Masts 'Knocking Racing Pigeons off Track.'" *The Press Association (UK)*, January 23.

Haughey, Nuala. 1997. "Mobile Phones Blamed for Poor Pigeon Performance." *Irish Times*, July 21.

Hummell, Steve. 2005. "Lost Pigeons Create Flap; Cellphone Signals Responsible for Sending Birds off Course, Racers Say." *Vancouver Sun*, October 3.

Indian Express. 2010. "Cellphone Towers Disorient Homer Pigeons." December 27.

Keeton, William T. 1972. "Effects of Magnets on Pigeon Homing." In: S. R. Galler, K. Schmidt-Koenig, G. J. Jacobs, and R. E. Belleville, eds., *Animal Orientation and Navigation* (Washington, DC: Government Printing Office), NASA SP-262, pp. 579-94.

———. 1979. "Avian Orientation and Navigation: A Brief Overview." *British Birds* 72(10): 451-70.

Keeton, William T., Timothy S. Larkin, and Donald M. Windsor. 1974. "Normal Fluctuations in the Earth's Magnetic Field Influence Pigeon Orientation." *Journal of Comparative Physiology* 95: 95-103.

New York Post. 1998. "2,400 Homing Pigeons Fly the Coop in Race." October 8.

Wee, Eric L. 1998. "Homing Pigeons Race Off to Oblivion." *Washington Post*, October 8.

———. 1998. "Some Birds Lost During Races Are Turning Up at Area Homes, Barns and Feeders." *Washington Post*, October 9.

Honey Bees

Anderson, John. 1930a. "'Isle of Wight Disease' in Bees. I." *Bee World* 11(4): 37-42.

———. 1930b. "'Isle of Wight Disease' in Bees – II. A Check to the Immunity Hypothesis." *Bee World* 11(5): 50-53.

Bailey, Leslie 1958. "The Epidemiology of the Infestation of the Honeybee, *Apis mellifera* L., by the Mite *Acarapis woodi* Rennie and the Mortality of Infested Bees." *Parasitology* 48(3-4): 493-506.

———. 1964. "The 'Isle of Wight disease': The Origin and Significance of the Myth." *Bee World* 45(1): 32-37, 18.

Bailey, Leslie and D. C. Lee. 1959. "The Effect of Infestation with *Acarapis woodi* (Rennie) on the Mortality of Honey Bees." *Journal of Insect Pathology* 1(1): 15-24.

Bailey, Leslie and Brenda V. Ball. 1991. *Honey Bee Pathology*. London: Academic.

Barrionuevo, Alexei. 2007. "Honeybees, Gone with the Wind, Leave Crops and Keepers in Peril." *New York Times*, February 27, p. A1.

Boecking O. and W. Ritter. 1993. "Grooming and Removal Behaviour of *Apis mellifera intermissa* in Tunisia against *Varroa jacobsoni*." *Journal of Apicultural Research* 32: 127-34.

Borenstein, Seth. 2007. "Honeybee Die-off Threatens Food Supply." *Washington Post*, May 2.

Calderón Rafael A., Natalia Fallas, Luis G. Zamora, Johan W. van Veen, and Luis A. Sánchez. 2009. "Behavior of Varroa Mites in Worker Brood Cells of Africanized Honey Bees." *Experimental and Applied Acarology* 49(4): 329-38.

Carr, Elmer G.. 1918. "An Unusual Disease of Honey Bees." *Journal of Economic Entomology* 11(4): 347-51.

Dahlen, Sage. 2007. "Colony Collapse Disorder." *The Wake*, Summer 2007, p. 15.

Favre, Daniel. 2011. "Mobile Phone-induced Honeybee Worker Piping." *Apidologie* 42: 270-79.

Finley, Jennifer, Scott Camazine, and Maryann Frazier. 1996. "The Epidemic of Honey Bee Colony Losses during the 1995-1996 Season." *American Bee Journal* 136(11): 805-8.

Fries, Ingemar, Anton Imdorf, and Peter Rosenkranz. 2006. "Survival of Mite Infested (*Varroa destructor*) Honey Bee (*Apis mellifera*) Colonies in a Nordic Climate. *Apidologie* 37: 1-7.

Hamzelou, Jessica. 2007. "Where Have All the Bees Gone?" *Lancet* 370: 639.

Henderson, Colin, Jerry Bromenshenk, Larry Tarver, and Dave Plummer. 2007. *National Honey Bee Loss Survey*. Missoula, MT: Bee Alert Technology, Inc.

Imms, Augustus D. 1907. "Report on a Disease of Bees in the Isle of Wight." *Journal of the Board of Agriculture* 14(3): 129-40.

Kauffeld, Norbert M., James H. Everitt, and Edgar A. Taylor. 1976. "Honey Bee Problems in the Rio Grande Valley of Texas." *American Bee Journal* 116: 220, 222, 232.

Kraus, Bernhard and Robert E. Page, Jr. 1995. "Effect of *Varroa jacobsoni* (Mesostigmata: Varroidae) on Feral *Apis mellifera* (Hymenoptera: Apidae) in California." *Environmental Entomology* 24(6): 1474-80.

Kumar, Neelima R., Sonika Sangwan, and Pooja Badotra. 2011. "Exposure to Cell Phone Radiations Produces Biochemical Changes in Worker Honey Bees." *Toxicology International* 18(1): 70-72.

Le Conte, Yves, Marion Ellis, and Wolfgang Ritter. 2010. "*Varroa* Mites and Honey Bee Health: Can *Varroa* Explain Part of the Colony Losses?" *Apidologie* 41(3): 353-63.

Lee, Kathleen V., Nathalie A. Steinhauer, Karen Rennich, Michael E. Wilson, David R. Tarpy, Dewey M. Caron, Robyn Rose, Keith S. Delaplane, Kathy Baylis, Eugene J. Lengerich, Jeffery Pettis, John A. Skinner, James T. Wilkes, Ramesh Sagili, and Dennis vanEngelsdorp. 2015. "A National Survey of Managed Honey Bee 2013-2014 Annual Colony Losses in the USA." *Apidologie* 46: 292-305.

Lindauer, Martin and Herman Martin. 1972. "Magnetic Effect on Dancing Bees." In: Sidney R. Galler, Klaus Schmidt-Koenig, G. J. Jacobs, and Richard E. Belleville, eds., *Animal Orientation and Navigation*, (Washington, DC: Government Printing Office), NASA SP-262, pp. 559-67.

McCarthy, Michael. 2011. "Decline of Honey Bees Now a Global Phenomenon, Says United Nations." *The Independent*, March 10.

O'Hanlon, Kevin. 1997. "Few Honeybees Means Poorer Fruit, Vegetables." *Associated Press*, May 28.

Oldroyd, Benjamin P. 1999. "Coevolution While You Wait: *Varroa jacobsoni*, a New Parasite of Western Honeybees." *Trends in Ecology and Evolution* 14(8): 312-15, 1999.

———. 2007. "What's Killing American Honey Bees?" *PLoS Biology* 5(6): 1195-99.

Page, Robert E. 1998. "Blessing or Curse? Varroa Mite Impacts Africanized Bee Spread and Beekeeping." *California Agriculture* 52(2): 9-13.

Pattazhy, Sainudeen. 2011a. *Impact of Electromagnetic Radiation on the Density of Honeybees: A Case Study.* Saarbrücken, Germany: Lambert Academic.

———. 2011b. "Impact of Mobile Phones on the Density of Honey Bees." *Munis Entomology and Zoology* 6(1): 396-99.

———. 2012. "Electromagnetic Radiation (EMR) Clashes with Honeybees." *Journal of Entomology and Nematology* 4(10): 1-3.

Phillips, Ernest F. 1925. "The Status of Isle of Wight Disease in Various Countries." *Journal of Economic Entomology* 18: 391-95.

Rennie, John, Philip Bruce White, and Elsie J. Harvey. 1921. "Isle of Wight Disease in Hive Bees: The Etiology of the Disease." *Transactions of the Royal Society of Edinburgh*, vol. 52, part 4, no. 29, pp. 737-79.

Rinderer, Thomas E., Lilia I. de Guzman, G. T. Delatte, J. A. Stelzer, V. A. Lancaster, V. Kuznetsov, L. Beaman, R. Watts, and J. W. Harris. 2001. "Resistance to the Parasitic Mite *Varroa destructor* in Honey Bees from Far-Eastern Russia." *Apidologie* 32: 381-94.

Ruzicka, Ferdinand. 2003. "Schäden Durch Elektrosmog." *Bienenwelt* 10: 34-35.

———. 2006. "Schäden an Bienenvölkern." *Diagnose: Funk* 2006.

Sanford, Malcolm T. 2004. "Mite Tolerance in Honey Bees." *Bee Culture* 132(10): 23-26.

Science Daily. 1998. "Where Have All the Honeybees Gone?" July 6.

———. 2010. "Survey Reports Latest Honey Bee Losses." May 3.

Seeley, Thomas D. 2004. "Forest Bees and Varroa Mites." *Bee Culture*, July, pp. 22-23.

———. 2007. "Honey Bees of the Arnot Forest: A Population of Feral Colonies Persisting with *Varroa destructor* in the Northeastern United States." *Apidologie* 38: 19-29.

Sharma, Ved Parkash and Neelima R. Kumar. 2010. "Changes in Honeybee Behaviour and Biology under the Influence of Cellphone Radiations." *Current Science* 98(10): 1376-78.

Spleen, Angela M., Eugene J. Lengerich, Karen Rennich, Dewey Caron, Robyn Rose, Jeffery S. Pettis, Mark Henson, James T. Wilkes, Michael Wilson, Jennie Stitzinger, Kathleen Lee, Michael Andree, Robert Snyder, and Dennis vanEngelsdorp, for the Bee Informed Partnership. 2013. "A National Survey of Managed Honey Bee 2011-12 Winter Losses in the United States: Results from the Bee Informed Partnership." *Journal of Apicultural Research* 52(2): 44-53.

Steinhauer, Nathalie A., Karen Rennich, Michael E. Wilson, Dewey M. Caron, Eugene J. Lengerich, Jeffery S. Pettis, Robyn Rose, John A. Skinner, David R. Tarpy, James T. Wilkes, and Dennis vanEngelsdorp. 2014. "A National Survey of Managed Honey Bee 2012-2013 Annual Colony Losses in the USA: Results from the Bee Informed Partnership." *Journal of Apicultural Research* 53(1): 1-18.

Steinhauer, Nathalie, Karen Rennich, Kathleen Lee, Jeffery Pettis, David R. Tarpy, Juliana Rangel, Dewey Caron, Ramesh Sagili, John A. Skinner, Michael E. Wilson, James T. Wilkes, Keith S. Delaplane, Robyn Rose, and Dennis vanEngelsdorp. 2015. "Colony Loss 2014-2015: Preliminary Results." Bee Informed Partnership, UK.

Svensson, Börje. 2003. "Silent Spring in Northern Europe?" *Bees for Development Journal* 71: 3-4.

United States Dept of Agriculture, National Agricultural Statistics Service. 2010. *Honey*, February.

———. 2011. *Honey*, February.

Underwood, Robyn M. and Dennis vanEngelsdorp. 2007. "Colony Collapse Disorder: Have We Seen This Before?" *Bee Culture* 35(7): 13-18.

vanEngelsdorp, Dennis, Jay D. Evans, Claude Saegerman, Chris Mullin, Eric Haubruge, Bach Kim Nguyen, Maryann Frazier, Jim Frazier, Diana Cox-Foster, Yanping Chen, Robyn Underwood, David R. Tarpy, and Jeffery S. Pettis. 2009. "Colony Collapse Disorder: A Descriptive Study." *PLoS ONE* 4(8): e6481.

Warnke, Ulrich. 1976. "Effects of Electric Charges on Honeybees." *Bee World* 57(2): 50-56.

———. 2009. *Bienen, Vögel und Menschen: Die Zerstörung der Natur durch "Elektrosmog."* Published in English as *Bees, Birds and Mankind: Destroying Nature by "Electrosmog."* Kempten, Germany: Kompetenzinitiative.

Wilson, William T. and Diana M. Menapace. 1979. "Disappearing Disease of Honey Bees: A Survey of the United States." *American Bee Journal*, February, pp. 118-19; March, pp. 184-86, 217.

House Sparrows

ASPO/BirdLife Suisse. 2015. "Oiseau de l'année 2015: Moineau domestique" ("Bird of the Year 2015: House Sparrow").

Balmori, Alfonso and Örjan Hallberg. 2007. "The Urban Decline of the House Sparrow (*Passer domesticus*): A Possible Link with Electromagnetic Radiation." *Electromagnetic Biology and Medicine* 26: 141-51.

Bokotey, Andrei A. and Igor M. Gorban. 2005. "Numbers, Distribution, and Ecology of the House Sparrow in Lvov (Ukraine)." *International Studies on Sparrows* 30: 7-22.

De Laet, Jenny and James Denis Summers-Smith. 2007. "The Status of the Urban House Sparrow *Passer domesticus* in North-western Europe: A Review." *Journal of Ornithology* 148 (suppl. 2): S275-78.

Deccan Herald. 2010. "House Sparrow Listed as an Endangered Species." June 24.

Dott, Harry E. M. and Allan W. Brown. 2000. "A Major Decline in House Sparrows in Central Edinburgh." *Scottish Birds* 21: 61-68.

Eaton, Mark A., Andy F. Brown, David G. Noble, Andy J. Musgrove, Richard D. Hearn, Nicholas J. Aebischer, David W. Gibbons, Andy Evans, and Richard D. Gregory. 2009. "Birds of Conservation Concern 3." *British Birds* 102: 296-341.

Everaert, Joris and Dirk Bauwens. 2007. "A Possible Effect of Electromagnetic Radiation from Mobile Phone Base Stations on the Number of Breeding House Sparrows (*Passer domesticus*)." *Electromagnetic Biology and Medicine* 26: 63-72.

Galbraith, Colin. 2002. "The Population Status of Birds in the U.K: Birds of Conservation Concern: 2002-2007." *Bird Populations* 7: 173-79.

Gregory, Richard D., Nicholas I. Wilkinson, David G. Noble, James A. Robinson, Andrew F. Brown, Julian Hughes, Deborah Procter, David W. Gibbons, and Colin A. Galbraith. 2002. "The Population Status of Birds in the United Kingdom, Channel Islands and Isle of Man: An Analysis of Conservation Concern 2002-2007." *British Birds* 95: 410-48.

Longino, Libby. 2013. "Researchers Stumped over Decline of Sparrow Populations." *USA Today*, October 5.

Pattazhy, Sainudeen. 2012. "Dwindling Number of Sparrows." *Karala Calling*, March, pp. 32-33.

Prowse, Alan. 2002. "The Urban Decline of the House Sparrow." *British Birds* 95: 143-46.

Robinson, Robert A., Gavin M. Siriwardena, and Humphrey Q. P. Crick. 2005. "Size and Trends of the House Sparrow *Passer domesticus* Population in Great Britain." *Ibis* 147(2): 552-62.

Sanderson, Roy F. 1995. "Autumn Bird Counts in Kensington Gardens, 1925-1995." *London Bird Report* 60: 170-76.

Sanderson, Roy F. 2001. "Further Declines in an Urban Population of House Sparrows." *British Birds* 94: 507-8.

Scott, Bob and Adrian Pitches. 2002. "Demise of the Cockney Sparrow." *British Birds* 95: 468-70.

Sen, Benita. 2012. "Calling Back the Sparrow." *Deccan Herald*, November 26.

Sherry, Kate. 2003. "Are Mobile Phones Behind the Decline of House Sparrows?" *Daily Mail*, January 13.

Škorpilová, Jana, Petr Voříšek, and Alena Klvaňová. 2010. "Trends of Common Birds in Europe, 2010 Update." European Bird Census Council.

Summers-Smith, James Denis. 2000. "Decline of House Sparrows in Large Towns." *British Birds* 93: 256-57.

———. 2003. "Decline of the House Sparrow: A Review." *British Birds* 96: 439-46.

———. 2005. "Changes in the House Sparrow Population in Britain." *International Studies on Sparrow* 30: 23-37.

Times of India. 2005. "Even Sparrows Don't Want to Live in Cities Anymore." June 13.

Townsend, Mark. 2003. "Mobile Phones Blamed for Sparrow Deaths." *The Observer*, January 12.

Insects

Balmori, Alfonso. 2006. "Efectos de las radiaciones electromagnéticas de la telefonía móvil sobre los insectos." *Ecosistemas* 15(1): 87-95.

Barbassa, Juliana. 2006. "The Plight of the Butterfly." *New Mexican*, May 11, p. D1.

Becker, Günther. 1977. "Communication Between Termites by Biofields." *Biological Cybernetics* 26: 41-44.

Cammaerts, Marie-Claire and Olle Johansson. 2014. "Ants Can Be Used as Bio-indicators to Reveal Biological Effects of Electromagnetic Waves from Some Wireless Apparatus." *Electromagnetic Biology and Medicine* 33(4): 282-88.

Evans, Elaine, Robbin Thorp, Sarina Jepsen, and Scott Hoffman Black. 2008. *Status Review of Three Formerly Common Species of Bumble Bee in the Subgenus* Bombus. Portland, OR: Xerces Society for Invertebrate Conservation.

Kluser, Stéphane and Pascal Peduzzi. 2007. *Global Pollinator Decline: A Literature Review.* Geneva: United Nations Environment Programme/GRID-Europe.

Margaritis, Lukas H., Areti K. Manta, Konstantinos D. Kokkaliaris, Dimitra Schiza, Konstantinos Alimisis, Georgios Barkas, Eleana Georgiou, Olympia Giannakopoulou, Ioanna Kollia, Georgia Kontogianni, Angeliki Kourouzidou, Angeliki Myari, Fani Roumelioti, Aikaterini Skouroliakou, Vasia Sykioti, Georgia Varda, Konstantinos Xenos, and Konstantinos Ziomas. 2014. "Drosophila Oogenesis as a Bio-marker Responding to EMF Sources." *Electromagnetic Biology and Medicine* 33(3): 165-89.

Massachusetts Division of Fisheries and Wildlife, Department of Fish and Game. 2015. *Massachusetts List of Endangered, Threatened and Special Concern Species.* Westborough, MA.

Ministry of Environment and Forests. 2011. *Report on Possible Impacts of Communication Towers on Wildlife Including Birds and Bees.* New Delhi.

National Research Council, Committee on the Status of Pollinators in North America. 2007. *Status of Pollinators in North America.* Washington, DC: National Academies Press.

Panagopoulos, Dimitris J. 2011. "Analyzing the Health Impacts of Modern Telecommunications Microwaves." *Advances in Medicine and Biology* 17: 1-55.

———. 2012a. "Effect of Microwave Exposure on the Ovarian Development of *Drosophila melanogaster*." *Cell Biochemistry and Biophysics* 63: 121-32.

———. 2012b. "Gametogenesis, Embryonic and Post-Embryonic Development of Drosophila Melanogaster, as a Model System for the Assessment of Radiation and Environmental Genotoxicity." In: M. Spindler-Barth, ed., *Drosophila Melanogaster: Life Cycle, Genetics, and Development* (New York: Nova Science), pp. 1-38.

Panagopoulos, Dimitris J., Evangelia D. Chavdoula, Andreas Karabarbounis, and Lukas H. Margaritis. 2007. "Comparison of Bioactivity between GSM 900 MHz and DCS 1800 MHz Mobile Telephony Radiation." *Electromagnetic Biology and Medicine* 26: 33-44.

Panagopoulos, Dimitris J., Evangelia D. Chavdoula, and Lukas H. Margaritis. 2010. "Bioeffects of Mobile Telephony Radiation in Relation to Its Intensity or Distance from the Antenna." *International Journal of Radiation Biology* 86(5): 345-57.

Panagopoulos, Dimitris J., Evangelia D. Chavdoula, Ioannis P. Nezis, and Lukas H. Margaritis. 2007. "Cell Death Induced by GSM 900-MHz and DCS 1800-MHz Mobile Telephony Radiation." *Mutation Research* 626: 69-78.

Panagopoulos, Dimitris J., Andreas Karabarbounis, and Lukas H. Margaritis. 2004. "Effect of GSM 900-MHz Mobile Phone Radiation on the Reproductive Capacity of *Drosophila melanogaster*." *Electromagnetic Biology and Medicine* 23(1): 29-43.

Panagopoulos, Dimitris J. and Lukas H. Margaritis. 2008. "Mobile Telephony Radiation Effects on Living Organisms." In: A. C. Harper and R. V. Buress, eds., *Mobile Telephones, Networks, Applications, and Performance* (New York: Nova Science), pp. 107-49.

———. 2010. "The Identification of an Intensity 'Window' on the Bioeffects of Mobile Telephony Radiation." *International Journal of Radiation Biology* 86(5): 358-66.

Serant, Claire. 2004. "A Human Science Experiment." *New York Newsday*, May 10.

Warnke, Ulrich. 1989. "Information Transmission by Means of Electrical Biofields." In: Fritz Albert Popp, Ulrich Warnke, Herbert L. König, and Walter Peschka, eds., *Electromagnetic Bio-Information* (München: Urban & Schwarzenberg), pp. 74-101.

Williams, Paul H., Miguel B Araújo, and Pierre Rasmont. 2007. "Can Vulnerability among British Bumblebee (*Bombus*) Species be Explained by Niche Position and Breadth?" *Biological Conservation* 138: 493-505.

Xerces Society for Invertebrate Conservation. 2015. *Red List of Bees: Native Bees in Decline*. Portland, OR.

———. 2015. *Red List of Butterflies and Moths*. Portland, OR.

Konstantynów

Flakiewicz, Wiesław and Antonina Cebulska-Wasilewska. 1992. "Biological Effects of EM Field on Randomly Selected Human Population Residing Permanently Close to the High Power, Long Wave Radio Transmitter, and Tradescantia Plant Model System In Situ." *EMC 92, Eleventh International Wrocław Symposium and Exhibition on Electromagnetic Compatibility, September 2-4, 1992*, pp. 72-76.

Mammals

Balmori, Alfonso. 2009. "Electromagnetic Pollution from Phone Masts. Effects on Wildlife." *Pathophysiology* 16(2-3): 191-99.

———. 2010. "The Incidence of Electromagnetic Pollution on Wild Mammals: A New 'Poison' with a Slow Effect on Nature?" *Environmentalist* 30: 90-97.

Magras, Ioannis N. and Thomas D. Xenos. 1997. "RF Radiation-Induced Changes in the Prenatal Development of Mice." *Bioelectromagnetics* 18: 455-61.

Radio Tagging Animals

Altonn, Helen. 2002. "High-tech Tags Give Scientists Tools to Track Sea Animal Movement." *Honolulu Star-Bulletin*, Feb 18.

Balmori, Alfonso. 2016. "Radiotelemetry and Wildlife: Highlighting a Gap in the Knowledge on Radiofrequency Radiation Effects." *Science of the Total Environment* 543: 662-69.

Burrows, Roger, Heribert Hofer, and Marion L. East. 1994. "Demography, Extinction and n a Small Population: the Case of the Serengeti Wild Dogs." *Proceedings of the Royal Society of London B* 256: 281-92.

———. 1995. "Population Dynamics, Intervention and Survival in African Wild Dogs *(Lycaon pictus)*." *Proceedings of the Royal Society of London B*: 235-45.

Caldwell, Mark. 1997. "The Wired Butterfly." *Discover Magazine*, February 1.

Godfrey, Jason D. and David M. Bryant. 2003. "Effects of Radio Transmitters: Review of Recent Radio-tracking Studies." In: Williams, M., ed., *Conservation Applications of Measuring Energy Expenditure of New Zealand Birds: Assessing Habitat Quality and Costs of Carrying Radio Transmitters* (Wellington, New Zealand: Dept. of Conservation), pp. 83-95.

Mech, L. David and Shannon M. Barber. 2002. *A Critique of Wildlife Radio-Tracking and Its Use in National Parks*. Jamestown, ND: U.S. Geological Survey, Northern Prairie Wildlife Research Center.

Moorhouse, Tom P. and David W. Macdonald. 2005. "Indirect Negative Impacts of Radio-collaring: Sex Ratio Variation in Water Voles." *Journal of Applied Ecology* 42: 91-98.

Roberts, Greg. 2000. "Sick as a Parrot: Deaths Halt DNA Program." *The Age*, February 8.

Swenson, Jon E., Kjell Wallin, Göran Ericsson, Göran Cederlund, and Finn Sandegren. 1999. "Effects of Ear-tagging with Radiotransmitters on Survival of Moose Calves." *Journal of Wildlife Management* 63(1): 354-58.

Reader's Digest. 1998. "The Snow Tiger's Last Stand." November.

Webster, A. Bruce and Ronald J. Brooks. 1980. "Effects of Radiotransmitters on the Meadow Vole, *Microtus pennsylvanicus*." *Canadian Journal of Zoology* 58: 997-1001.

Withey, John C., Thomas D. Bloxton, and John M. Marzluff. 2001. "Effects of Tagging and Location Error in Wildlife Radiotelemetry Studies." In: Joshua J. Millspaugh and John M. Marzluff, eds., *Radio Tracking and Animal Populations* (San Diego: Academic), pp. 43-75.

Schwarzenburg

Abelin, Theodor, Ekkehardt Altpeter, and Martin Röösli. 2005. "Sleep Disturbances in the Vicinity of the Short-Wave Broadcast Transmitter Schwarzenburg." *Somnologie* 9: 203-9.

Altpeter, Ekkehardt-Siegfried, Katharina Sprenger, Katrin Madarasz, and Theodor Abelin. 1997. "Do Radiofrequency Electromagnetic Fields Cause Sleep Disorders?" European Regional Meeting of the International Epidemiological Association, Münster, Germany, September. Abstract no. 351.

Altpeter, Ekkehardt-Siegfried, Martin Röösli. Markus Battaglia, Dominik H. Pfluger, Christoph E. Minder, and Theodor Abelin. 2006. "Effect of Short-Wave (6-22

MHz) Magnetic Fields on Sleep Quality and Melatonin Cycle in Humans: The Schwarzenburg Shut-Down Study." *Bioelectromagnetics* 27: 142-50.

Altpeter, Ekkehardt-Siegfried, Thomas Krebs, Dominik H. Pfluger, J. von Känel, R. Blattmann, D. Emmenegger, B. Cloetta, U. Rogger, H. Gerber, Bernhard Manz, R. Coray, R. Baumann, Katharina Staerk, Christian Griot, and Theodor Abelin. 1995. *Study on Health Effects of the Shortwave Transmitter Station of Schwarzenburg, Berne, Switzerland.* BEW Publication Series, Study no. 55. Federal Office of Energy, August 1995.

Jakob, Hans-U. 2006. "Schwarzenburg – Nach 8 Jahren Geheimhaltung." Basel: Diagnose-Funk, June 25.

———. 2000. "State of Health after Shutdown of the Schwarzenburg Transmitter." *No Place To Hide* 2(4): 21-22.

Roch, Phillippe. 1996. "Health Effects of the Schwarzenburg Shortwave Transmitter," Letter of May 29, 1996, Bern: Federal Office of Environment, Forests and Landscape. English translation in *No Place To Hide* 1(3): 7-8.

Stärk, Katharina D. C., Thomas Krebs, Ekkehardt Altpeter, Bernhard Manz, Christian Griot, and Theodor Abelin. 1997. "Absence of Chronic Effect of Exposure to Short-wave Radio Broadcast Signal on Salivary Melatonin Concentrations in Dairy Cattle." *Journal of Pineal Research* 22: 171-76.

Skrunda

Balode, Zanda. 1996. "Assessment of Radio-Frequency Radiation by the Micronucleus Test in Bovine Peripheral Erythrocytes." *Science of the Total Environment* 180: 81-85.

Balodis, Valdis, Guntis Brūmelis, Kārlis Kalviškis, Oļģerts Nikodemus, Didzis Tjarve, and Vija Znotiņa. 1996. "Does the Skrunda Radio Location Station Diminish the Radial Growth of Pine Trees?" *Science of the Total Environment* 180: 57-64.

Brūmelis, Guntis, Valdis Balodis, and Zanda Balode. 1996. "Radio-frequency Electromagnetic Fields: The Skrunda Radio Location Station Case." *Science of the Total Environment* 180: 49-50.

Goldsmith, John R. 1995. "Epidemiologic Evidence of Radiofrequency Radiation (Microwave) Effects on Health in Military, Broadcasting, and Occupational Studies." *International Journal of Occupational and Environmental Health* 1: 47-57.

Kalniņš, T., R. Križbergs, and A. Romančuks. 1996. "Measurement of the Intensity of Electromagnetic Radiation from the Skrunda Radio Location Station, Latvia." *Science of the Total Environment* 180: 51-56.

Kolodynski, Anton and Valda Kolodynska. 1996. "Motor and Psychological Functions of School Children Living in the Area of the Skrunda Radio Location Station in Latvia." *Science of the Total Environment* 180: 87-93.

Liepa, V. and Valdis Balodis. 1994. "Monitoring of Bird Breeding near a Powerful Radar Station." *The Ring* 16(1-2): 100. Abstract.

Magone, I. 1996. "The Effect of Electromagnetic Radiation from the Skrunda Radio Location Station on *Spirodela polyrhiza* (L.) Cultures." *Science of the Total Environment* 180: 75-80.

Microwave News. 1994. "Latvia's Russian Radar May Yield Clues to RF Health Risks." September/October, pp. 12-13.

Science of the Total Environment. 1996. "Special Issue: Effects of RF Electromagnetic Radiation on Organisms. A Collection of Papers Presented at The International Conference on the Effect of Radio Frequency Electromagnetic Radiation on Organisms, Skrunda, Latvia, June 17-21, 1994." 180: 277-78.

Selga, Turs and Maija Selga. 1996. "Response of *Pinus sylvestris L.* needles to Electromagnetic Fields: Cytological and Ultrastructural Aspects." *Science of the Total Environment* 180: 65-73.

Chapter 17

Adey, William Ross. 1993. "Effects of Electromagnetic Fields. *Journal of Cellular Biochemistry* 51: 410-16.

———. 1993. "Whispering Between Cells: Electromagnetic Fields and Regulatory Mechanisms in Tissue." *Frontier Perspectives* 3(2): 21-25.

Baş, Orhan, Osman Fikret Sönmez, Ali Aslan, Ayşe İkinci, Hatice Hancı, Mehmet Yıldırım, Haydar Kaya, Metehan Akça, and Ersan Odacı. 2013. "Pyramidal Cell Loss in the Cornu Ammonis of 32-day-old Female Rats Following Exposure to a 900 Megahertz Electromagnetic Field during Prenatal Days 13-21." *NeuroQuantology* 11(4): 591-99.

Bejot, Yannick, Benoit Daubail, Agnès Jacquin, Jérôme Durier, Guy-Victor Osseby, Olivier Rouaud, and Maurice Giroud. 2014. "Trends in the Incidence of Ischaemic Stroke in Young Adults Between 1985 and 2011: the Dijon Stroke Registry." *Journal of Neurology, Neurosurgery, and Psychiatry* 85: 509-13.

Blue Cross Blue Shield. 2019. *The Health of Millennials.* Washington, DC.

Broomhall, Mark. 2017. *Report Detailing the Exodus of Species from the Mt. Nardi Area of the Nightcap National Park World Heritage Area During a 15-Year Period (2000-2015).* Report for the United Nations Educational Scientific and Cultural Organization (UNESCO). New South Wales, Australia.

Byun, Yoon-Hwan, Mina Ha, Ho-Jang Kwon, Yun-Chul Hong, Jong-Han Leem, Joon Sakong, Su Young Kim, Chul Gab Lee, Dongmug Kang, Hyung-Do Choi, and Nam Kim. 2013. "Mobile Phone Use, Blood Lead Levels, and Attention Deficit Hyperactivity Symptoms in Children: A Longitudinal Study." *PLoS ONE* 8(3): e59742.

Centola, G. M., A. Blanchard, J. Demick, S. Li, and M. L. Eisenberg. 2016. "Decline in Sperm Count and Motility in Young Adult Men from 2003 to 2013: Observations from a U.S. Sperm Bank." *Andrology* 4: 270-76.

Cherry, Neil. 2000. *Safe Exposure Levels.* Lincoln University, Lincoln, New Zealand.

———. 2002. "Schumann Resonances, a Plausible Biophysical Mechanism for the Human Health Effects of Solar/Geomagnetic Activity." *Natural Hazards Journal* 26(3): 279-331.

Dalsegg, Aud. 2002. "Får hodesmerter av mobilstråling" ("She Gets Headaches from Mobile Radiation"). *Dagbladet,* March 9.

Grigoriev, Yury Grigorievich. 2005. "Elektromagnitnye polya sotovykh telefonov i zdorovye detey i podrostkov: Situatsiya, trebuyushchaya prinyatiya neotlozhnykh mer" ("The Electromagnetic Field of Mobile Phones and the Health of Children and Adolescents: This Situation Requires Urgent Action"). *Radiatsionnaya biologiya. Radioekologiya* 45(4): 442-50.

————. 2012. "Mobile Communications and Health of Population: The Risk Assessment, Social and Ethical Problems." *The Environmentalist* 32(2): 193-200.

Grigoriev, Yury Grigorievich and Oleg Aleksandrovich Grigoriev. 2011. "Mobil'naya svyaz' i zdorovye naseleniya: Otsenka opasnosti, sotsial'nye i eticheskiye problemi" ("Mobile Communication and Health of Population: Estimation of Danger, Social and Ethical Problems"). *Radiatsionnaya biologiya. Radioekologiya* 51(3): 357-68.

————. 2013. *Sotovaya Svyaz' i Zdorov'e* ("Cellular Communication and Health"). Moscow: Ekonomika.

Grigoriev, Yury Grigorievich and Nataliya Igorevna Khorseva. 2014. *Mobil'naya Svyaz' i Zdorov'e Detey* ("Mobile Communication and Children's Health"). Moscow: Ekonomika.

Hallberg, Örjan and Olle Johansson. 2009. "Apparent Decreases in Swedish Public Health Indicators after 1997 – Are They Due to Improved Diagnostics or to Environmental Factors?" *Pathophysiology* 16(1): 43-46.

Hallberg, Örjan and Olle Johansson. 2004. *Glesbygd är en sjuk miljö, nu börjar även friska dö* ("Say To Countryside Goodbye, When Even Healthy People Die"). Stockholm: Karolinska Institute, Experimental Dermatology Unit. Report no. 6.

Hallberg, Örjan and Gerd Oberfeld. 2006. "Letter to the Editor: Will We All Become Electrosensitive?" *Electromagnetic Biology and Medicine* 25(3): 189-91.

Hallman, Caspar A., Martin Sorg, Eelke Jongejans, Hank Siepel, Nick Hofland, Heinz Schwan, Werner Stenmans, Andreas Müller, Hubert Sumser, Thomas Hörren, Dave Goulson, Hans de Kroon. 2017. "More than 75 Percent Decline over 27 Years in Total Flying Insect Biomass in Protected Areas." *PLoS ONE* 12(10): e0185809.

Hancı, Hatice, Ersan Odacı, Haydar Kaya, Yüksel Aliyazıcıoğlu, İbrahim Turan, Selim Demir, and Serdar Çolakoğlu. 2013. "The Effect of Prenatal Exposure to 900-MHz Electromagnetic Field on the 21-old-day Rat Testicle." *Reproductive Toxicology* 42: 203-9.

Hancı, Hatice, Sibel Türedi, Zehra Topal, Tolga Mercantepe, İlyas Bozkurt, Haydar Kaya, Safak Ersöz, Bünyami Ünal, and Ersan Odacı. 2015. "Can Prenatal Exposure to a 900 MHz Electromagnetic Field Affect the Morphology of the Spleen and Thymus, and Alter Biomarkers of Oxidative Damage in 21-day-old Male Rats?" *Biotechnic & Histochemistry* 90(7). 535-43.

Hutton, John S., Jonathan Dudley, Tzipi Horowitz-Kraus, Tom DeWitt, and Scott K. Holland. 2019. "Associations Between Screen-Based Media Use and Brain White Matter Integrity in Preschool-Aged Children." *JAMA Pediatrics* 2019 Nov. 4: e193869.

İkinci, Ayşe, Ersan Odacı, Mehmet Yıldırım, Haydar Kaya, Metehan Akça, Hatice Hancı, Ali Aslan, Osman Fikret Sönmez, and Orhan Baş. 2013. "The Effects of Prenatal Exposure to a 900 Megahertz Electromagnetic Field on Hippocampus Morphology and Learning Behavior in Rat Pups." *Journal of Experimental and Clinical Medicine* 30: 278. Abstract.

İkinci, Ayşe, Tolga Mercantepe, Deniz Unal, Hüseyin Serkan Erol, Arzu Şahin, Ali Aslan, Orhan Baş, Havva Erdem, Osman Fikret Sönmez, Haydar Kaya, and Ersan Odacı. 2015. "Morphological and Antioxidant Impairments in the Spinal Cord of Male Offspring Rats Following Exposure to a Continuous 900 MHz Electromagnetic Field During Early and Mid-Adolescence." *Journal of Chemical Neuroanatomy* [Epub ahead of print].

Kimata, Hajime. 2002. "Enhancement of Allergic Skin Wheal Responses by Microwave Radiation from Mobile Phones in Patients with Atopic Eczema/Dermatitis Syndrome." *International Archives of Allergy and Immunology* 129(4): 348-50.

Li, De-Kun, Hong Chen, and Roxana Odouli. 2011. "Maternal Exposure to Magnetic Fields during Pregnancy in Relation to the Risk of Asthma in Offspring." *Archives of Pediatrics & Adolescent Medicine* 165(10): 945-50.

Lister, Bradford C. and Andres Garcia. 2018. "Climate-driven Declines in Arthropod Abundance Restructure a Rainforest Food Web." *Proceedings of the National Academy of Sciences* 115(44): E10397–E10406.

Mild, Kjell Hansson, Gunnhild Oftedal, Monica Sandström, Jonna Wilén, Tore Tynes, Bjarte Haugsdal, and Egil Hauger. 1998. *Comparison of Symptoms Experienced by Users of Analogue and Digital Mobile Phones. A Swedish-Norwegian Epidemiological Study.* Umeå, Sweden: National Institute for Working life. Arbetslivsrapport 23.

Mishra, Lata. 2011. "Heard This? Talking on the Phone Makes You Deaf." *Mumbai Mirror*, October 26.

Mishra, Srikanta Kumar. 2010. "Otoacoustic Emission (OAE)-Based Measurement of the Functioning of the Human Cochlea and the Efferent Auditory System." Ph.D. thesis, University of Southampton.

Nittby, Henrietta, Gustav Grafström, Dong Ping Tian, Lars Malmgren, Arne Brun, Bertil R. R. Persson, Leif G. Salford, and Jacob Eberhardt. 2008. "Cognitive Impairment in Rats After Long-Term Exposure to GSM-900 Mobile Phone Radiation." *Bioelectromagnetics* 29: 219-32.

Odacı, Ersan, Hatice Hancı, Ayşe İkinci, Osman Fikret Sönmez, Ali Aslan, Arzu Şahin, Haydar Kaya, Serdar Çolakoğlu, and Orhan Baş. 2015. "Maternal Exposure to a Continuous 900-MHz Electromagnetic Field Provokes Neuronal Loss and Pathological Changes in Cerebellum of 32-day-old Female Rat Offspring." *Journal of Chemical Neuroanatomy* [Epub ahead of print].

Odacı, Ersan, Hatice Hancı, Esin Yuluğ, Sibel Türedi, Yüksel Aliyazıcıoğlu, Haydar Kaya, and Serdar Çolakoğlu. 2016. "Effects of Prenatal Exposure to a 900 MHz Electromagnetic Field on 60-day-old Rat Testis and Epididymal Sperm Quality." *Biotechnic & Histochemistry* 91(1): 9-19.

Odacı, Ersan, Ayşe İkinci, Mehmet Yıldırım, Haydar Kaya, Metehan Akça, Hatice Hancı, Osman Fikret Sönmez, Ali Aslan, Mukadder Okuyan, and Orhan Baş. 2013. "The Effects of 900 Megahertz Electromagnetic Field Applied in the Prenatal Period on Spinal Cord Morphology and Motor Behavior in Female Rat Pups." *NeuroQuantology* 11(4): 573-81.

Odacı, Ersan and Cansu Özyılmaz. 2015. "Exposure to a 900 MHz Electromagnetic Field for 1 Hour a Day over 30 Days Does Change the Histopathology and Biochemistry of the Rat Testis." *International Journal of Radiation Biology* 91: 547-54.

Odacı, Ersan, Deniz Ünal, Tolga Mercantepe, Zehra Topal, Hatice Hancı, Sibel Türedi, Hüseyin Serkan Erol, Sevdegül Mungan, Haydar Kaya, and Serdar Çolakoğlu. 2015. "Pathological Effects of Prenatal Exposure to a 900 MHz Electromagnetic Field on the 21-day-old Male Rat Kidney." *Biotechnic & Histochemistry* 90(2): 93-101.

Oktay, M. Faruk and Suleyman Dasdag. 2006. "Effects of Intensive and Moderate Cellular Phone Use on Hearing Function." *Electromagnetic Biology and Medicine* 25: 13-21.

Panda, Naresh K., Rahul Modi, Sanjay Munjal, and Ramandeep S. Virk. 2011. "Auditory Changes in Mobile Users: Is Evidence Forthcoming?" *Otolaryngology – Head and Neck Surgery* 144(4): 581-85.

Putaala, Jukka, Antti J. Metso, Tiina M. Metso, Nina Konkola, Yvonn Kraemer, Elena Haapaniemi, Markku Kaste, and Turgut Tatlisumak. 2009. "Analysis of 1008 Consecutive Patients Aged 15 to 49 with First-Ever Ischemic Stroke: the Helsinki Young Stroke Registry." *Stroke* 40: 1195-1203.

Rosengren, Annika, Kok Wai Giang, Georgios Lappas, Christina Jern, Kjell Torén, and Lena Björck. 2013. "Twenty-four-year Trends in the Incidence of Ischemic Stroke in Sweden from 1987 to 2010." *Stroke* 44: 2388-93.

Şahin, Arzu, Ali Aslan, Orhan Baş, Ayşe İkinci, Cansu Özyılmaz, Osman Fikret Sönmez, Serdar Çolakoğlu, and Ersan Odacı. 2015. "Deleterious Impacts of a 900-MHz Electromagnetic Field on Hippocampal Pyramidal Neurons of 8-week-old Sprague Dawley Male Rats." *Brain Research* 1624: 232-38.

Salford, Leif G., Arne E. Brun, Jacob L. Eberhardt, Lars Malmgren, and Bertil R.R. Persson. 2003. "Nerve Cell Damage in Mammalian Brain after Exposure to Microwaves from GSM Mobile Phones." *Environmental Health Perspectives* 111(7): 881-83.

Sánchez-Bayo, Francisco and Kris A. G. Wyckhuys. 2019. "Worldwide Decline of the Entomofauna: A Review of Its Drivers. *Biological Conservation* 232: 8-27.

Shinjyo, Tetsuharu and Akemi Shinjyo. 2014. "Signifikanter Rückgang klinischer Symptome nach Senderabbau – eine Interventionsstudie." *Umwelt-Medizin-Gesellschaft* 27(4): 294-301.

Siegel, Rebecca L., Stacey A. Fedewa, William F. Anderson, Kimberly D. Miller, Jiemin Ma, Philip S. Rosenberg, and Ahmedin Jemal. 2017. "Colorectal Cancer Incidence Patterns in the United States, 1974-2013." *Journal of the National Cancer Institute* 109(8): djw322.

Tatemichi, Masayuki, Tadashi Nakano, Katsutoshi Tanaka, Takeshi Hayashi, Takeshi Nawa, Toshiaki Miyamoto, Hisanori Hiro, and Minoru Sugita. 2004. "Possible Association between Heavy Computer Users and Glaucomatous Visual Field Abnormalities: A Cross Sectional Study in Japanese Workers." *Journal of Epidemiology and Community Health* 58: 1021-27.

Tibæk, Maiken, Christian Dehlendorff, Henrik S. Jørgensen, Hysse B. Forchhammer, Søren P. Johnsen, and Lars P. Kammersgaard. 2016. "Increasing Incidence of Hospitalization for Stroke and Transient Ischemic Attack in Young Adults: A Registry-Based Study." *Journal of the American Heart Association* 5(5): e003158.

Topal, Zehra, Hatice Hancı, Tolga Mercantepe, Hüseyin Serkan Erol, Osman Nuri Keleş, Haydar Kaya, Sevdegül Mungan, and Ersan Odacı. 2015. "The Effects of Prenatal Long-duration Exposure to 900-MHz Electromagnetic Field on the 21-day-old Newborn Male Rat Liver." *Turkish Journal of Medical Sciences* 45(2): 291-97.

Türedi, Sibel, Hatice Hancı, Zehra Topal, Deniz Ünal, Tolga Mercantepe, İlyas Bozkurt, Haydar Kaya, and Ersan Odacı. 2015. "The Effects of Prenatal Exposure to a

900-MHz Electromagnetic Field on the 21-day-old Male Rat Heart." *Electromagnetic Biology and Medicine* 34(4): 390-97.

Velayutham, P., Gopala Krishnan Govindasamy, R. Raman, N. Prepageran, and K. H. Ng. 2014. "High-frequency Hearing Loss Among Mobile Phone Users." *Indian Journal of Otolaryngology and Head & Neck Surgery* 66: S169-S172.

Weiner, A. B., R. S. Matulewicz, S. E. Eggener, and E. M. Schaeffer. 2016. "Increasing Incidence of Metastatic Prostate Cancer in the United States (2004-2013). *Prostate Cancer and Prostatic Diseases* 19: 395-97.

West, John G., Nimmi S. Kapoor, Shu-Yuan Liao, June W. Chen, Lisa Bailey, and Robert A. Nagourney. 2013. "Multifocal Breast Cancer in Young Women with Prolonged Contact between Their Breasts and Their Cellular Phones. *Case Reports in Medicine*, article ID 354682.

Wiedbrauk, Danny L. 1997. "The 1996-1997 Influenza Season – A View from the Benches." *Pan American Society for Clinical Virology Newsletter* 23(1): 1 ff.

Wolford, Monica L., Kathleen Palso, and Anita Bercovitz. 2015. "Hospitalization for Total Hip Replacement Among Inpatients Aged 45 and Over: United States, 2000-2010." *NCHS Data Brief* no. 186.

Wong, Martin C. S., William B. Goggins, Harry H. X. Wang, Franklin D. H. Fung, Colette Leung, Samuel Y. S. Wonga, Chi Fai Ng, and Joseph J. Y. Sung. 2016. "Global Incidence and Mortality for Prostate Cancer: Analysis of Temporal Patterns and Trends in 36 Countries." *European Urology* 70: 862-74.

Yakymenko, I. L., E. P. Sidorik, A. S. Tsybulin, and V. F. Chekhun. 2011. "Potential Risks of Microwaves from Mobile Phones for Youth Health." *Environment & Health* 56(1): 48-51.

Ye, Juan, Ke Yao, Dequiang Lu, Renyi Wu, and Huai Jiang. 2001. "Low Power Density Microwave Radiation Induced Early Changes in Rabbit Lens Epithelial Cells." *Chinese Medical Journal* 114(12): 1290-94.

Index

About the Author

Arthur Firstenberg is a scientist and journalist who is at the forefront of a global movement to tear down the taboo surrounding this subject. After graduating Phi Beta Kappa from Cornell University with a degree in mathematics, he attended the University of California, Irvine School of Medicine from 1978 to 1982. Injury by X-ray overdose cut short his medical career. For the past thirty-eight years he has been a researcher, consultant, and lecturer on the health and environmental effects of electromagnetic radiation, as well as a practitioner of several healing arts.